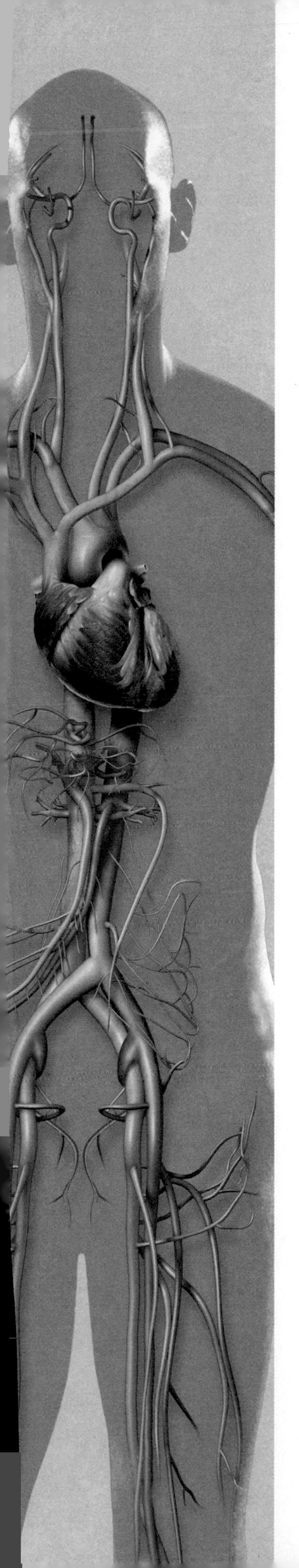

PRINCIPLES OF ANATOMY AND PHYSIOLOGY

Tenth Edition

Volume 2

Support and Movement of the Human Body

Gerard J. Tortora

Bergen Community College

Sandra Reynolds Grabowski

Purdue University

John Wiley & Sons, Inc.

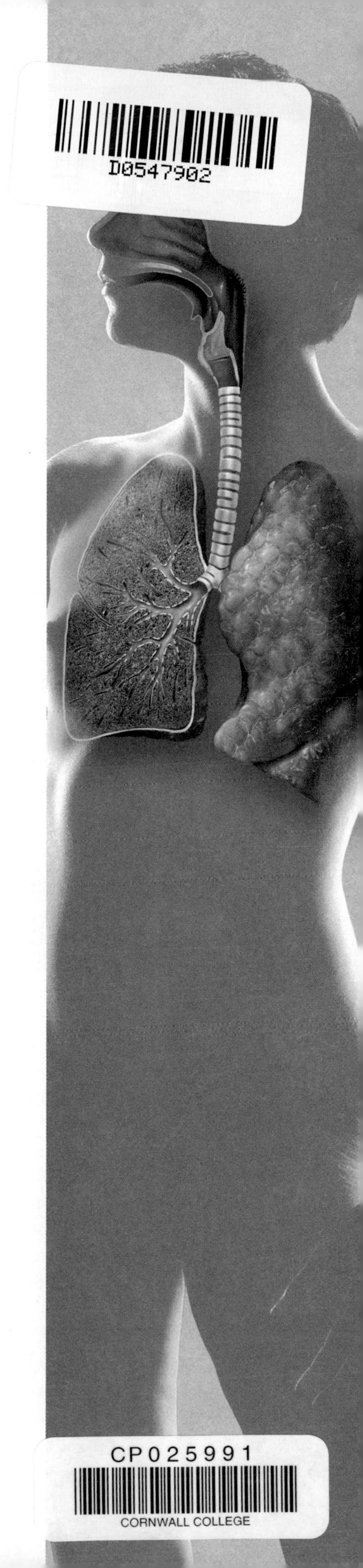

Senior Editor	**Bonnie Roesch**
Marketing Manager	**Clay Stone**
Developmental Editor	**Ellen Ford**
Production Director	**Pamela Kennedy**
Associate Production Manager	**Kelly Tavares**
Text and Cover Designer	**Karin Gerdes Kincheloe**
Art Coordinator	**Claudia Durrell**
Photo Editor	**Hilary Newman**
Chapter Opener Illustrations	**Keith Kasnot**

Cover photos: Photo of woman: ©PhotoDisc, Inc.
Photo of man: ©Ray Massey/Stone
Illustration and photo credits follow the Glossary.

This book was typeset by Progressive Information Technologies. It was printed and bound by Von Hoffmann Press, Inc. The cover was also printed by Von Hoffman Press, Inc.

The paper in this book was manufactured by a mill whose forest management programs include sustained yield harvesting of its timberlands. Sustained yield harvesting principles ensure that the number of trees cut each year does not exceed the amount of new growth.

This book is printed on acid-free paper. ♾

This title published by John Wiley & Sons, Inc.

USA ISBN: 0-471-22932-6

Printed in the United States of America.
10 9 8 7 6 5 4 3 2 1

About The Cover

In choosing the cover illustration for the tenth edition of ***Principles of Anatomy and Physiology,*** we wanted to reflect the underlying principle in the textbook as well as our approach to presentation of the content. In a word, we wanted to suggest the idea of balance. Successfully practicing yoga requires finding the balance between the physical, mental, and spiritual aspects of the human body. Achieving this balance produces flexibility and strength, as demonstrated by the figures on the cover.

Similarly, balance has always been at the heart of the study of anatomy and physiology. Homeostasis, the underlying principle discussed throughout this textbook, is the state of equilibrium of the internal environment of the human body. During your study of anatomy and physiology you will learn about the structures and functions needed to regulate and maintain this balanced state and its importance in promoting optimal well being.

Balance also applies to the way we present the content in the tenth edition of ***Principles of Anatomy & Physiology***. We offer a balanced coverage of structure and function; between normal and abnormal anatomy and physiology; between concepts and applications; between narrative and illustrations; between print and media; and among supplemental materials developed to meet the variety of teaching and learning styles. We hope that this balanced approach will provide you with the flexibility and strength to succeed in this course and in your future careers.

About the Authors

Gerard J. Tortora is Professor of Biology and former Coordinator at Bergen Community College in Paramus, New Jersey, where he teaches human anatomy and physiology as well as microbiology. He received his bachelor's degree in biology from Fairleigh Dickinson University and his master's degree in science education from Montclair State College. He is a member of many professional organizations, such as the Human Anatomy and Physiology Society (HAPS), the American Society of Microbiology (ASM), American Association for the Advancement of Science (AAAS), National Education Association (NEA), and the Metropolitan Association of College and University Biologists (MACUB).

Above all, Jerry is devoted to his students and their aspirations. In recognition of this commitment, Jerry was the recipient of MACUB's 1992 President's Memorial Award. In 1996, he received a National Institute for Staff and Organizational Development (NISOD) excellence award from the University of Texas and was selected to represent Bergen Community College in a campaign to increase awareness of the contributions of community colleges to higher education.

Jerry is the author of several best-selling science textbooks and laboratory manuals, a calling that often requires an additional 40 hours per week beyond his teaching responsibilities. Nevertheless, he still makes time for four or five weekly aerobic workouts that include biking and running. He also enjoys attending college basketball and professional hockey games and performances at the Metropolitan Opera House.

To my children, Lynne, Gerard, Kenneth, Anthony, and Andrew, who make it all worthwhile. — *G.J.T.*

Sandra Reynolds Grabowski is an instructor in the Department of Biological Sciences at Purdue University in West Lafayette, Indiana. For 25 years, she has taught human anatomy and physiology to students in a wide range of academic programs. In 1992 students selected her as one of the top ten teachers in the School of Science at Purdue.

Sandy received her BS in biology and her PhD in neurophysiology from Purdue. She is a founding member of the Human Anatomy and Physiology Society (HAPS) and has served HAPS as President and as editor of *HAPS News*. In addition, she is a member of the American Anatomy Association, the Association for Women in Science (AWIS), the National Science Teachers Association (NSTA), and the Society for College Science Teachers (SCST).

To students around the world, whose questions and comments continue to inspire the fine-tuning of this textbook. — *S.R.G.*

Preface

An anatomy and physiology course can be the gateway to a gratifying career in a host of health-related professions. As active teachers of the course, we recognize both the rewards and challenges in providing a strong foundation for understanding the complexities of the human body to an increasingly diverse population of students. Building on the unprecedented success of previous editions, the tenth edition of *Principles of Anatomy and Physiology* continues to offer a balanced presentation of content under the umbrella of our primary and unifying theme of homeostasis, supported by relevant discussions of disruptions to homeostasis. In addition, years of student feedback have convinced us that readers learn anatomy and physiology more readily when they remain mindful of the relationship between structure and function. As a writing team—an anatomist and a physiologist—our very different specializations offer practical advantages in fine-tuning the balance between anatomy and physiology.

Most importantly, our students continue to remind us of their needs for—and of the power of—simplicity, directness, and clarity. To meet these needs each chapter has been written and revised to include:

- clear, compelling, and up-to-date discussions of anatomy and physiology
- expertly executed and generously sized art
- classroom-tested pedagogy
- outstanding student study support.

As we revised the content for this edition we kept our focus on these important criteria for success in the anatomy and physiology classroom and have refined or added new elements to enhance the teaching and learning process.

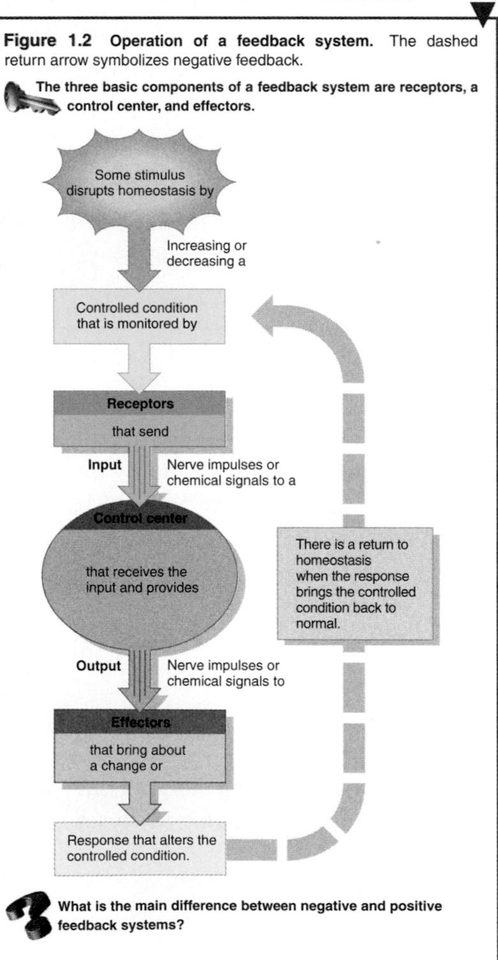

Homeostasis: A Unifying Theme

The dynamic physiological constancy known as homeostasis is the prime theme in *Principles of Anatomy and Physiology*. We immediately introduce this unifying concept in Chapter 1 and describe how various feedback mechanisms work to maintain physiological processes within the narrow range that is compatible with life. Homeostatic mechanisms are discussed throughout the book, and homeostatic processes are clarified and reinforced through our well-received series of homeostasis feedback illustrations.

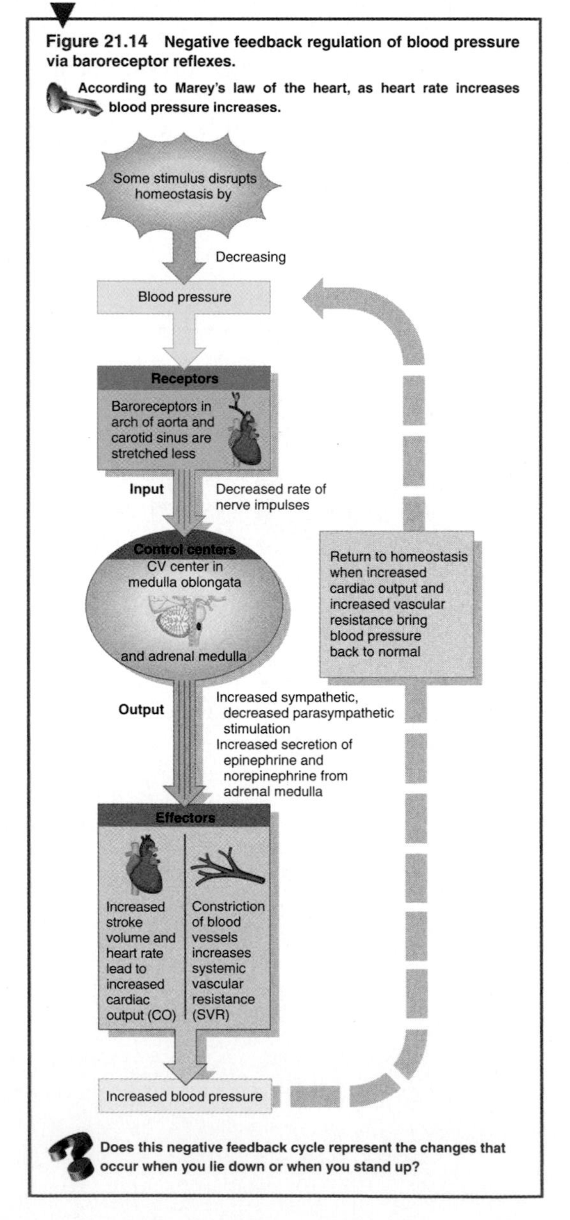

New to this edition are ten *Focus on Homeostasis* pages, one each for the integumentary, skeletal, muscular, nervous, endocrine, cardiovascular, lymphatic and immune, respiratory, digestive, and urinary systems. Incorporating both graphic and narrative elements, these pages explain, clearly and succinctly, how the system under consideration contributes to the homeostasis of each of the other body systems. Use of this feature will enhance student understanding of the links between body systems and how interactions among systems contribute to the homeostasis of the body as a whole

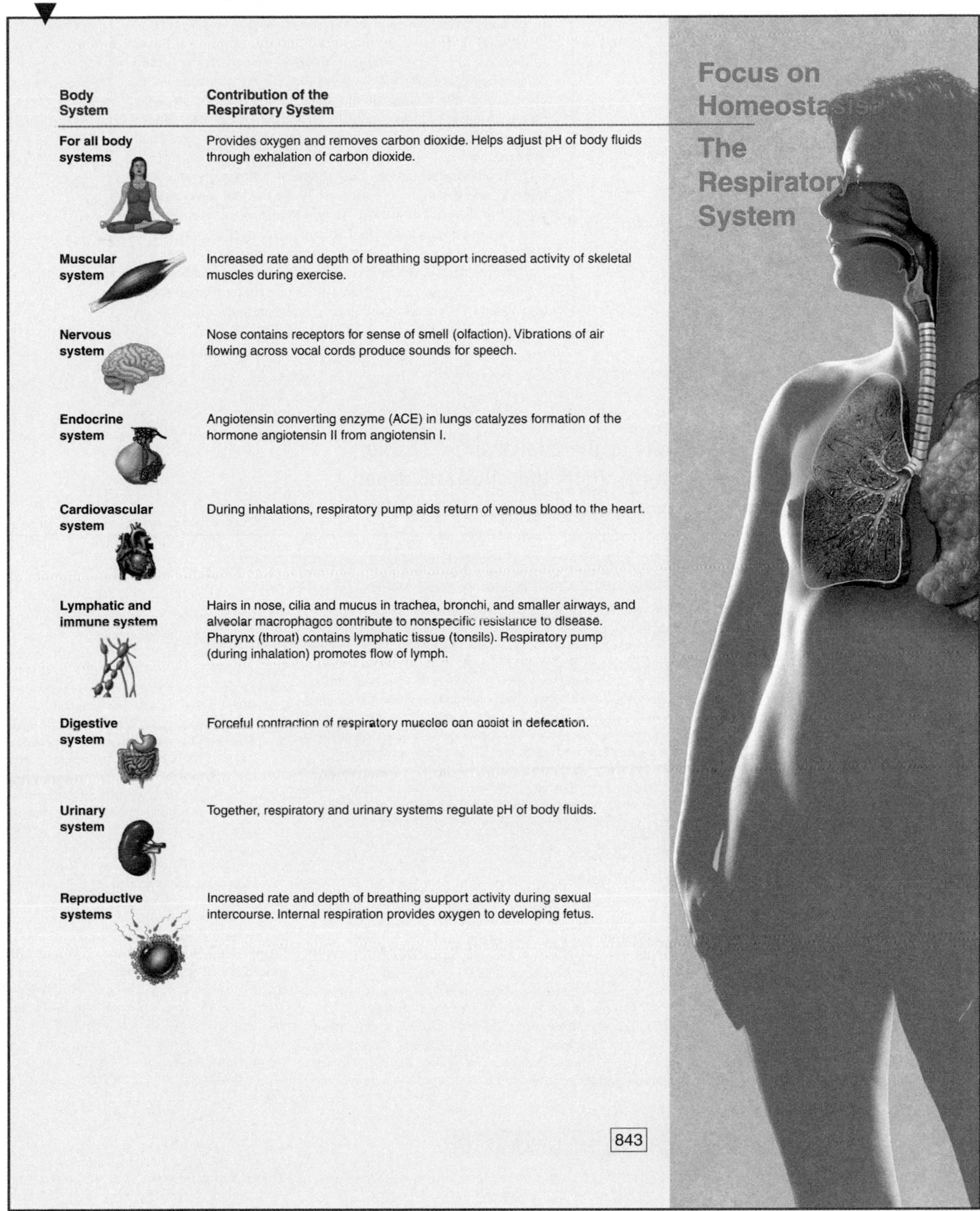

Focus on Homeostasis

The Respiratory System

Body System	Contribution of the Respiratory System
For all body systems	Provides oxygen and removes carbon dioxide. Helps adjust pH of body fluids through exhalation of carbon dioxide.
Muscular system	Increased rate and depth of breathing support increased activity of skeletal muscles during exercise.
Nervous system	Nose contains receptors for sense of smell (olfaction). Vibrations of air flowing across vocal cords produce sounds for speech.
Endocrine system	Angiotensin converting enzyme (ACE) in lungs catalyzes formation of the hormone angiotensin II from angiotensin I.
Cardiovascular system	During inhalations, respiratory pump aids return of venous blood to the heart.
Lymphatic and immune system	Hairs in nose, cilia and mucus in trachea, bronchi, and smaller airways, and alveolar macrophages contribute to nonspecific resistance to disease. Pharynx (throat) contains lymphatic tissue (tonsils). Respiratory pump (during inhalation) promotes flow of lymph.
Digestive system	Forceful contraction of respiratory muscles can assist in defecation.
Urinary system	Together, respiratory and urinary systems regulate pH of body fluids.
Reproductive systems	Increased rate and depth of breathing support activity during sexual intercourse. Internal respiration provides oxygen to developing fetus.

843

In addition, we believe students can better understand normal physiological processes by examining situations in which diseases or disorders impair those processes. We offer three features that highlight these disruptions to homeostasis.

A perennial favorite among students, the intriguing **Clinical Applications** in every chapter explore the clinical, professional, or everyday relevance of a particular anatomical structure or its related function. Many are new to this edition, and all have been reviewed for accuracy and relevance. Each application directly follows the discussion to which it relates.

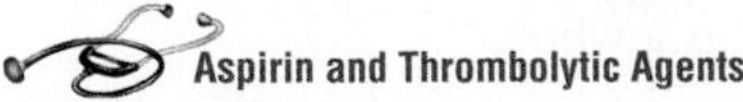

Aspirin and Thrombolytic Agents

In patients with heart and blood vessel disease, the events of hemostasis may occur even without external injury to a blood vessel. At low doses, **aspirin** inhibits vasoconstriction and platelet aggregation by blocking synthesis of thromboxane A2. It also reduces the chance of thrombus formation. Due to these effects, aspirin reduces the risk of transient ischemic attacks (TIA), strokes, myocardial infarction, and blockage of peripheral arteries.

Thrombolytic agents are chemical substances that are injected into the body to dissolve blood clots that have already formed to restore circulation. They either directly or indirectly activate plasminogen. The first thrombolytic agent, approved in 1982 for dissolving clots in the coronary arteries of the heart, was **streptokinase,** which is produced by streptococcal bacteria. A genetically engineered version of human **tissue plasminogen activator (t-PA)** is now used to treat both heart attacks and brain attacks (strokes) that are caused by blood clots. ■

The **Disorders: Homeostatic Imbalances** sections at the end of most chapters include concise discussions of major diseases and disorders that illustrate departures from normal homeostasis. They provide answers to many questions that students ask about medical problems.

DISORDERS: HOMEOSTATIC IMBALANCES

Disorders of epithelial tissues are mainly specific to individual organs, for example, peptic ulcer disease (PUD), which erodes the epithelial lining of the stomach or small intestine. For this reason, epithelial disorders are described together with the relevant body system throughout the text. The most prevalent disorders of connective tissues are **autoimmune diseases**—diseases in which antibodies produced by the immune system fail to distinguish what is foreign from what is self and attack the body's own tissues. The most common autoimmune disorder is rheumatoid arthritis, which attacks the synovial membranes of joints. Because connective tissue is one of the most abundant and widely distributed of the four main types of tissues, its disorders often affect multiple body systems. Common disorders of muscle tissue and nervous tissue are described at the ends of Chapters 10 and 12, respectively.

Sjögren's Syndrome

Sjögren's syndrome (SHŌ-grenz) is a common autoimmune disorder that causes inflammation and destruction of exocrine glands, especially the lacrimal (tear) glands and salivary glands. Signs include dryness of the mucous membranes in the eyes and mouth and salivary gland enlargement. Systemic effects include arthritis, difficulty swallowing, pancreatitis (inflammation of the pancreas), pleuritis (inflammation of the pleurae of the lungs), and migraine headaches. The disorder affects more females than males by a ratio of 9 to 1. About 20% of older adults experience some signs of Sjögren's. Treatment is supportive, including using artificial tears to moisten the eyes, sipping fluids, chewing sugarless gum, and using a saliva substitute to moisten the mouth.

Systemic Lupus Erythematosus

Systemic lupus erythematosus (er-i-thē-ma-TŌ-sus), **SLE,** or simply lupus, is a chronic inflammatory disease of connective tissue occurring mostly in nonwhite women during their childbearing years. It is an autoimmune disease that can cause tissue damage in every body system. The disease, which can range from a mild condition in most patients to a rapidly fatal disease, is marked by periods of exacerbation and remission. The prevalence of SLE is about 1 in 2000 persons, with females more likely to be afflicted than males by a ratio of 8 or 9 to 1.

Although the cause of SLE is not known, genetic, environmental, and hormonal factors are implicated. The genetic component is suggested by studies of twins and family history. Environmental factors include viruses, bacteria, chemicals, drugs, exposure to excessive sunlight, and emotional stress. With regard to hormones, sex hormones, such as estrogens, may trigger SLE.

Signs and symptoms of SLE include painful joints, low-grade fever, fatigue, mouth ulcers, weight loss, enlarged lymph nodes and spleen, sensitivity to sunlight, rapid loss of large amounts of scalp hair, and anorexia. A distinguishing feature of lupus is an eruption across the bridge of the nose and cheeks called a "butterfly rash." Other skin lesions may occur, including blistering and ulceration. The erosive nature of some SLE skin lesions was thought to resemble the damage inflicted by the bite of a wolf—thus, the term *lupus* (= wolf). The most serious complications of the disease involve inflammation of the kidneys, liver, spleen, lungs, heart, brain, and gastrointestinal tract. Because there is no cure for SLE, treatment is supportive, including anti-inflammatory drugs, such as aspirin, and immunosuppressive drugs.

Vocabulary-building glossaries of selected **Medical Terminology** and conditions also appear at the end of appropriate chapters. This feature has been expanded and updated for this edition.

MEDICAL TERMINOLOGY

Atrophy (AT-rō-fē; *a-* = without; *-trophy* = nourishment) A decrease in the size of cells, with a subsequent decrease in the size of the affected tissue or organ.

Hypertrophy (hī-PER-trō-fē; *hyper-* = above or excessive) Increase in the size of a tissue because its cells enlarge without undergoing cell division.

Tissue rejection An immune response of the body directed at foreign proteins in a transplanted tissue or organ; immunosuppressive drugs, such as cyclosporine, have largely overcome tissue rejection in heart-, kidney-, and liver-transplant patients.

Tissue transplantation The replacement of a diseased or injured tissue or organ. The most successful transplants involve use of a person's own tissues or those from an identical twin.

Xenotransplantation (zen′-ō-trans-plan-TĀ-shun; *xeno-* = strange, foreign) The replacement of a diseased or injured tissue or organ with cells or tissues from an animal. Porcine (from pigs) and bovine (from cows) heart valves are used for some heart-valve replacement surgeries.

Organization, Special Topics, and Content Improvements

The book follows the same unit and topic sequence as its nine earlier editions. It is divided into five principal sections: Unit 1, "Organization of the Human Body," provides an understanding of the structural and functional levels of the body, from molecules to organ systems. Unit 2, "Principles of Support and Movement," analyzes the anatomy and physiology of bones, joints, and muscles. Unit 3, "Control Systems of the Human Body," emphasizes the importance of neural communication in the immediate maintenance of homeostasis, the role of sensory receptors in providing information about the internal and external environment, and the significance of hormones in maintaining long-term homeostasis. Unit 4, "Maintenance of the Human Body," explains how body systems function to maintain homeostasis on a moment-to-moment basis through the processes of circulation, respiration, digestion, cellular metabolism, urinary functions, and buffer systems. Unit 5, "Continuity," covers the anatomy and physiology of the reproductive systems, development, and the basic concepts of genetics and inheritance.

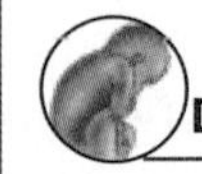

DEVELOPMENT OF THE HEART

▶ OBJECTIVE

- **Describe the development of the heart.**

The *heart,* a derivative of **mesoderm,** begins to develop early in the third week after fertilization. In the ventral region of the embryo inferior to the foregut, the heart develops from a group of mesodermal cells called the **cardiogenic area.** The next step in its development is the formation of a pair of tubes, the **endocardial tubes,** which develop from the cardiogenic area (Figure 20.18). These tubes then unite to form a common tube, referred to as the **primitive heart tube.** Next, the primitive heart tube develops into five distinct regions: (1) **truncus arteriosus,** (2) **bulbus cordis,** (3) **ventricle,** (4) **atrium,** and (5) **sinus venosus.** Because the bulbus cordis and ventricle grow most rapidly, and because the heart enlarges more rapidly than its superior and inferior attachments, the heart first assumes a U-shape and then an S-shape. The flexures of the heart reorient the regions so that the atrium and sinus venosus eventually come to lie superior to the bulbus cordis, ventricle, and truncus arteriosus. Contractions

Development We often tell our students that they can better appreciate the "logic" of human anatomy by becoming aware of how various structures developed in the first place. As in previous editions, illustrated discussions of development are found near the conclusion of most body system chapters. Placing this coverage at the end of chapters enables students to master the anatomical terminology they need before attempting to learn about embryonic and fetal structures. The fetus icon designates the start of each development discussion.

AGING AND THE IMMUNE SYSTEM

▶ OBJECTIVE

- **Describe the effects of aging on the immune system.**

With advancing age, most people become more susceptible to all types of infections and malignancies. Their response to vaccines is decreased, and they tend to produce more autoantibodies (antibodies against their body's own molecules). In addition, the immune system exhibits lowered levels of function. For example, T cells become less responsive to antigens, and fewer T cells respond to infections. This may result from age-related atrophy of the thymus or decreased production of thymic hormones. Because the T cell population decreases with age, B cells are also less responsive. Consequently, antibody levels do not increase as rapidly in response to a challenge by an antigen, resulting in increased susceptibility to various infections. It is for this key reason that elderly individuals are encouraged to get influenza (flu) vaccinations each year.

Aging Students need to be reminded from time to time that anatomy and physiology is not static. As the body ages, its structures and related functions subtly change. Moreover, aging is a professionally relevant topic for the majority of this book's readers, who will go on to careers in *health-related* fields in which the average age of the client population is steadily advancing. For these reasons, age-related changes in anatomy and physiology are discussed at the end of sixteen chapters.

EXERCISE AND BONE TISSUE

▶ OBJECTIVE

- **Describe how exercise and mechanical stress affect bone tissue.**

Within limits, bone has the ability to alter its strength in response to changes in mechanical stress. When placed under stress, bone tissue adapts by becoming stronger through increased deposition of mineral salts and production of collagen fibers. Another effect of stress is to increase the production of calcitonin, a hormone that inhibits bone resorption. Without mechanical stress, bone does not remodel normally because bone resorption outstrips bone formation. Removal of mechanical stress weakens bone through demineralization (loss of bone minerals) and decreased numbers of collagen fibers.

The main mechanical stresses on bone are those that result from the pull of skeletal muscles and the pull of gravity. If a person is bedridden or has a fractured bone in a cast, the strength of the unstressed bones diminishes. Astronauts subjected to the microgravity of space also lose bone mass. In both cases, bone loss can be dramatic—as much as 1% per week. Bones of athletes, which are repetitively and highly stressed, become notably thicker and stronger than those of nonathletes. Weight-bearing activities, such as walking or moderate weight lifting, help build and retain bone mass. Adolescents and young adults should engage in regular weight-bearing exercise prior to the closure of the epiphyseal plates to help build total mass prior to its inevitable reduction with aging. Even elderly people can strengthen their bones by engaging in weight-bearing exercise.

Exercise Physical exercise can produce favorable changes in some anatomical structures and enhance many physiological functions, most notably those associated with the muscular, skeletal, and cardiovascular systems. This information is especially relevant to readers embarking on careers in physical education, sports training, and dance. Hence, key chapters include brief discussions of exercise-related considerations, which are signaled by a distinctive running shoe icon.

Every chapter in the tenth edition of *Principles of Anatomy and Physiology* incorporates a host of improvements to both the text and the art, many suggested by reviewers, educators, and students. In addition, most chapters offer new Clinical Applications. Here are some of the more noteworthy changes.

Chapter 2 / The Chemical Level of Organization
A new Clinical Application on "Saturated Fats, Cholesterol, and Atherosclerosis." Some sections are more concise, and the complete discussion of oxidation–reduction reactions is now in chapter 25.

Chapter 3 / The Cellular Level of Organization
Description of a newly discovered cellular organelle, the proteasome. New Clinical Applications on Medical Uses of Isotonic, Hypertonic, and Hypotonic Solutions; Proteasomes and Disease; Genomics; and Antisense Therapy. Transport through the Golgi complex, tumor-suppressor genes, and causes of cancer updated and revised.

Chapter 4 / The Tissue Level of Organization
New illustrations of cell junctions and types of multicellular exocrine glands; several new photomicrographs. Revisions in sections on cell junctions, the basement membrane, and ground substance of connective tissue.

Chapter 5 / The Integumentary System
New section on types of hairs. New Clinical Applications on Photosensitivity, Hair Removal, and Hair and Hormones; New illustrations of the rule of nines, types of epidermal cells, epidermal wound healing, and deep wound healing. Structure of epidermis and development of the integumentary system revised.

Chapter 6 / The Skeletal System: Bone Tissue
New illustrations of the histology of bone, endochondral ossification, and fracture repair.

Chapter 8 / The Skeletal System: The Appendicular Skeleton
New illustration of the hip bone. Addition of Hip Fracture to the disorders section at the end of the chapter.

Chapter 9 / Joints
New Clinical Applications on Rotator Cuff Injuries, Separated Shoulder, Tennis Elbow, Little League Elbow, Dislocated Radius, Swollen Knee, Dislocated Knee, and Lyme Disease.

Chapter 10 / Muscle Tissue
New coverage of exercise and skeletal muscles, fibromyalgia, and benefits of stretching. New illustrations on microscopic organization of skeletal muscle tissue, the neuromuscular junction, and development of the muscular system. Proteins in skeletal muscle and uses of Botox revised.

Chapter 11 / The Muscular System
New illustration of a newly discovered skeletal muscle, the sphenomandibularis muscle (see page 325).

Chapter 12 / Nervous Tissue New Medical Terminology list. New illustrations of the structure of a neuron, myelinated and unmyelinated axons, and signal transmission at synapses. Updated definitions of sensory neuron, interneuron, and motor neuron.

Chapter 13 / The Spinal Cord and Spinal Nerves
New Medical Terminology list. New illustration of the connective tissue coverings of a spinal nerve.

Chapter 14 / The Brain and Cranial Nerves
New illustrations of the parts of the brain, protective coverings of the brain, pathways of cerebrospinal fluid flow, thalamus, development of the brain, seven figures on cranial nerves. Revised discussions of the thalamus and cranial nerves.

Chapter 15 / Sensory, Motor and Integrative Systems
New illustrations of muscle spindles, tendon organs, somatic sensory pathways, and motor pathways. New Clinical Applications on Phantom Limb Sensation, Amyotrophic Lateral Sclerosis, Damage to Basal Ganglia, and Damage to the Cerebellum. Section on Muscle Spindles rewritten to simplify terminology but more clearly explain function; sections on somatic motor pathways, memory, and sleep updated.

Chapter 16 / The Special Senses
New illustrations of taste buds, effects of tastants on taste neurons, and development of the eyes and ears. New Clinical Applications on Hyposmia, Macular Degeneration, Taste Aversion, Presbyopia, and Color and Night Blindness. New section on Development of the Eyes and Ears. Expanded Medical Terminology list.

Chapter 17 / The Autonomic Nervous System
New Table comparing sympathetic and parasympathetic divisions of the autonomic nervous system; new illustrations of sympathetic and parasympathetic divisions.

Chapter 18 / The Endocrine System
New illustrations of the location of endocrine glands, thyroid gland, parathyroid glands, adrenal glands, pancreas, and development of the endocrine system. New Clinical Applications on Administering Hormones, Oxytocin and Childbirth, Congenital Adrenal Hyperplasia, and Posttraumatic Stress Disorder. Photographs of people with endocrine disorders, including an interesting case of a twin with giantism (see page 625). New Medical Terminology list.

Chapter 19 / The Blood
New description of red bone marrow. New Clinical Applications on Iron Overload and Tissue Damage, Induced Polycythemia in Athletes, and Aspirin and Thrombolytic Agents.

Chapter 20 / The Heart
New section on ATP production in cardiac muscle fibers. New Clinical Applications on Heart Valve Disorders, Reperfusion Damage, Ectopic Pacemaker, Regeneration of Cardiac Muscle Tissue, Artificial Hearts. New illustrations of the serous pericardium, structure of the heart, coronary circulation, pacemaker potentials, route of action potential depolarization and repolarization, and development of the heart. Arrhythmias added to disorders section. Expanded Medical Terminology list.

Chapter 21 / Blood Vessels and Hemodynamics
New illustrations of the structure of blood vessels, types of capillaries, action of skeletal muscles, ANS innervation of the heart,

circulatory routes, principal branches of the aorta, ascending aorta, branches of the brachiocephalic trunk in the neck, cerebral arterial circle, branches of the abdominal aorta, arteries of the pelvis and lower limbs, pulmonary circulation, fetal circulation, and development of blood vessels and blood cells.

Chapter 22 / The Lymphatic and Immune System and Resistance to Disease
New section on stress and immunity. New illustrations of lymphatic tissue, routes for lymph drainage, lymph node structure, development of the lymphatic system, phagocytosis, and stages of inflammation. New Clinical Applications on Ruptured Spleen, Edema and Lymph Flow, and Microbial Evasion of Phagocytosis. Autoimmune diseases, severe combined immunodeficiency disease, SLE, and lymphomas added to the disorders section. Descriptions of thymus and lymph nodes revised.

Chapter 23 / The Respiratory System
New illustrations of the structures of the respiratory system, location of peripheral chemoreceptors that help regulate respiration, and development of the bronchial tubes and lungs. New Clinical Applications on Laryngitis, Cancer of the Larynx, and Nebulization. New discussion and explanation of ventilation-perfusion coupling, nitrogen narcosis, and decompression sickness. Addition of ARDS to disorders section. Updated percentages of carbon dioxide dissolved in blood plasma and carried by bicarbonate ions and hemoglobin.

Chapter 24 / The Digestive System
New illustrations of organs of the gastrointestinal tract, peritoneal folds, salivary glands, histology of the stomach, secretion of HCl by stomach cells, histology of the small intestine, anatomy of the large intestine, and histology of the large intestine. New Clinical Applications on Root Canal Therapy, Jaundice, Occult Blood, and Absorption of Alcohol.

Chapter 25 / Metabolism
New section on energy homeostasis. New Clinical Application on Emotional Eating and new Medical Terminology list. Updated sections on regulation of food intake and obesity.

Chapter 26 / The Urinary System
New illustrations of a sagittal section through the kidney, blood supply of the kidneys, anatomy of the ureter, urinary bladder, and urethra, and development of the urinary system. More concise sections on reabsorption and secretion. New Clinical Applications on Nephroptosis and Loss of Plasma Proteins. Addition of Urinary Bladder Cancer to the disorders section.

Chapter 27 / Fluid, Electrolyte, and Acid–Base Homeostasis
New Summary table of factors that maintain body water balance. Discussion and related figure describing how kidneys contribute to acid–base balance by secreting hydrogen ions moved from Chapter 26 to this chapter.

Chapter 28 / The Reproductive Systems
New illustrations to compare mitosis and meiosis and to show development of internal reproductive organs. New table reviewing oogenesis and development of ovarian follicles; new explanation of mittelschmerz and the role of leptin in puberty. New Clinical Applications on Uterine Prolapse, Episiotomy, and Female Athlete Triad. Addition of genital warts and premenstrual dysphoric disorder to the disorders section.

Chapter 29 / Development and Inheritance
New illustrations of fertilization, cleavage, formation of primary germ layers, gastrulation, development of the notochordal process, neurulation, development of chorionic villi, embryo folding, and development of pharyngeal arches. Beautiful new photographs of embryonic and fetal development. New Clinical Applications on Anencephaly and Stem Cell Research. Addition of Trinucleotide Repeat Diseases to the disorders section. Updated and revised sections on embryonic and fetal development, noninvasive prenatal tests, and inheritance.

Enhancement to the Illustration Program

New Design A textbook with beautiful illustrations or photographs on most pages requires a carefully crafted and functional design. The new design for the tenth edition has been transformed to assist students in making the most of the text's many features and outstanding art. A larger trim size provides even more space for the already large and highly praised illustrations and accommodates the larger font size for easier reading. Each page is carefully laid out to place related text, figures, and tables near one another, minimizing the need for page turning while reading a topic. New to this edition is the red print used to indicate the first mention of a figure or table. Not only is the reader alerted to refer to the figure or table, but the color print also serves as a place locator for easy return to the narrative.

Distinctive icons incorporated throughout the chapters signal special features and make them easy to find during review. These include the **key** with Key Concepts Statements; the **question mark** with the applicable questions that enhance every figure; the **stethoscope** indicating a clinical application within the chapter narrative; the **fetus icon** announcing the developmental anatomy section; the **running shoe** highlighting content relevant to exercise and the icons that indicate the study outline and distinctive types of **chapter-ending questions.**

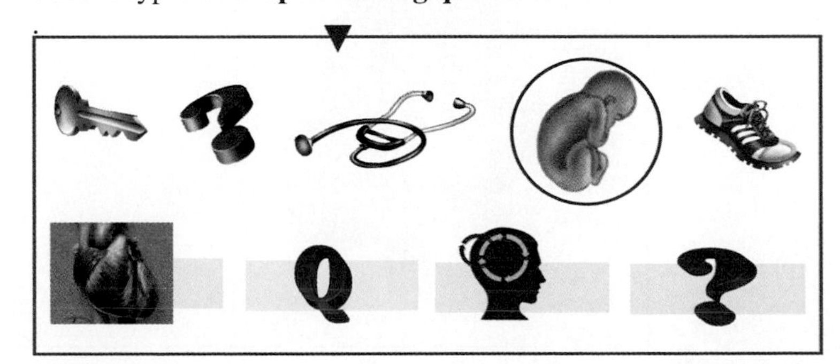

New Art Studying human anatomy and physiology is both a visual and a descriptive enterprise. Countless students have now benefited from the detailed and generously proportioned bone and muscle art that characterizes our book. Continuing in this tradition, you will find exciting new three-dimensional illustrations gracing the pages of virtually every chapter in the book. More than 90 figures are new to this edition, and nearly every figure has been revised or improved in some way.

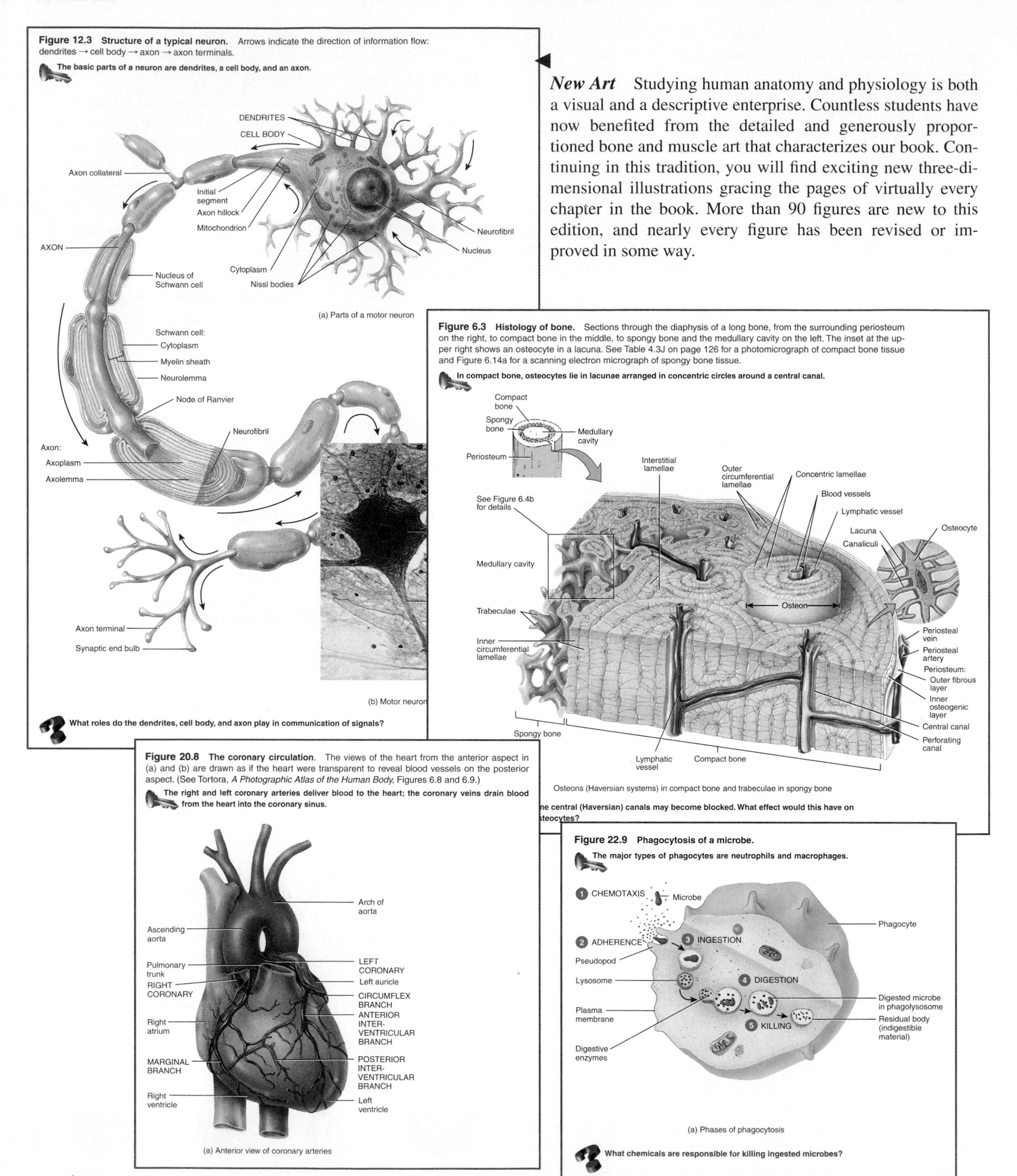

New Histology-Based Art As part of our goal for continuous improvement, many of the anatomical illustrations based on histological preparations have been revised and redrawn for this edition.

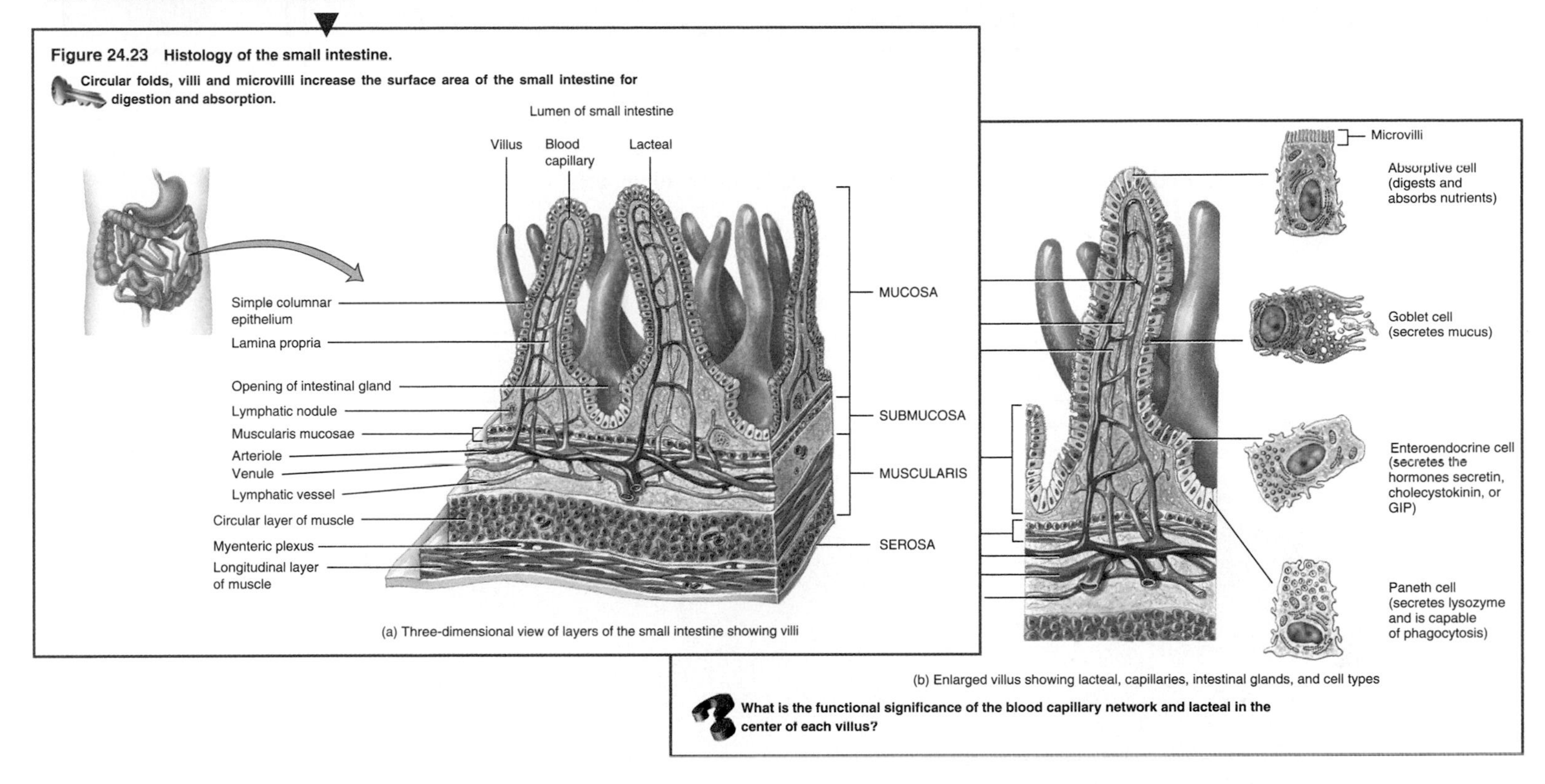

Figure 24.23 Histology of the small intestine.

Circular folds, villi and microvilli increase the surface area of the small intestine for digestion and absorption.

(a) Three-dimensional view of layers of the small intestine showing villi

(b) Enlarged villus showing lacteal, capillaries, intestinal glands, and cell types

What is the functional significance of the blood capillary network and lacteal in the center of each villus?

New Photomicrographs Dr. Michael Ross of the University of Florida has again provided us with more beautiful, customized photomicrographs of various tissues of the body. We have always considered Dr. Ross' photos among the best available and their inclusion in this edition greatly enhances the illustration program.

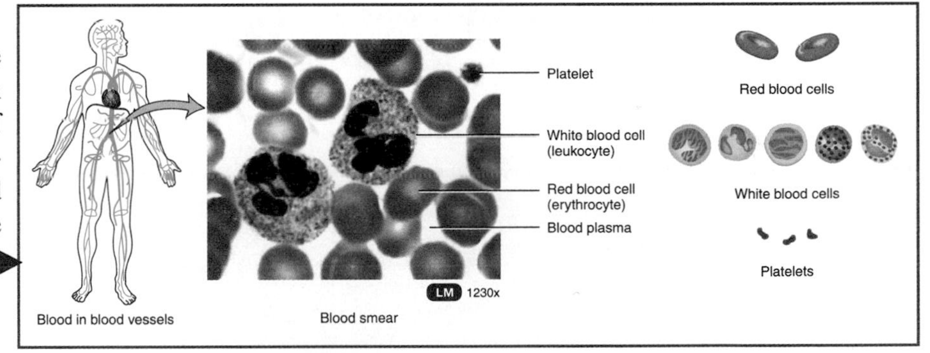

Cadaver Photos As before, we provide an assortment of large, clear cadaver photos at strategic points in many chapters. What is more, many anatomy illustrations are keyed to the large cadaver photos available in a companion text, *A Photographic Atlas of the Human Body*, by Gerard J. Tortora.

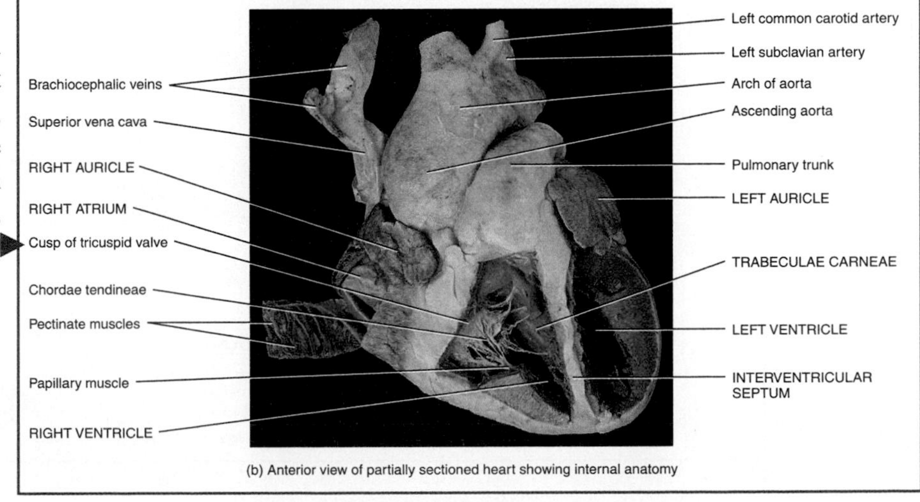

(b) Anterior view of partially sectioned heart showing internal anatomy

Helpful Orientation Diagrams Students sometimes need help figuring out the plane of view of anatomy illustrations — descriptions alone do not always suffice. An orientation diagram that depicts and explains the perspective of the view represented in the figure accompanies every major anatomy illustration. There are three types of diagrams: (1) planes used to indicate where certain sections are made when a part of the body is cut; (2) diagrams containing a directional arrow and the word "View" to indicate the direction from which the body part is viewed, and (3) diagrams with arrows leading from or to them that direct attention to enlarged and detailed parts of illustrations.

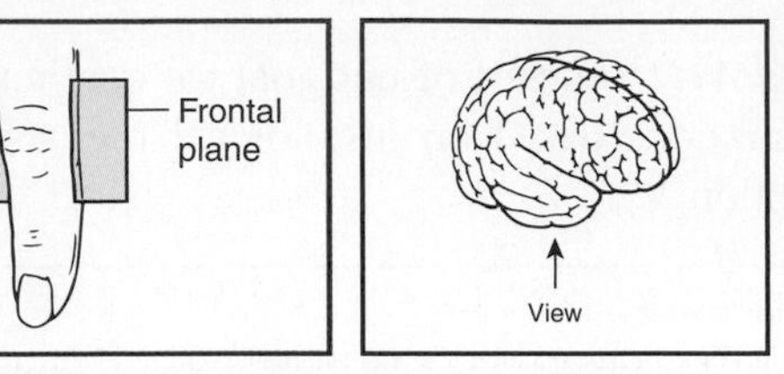

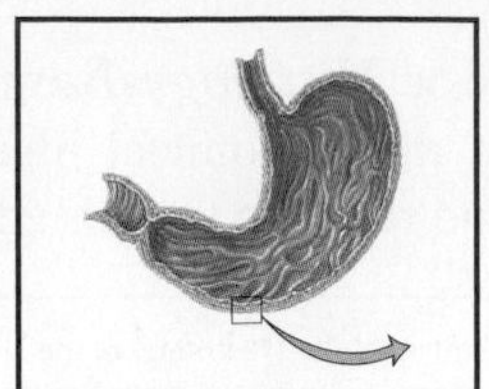

Key Concept Statements This art-related feature summarizes an idea that is discussed in the text and demonstrated in a figure. Each Key Concept Statement is positioned adjacent to its figure and is denoted by a distinctive key icon.

Figure 21.1 Comparative structure of blood vessels. The relative size of the capillary in (c) is enlarged.

Arteries carry blood from the heart to tissues; veins carry blood from tissues to the heart.

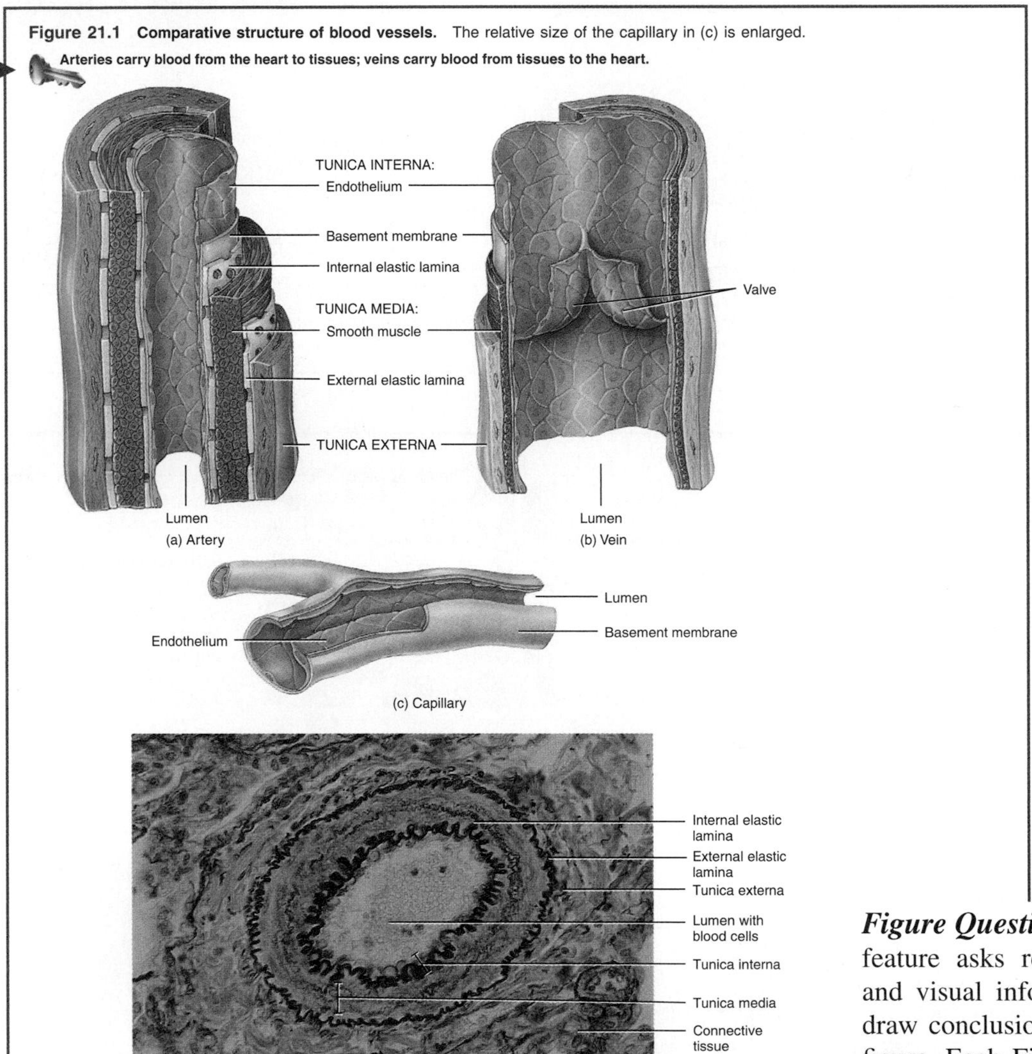

Figure Questions This highly applauded feature asks readers to synthesize verbal and visual information, think critically, or draw conclusions about what they see in a figure. Each Figure Question appears adjacent to its illustration and is highlighted in this edition by the purple question mark icon. Answers are located at the end of each chapter.

Which vessel—the femoral artery or the femoral vein—has a thicker wall? Which has a wider lumen?

Correlation of Sequential Processes Correlation of sequential processes in text and art is achieved through the use of numbered lists in the narrative that correspond to numbered segments in the accompanying figure. This approach is used extensively throughout the book to lend clarity to the flow of complex processes.

Figure 20.10 The conduction system of the heart. Autorhythmic fibers in the SA node, located in the right atrial wall (a), act as the heart's pacemaker, initiating cardiac action potentials (b) that cause contraction of the heart's chambers.

The conduction system ensures that the chambers of the heart contract in a coordinated manner.

Frontal plane

Left atrium

Right atrium

1 SINOATRIAL (SA) NODE

2 ATRIOVENTRICULAR (AV) NODE

3 ATRIOVENTRICULAR (AV) BUNDLE (BUNDLE OF HIS)

4 RIGHT AND LEFT BUNDLE BRANCHES

Left ventricle

Right ventricle

5 PURKINJE FIBERS

(a) Anterior view of frontal section

+ 10 mV

Membrane potential

Threshold

– 60 mV

Action potential

Pacemaker potential

0 0.8 1.6 2.4

Time (sec)

(b) Pacemaker potentials and action potentials in autorhythmic fibers of SA node

Which component of the conduction system provides the only electrical connection between the atria and the ventricles?

Functions Overview This feature succinctly lists the functions of an anatomical structure or body system depicted within a figure. The juxtaposition of text and art further reinforces the connection between structure and function.

OVERVIEW OF KIDNEY FUNCTIONS

OBJECTIVE

- **List the functions of the kidneys.**

The kidneys do the major work of the urinary system. The other parts of the system are mainly passageways and storage areas. Functions of the kidneys include:

- *Excreting wastes and foreign s*... urine, the kidneys help excrete ... have no useful function in the bod... in urine result from metabolic reac... include ammonia and urea from th... acids; bilirubin from the catabolism of hemoglobin; creatinine from the breakdown of creatine phosphate in muscle fibers; and uric acid from the catabolism of nucleic acids.

Figure 26.1 Organs of the urinary system in a female. (See Tortora, *A Photographic Atlas of the Human Body,* Figure 13.2.)

Urine formed by the kidneys passes first into the ureters, then to the urinary bladder for storage, and finally through the urethra for elimination from the body.

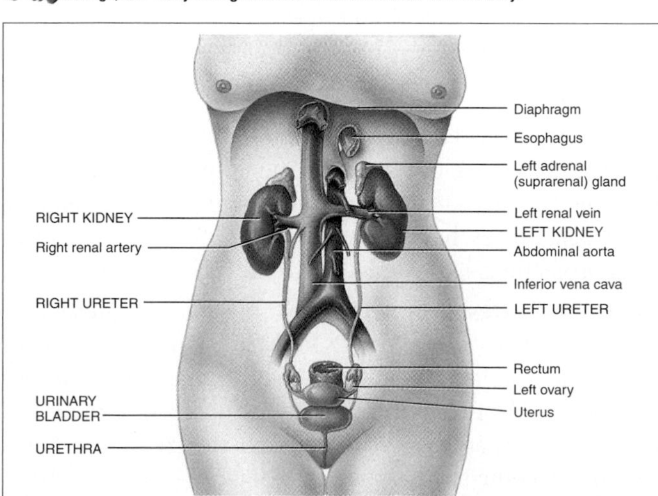

Functions of the Urinary System

1. **The kidneys excrete wastes in urine, regulate blood volume and composition, help regulate blood pressure, synthesize glucose, release erythropoietin, and participate in vitamin D synthesis.**
2. **The ureters transport urine from the kidneys to the urinary bladder.**
3. **The urinary bladder stores urine.**
4. **The urethra discharges urine from the body.**

Which organs constitute the urinary system?

19

Hallmark Features

The tenth edition of *Principles of Anatomy and Physiology* builds on the legacy of thoughtfully designed and class-tested pedagogical features that provide a complete learning system for students as they navigate their way through the text and course. All have been revised to reflect the enhancements to the text.

Chapter-opening Pages Each chapter now begins with a chapter-opening page that includes a beautiful new piece of art related to the body system under consideration. A perspective on the chapter's content from either a recent student of the course or a practitioner in a health related field introduces current users to the relevance of the material or particularly interesting portions of the chapter. Suggestions for energizing their study with dynamic multimedia activities on the Foundations CD included with the text and specially developed internet activities on the companion website are highlighted.

Helpful Exhibits Students of anatomy and physiology need extra help learning the many structures that constitute certain body systems—most notably skeletal muscles, articulations, blood vessels, and nerves. As in previous editions, the chapters that present these topics are organized around **Exhibits,** each of which consists of an overview, a tabular summary of the relevant anatomy, and an associated suite of illustrations or photographs. Each Exhibit is prefaced by an Objective and closes with a Checkpoint activity. We trust you will agree that our spaciously designed Exhibits are ideal study vehicles for learning anatomically complex body systems.

Student Objectives and Checkpoints **Objectives** are found at the beginning of major sections throughout each chapter. Complementing this format, **Checkpoint** questions appear at strategic intervals within chapters to give students the chance to validate their understanding as they read.

Tools for Mastering Vocabulary Students—even the best ones—generally find it difficult at first to read and pronounce anatomical and physiological terms. Moreover, as teachers we are sympathetic to the needs of the growing ranks of college students who speak English as a second language. For these reasons, we have endeavored to ensure that this book has a strong and helpful vocabulary component. The key terms in every chapter are emphasized by use of **boldface type.** We include **pronunciation guides** when major, or especially hard-to-pronounce, structures and functions are introduced in the discussions, Tables, or Exhibits. **Word roots** citing the Greek or Latin derivations of anatomical terms are offered as an additional aid. As a further service to readers, we provide a list of **Medical Terminology** at the conclusion of most chapters, and a comprehensive **Glossary** at the back of the book. Also included at the end of the book is a list of the basic building blocks of medical terminology—**Combining Forms, Word Roots, Prefixes, and Suffixes.**

Study Outline As always, readers will benefit from the popular end-of-chapter **Study Outline** that is page referenced to the chapter discussions.

End-of-chapter Questions The **Self-Quiz Questions** are written in a variety of styles that are calculated to appeal to readers' different testing preferences. **Critical Thinking Questions** challenge readers to apply concepts to real-life situations. The style of these questions ought to make students smile on occasion as well as think! Answers to the Self-Quiz and Critical Thinking Questions are located in Appendix E.

Complete Teaching And Learning Package

Continuing the tradition of providing a complete teaching and learning package, the tenth edition of *Principles of Anatomy and Physiology* is available with a host of carefully planned supplementary materials that will help you and your students attain the maximal benefit from our textbook. Please contact your Wiley sales representative for additional information about any of these resources.

Media

Dynamic CD-ROMs and an enhanced Companion Website have been developed to complement the tenth edition of *Principles of Anatomy and Physiology.* These are fully described on the inside front cover and include:

- **Student Companion CD-ROM,** provided free with the purchase of a new textbook
- **Dedicated Book Companion Website** free to students who purchase a new textbook
- **Interactions: Exploring the Functions of the Human Body.** The first CD in this dramatic new series—Foundations—is provided free to adopters of the text and to students purchasing a new text.

For Instructors

- **Full-color Overhead Transparencies** This set of full-color acetates includes over 650 figures, including histology micrographs, from the text — complete with figure numbers and captions. Transparencies have enlarged labels for effective use as overhead projections in the classroom. (0-471- 25425-8)
- **Instructor's Resource CD-ROM** For lecture presentation purposes, this cross-platform CD-ROM includes all of the line art from the text in labeled and unlabeled formats. In addition, a pre-designed set of Powerpoint Slides is included with text and images. The slides are easily edited to customize your presentations to your specific needs. (0-471- 25350-2)

• **Professor's Resource Manual** by Lee Famiano of Cuyahoga CC. This on-line manual provides instructors with tools to enhance their lectures. Key features include: Lecture Outlines, What's New and Different in this Chapter, Critical Thinking Questions, Teaching Tips, and a guide to other educational resources. The entire manual is found on the text's dedicated companion website and is fully downloadable. The electronic format allows you to customize lecture outlines or activities for delivery either in print or electronically to your students. (0-471-25355-3)

• **Printed Test Bank** by P. James Nielsen of Western Illinois University. A testbank of nearly 3,000 questions, many new to the tenth edition, is available. A variety of formats—multiple choice, short answer, matching and essay—are provided to accommodate different testing preferences. (0-471-25421-5)

• **Computerized Test Bank** An electronic version of the printed test bank is available on a cross-platform CD-ROM. (0-471-25146-1)

• **Interactions: Exploring the Functions of the Human Body**

• **WebCT** or **Blackboard** Course Management Systems with content prepared by Sharon Simpson are available.

• **Faculty Resource Network –** New from Wiley is a support structure to help instructors implement the dynamic new media that supports this text into their classrooms, laboratories, or online courses. Consult with your Wiley representative for details about this program.

For Students

• **Student Companion CD-ROM**

• **Interactions: Exploring the Functions of the Human Body**

• **Book Companion Website**

• **Learning Guide (0-471-43447-7)** by Kathleen Schmidt Prezbindowski, College of Mount St. Joseph. Designed specifically to fit the needs of students with different learning styles, this well-received guide helps students to more closely examine important concepts through a variety of activities and exercises. The 29 chapters in the *Learning Guide* parallel those of the textbook and include many activities, quizzes and tests for review and study.

• **Illustrated Notebook (0-471-25150-X)** A true companion to the text, this unique notebook is a tool for organized note taking in class and for review during study. Following the sequence in the textbook, each left-handed page displays an unlabeled black and white copy of every text figure. Students can fill in the labels during lecture or lab at the instructor's directions and take additional notes on the lined right-handed pages.

• **Anatomy and Physiology: A Companion Coloring Book (0-471-39515-3)** This helpful study aid features over 500 striking, original illustrations that, by actively coloring them, give students a clear understanding of key anatomical structures and physiological processes.

For the Laboratory

• **A Brief Atlas of the Human Skeleton** This brief photographic review of the human skeleton is provided free with every new copy of the text.

• **Laboratory Manual for Anatomy and Physiology** by Connie Allen and Valerie Harper, Edison Community College **(0-471-39464-5)** This new laboratory manual presents material covered in the 2-semester undergraduate anatomy & physiology laboratory course in a clear and concise way, while maintaining a student-friendly tone. The manual is very interactive and contains activities and experiments that enhance students' ability to both visualize anatomical structures and understand physiological topics.

• **Cat Dissection Manual** by Connie Allen and Valerie Harper, Edison Community College **(0-471-26457-1)** This manual includes photographs and illustrations of the cat along with guidelines for dissection. It is available independently as well as bundled with the main manual depending upon your an adoption needs.

• **Fetal Pig Dissection Manual** by Connie Allen and Valerie Harper, Edison Community College **(0-471-26458-X)** This manual includes photographs and illustrations of the fetal pig along with guidelines for dissection. It is available independently as well as bundled with the main manual depending upon your an adoption needs.

• **Photographic Atlas of the Human Body with Selected Cat, Sheep, and Cow Dissections** by Gerard Tortora **(0-471-37487-3)** This four-colored atlas is designed to support both study and laboratory experiences. Organized by body systems, the clearly labeled photographs provide a stunning visual reference to gross anatomy. Histological micrographs are also included. Many of the illustrations within *Principles of Anatomy and Physiology*, 10e are cross-referenced to this atlas.

Like each of our students, this book has a life of its own. The structure, content, and production values of *Principles of Anatomy and Physiology* are shaped as much by its relationship with educators and readers as by the vision that gave birth to the book ten editions ago. Today you, our readers, are the "heart" of this book. We invite you to continue the tradition of sending in your suggestions to us so that we can include them in the eleventh edition.

Gerard J. Tortora
Department of Science and Health, S229
Bergen Community College
400 Paramus Road
Paramus, NJ 07652

Sandra Reynolds Grabowski
Department of Biological Sciences
1392 Lilly Hall of Life Sciences
Purdue University
West Lafayette, IN 47907-1392
email: Sgrabows@bilbo.bio.purdue.edu

Acknowledgements

For the tenth edition of *Principles of Anatomy and Physiology*, we have enjoyed the opportunity of collaborating with a group of dedicated and talented professionals. Accordingly, we would like to recognize and thank the members of our book team, who often worked evenings and weekends, as well as days, to bring this book to you. At John Wiley & Sons, Inc., our Editor Bonnie Roesch again illuminates the path toward ever better books with her creative ideas and dedication. Bonnie was our valued editor during the seventh and eighth editions—welcome back and thanks! Karin Kincheloe is the Wiley designer whose vision for the tenth edition is a larger, more colorful, more user-friendly new style. Moreover, Karin laid out each page of the book to achieve the best possible placement of text, figures, and other elements. Both instructors and students will appreciate and benefit from the pedagogically effective and visually pleasing design elements that augment the content changes made to this edition. Danke sehr! Claudia Durrell, our Art Coordinator, has collaborated on this text since its seventh edition. Her artistic ability, organizational skills, attention to detail, and understanding of our illustration preferences greatly enhance the visual appeal and style of the figures. She remains a cornerstone of our projects—thank you, Claudia, for all your contributions. Kelly Tavares, Senior Production Editor, demonstrated her untiring expertise during each step of the production process. She coordinated all aspects of actually making and manufacturing the book. Kelly also was "on press" as the book was being printed to ensure the highest possible quality. Muito obrigada, Kelly for all the extra hours you spent to implement book-improving changes! Ellen Ford, Developmental Editor, shepherded the manuscript and electronic files during the revision process. She also contacted many reviewers for their comments and sought out former students to obtain their perspectives on learning anatomy and physiology for the new chapter-opening pages. Hillary Newman, Photo Editor, provided us with all of the photos we requested and did it with efficiency, accuracy, and professionalism. Mary O'Sullivan, Assistant Editor, coordinated the development of many of the supplements that support this text. We are most appreciative! Wiley Editorial Assistants Justin Bow and Kelli Coaxum helped with various aspects of the project and took care of many details. Thanks to all of you!

Jerri K. Lindsey, Tarrant County Junior College, Northeast and Caryl Tickner, Stark State College wrote the end-of-chapter Self-Quiz Questions. Joan Barber of Delaware Technical and Community College contributed the Critical Thinking Questions. Thanks to all three for writing questions that students will appreciate. And a special thank you to Kathleen Prezbindowski, who has authored the *Learning Guide* for many editions. The high quality of her study activities ensures student success.

Outstanding illustrations and photographs have always been a signature feature of *Principles of Anatomy and Physiology*. Respected scientific and medical illustrators on our team of exceptional artists include Mollie Borman, Leonard Dank, Sharon Ellis, Wendy Hiller Gee, Jean Jackson, Keith Kasnot, Lauren Keswick, Steve Oh, Lynn O'Kelley, Hilda Muinos, Tomo Narashima, Nadine Sokol, and Kevin Somerville. Mark Nielsen of the University of Utah provided many of the cadaver photos that appear in this edition. Artists at Imagineering created the amazing computer graphics images and provided all labeling of figures.

Reviewers

We are extremely grateful to our colleagues who reviewed the manuscript and offered insightful suggestions for improvement. The contributions of all these people, who generously provided their time and expertise to help us maintain the book's accuracy and clarity, are acknowledged in the list that follows.

Patricia Ahanotu, Georgia Perimeter College
Cynthia L. Beck, George Mason University
Clinton L. Benjamin, Lower Columbia College
Anna Berkovitz, Purdue University
Charles J. Biggers, University of Memphis
Mark Bloom, Tyler Junior College
Michele Boiani, University of Pennsylvania
Bruce M. Carlson, University of Michigan
Barbara Janson Cohen, Delaware County Community College
Matthew Jarvis Cohen, Delaware County Community College
Marcia Carol Coss, George Mason University
Victor P. Eroschenko, University of Idaho
Lorraine Findlay, Nassau Community College
Candice Francis, Palomar College
Christina Gan, Rogue Community College
Gregory Garman, Centralia College

Alan Gillen, Pensacola Christian College
Chaya Gopalan, St. Louis Community College
Janet Haynes, Long Island University
Clare Hayes, Metropolitan State College of Denver
James Junker, Campbell University
Gerald Karp, University of Florida
William Kleinelp, Middlesex County College
John Langdon, University of Indianapolis
John Lepri, University of North Carolina at Greensboro
Jerri K. Lindsey, Tarrant County College
Mary Katherine K. Lockwood, University of New Hampshire
Jennifer Lundmark, California State University, Sacramento
Paul Malven, Purdue University
G. K. Maravelas, Bristol Community College
Jane Marks, Paradise Valley Community College
Lee Meserve, Bowling Green State University
Javanika Mody, Anne Arundel Community College
Robert L. Moskowitz, Community College of Philadelphia
Shigihiro Nakajima, University of Illinois at Chicago
Jerry D. Norton, Georgia State University
Justicia Opoku, University of Maryland
Weston Opitz, Kansas Wesleyan University
Joann Otto, Purdue University
David Parker, Northern Virginia Community College
Karla Pouillon, Everett Community College
Linda Powell, Community College of Philadelphia
C. Lee Rocket, Bowling Green State University
Esmond J. Sanders, University of Alberta
Louisa Schmid, Tyler Junior College
Hans Schöler, University of Pennsylvania
Charles Sinclair, University of Indianapolis
Dianne Snyder, Augusta State University
Dennis Strete, McLennan Community College
Eric Sun, Macon State College
Antonietto Tan, Worcester State College
Jim Van Brunt, Rogue Community College
Jyoti Wagle, Houston Community College
Curt Walker, Dixie State College
DeLoris M. Wenzel, The University of Georgia
David Westmoreland, US Air Force Academy
Frederick E. Williams, University of Toledo

In addition, every chapter was read and reviewed by either an Anatomy & Physiology student or a health care professional now working in his or her chosen career. Their comments on the relevance and interest of the chapter topics is presented on each chapter opening page. We appreciate their time and effort, and especially their enthusiasm for the subject matter. We think that enthusiasm is catching! Thanks go out to the following:

Tamatha Adkins, RN
Molly Causby, Georgia Perimeter College
Margaret Chambers, George Mason University
RoNell Coco, Clark College
John Curra, Respiratory Technologist
Stephanie Hall Ford, Barry University
Helen Hart Ford, Radiological Technologist
Elizabeth Garrison, Modesto Junior College
Emily Gordon, Edison Community College
Kim Green, Edison Community College
Mike Grosse, Licensed Physical Therapy Assistant
Caroline Guerra, Broward Community College
Jill Haan, Cardiographic Technician
Dorie Hart, RN
Lisa P. Hubbard, Macon State College
Mark Johnson, Edison Community College
Wendy Lawrence, Troy State University
Candice Machado, Modesto Junior College
Susan Mahoney, Physical Therapist
Christine McGrellis, Mohawk Valley Community College
Jacqueline Opera, Medical Laboratory Technologist
Denise Pacheco, Modesto Junior College
Joan Petrokovsky, Licensed Massage Therapist
Toni Sheridan, Pharmacist
Barbara Simone, RN, FNP
Tiffany Smith, Stark State College of Technology
Sabrina von Brueckwitz, Edison Community College

To the Student

Your book has a variety of special features that will make your time studying anatomy and physiology a more rewarding experience. These have been developed based on feedback from students – like you – who have used previous editions of the text. A review of the preface will give you insight, both visually and in narrative, to all of the text's distinctive features.

Our experience in the classroom has taught us that student's appreciate a hint – both visually and verbally – at the beginning of each chapter about what to expect from its contents. Each chapter of your book begins with a stunning illustration depicting the system or main content being covered in the chapter. In addition, a short introduction to the chapter contents – usually written by a student who has recently completed the course, but occasionally by a practitioner in an allied health field – offers you an insight into some of the most intriguing or relevant aspects of the chapter. Links to activities on your **Foundations** CD, and special web-based activities called **Insights and Explorations** are suggested to make your study time worthwhile and interesting.

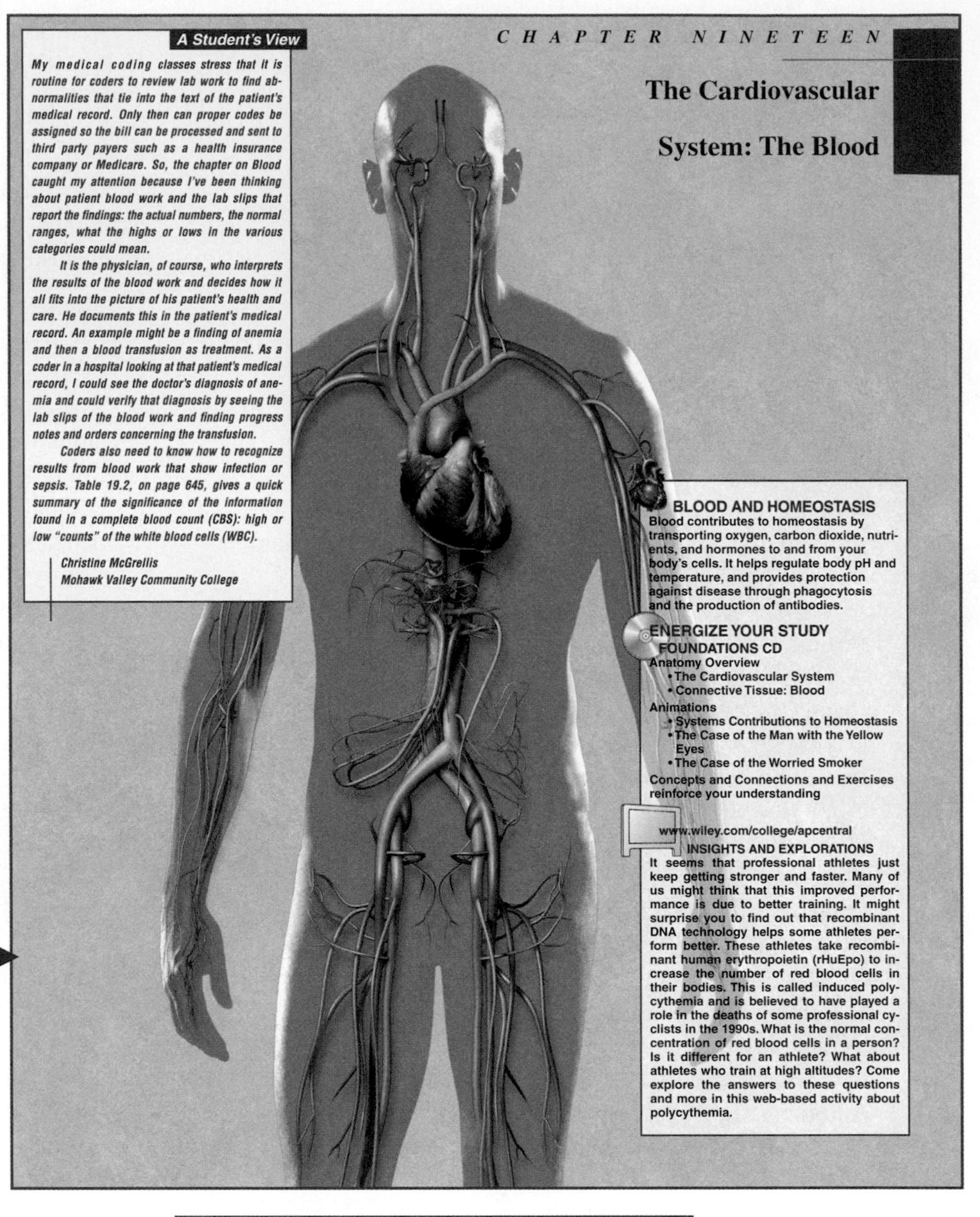

A Student's View

My medical coding classes stress that it is routine for coders to review lab work to find abnormalities that tie into the text of the patient's medical record. Only then can proper codes be assigned so the bill can be processed and sent to third party payers such as a health insurance company or Medicare. So, the chapter on Blood caught my attention because I've been thinking about patient blood work and the lab slips that report the findings: the actual numbers, the normal ranges, what the highs or lows in the various categories could mean.

It is the physician, of course, who interprets the results of the blood work and decides how it all fits into the picture of his patient's health and care. He documents this in the patient's medical record. An example might be a finding of anemia and then a blood transfusion as treatment. As a coder in a hospital looking at that patient's medical record, I could see the doctor's diagnosis of anemia and could verify that diagnosis by seeing the lab slips of the blood work and finding progress notes and orders concerning the transfusion.

Coders also need to know how to recognize results from blood work that show infection or sepsis. Table 19.2, on page 645, gives a quick summary of the significance of the information found in a complete blood count (CBS): high or low "counts" of the white blood cells (WBC).

Christine McGrellis
Mohawk Valley Community College

CHAPTER NINETEEN

The Cardiovascular System: The Blood

BLOOD AND HOMEOSTASIS
Blood contributes to homeostasis by transporting oxygen, carbon dioxide, nutrients, and hormones to and from your body's cells. It helps regulate body pH and temperature, and provides protection against disease through phagocytosis and the production of antibodies.

ENERGIZE YOUR STUDY
FOUNDATIONS CD
Anatomy Overview
- The Cardiovascular System
- Connective Tissue: Blood

Animations
- Systems Contributions to Homeostasis
- The Case of the Man with the Yellow Eyes
- The Case of the Worried Smoker

Concepts and Connections and Exercises reinforce your understanding

www.wiley.com/college/apcentral
INSIGHTS AND EXPLORATIONS
It seems that professional athletes just keep getting stronger and faster. Many of us might think that this improved performance is due to better training. It might surprise you to find out that recombinant DNA technology helps some athletes perform better. These athletes take recombinant human erythropoietin (rHuEpo) to increase the number of red blood cells in their bodies. This is called induced polycythemia and is believed to have played a role in the deaths of some professional cyclists in the 1990s. What is the normal concentration of red blood cells in a person? Is it different for an athlete? What about athletes who train at high altitudes? Come explore the answers to these questions and more in this web-based activity about polycythemia.

As you begin each narrative section of the chapter, be sure to take note of the **objectives** at the beginning of the section to help you focus on what is important as you read it.

FUNCTIONS AND PROPERTIES OF BLOOD

OBJECTIVES
- **Describe the functions of blood.**
- **Describe the physical characteristics and principal components of blood.**

Most cells of a multicellular organism cannot move around to obtain oxygen and nutrients or eliminate carbon dioxide and other wastes. Instead, these needs are met by two fluids: blood and interstitial fl... a liquid matrix c...

At the end of the section, take time to try and answer the **Checkpoint** questions placed there. If you can, then you are ready to move on to the next section. If you experience difficulty answering the questions, you may want to re-read the section before continuing.

CHECKPOINT
1. In what ways is blood plasma similar to and different from interstitial fluid?
2. How many kilograms or pounds of blood is present in your body?
3. How does the volume of blood plasma in your body compare to the volume of fluid in a two-liter bottle of Coke?
4. What are the major solutes in blood plasma? What does each do?
5. What is the significance of lower-than-normal or higher-than-normal hematocrit?

Studying the figures (illustrations that include artwork and photographs) in this book is as important as reading the text. To get the most out of the visual parts of this book, use the tools we have added to the figures to help you understand the concepts being presented. Start by reading the **legend**, which explains what the figure is about. Next, study the **key concept statement**, which reveals a basic idea portrayed in the figure. Added to many figures you will also find an **orientation diagram** to help you understand the perspective from which you are viewing a particular piece of anatomical art. Finally, at the bottom of each figure you will find a **figure question**. If you try to answer these questions as you go along, they will serve as self-checks to help you understand the material. Often it will be possible to answer a question by examining the figure itself. Other questions will encourage you to integrate the knowledge you've gained by carefully reading the text associated with the figure. Still other questions may prompt you to think critically about the topic at hand or predict a consequence in advance of its description in the text.

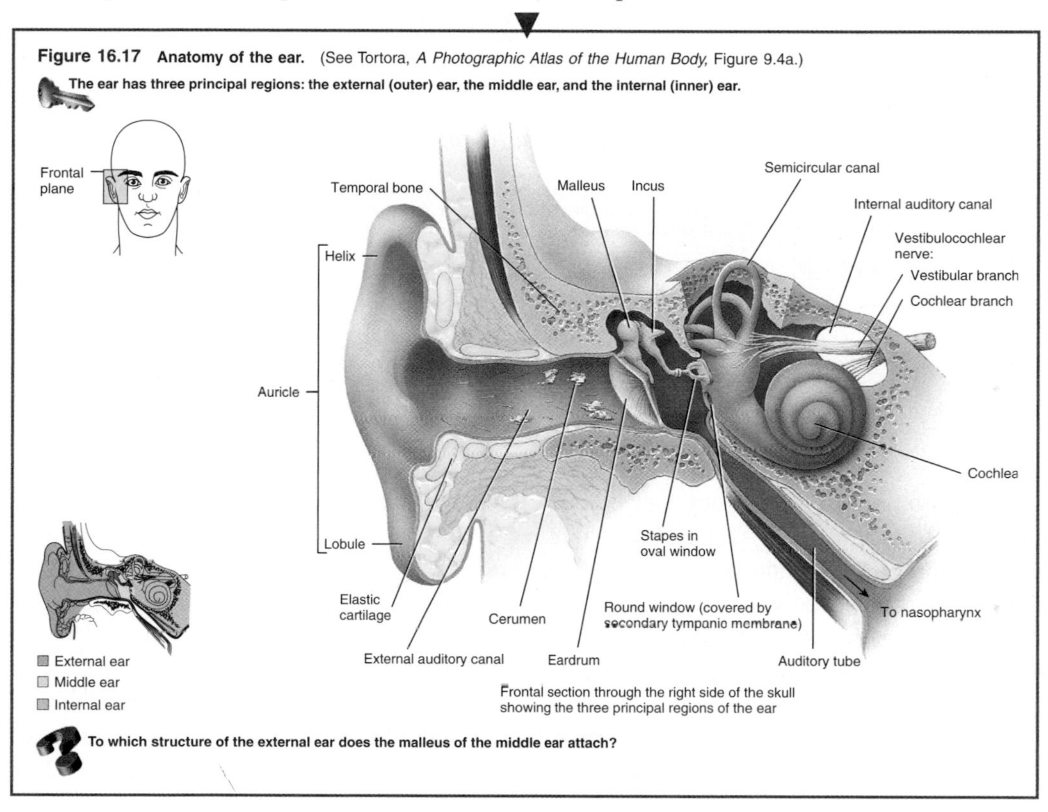

Figure 16.17 Anatomy of the ear. (See Tortora, *A Photographic Atlas of the Human Body,* Figure 9.4a.)

The ear has three principal regions: the external (outer) ear, the middle ear, and the internal (inner) ear.

To which structure of the external ear does the malleus of the middle ear attach?

At the end of each chapter are other resources that you will find useful. **The Study Outline** is a concise statement of important topics discussed in the chapter. Page numbers are listed next to key concepts so you can easily refer to the specific passages in the text for clarification or amplification.

STUDY OUTLINE

SPINAL CORD ANATOMY (p. 420)

1. The spinal cord is protected by the vertebral column, the meninges, cerebrospinal fluid, and denticulate ligaments.
2. The three meninges are coverings that run continuously around the spinal cord and brain. They are the dura mater, arachnoid mater, and pia mater.
3. The spinal cord begins as a continuation of the medulla oblongata and ends at about the second lumbar vertebra in an adult.
4. The spinal cord contains cervical and lumbar enlargements that serve as points of origin for nerves to the limbs.
5. The tapered inferior portion of the spinal cord is the conus medullaris, from which arise the filum terminale and cauda equina.
6. Spinal nerves connect to each segment of the spinal cord by two roots. The posterior or dorsal root contains sensory axons, and the anterior or ventral root contains motor neuron axons.
7. The anterior median fissure and the posterior median sulcus partially divide the spinal cord into right and left sides.
8. The gray matter in the spinal cord is divided into horns, and the white matter into columns. In the center of the spinal cord is the central canal, which runs the length of the spinal cord.
9. Parts of the spinal cord observed in transverse section are the gray commissure; central canal; anterior, posterior, and lateral gray horns; and anterior, posterior, and lateral white columns, which contain ascending and descending tracts. Each part has specific functions.
10. The spinal cord conveys sensory and motor information by way of ascending and descending tracts, respectively.

The **Self-quiz Questions** are designed to help you evaluate your understanding of the chapter contents. **Critical Thinking Questions** are word problems that allow you to apply the concepts you have studied in the chapter to specific situations.

SELF-QUIZ QUESTIONS

Fill in the blanks in the following statements.

1. _______ are fast, predictable, automatic responses to changes in the environment.
2. Because they contain both sensory and motor axons, spinal nerves are considered to be _______ nerves.

Indicate whether the following statements are true or false.

3. Reflexes permit the body to make exceedingly rapid adjustments to homeostatic imbalances.
4. Autonomic reflexes involve responses of smooth muscle, cardiac muscle, and glands.

Choose the one best answer to the following questions.

(a) endoneurium, (b) fascicle, (c) perineurium, (d) epineurium, (e) neurolemma.

6. Which of the following is *not* a function of the spinal cord? (a) reflex center, (b) integration of EPSPs and IPSPs, (c) conduction pathway for sensory impulses, (d) conduction pathway for motor impulses, (e) interpretation of sensory stimuli.
7. Which of the following are true? (1) The anterior (ventral) gray horns contain cell bodies of neurons that cause skeletal muscle contraction. (2) The gray commissure connects the white matter of the right and left sides of the spinal cord. (3) Cell bodies of autonomic motor neurons are located in the lateral gray horns. (4) Sensory (ascending) tracts conduct motor impulses down the spinal ...) Gray matter in the spinal cord consists of cell bodies of

CRITICAL THINKING QUESTIONS

1. Pearl, a lifeguard at the local beach, stepped on a discarded lit cigarette with her bare foot. Trace the reflex arcs set in motion by her accident. Name the reflex types.
   HINT ***How will she remain standing when she picks up her foot?***
2. Why doesn't your spinal cord creep up toward your h... time you bend over? Why doesn't it get all twisted out o... when you exercise?
   HINT ***How would you prevent a boat from floating aw... dock?***
3. Jose's severe headaches and other symptoms were suggestive of meningitis, so his physician ordered a spinal tap. List the structures that the needle will pierce from the most superficial to the deepest. Why would the physican order a test in the spinal region to check a problem in Jose's head?

You will also find the **Answers to Figure Questions** at the end of chapters.

ANSWERS TO FIGURE QUESTIONS

13.1 The superior boundary of the spinal dura mater is the foramen magnum of the occipital bone. The inferior boundary is the second sacral vertebra.

13.2 The cervical enlargement connects with sensory and motor nerves of the upper limbs.

13.3 A horn is an area of gray matter, and a column is a region of white matter in the spinal cord.

13.4 The anterior corticospinal tract is located on the anterior side of the spinal cord, originates in the cortex of the cerebrum, and

the motor impulses leave the spinal cord on the side opposite the entry of sensory impulses.

13.10 All spinal nerves are mixed (have sensory and motor components) because the posterior root containing sensory axons and the anterior root containing motor axons unite to form the spinal nerve.

13.11 The anterior rami serve the upper and lower limbs.

13.12 Severing the spinal cord at level C2 causes respiratory arrest because it prevents descending nerve impulses from reaching the

Learning the language of anatomy and physiology can be one the more challenging aspects of taking this course. Throughout the text we have included pronunciations, and sometimes, Word Roots, for many terms that may be new to you. These appear in parentheses immediately following the new words, and the pronunciations are repeated in the glossary at the back of the book. You companion CD includes the complete glossary and offers you the opportunity to hear the words pronounced. In addition, the Companion Website offers you a review of these terms by chapter, pronounces them for you, and allows you the opportunity to create flash cards or quiz yourself on the many new terms.

Look at the words carefully and say them out loud several times. Learning to pronounce a new word will help you remember it and make it a useful part of your medical vocabulary. Take a few minutes to review the following pronunciation key, so it will be familiar to you when you encounter new words. The key is repeated at the beginning of the Glossary, as well.

Pronounciation Key

1. The most strongly accented syllable appears in capital letters, for example, bilateral (bī-LAT-er-al) and diagnosis (dī-ag-NŌ-sis).
2. If there is a secondary accent, it is noted by a prime (′), for example, constitution (kon′-sti-TOO-shun) and physiology (fiz′-ē-OL-ō-jē). Any additional secondary accents are also noted by a prime, for example, decarboxylation (dē′-kar-bok′-si-LĀ-shun).
3. Vowels marked by a line above the letter are pronounced with the long sound, as in the following common words:
   ā as in *māke* *ō* as in *pōle*
   ē as in *bē* *ū* as in *cute*
   ī as in *īvy*
4. Vowels not marked by a line above the letter are pronounced with the short sound, as in the following words:
   a as in *above* or *at* *o* as in *not*
   e as in *bet* *u* as in *bud*
   i as in *sip*
5. Other vowel sounds are indicated as follows:
   oy as in *oil*
   oo as in *root*
6. Consonant sounds are pronounced as in the following words:
   b as in *bat* *m* as in *mother*
   ch as in *chair* *n* as in *no*
   d as in *dog* *p* as in *pick*
   f as in *father* *r* as in *rib*
   g as in *get* *s* as in *so*
   h as in *hat* *t* as in *tea*
   j as in *jump* *v* as in *very*
   k as in *can* *w* as in *welcome*
   ks as in *tax* *z* as in *zero*
   kw as in *quit* *zh* as in *lesion*
   l as in *let*

Brief Table of Contents

Contents

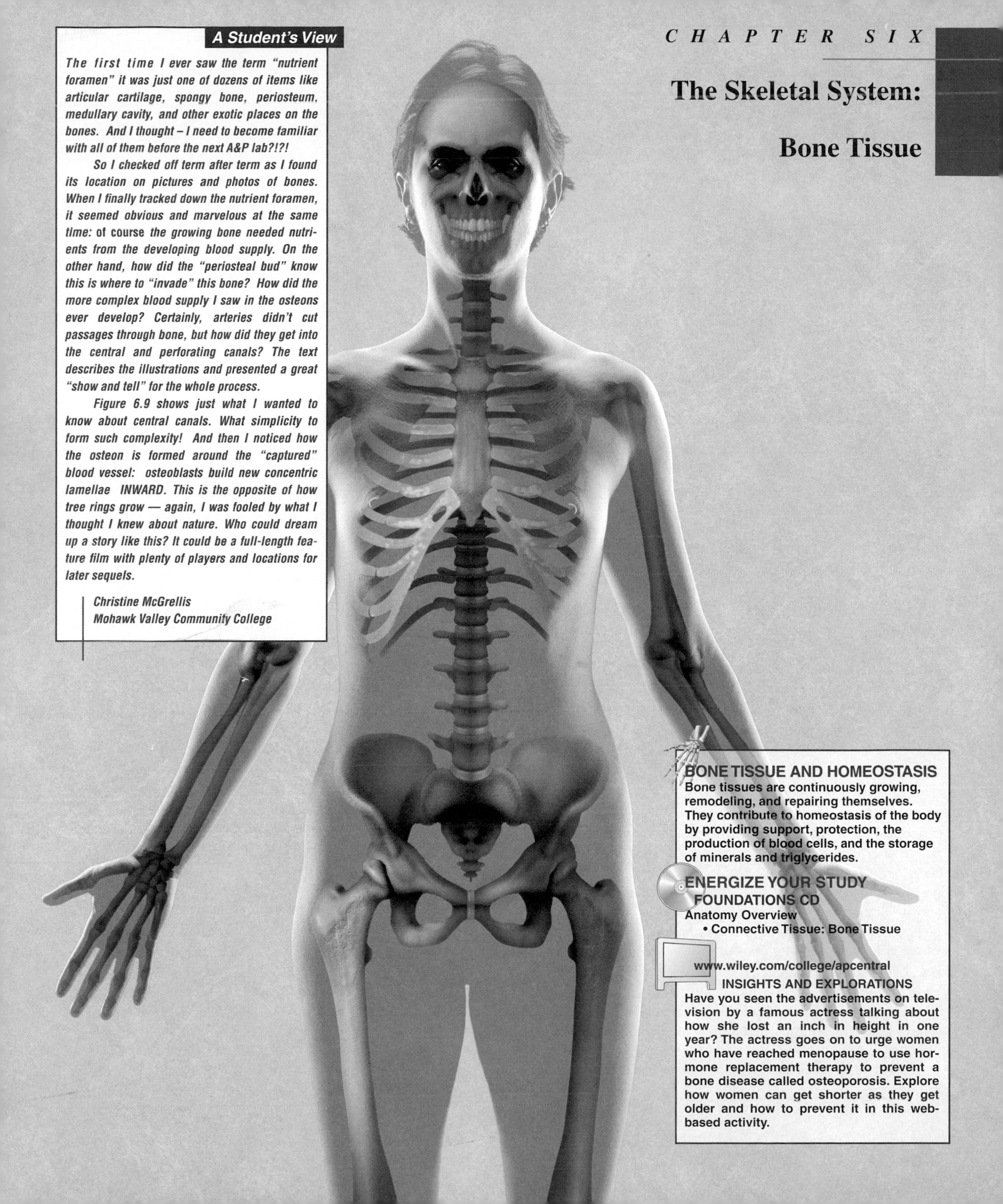

A Student's View

The first time I ever saw the term "nutrient foramen" it was just one of dozens of items like articular cartilage, spongy bone, periosteum, medullary cavity, and other exotic places on the bones. And I thought – I need to become familiar with all of them before the next A&P lab?!?!

So I checked off term after term as I found its location on pictures and photos of bones. When I finally tracked down the nutrient foramen, it seemed obvious and marvelous at the same time: of course the growing bone needed nutrients from the developing blood supply. On the other hand, how did the "periosteal bud" know this is where to "invade" this bone? How did the more complex blood supply I saw in the osteons ever develop? Certainly, arteries didn't cut passages through bone, but how did they get into the central and perforating canals? The text describes the illustrations and presented a great "show and tell" for the whole process.

Figure 6.9 shows just what I wanted to know about central canals. What simplicity to form such complexity! And then I noticed how the osteon is formed around the "captured" blood vessel: osteoblasts build new concentric lamellae INWARD. This is the opposite of how tree rings grow — again, I was fooled by what I thought I knew about nature. Who could dream up a story like this? It could be a full-length feature film with plenty of players and locations for later sequels.

Christine McGrellis
Mohawk Valley Community College

CHAPTER SIX

The Skeletal System: Bone Tissue

BONE TISSUE AND HOMEOSTASIS

Bone tissues are continuously growing, remodeling, and repairing themselves. They contribute to homeostasis of the body by providing support, protection, the production of blood cells, and the storage of minerals and triglycerides.

ENERGIZE YOUR STUDY
FOUNDATIONS CD

Anatomy Overview
- Connective Tissue: Bone Tissue

www.wiley.com/college/apcentral

INSIGHTS AND EXPLORATIONS

Have you seen the advertisements on television by a famous actress talking about how she lost an inch in height in one year? The actress goes on to urge women who have reached menopause to use hormone replacement therapy to prevent a bone disease called osteoporosis. Explore how women can get shorter as they get older and how to prevent it in this web-based activity.

A bone is made up of several different tissues working together: bone or osseous tissue, cartilage, dense connective tissues, epithelium, adipose tissue, and nervous tissue. For this reason, each individual bone is an organ. Bone tissue is a complex and dynamic living tissue. It continually engages in a process called remodeling—building new bone tissue and breaking down old bone tissue. The entire framework of bones and their cartilages together constitute the **skeletal system.** This chapter will survey the various components of bones so you can understand how bones form, how they age, and how exercise affects the density and strength of bones. The study of bone structure and the treatment of bone disorders is called **osteology** (os-tē-OL-ō-jē; *osteo-* = bone; *-logy* = study of).

FUNCTIONS OF BONE AND THE SKELETAL SYSTEM

OBJECTIVE

- **Discuss the functions of the skeletal system.**

Bone tissue and the skeletal system perform several basic functions:

1. ***Support.*** The skeleton serves as the structural framework for the body by supporting soft tissues and providing attachment points for the tendons of most skeletal muscles.
2. ***Protection.*** The skeleton protects many internal organs from injury. For example, cranial bones protect the brain, vertebrae protect the spinal cord, and the rib cage protects the heart and lungs.
3. ***Assistance in movement.*** Because skeletal muscles attach to bones, when muscles contract, they pull on bones. Together, bones and muscles produce movement. This function is discussed in detail in Chapter 10.
4. ***Mineral homeostasis.*** Bone tissue stores several minerals, especially calcium and phosphorus. On demand, bone releases minerals into the blood to maintain critical mineral balances (homeostasis) and to distribute the minerals to other parts of the body.
5. ***Blood cell production.*** Within certain bones, a connective tissue called **red bone marrow** produces red blood cells, white blood cells, and platelets, a process called **hemopoiesis** (hēm-ō-poy-Ē-sis; *hemo-* = blood; *poiesis* = making). Red bone marrow consists of developing blood cells, adipocytes, fibroblasts, and macrophages within a network of reticular fibers. It is present in developing bones of the fetus and in some adult bones, such as the pelvis, ribs, breastbone, backbones, skull, and ends of the arm bones and thighbones.
6. ***Triglyceride storage.*** Triglycerides stored in the adipose cells of **yellow bone marrow** are an important chemical energy reserve. Yellow bone marrow consists mainly of adipose cells, which store triglycerides, and a few blood cells. In the newborn, all bone marrow is red and is involved in hemopoiesis. With increasing age, much of the bone marrow changes from red to yellow.

▶ CHECKPOINT

1. What kinds of tissues make up the skeletal system?
2. How do red and yellow bone marrow differ in composition, location, and function?

STRUCTURE OF BONE

OBJECTIVE

- **Describe the parts of a long bone.**

The structure of a bone may be analyzed by considering the parts of a long bone, for instance, the humerus (the arm bone), as shown in Figure 6.1. A long bone is one that has greater length than width. A typical long bone consists of the following parts:

1. The **diaphysis** (dī-AF-i-sis = growing between) is the bone's shaft or body—the long, cylindrical, main portion of the bone.
2. The **epiphyses** (e-PIF-i-sēz = growing over; singular is *epiphysis*) are the distal and proximal ends of the bone.
3. The **metaphyses** (me-TAF-i-sēz; *meta-* = between; singular is *metaphysis*) are the regions in a mature bone where the diaphysis joins the epiphyses. In a growing bone, each metaphysis includes an **epiphyseal plate** (ep′-i-FIZ-ē-al), a layer of hyaline cartilage that allows the diaphysis of the bone to grow in length (described later in the chapter). When bone growth in length stops, the cartilage in the epiphyseal plate is replaced by bone and the resulting bony structure is known as the **epiphyseal line.**
4. The **articular cartilage** is a thin layer of hyaline cartilage covering the epiphysis where the bone forms an articulation (joint) with another bone. Articular cartilage reduces friction and absorbs shock at freely movable joints. Because articular cartilage lacks a perichondrium, repair of damage is limited.
5. The **periosteum** (per′-ē-OS-tē-um; *peri-* = around) is a tough sheath of dense irregular connective tissue that surrounds the bone surface wherever it is not covered by articular cartilage. The periosteum contains bone-forming cells that enable bone to grow in diameter or thickness, but not in length. It also protects the bone, assists in fracture repair, helps nourish bone tissue, and serves as an attachment point for ligaments and tendons.

Figure 6.1 Parts of a long bone. The spongy bone tissue of the epiphysis and metaphysis contains red bone marrow, whereas the medullary cavity of the diaphysis contains yellow bone marrow (in adults).

A long bone is covered by articular cartilage at its proximal and distal epiphyses and by periosteum around the diaphysis.

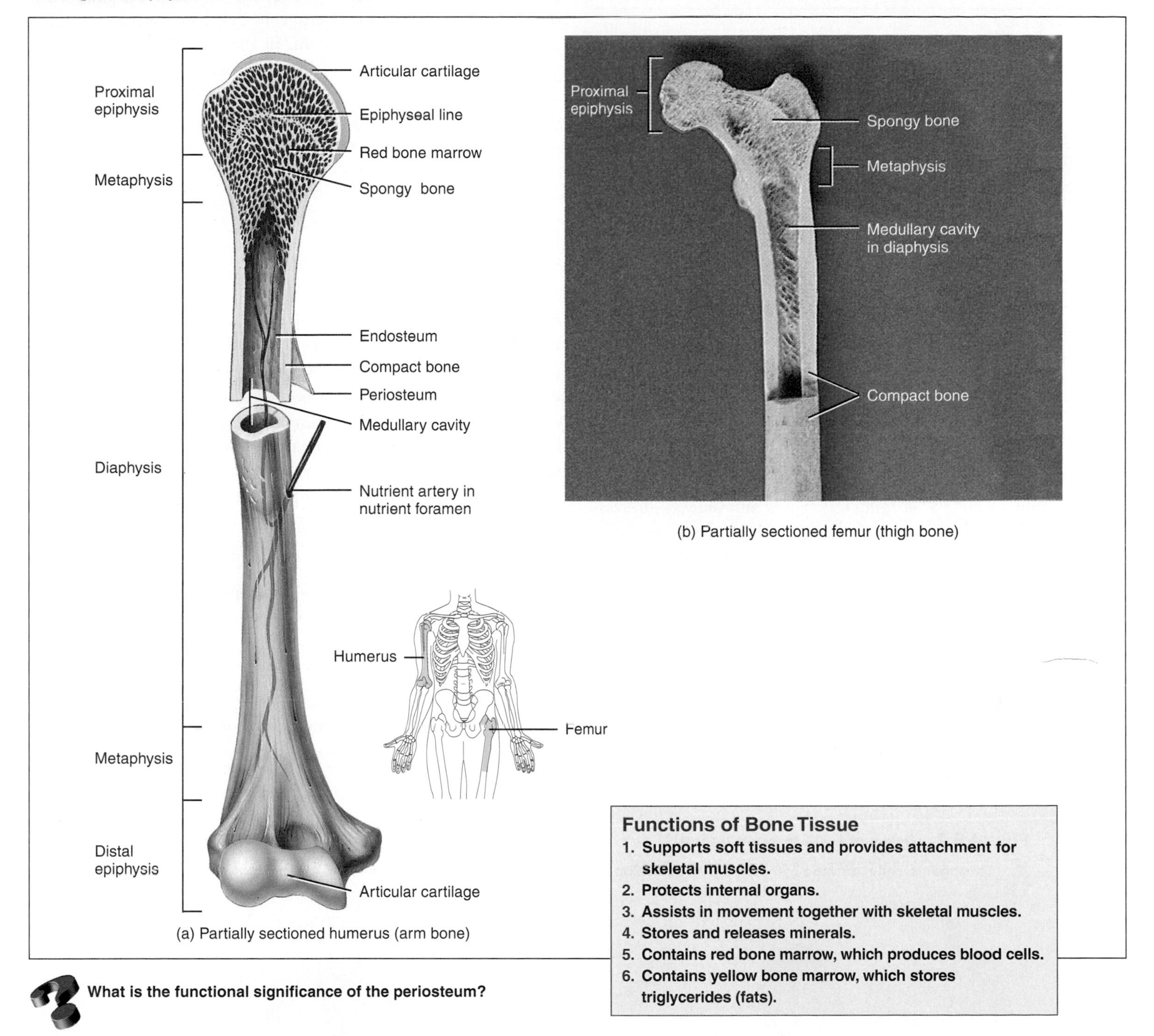

(a) Partially sectioned humerus (arm bone)

(b) Partially sectioned femur (thigh bone)

Functions of Bone Tissue

1. **Supports soft tissues and provides attachment for skeletal muscles.**
2. **Protects internal organs.**
3. **Assists in movement together with skeletal muscles.**
4. **Stores and releases minerals.**
5. **Contains red bone marrow, which produces blood cells.**
6. **Contains yellow bone marrow, which stores triglycerides (fats).**

What is the functional significance of the periosteum?

6. The **medullary cavity** (MED-ū-lar′-ē; *medulla-* = marrow, pith) or **marrow cavity** is the space within the diaphysis that contains fatty yellow bone marrow in adults.

7. The **endosteum** (end-OS-tē-um; *endo-* = within) is a thin membrane that lines the medullary cavity. It contains a single layer of bone-forming cells and a small amount of connective tissue.

▶ CHECKPOINT

3. Diagram the parts of a long bone, and list the functions of each part.

HISTOLOGY OF BONE TISSUE

OBJECTIVE

- **Describe the histological features of bone tissue.**

Like other connective tissues, **bone,** or **osseous tissue** (OS-ē -us), contains an abundant matrix of intercellular materials that surround widely separated cells. The matrix is about 25% water, 25% collagen fibers, and 50% crystallized mineral salts. The abundant inorganic mineral salts are mainly *hydroxyapatite* (calcium phosphate and calcium carbonate). In addition, bone matrix includes small amounts of magnesium hydroxide, fluoride, and sulfate. As these mineral salts are deposited in the framework formed by the collagen fibers of the matrix, they crystallize and the tissue hardens. This process of **calcification** is initiated by osteoblasts, the bone-building cells.

It was once thought that calcification simply occurred when enough mineral salts were present to form crystals. Now, however, we know that the process occurs only in the presence of collagen fibers. Mineral salts begin to crystallize in the microscopic spaces between collagen fibers. After the spaces are filled, mineral crystals accumulate around the collagen fibers. The combination of crystallized salts and collagen fibers is responsible for the hardness that is characteristic of bone.

Although a bone's *hardness* depends on the crystallized inorganic mineral salts, a bone's *flexibility* depends on its collagen fibers. Like reinforcing metal rods in concrete, collagen fibers and other organic molecules provide *tensile strength,* which is resistance to being stretched or torn apart. Soaking a bone in an acidic solution, such as vinegar, dissolves its mineral salts, causing the bone to become rubbery and flexible. As you will see shortly, bone cells called osteoclasts secrete enzymes and acids that break down bone matrix.

Four types of cells are present in bone tissue: osteogenic cells, osteoblasts, osteocytes, and osteoclasts (Figure 6.2).

1. **Osteogenic cells** (os′-tē-ō-JEN-ik; *-genic* = producing) are unspecialized stem cells derived from mesenchyme, the tissue from which all connective tissues are formed. They are the only bone cells to undergo cell division; the resulting daughter cells develop into osteoblasts. Osteogenic cells are found along the inner portion of the periosteum, in the endosteum, and in the canals within bone that contain blood vessels.
2. **Osteoblasts** (OS-tē-ō-blasts′; *-blasts* = buds or sprouts) are bone-building cells. They synthesize and secrete collagen fibers and other organic components needed to build the matrix of bone tissue, and they initiate calcification (described shortly). As osteoblasts surround themselves with matrix, they become trapped in their secretions and become osteocytes. (Note: *Blasts* in bone or any other connective tissue secrete matrix.)
3. **Osteocytes** (OS-tē-ō-sīts′; *-cytes* = cells), mature bone cells, are the main cells in bone tissue and maintain its daily metabolism, such as the exchange of nutrients and wastes with the blood. Like osteoblasts, osteocytes do not undergo cell division. (Note: *Cytes* in bone or any other tissue maintain the tissue.)
4. **Osteoclasts** (OS-tē-ō-clasts′; *-clast* = break) are huge cells derived from the fusion of as many as 50 monocytes (a type of white blood cell) and are concentrated in the endosteum. On the side of the cell that faces the bone surface, the osteoclast's plasma membrane is deeply folded into a *ruffled border.* Here the cell releases powerful lysosomal enzymes and acids that digest the protein and mineral components of the underlying bone matrix. This breakdown of bone matrix, termed **resorption,** is part of the normal development, growth, maintenance, and repair of bone. (Note: *Clasts* in bone break down matrix.)

Bone is not completely solid but has many small spaces between its cells and matrix components. Some spaces are channels for blood vessels that supply bone cells with nutrients. Other spaces are storage areas for red bone marrow. Depending on the size and distribution of the spaces, the regions of a bone may be

Figure 6.2 Types of cells in bone tissue.

Osteogenic cells undergo cell division and develop into osteoblasts, which secrete bone matrix.

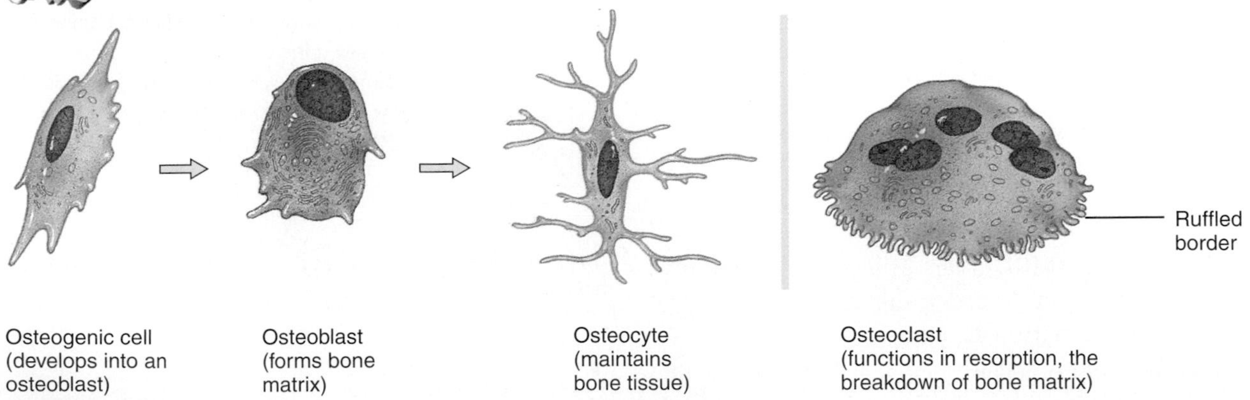

Why is bone resorption important?

categorized as compact or spongy (see Figure 6.1). Overall, about 80% of the skeleton is compact bone and 20% is spongy bone.

Compact Bone Tissue

Compact bone tissue contains few spaces. It forms the external layer of all bones and makes up the bulk of the diaphyses of long bones. Compact bone tissue provides protection and support and resists the stresses produced by weight and movement.

Compact bone tissue is arranged in units called **osteons** or **Haversian systems** (Figure 6.3). Blood vessels, lymphatic vessels, and nerves from the periosteum penetrate the compact bone through transverse **perforating (Volkmann's) canals.** The vessels and nerves of the perforating canals connect with those of the medullary cavity, periosteum, and **central (Haversian) canals.** The central canals run longitudinally through the bone. Around the canals are **concentric lamellae** (la-MEL-ē)—rings of hard, calcified matrix. Between the lamellae are small spaces called **lacunae** (la-KOO-nē = little lakes; singular is *lacuna*), which contain osteocytes. Radiating in all directions from the lacunae are tiny **canaliculi** (kan′-a-LIK-ū-lī = small channels), which are filled with extracellular fluid. Inside the canaliculi are slender fingerlike processes of osteocytes (see inset at right

Figure 6.3 Histology of bone. Sections through the diaphysis of a long bone, from the surrounding periosteum on the right, to compact bone in the middle, to spongy bone and the medullary cavity on the left. The inset at the upper right shows an osteocyte in a lacuna. See Table 4.3J on page 126 for a photomicrograph of compact bone tissue and Figure 6.14a for a scanning electron micrograph of spongy bone tissue.

In compact bone, osteocytes lie in lacunae arranged in concentric circles around a central canal.

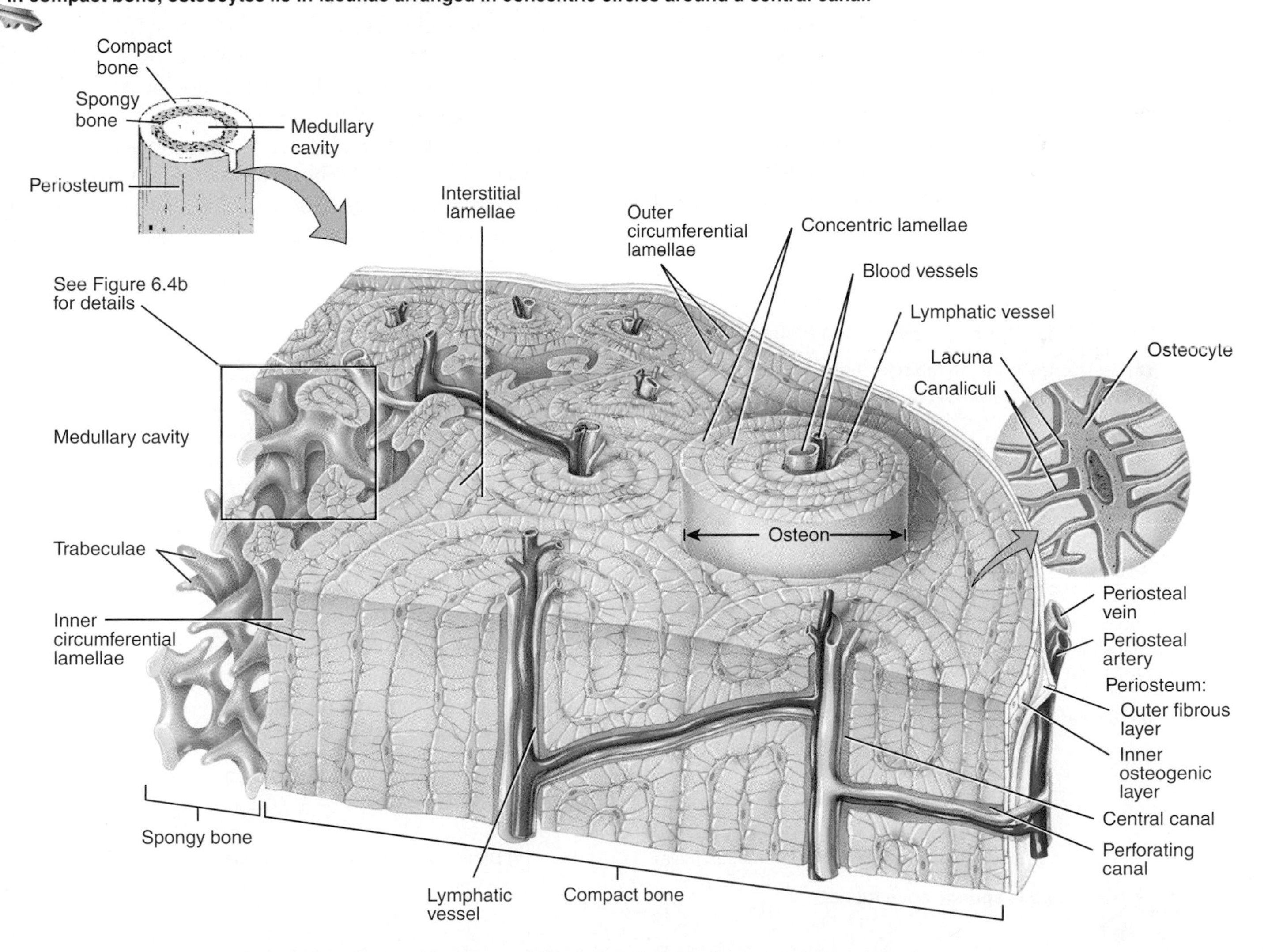

Osteons (Haversian systems) in compact bone and trabeculae in spongy bone

As people age, some central (Haversian) canals may become blocked. What effect would this have on the surrounding osteocytes?

of Figure 6.3). Neighboring osteocytes communicate via gap junctions. The canaliculi connect lacunae with one another and with the central canals. Thus, an intricate, miniature canal system throughout the bone provides many routes for nutrients and oxygen to reach the osteocytes and for wastes to diffuse away. This is very important because diffusion through the lamellae is extremely slow.

Osteons in compact bone tissue are aligned in the same direction along lines of stress. In the shaft, for example, they are parallel to the long axis of the bone. As a result, the shaft of a long bone resists bending or fracturing even when considerable force is applied from either end. Compact bone tissue tends to be thickest in those parts of a bone where stresses are applied in relatively few directions. The lines of stress in a bone are not static. They change as a person learns to walk and in response to repeated strenuous physical activity, such as occurs when a person undertakes weight training. The lines of stress in a bone also can change because of fractures or physical deformity. Thus, the organization of osteons changes over time in response to the physical demands placed on the skeleton.

The areas between osteons contain **interstitial lamellae** (in′-ter-STISH-al), which also have lacunae with osteocytes and canaliculi. Interstitial lamellae are fragments of older osteons that have been partially destroyed during bone rebuilding or growth. Lamellae that encircle the bone just beneath the periosteum are called **outer circumferential lamellae;** those that encircle the medullary cavity are called **inner circumferential lamellae.**

Spongy Bone Tissue

In contrast to compact bone tissue, **spongy bone tissue** does not contain osteons. As shown in Figure 6.4, it consists of **trabeculae** (tra-BEK-ū-lē = little beams; singular is *trabecula*), an irregular latticework of thin columns of bone. The spaces between the trabeculae of some bones are filled with red bone marrow. Within each trabecula are osteocytes that lie in lacunae. Radiating from the lacunae are canaliculi. Because osteocytes are located on the superficial surfaces of trabeculae, they receive nourishment directly from the blood circulating through the medullary cavities.

Spongy bone tissue makes up most of the bone tissue of short, flat, and irregularly shaped bones. It also forms most of the epiphyses of long bones and a narrow rim around the medullary cavity of the diaphysis of long bones.

At first glance, the structure of the osteons of compact bone tissue may appear to be highly organized, whereas the trabeculae of spongy bone tissue may appear to be far less well organized. However, the trabeculae in spongy bone tissue are precisely oriented along lines of stress, a characteristic that helps bones resist stresses and transfer force without breaking. Spongy bone tissue tends to be located where bones are not heavily stressed or where stresses are applied from many directions.

Spongy bone tissue is different from compact bone tissue in two respects. First, spongy bone tissue is light, which reduces the overall weight of a bone so that it moves more readily when pulled by a skeletal muscle. Second, the trabeculae of spongy

Figure 6.4 Spongy bone. See Figure 6.14a for a scanning electron micrograph of spongy bone tissue.

In spongy bone, osteocytes lie in lacunae arranged irregularly in trabeculae.

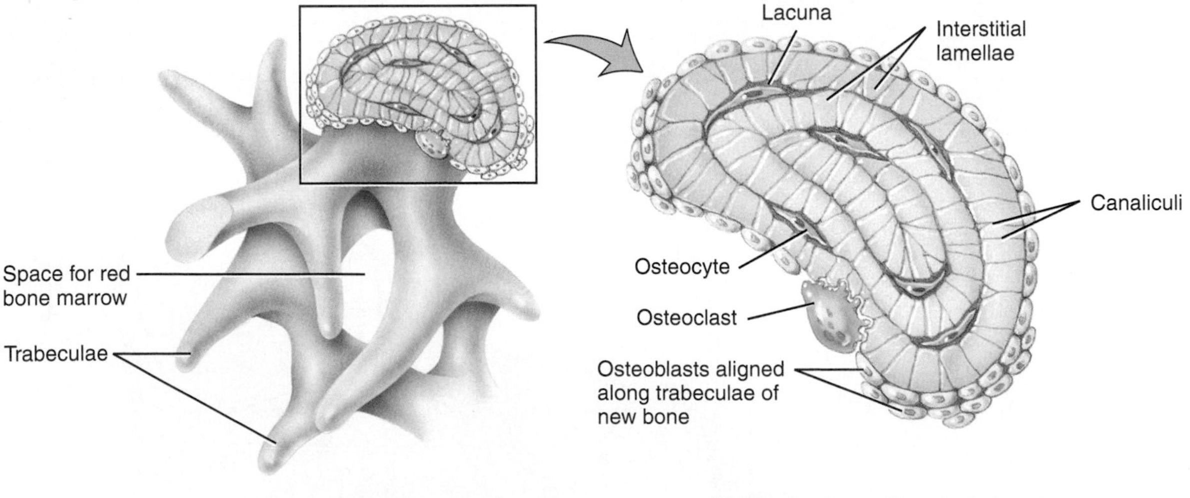

Space for red bone marrow

Trabeculae

(a) Enlarged aspect of spongy bone trabeculae

Lacuna

Interstitial lamellae

Canaliculi

Osteocyte

Osteoclast

Osteoblasts aligned along trabeculae of new bone

(b) Details of a section of a trabecula

Where is spongy bone found?

bone tissue support and protect the red bone marrow. The spongy bone tissue in the hip bones, ribs, breastbone, backbones, and the ends of long bones is the only site of red bone marrow and, thus, of hemopoiesis in adults.

Bone Scan

A **bone scan** is a diagnostic procedure that takes advantage of the fact that bone is living tissue. A small amount of a radioactive tracer compound that is readily absorbed by bone is injected intravenously. The degree of uptake of the tracer is related to the amount of blood flow to the bone. A scanning device measures the radiation emitted from the bones, and the information is translated into a photograph or diagram that can be read like an x ray. Normal bone tissue is identified by a consistent gray color throughout because of its uniform uptake of the radioactive tracer. Darker or lighter areas, however, may indicate bone abnormalities. Darker areas are called "hot spots," areas of increased metabolism that absorb more of the radioactive tracer. Hot spots may indicate bone cancer, abnormal healing of fractures, or abnormal bone growth. Lighter areas are called "cold spots," areas of decreased metabolism that absorb less of the radioactive tracer. Cold spots may indicate problems such as degenerative bone disease, decalcified bone, fractures, bone infections, Paget's disease, and rheumatoid arthritis. A bone scan detects abnormalities 3 to 6 months sooner than standard x-ray procedures and exposes the patient to less radiation. ■

▶ CHECKPOINT

4. Why is bone considered a connective tissue?
5. What are the four types of cells in bone tissue?
6. What is the composition of the matrix of bone tissue?
7. How are spongy and compact bone tissue different in terms of their microscopic appearance, location, and function?

BLOOD AND NERVE SUPPLY OF BONE

▶ OBJECTIVE

- **Describe the blood and nerve supply of bone.**

Bone is richly supplied with blood. Blood vessels, which are especially abundant in portions of bone containing red bone marrow, pass into bones from the periosteum. We will consider the blood supply to a long bone such as the mature tibia (shin bone) shown in Figure 6.5.

Figure 6.5 Blood supply of a mature long bone, the tibia (shin bone).

Bone is richly supplied with blood vessels.

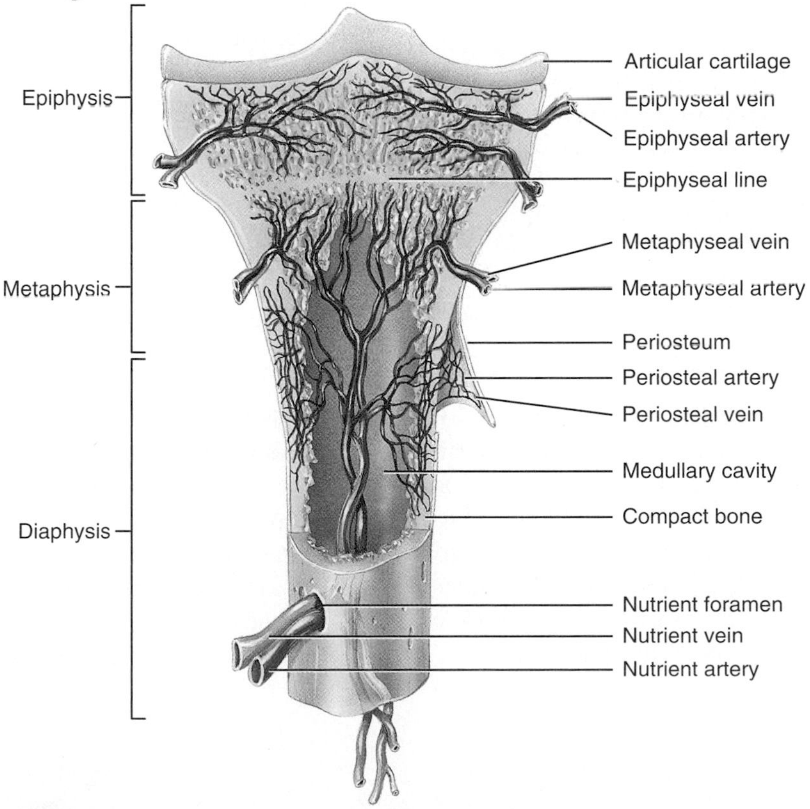

Epiphysis
Metaphysis
Diaphysis
Articular cartilage
Epiphyseal vein
Epiphyseal artery
Epiphyseal line
Metaphyseal vein
Metaphyseal artery
Periosteum
Periosteal artery
Periosteal vein
Medullary cavity
Compact bone
Nutrient foramen
Nutrient vein
Nutrient artery

Where do periosteal arteries enter bone tissue?

Periosteal arteries accompanied by nerves enter the diaphysis through many perforating (Volkmann's) canals and supply the periosteum and outer part of the compact bone (see Figure 6.3). Near the center of the diaphysis, a large **nutrient artery** passes through a hole in compact bone called the **nutrient foramen.** On entering the medullary cavity, the nutrient artery divides into proximal and distal branches that supply both the inner part of compact bone tissue of the diaphysis and the spongy bone tissue and red marrow as far as the epiphyseal plates (or lines). Some bones, like the tibia, have only one nutrient artery; others like the femur (thigh bone) have several. The ends of long bones are supplied by the metaphyseal and epiphyseal arteries, which arise from arteries that supply the associated joint. The **metaphyseal arteries** enter the metaphyses of a long bone and, together with the nutrient artery, supply the red bone marrow and bone tissue of the metaphyses. The **epiphyseal arteries** enter the epiphyses of a long bone and supply the red bone marrow and bone tissue of the epiphyses.

Veins that carry blood away from long bones are evident in three places: (1) One or two **nutrient veins** accompany the nutrient artery in the diaphysis; (2) numerous **epiphyseal veins** and **metaphyseal veins** exit with their respective arteries in the epiphyses; and (3) many small **periosteal veins** exit with their respective arteries in the periosteum.

Nerves accompany the blood vessels that supply bones. The periosteum is rich in sensory nerves, some of which carry pain sensations. These nerves are especially sensitive to tearing or tension, which explains the severe pain resulting from a fracture or a bone tumor.

▶ CHECKPOINT

8. Explain the location and roles of the nutrient arteries, nutrient foramina, epiphyseal arteries, and periosteal arteries.

BONE FORMATION

▶ OBJECTIVE

- **Describe the steps of intramembranous and endochondral ossification.**

The process by which bone forms is called **ossification** (os′-i-fi-KĀ-shun; *ossi-* = bone; *-fication* = making) or **osteogenesis.** The "skeleton" of a human embryo is composed of fibrous connective tissue membranes and pieces of hyaline cartilage, which are shaped like bones and are the sites where ossification occurs. These embryonic tissues provide the template for subsequent ossification, which begins during the sixth or seventh week of embryonic development and follows one of two patterns.

The two methods of bone formation involve the replacement of a preexisting connective tissue with bone. These two methods of ossification do not lead to differences in the structure of mature bones, but are simply different methods of bone development. In the first type of ossification, called **intramembranous ossification** (in′-tra-MEM-bra-nus; *intra-* = within; *membran-* = membrane), bone forms directly on or within loose fibrous connective tissue membranes. In the second type, **endochondral ossification** (en′-dō-KON-dral; *endo-* = within; *-chondral* = cartilage), bone forms within hyaline cartilage.

Intramembranous Ossification

Intramembranous ossification is the simpler of the two methods of bone formation. The flat bones of the skull and mandible (lower jawbone) are formed in this way. Also, the "soft spots" that help the fetal skull pass through the birth canal later harden as they undergo intramembranous ossification, which occurs as follows (Figure 6.6):

1. ***Development of the center of ossification.*** At the site where the bone will develop, mesenchymal cells cluster together and differentiate, first into osteogenic cells and then into osteoblasts. (Recall that *mesenchyme* is the tissue from which all other connective tissues arise.) The site of such a cluster is called a **center of ossification.** Osteoblasts secrete the organic matrix of bone until they are surrounded by it.
2. ***Calcification.*** Then secretion of matrix stops and the cells, now called osteocytes, lie in lacunae and extend their narrow cytoplasmic processes into canaliculi that radiate in all directions. Within a few days, calcium and other mineral salts are deposited and the matrix hardens or calcifies.
3. ***Formation of trabeculae.*** As the bone matrix forms, it develops into trabeculae that fuse with one another to form spongy bone. Blood vessels grow into the spaces between the trabeculae and the mesenchyme along the surface of the newly formed bone. Connective tissue that is associated with the blood vessels in the trabeculae differentiates into red bone marrow.
4. ***Development of the periosteum.*** At the periphery of the bone, the mesenchyme condenses and develops into the periosteum. Eventually, a thin layer of compact bone replaces the surface layers of the spongy bone, but spongy bone remains in the center. Much of the newly formed bone is remodeled (destroyed and reformed) as the bone is transformed into its adult size and shape.

Figure 6.6 Intramembranous ossification. Follow along on this figure as you read the corresponding numbered paragraphs in the text. Parts 1 and 2 show a smaller field of view at higher magnification than parts 3 and 4.

Intramembranous ossification involves the formation of bone directly on or within fibrous connective tissue membranes formed by clusters of mesenchymal cells.

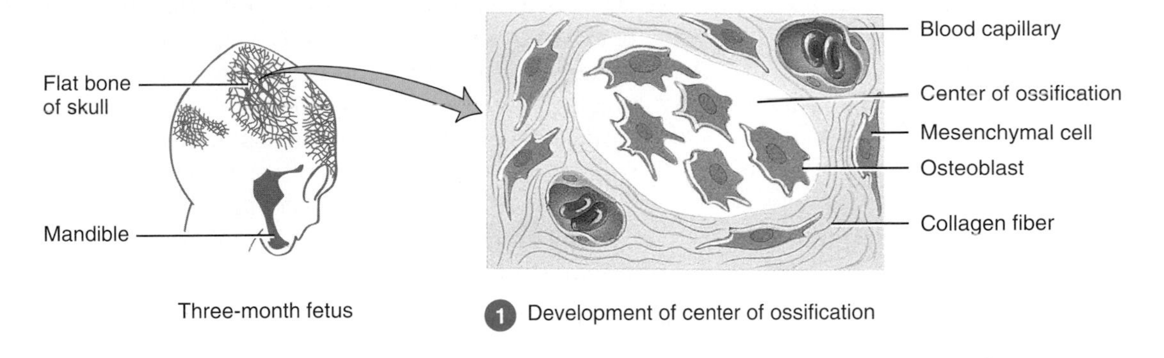

Three-month fetus

1 Development of center of ossification

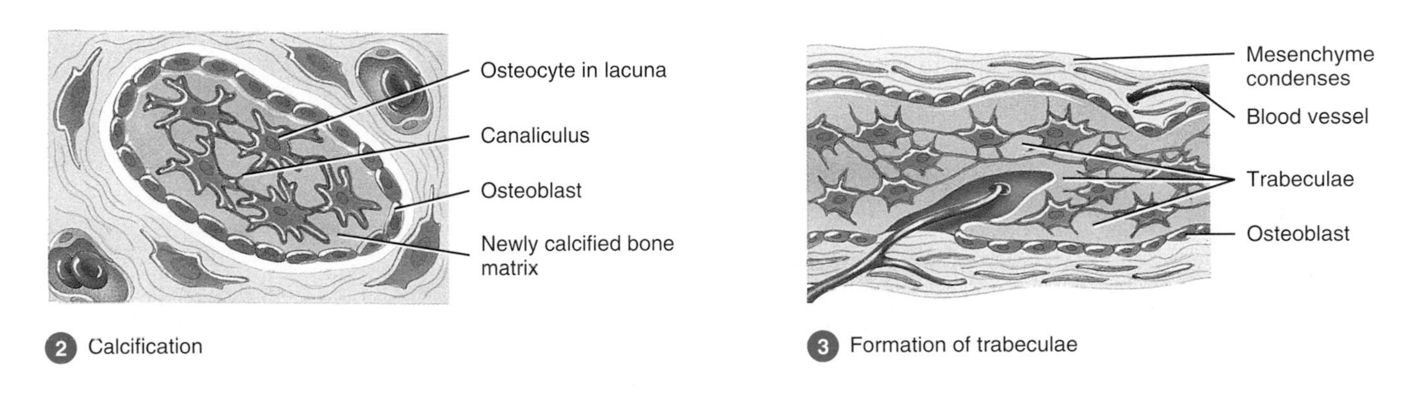

2 Calcification

3 Formation of trabeculae

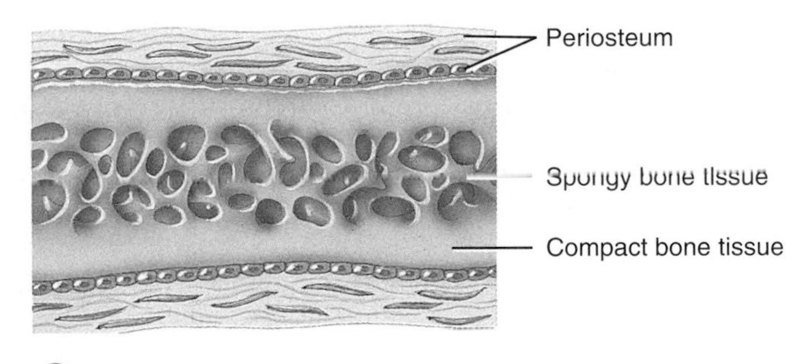

4 Development of the periosteum

Which bones of the body develop by intramembranous ossification?

Endochondral Ossification

The replacement of cartilage by bone is called **endochondral ossification.** Although most bones of the body are formed in this way, the process is best observed in a long bone. It proceeds as follows (Figure 6.7):

1. ***Development of the cartilage model.*** At the site where the bone is going to form, mesenchymal cells crowd together in the shape of the future bone, and then develop into chondroblasts. The chondroblasts secrete cartilage matrix, producing a **cartilage model** consisting of hyaline cartilage. A membrane called the **perichondrium** (per-i-KON-drē-um) develops around the cartilage model.
2. ***Growth of the cartilage model.*** Once chondroblasts become deeply buried in cartilage matrix, they are called chondrocytes. The cartilage model grows in length by continual cell division of chondrocytes accompanied by further secretion of the cartilage matrix. This type of growth is termed **interstitial growth** and results in an increase in length. In contrast, growth of the cartilage in thickness is due mainly to the addition of more matrix material to the periphery of the model by new chondroblasts that develop from the perichondrium. This growth pattern in which matrix is deposited on the cartilage surface is called **appositional growth** (a-pō-ZISH-i-nal).

   As the cartilage model continues to grow, chondrocytes in its mid-region hypertrophy (increase in size). Some hypertrophied cells burst and release their contents, which increases the pH of the surrounding matrix. This change in pH triggers calcification. Other chondrocytes within the calcifying cartilage die because nutrients can no longer diffuse quickly enough through the matrix. As chondrocytes die, lacunae form and eventually merge into small cavities.
3. ***Development of the primary ossification center.*** Primary ossification proceeds *inward* from the external surface of the bone. A nutrient artery penetrates the perichondrium and the calcifying cartilage model through a nutrient foramen in the mid-region of the cartilage model, stimulating osteogenic cells in the perichondrium to differentiate into osteoblasts. Deep to the perichondrium the osteoblasts secrete a thin shell of compact bone called the **periosteal bone collar.** Once the perichondrium starts to form bone, it is known as the **periosteum.** Near the middle of the model, periosteal capillaries grow into the disintegrating calcified cartilage. Upon growing into the cartilage model, the capillaries induce growth of a **primary ossification center,** a region where bone tissue will replace most of the cartilage. Osteoblasts then begin to deposit bone matrix over the remnants of calcified cartilage, forming spongy bone trabeculae. As the ossification center grows toward the ends of the bone, osteoclasts break down some of the newly formed spongy bone trabeculae. This activity leaves a cavity, the medullary (marrow) cavity, in the core of the model. The medullary cavity then fills with red bone marrow.
4. ***Development of the secondary ossification centers.*** The diaphysis (shaft), which was once a solid mass of hyaline cartilage, is replaced by compact bone, the core of which contains a red bone marrow–filled medullary cavity. When branches of the epiphyseal artery enter the epiphyses, **secondary ossification centers** develop, usually around the time of birth. Bone formation is similar to that in primary ossification centers. One difference, however, is that spongy bone remains in the interior of the epiphyses (no medullary cavities are formed there). Secondary ossification proceeds *outward* from the center of the epiphysis toward the outer surface of the bone.
5. ***Formation of articular cartilage and the epiphyseal plate.*** The hyaline cartilage that covers the epiphyses becomes the articular cartilage. Prior to adulthood, hyaline cartilage remains between the diaphysis and epiphysis as the **epiphyseal plate,** which is responsible for the lengthwise growth of long bones.

▶ CHECKPOINT

9. What are the major events of intramembranous and endochondral ossification and how are they different?

BONE GROWTH

▶ OBJECTIVES

- **Describe how bone grows in length and thickness.**
- **Explain the role of nutrients and hormones in regulating bone growth.**

During childhood, bones throughout the body grow in thickness by appositional growth, and long bones lengthen by the addition of bone material on the diaphyseal side of the epiphyseal plate. Bones stop growing in length at about age 25, although they may continue to thicken.

Growth in Length

To understand how a bone grows in length, you need to know some of the details of the structure of the epiphyseal plate (Figure 6.8 on page 172). The **epiphyseal plate** is a layer of hyaline cartilage in the metaphysis of a growing bone that consists of four zones (Figure 6.8b).

1. ***Zone of resting cartilage.*** This layer is nearest the epiphysis and consists of small, scattered chondrocytes. The term "resting" is used because the cells do not function in bone growth. Rather, they anchor the epiphyseal plate to the bone of the epiphysis.

Figure 6.7 Endochondral ossification.

During endochondral ossification, bone gradually replaces a cartilage model.

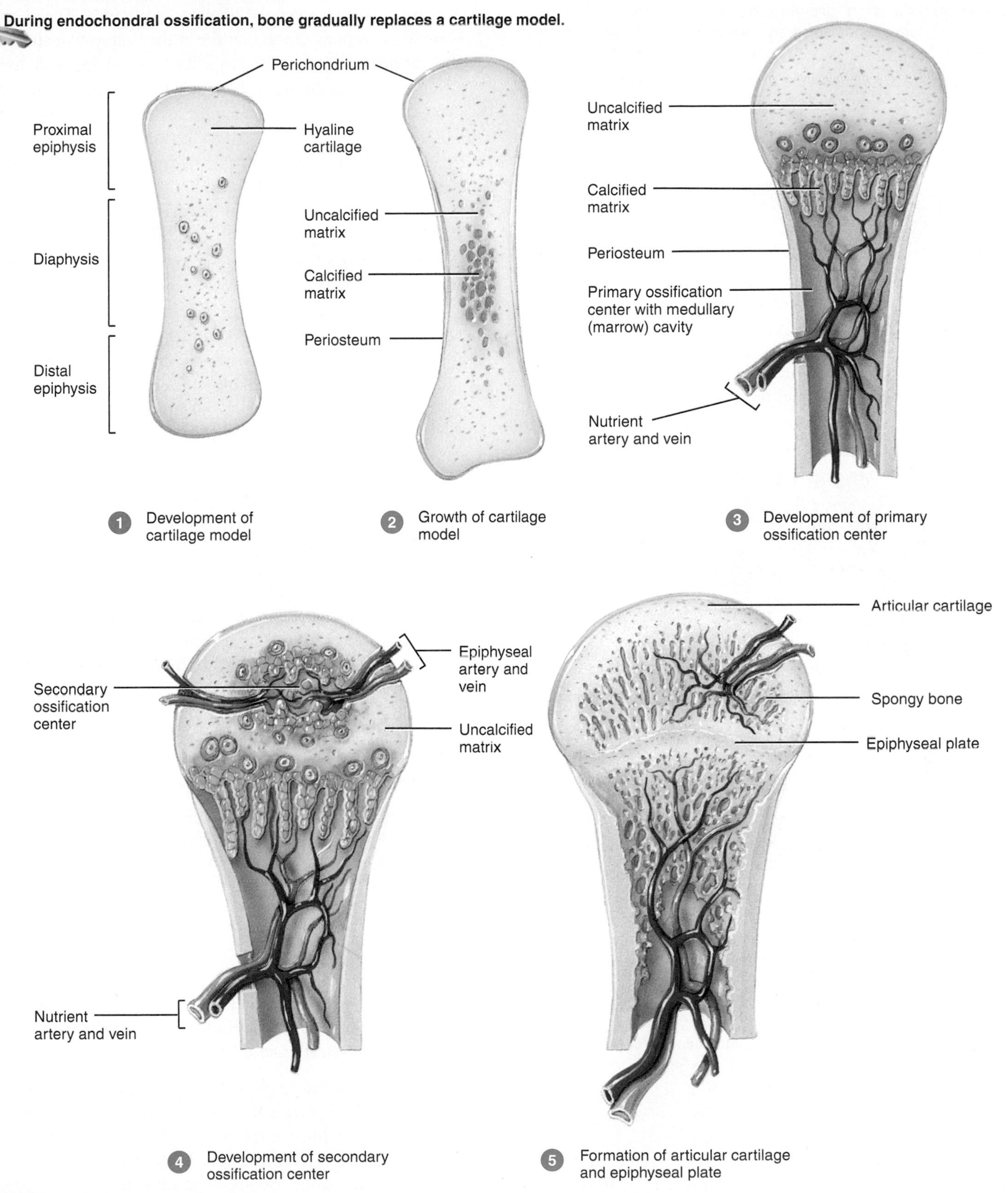

If radiographs of an 18-year-old basketball player show clear epiphyseal plates but no epiphyseal lines, is she likely to grow taller?

Figure 6.8 The epiphyseal plate is a layer of hyaline cartilage in the metaphysis of a growing bone. The epiphyseal plate appears as a dark band between whiter calcified areas in the radiograph.

The epiphyseal plate allows the diaphysis of a bone to increase in length.

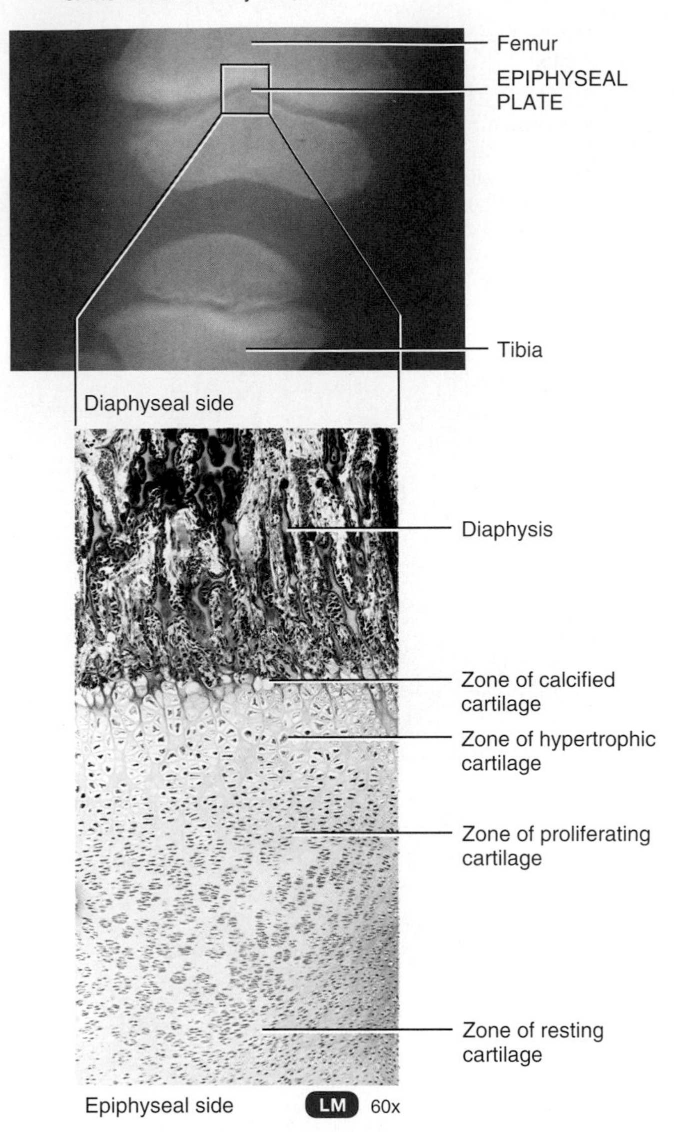

(b) Histology of the epiphyseal plate

What activities of the epiphyseal plate account for the lengthwise growth of the diaphysis?

2. ***Zone of proliferating cartilage.*** Slightly larger chondrocytes in this zone are arranged like stacks of coins. The chondrocytes divide to replace those that die at the diaphyseal side of the epiphyseal plate.
3. ***Zone of hypertrophic cartilage.*** In this layer, the chondrocytes are even larger and remain arranged in columns. The chondrocytes accumulate glycogen in their cytoplasm, and the matrix between lacunae narrows in response to the growth of lacunae. The lengthening of the diaphysis is the result of cell divisions in the zone of proliferating cartilage and maturation of the cells in the zone of hypertrophic cartilage.
4. ***Zone of calcified cartilage.*** The final zone of the epiphyseal plate is only a few cells thick and consists mostly of dead chrondrocytes because the matrix around them has calcified. Osteoclasts dissolve the calcified cartilage, and osteoblasts and capillaries from the diaphysis invade the area. The osteoblasts lay down bone matrix, replacing the calcified cartilage. As a result, the diaphyseal border of the epiphyseal plate is firmly cemented to the bone of the diaphysis.

The activity of the epiphyseal plate is the only way that the diaphysis can increase in length. As a bone grows, chondrocytes proliferate on the epiphyseal side of the plate. New chondrocytes cover older ones, which are then destroyed by the process of calcification. Thus, the cartilage is replaced by bone on the diaphyseal side of the plate. In this way the thickness of the epiphyseal plate remains relatively constant, but the bone on the diaphyseal side increases in length.

Between the ages of 18 and 25, the epiphyseal plates close; that is, the epiphyseal cartilage cells stop dividing, and bone replaces all the cartilage. The epiphyseal plate fades, leaving a bony structure called the **epiphyseal line.** The appearance of the epiphyseal line signifies that the bone has stopped growing in length. The clavicle is the last bone to stop growing. If a bone fracture damages the epiphyseal plate, the fractured bone may be shorter than normal once adult stature is reached. This is because damage to cartilage, which is avascular, accelerates closure of the epiphyseal plate, thus inhibiting lengthwise growth of the bone.

Growth in Thickness

Unlike cartilage, which can thicken by both interstitial and appositional growth, bone can grow in thickness or diameter only by **appositional growth** (Figure 6.9):

1. At the bone surface, periosteal cells differentiate into osteoblasts, which secrete the collagen fibers and other organic molecules that form bone matrix. The osteoblasts become surrounded by matrix and develop into osteocytes. This process forms bone ridges on either side of a periosteal blood vessel. The ridges slowly enlarge and create a groove for the periosteal blood vessel.

Figure 6.9 Bone growth in diameter: appositional growth.

Whereas cartilage can grow by both interstitial and appositional growth, bone can grow in diameter only by appositional growth.

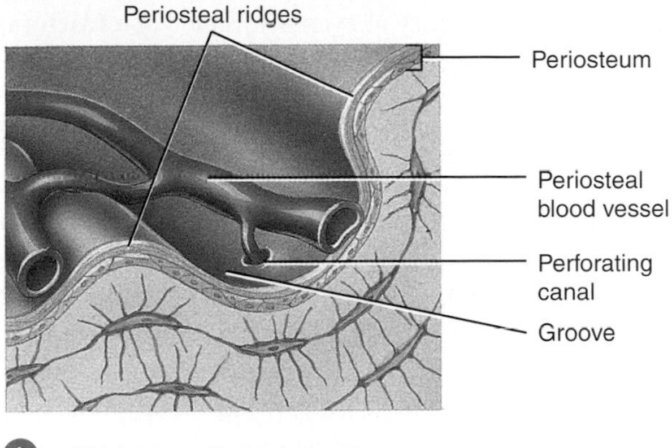

1 Ridges in periosteum create groove for periosteal blood vessel.

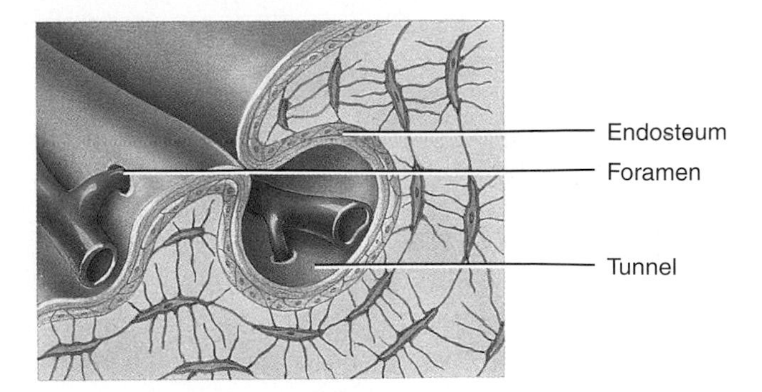

2 Periosteal ridges fuse, forming an endosteum-lined tunnel.

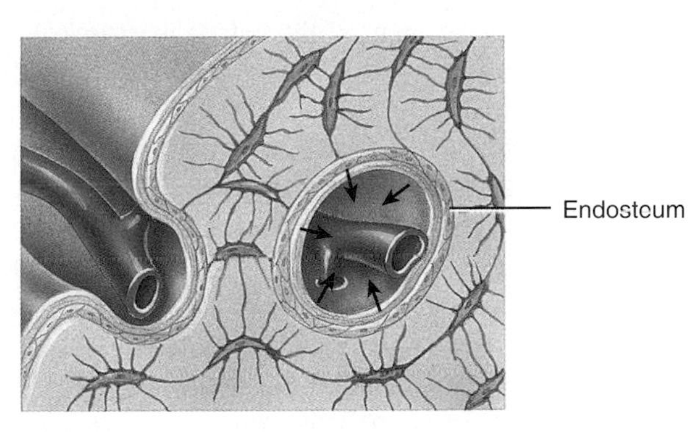

3 Osteoblasts in endosteum build new concentric lamellae inward toward center of tunnel, forming a new osteon.

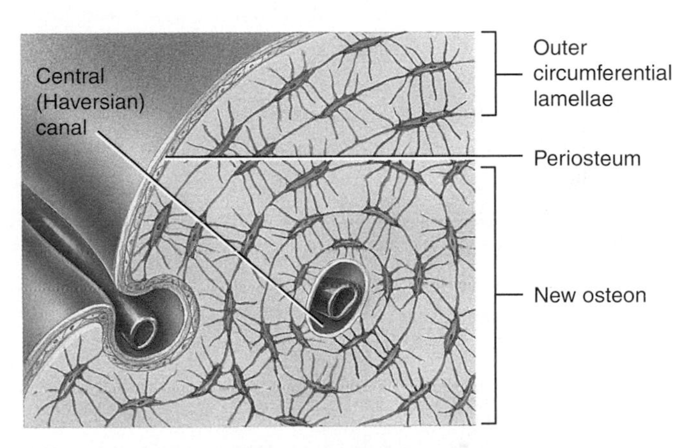

4 Bone grows outward as osteoblasts in periosteum build new outer circumferential lamellae. Osteon formation repeats as new periosteal ridges fold over blood vessels.

How does the medullary cavity enlarge during growth in diameter?

2 Eventually, the ridges fold together and fuse, and the groove becomes a tunnel that encloses the blood vessel. The former periosteum now becomes the endosteum that lines the tunnel.

3 Osteoblasts in the endosteum deposit bone matrix, forming new concentric lamellae. The formation of additional concentric lamellae proceeds inward toward the periosteal blood vessel. In this way, the tunnel fills in, and a new osteon is created.

4 As an osteon is forming, osteoblasts under the periosteum deposit new outer circumferential lamellae, further increasing the thickness of the bone. As additional periosteal blood vessels become enclosed as in step 1, the growth process continues.

Recall that as new bone tissue is being deposited on the outer surface of bone, the bone tissue lining the medullary cavity is destroyed by osteoclasts in the endosteum. In this way, the medullary cavity enlarges as the bone increases in diameter.

Factors Affecting Bone Growth

Growth and maintenance of bone depend on adequate dietary intake of minerals and vitamins, as well as on sufficient levels of several hormones. Large amounts of calcium and phosphorus are needed while bones are growing, as are smaller amount of fluoride, magnesium, iron, and manganese. Vitamin C is needed for synthesis of collagen, the main bone protein, and also for differentiation of osteoblasts into osteocytes. Vitamins K and B_{12} also are needed for protein synthesis, whereas vitamin A stimulates activity of osteoblasts.

During childhood, the most important hormones that stimulate bone growth are the insulinlike growth factors (IGFs), which are produced by bone tissue and the liver. IGFs promote cell division at the epiphyseal plate and in the periosteum and enhance synthesis of the proteins needed to build new bone. Production of IGFs, in turn, is stimulated by human growth hormone (hGH) from the anterior lobe of the pituitary gland. Thyroid hormones (T_3 and T_4), from the thyroid gland, and insulin, from the pancreas, also are needed for normal bone growth.

At puberty, hormones known as sex steroids start to be released in larger quantities. The **sex steroids** include estrogens (produced by the ovaries) and androgens (produced by the testes in males). Although females have much higher levels of estrogens and males have higher levels of androgens, females also have low levels of androgens, and males have low levels of estrogens. The adrenal glands of both sexes produce androgens, and other tissues, such as adipose tissue, can convert androgens to estrogens. These hormones are responsible for increased osteoblast activity and synthesis of bone matrix and the sudden "growth spurt" that occurs during the teenage years. Estrogens also promote changes in the skeleton that are typical of females, for example, widening of the pelvis. Ultimately, sex steroids, especially estrogens in both sexes, shut down growth at the epiphyseal plates. At this point, elongation of the bones ceases. Lengthwise growth of bones typically ends earlier in females than in males due to the higher levels of estrogens in females.

Hormonal Abnormalities that Affect Height

Excessive or deficient secretion of hormones that normally govern bone growth causes a person to be abnormally tall or short. During childhood, oversecretion of hGH produces giantism, in which a person becomes much taller and heavier than normal, whereas undersecretion of hGH or thyroid hormones produces short stature. Because estrogens terminate growth at the epiphyseal plates, both men and women who lack estrogens or receptors for estrogens grow taller than normal. ■

▶ CHECKPOINT

10. Describe the zones of the epiphyseal plate and their functions.
11. What is the significance of the epiphyseal line?
12. How is bone growth in length different from bone growth in thickness?

BONES AND HOMEOSTASIS

▶ OBJECTIVES

- **Describe the processes involved in bone remodeling.**
- **Describe the sequence of events in repair of a fracture.**
- **Describe the role of bone in calcium homeostasis.**

Bone Remodeling

Even after bones have reached their adult shapes and sizes, they continue to be renewed. **Bone remodeling** is an ongoing process whereby osteoclasts first carve out small tunnels in old bone tissue and then osteoblasts rebuild it anew. In different parts of the skeleton, a full cycle of bone remodeling may take as little as 2 to 3 months or last much longer. For example, remodeling of the distal portion of the femur takes about four months. Remodeling normally serves two purposes: It renews bone tissue before deterioration sets in, and it redistributes bone matrix along lines of mechanical stress. Remodeling also is the way that injured bone heals.

The breakdown of matrix by osteoclasts is called **bone resorption.** In this process, an osteoclast attaches tightly to the bone surface at the endosteum or periosteum and forms a leakproof seal at the edges of its ruffled border (see Figure 6.2). Then it releases protein-digesting lysosomal enzymes and several acids into the sealed pocket. The enzymes digest collagen fibers and other organic substances while the acids dissolve the bone minerals. Working together, several osteoclasts carve out a small tunnel in the old bone. The degraded bone proteins and matrix minerals, mainly calcium and phosphorus, enter an osteoclast by endocytosis, cross the cell in vesicles, and undergo exocytosis on the side opposite the ruffled border. Now in the interstitial fluid, the products of bone resorption diffuse into nearby blood capillaries. Once a small area of bone has been resorbed, osteoclasts depart and osteoblasts move in to rebuild the bone in that area.

To achieve homeostasis in bone, the bone-resorbing actions of osteoclasts must balance the bone-making actions of osteoblasts. If osteoblasts form too much new bone tissue, the bones become abnormally thick and heavy. If bone tissue becomes overly calcified, thick bumps called *spurs* may appear and may interfere with movement at joints. Alternatively, a loss of too much calcium or inadequate formation of new tissue weakens bone tissue.

The control of bone growth and bone remodeling is complex and not completely understood. Several circulating hormones as well as locally produced substances are involved. Sex steroids slow resorption of old bone and promote deposition of new bone. One way that estrogens slow resorption is by promoting apoptosis (programmed death) of osteoclasts. Other hormones that govern remodeling include parathyroid hormone and calcitriol, the active form of vitamin D, both of which also increase blood calcium level (see Figure 6.12).

Fracture and Repair of Bone

A **fracture** is any break in a bone. Fractures are named according to their severity, the shape or position of the fracture line, or even the physician who first described them. Among the common kinds of fractures are the following (Figure 6.10):

- **Open (compound) fracture:** The broken ends of the bone protrude through the skin (Figure 6.10a). Conversely, a **closed (simple) fracture** does not break the skin.
- **Comminuted fracture** (KOM-i-noo-ted; *com-* = together; *-minuted* = crumbled): The bone splinters at the site of impact, and smaller bone fragments lie between the two main fragments (Figure 6.10b).
- **Greenstick fracture:** A partial fracture in which one side of the bone is broken and the other side bends; occurs only in children, whose bones are not yet fully ossified and contain more organic material than inorganic material (Figure 6.10c).
- **Impacted fracture:** One end of the fractured bone is forcefully driven into the interior of the other (Figure 6.10d).
- **Pott's fracture:** A fracture of the distal end of the lateral leg bone (fibula), with serious injury of the distal tibial articulation (Figure 6.10e).
- **Colles' fracture** (KOL-ez): A fracture of the distal end of the lateral forearm bone (radius) in which the distal fragment is displaced posteriorly (Figure 6.10f).

In some cases, a bone may fracture without visibly breaking. For example, a **stress fracture** is a series of microscopic fissures in bone that forms without any evidence of injury to other tissues. In healthy adults, stress fractures result from repeated, strenuous activities such as running, jumping, or aerobic dancing. Stress fractures also result from disease processes that disrupt normal bone calcification, such as osteoporosis (discussed later). About 25% of stress fractures involve the tibia. Although standard x-ray images often fail to reveal the presence of stress fractures, they show up clearly in a bone scan.

The following steps occur in the repair of a bone fracture (Figure 6.11):

1. *Formation of fracture hematoma.* Because of the fracture, blood vessels crossing the fracture line are broken. These include vessels in the periosteum, osteons (Haversian systems), medullary cavity, and perforating canals. As blood leaks from the torn ends of the vessels, it forms a clot around the site of the fracture. This clot, called a **fracture hematoma** (hē′-ma-TŌ-ma; *hemat-* = blood; *-oma* = tumor), usually forms 6 to 8 hours after the injury. Because the circulation of blood stops at the site where the fracture hematoma forms, nearby bone cells die. Swelling and inflammation occur in response to dead bone cells, producing additional cellular debris. Blood capillaries grow into the blood clot, and phagocytes (neutrophils and macrophages)

Figure 6.10 Types of bone fractures.

A fracture is any break in a bone.

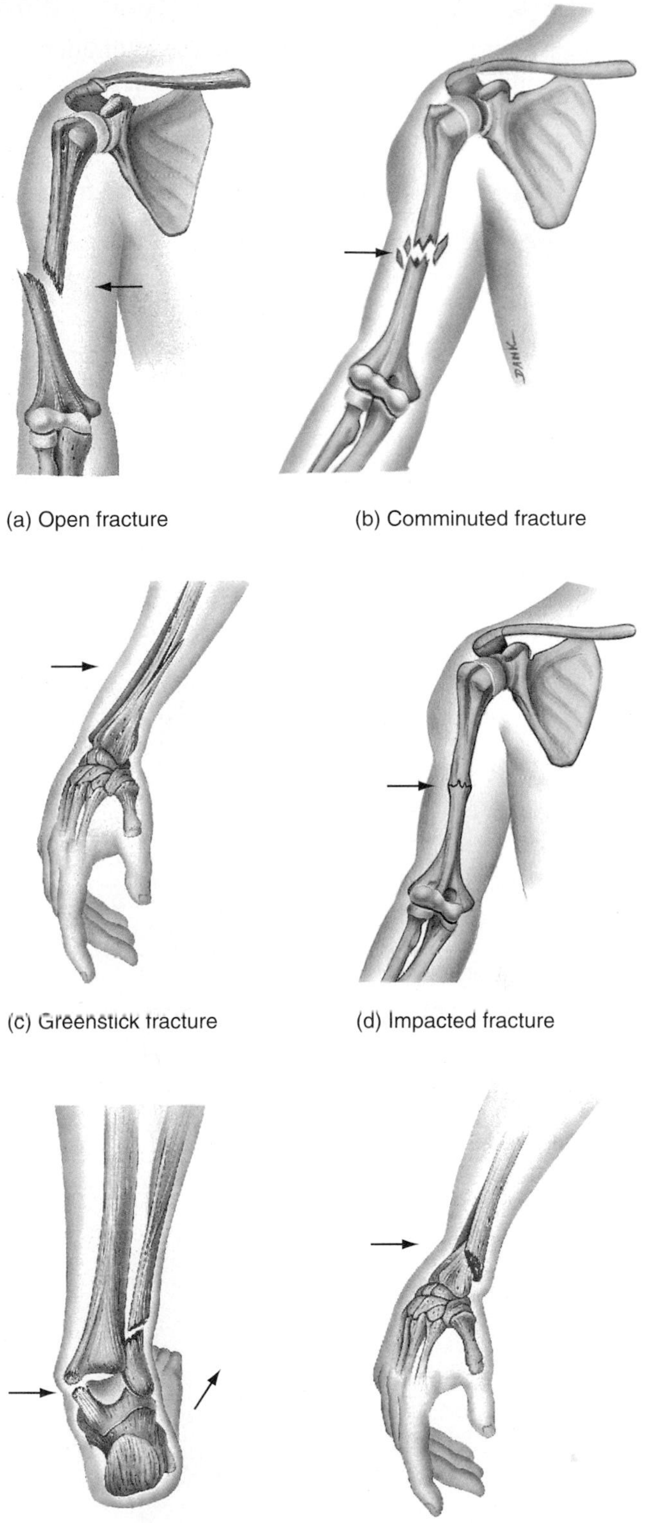

(a) Open fracture

(b) Comminuted fracture

(c) Greenstick fracture

(d) Impacted fracture

(e) Pott's fracture

(f) Colles' fracture

What is the difference between an open fracture and a closed fracture?

and osteoclasts begin to remove the dead or damaged tissue in and around the fracture hematoma. This stage may last up to several weeks.

2 ***Fibrocartilaginous callus formation.*** The infiltration of new blood capillaries into the fracture hematoma helps organize it into an actively growing connective tissue called a **procallus.** Next, fibroblasts from the periosteum and osteogenic cells from the periosteum, endosteum, and red bone marrow invade the procallus. The fibroblasts produce collagen fibers, which help connect the broken ends of the bone. Phagocytes continue to remove cellular debris. Osteogenic cells develop

Figure 6.11 Steps in repair of a bone fracture.

Bone heals more rapidly than cartilage because its blood supply is more plentiful.

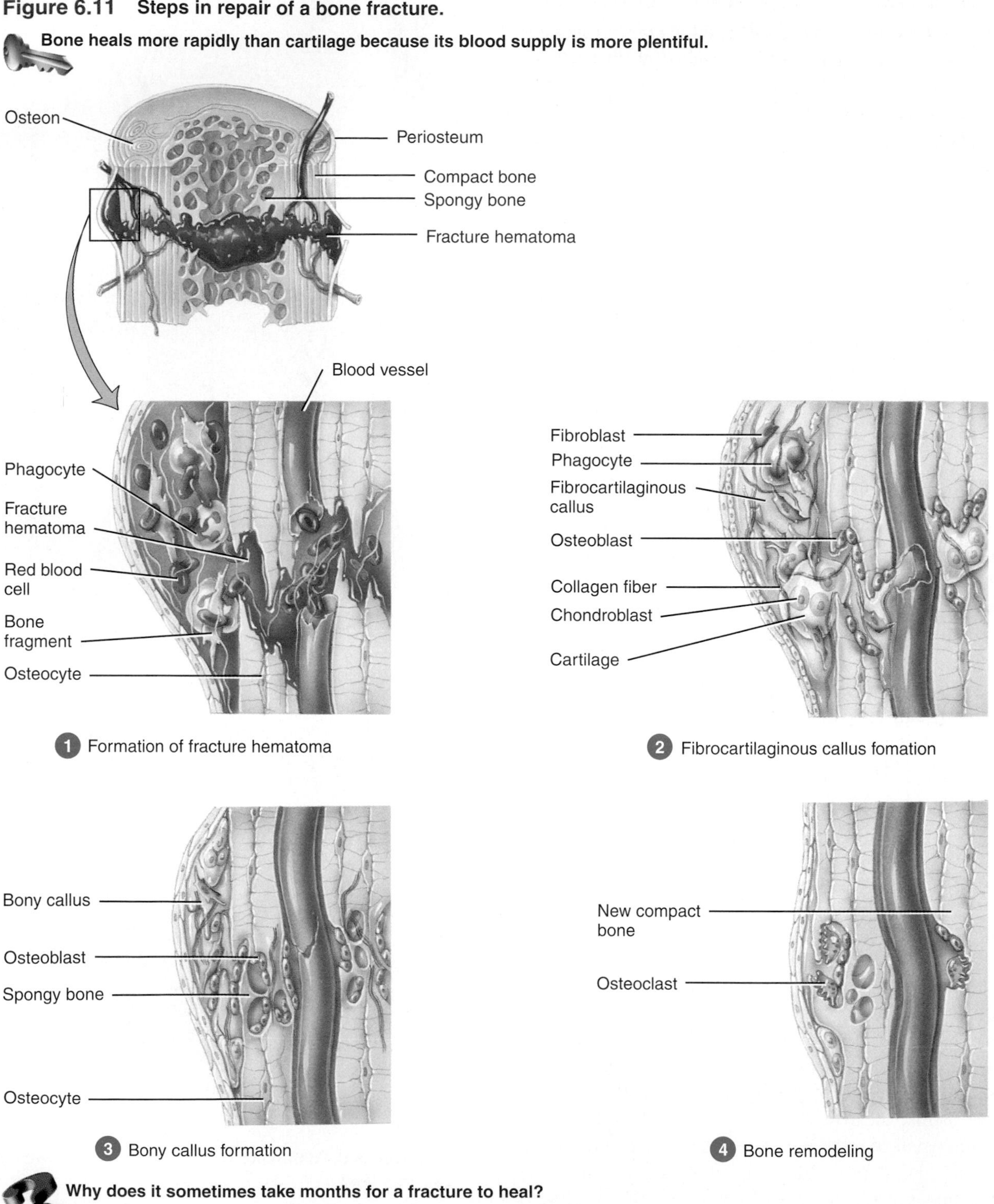

Why does it sometimes take months for a fracture to heal?

into chondroblasts in areas of avascular healthy bone tissue and begin to produce fibrocartilage. Eventually, the procallus is transformed into a **fibrocartilaginous callus,** a mass of repair tissue that bridges the broken ends of the bone. Formation of the fibrocartilaginous callus takes about 3 weeks.

3. ***Bony callus formation.*** In areas closer to well-vascularized healthy bone tissue, osteogenic cells develop into osteoblasts, which begin to produce spongy bone trabeculae. The trabeculae join living and dead portions of the original bone fragments. In time, the fibrocartilage is converted to spongy bone, and the callus is then referred to as a **bony callus.** The bony callus lasts about 3 to 4 months.
4. ***Bone remodeling.*** The final phase of fracture repair is **bone remodeling** of the callus. Dead portions of the original fragments of broken bone are gradually resorbed by osteoclasts. Compact bone replaces spongy bone around the periphery of the fracture. Sometimes, the repair process is so thorough that the fracture line is undetectable, even in a radiograph (x ray). Often, a thickened area on the surface of the bone remains as evidence of a healed fracture, and a healed bone may eventually be stronger than it was before the break.

Although bone has a generous blood supply, healing sometimes takes months. The calcium and phosphorus needed to strengthen and harden new bone are deposited only gradually, and bone cells generally grow and reproduce slowly. Moreover, the temporary disruption in its blood supply helps explain the slowness of healing of severely fractured bones.

Treatments for Fractures

Treatments for fractures vary according to the person's age, the type of fracture, and the bone involved. The ultimate goals of fracture treatment are realignment of the bone fragments, immobilization to maintain realignment, and restoration of function. For bones to unite, the fractured ends must be brought into alignment, a process called **reduction.** In **closed reduction,** the fractured ends of a bone are brought into alignment by manual manipulation, and the skin remains intact. In **open reduction,** the fractured ends of a bone are brought into alignment by a surgical procedure in which internal fixation devices such as screws, plates, pins, rods, and wires are used. Following reduction, a fractured bone may be kept immobilized by a cast, sling, splint, elastic bandage, external fixation device, or a combination of these devices. ■

Bone's Role in Calcium Homeostasis

Bone is the body's major calcium reservoir, storing 99% of total body calcium. The level of calcium in the blood can be regulated by controlling the rates of calcium resorption from bone into blood and of calcium deposition from blood into bone. Most functions of nerve cells depend on having a stable level of calcium ions (Ca^{2+}) in extracellular fluid. Also, many enzymes require Ca^{2+} as a cofactor (an additional substance needed for an enzymatic reaction to occur), and blood clotting requires Ca^{2+}. For this reason, the blood plasma level of Ca^{2+} is very closely regulated between 9 and 11 mg/100 mL. Even small changes in Ca^{2+} concentration outside this range may prove fatal—the heart may stop (cardiac arrest) if the concentration goes too high, or breathing may cease (respiratory arrest) if the level falls too low. The role of bone in calcium homeostasis is to "buffer" the blood Ca^{2+} level, releasing Ca^{2+} into blood plasma when the level decreases, and taking Ca^{2+} back when the level rises. Hormones regulate these exchanges.

The most important hormone that regulates Ca^{2+} exchange between bone and blood is **parathyroid hormone (PTH),** secreted by the parathyroid glands (see Figure 18.13 on page 607). PTH secretion operates via a negative feedback system (Figure 6.12). If some stimulus causes blood Ca^{2+} level to decrease, parathyroid gland cells (receptors) detect this change and increase their production of a molecule known as cyclic adenosine monophosphate (cyclic AMP). The gene for PTH within the nucleus of a parathyroid gland cell, which acts as the control center, detects the increased production of cyclic AMP (the input). As a result, PTH synthesis speeds up, and more PTH (the output) is released into the blood. The presence of higher levels of PTH increases the number and activity of osteoclasts (effectors), which step up the pace of bone resorption. The resulting release of Ca^{2+} from bone into blood returns the blood Ca^{2+} level to normal.

PTH also acts on the kidneys (effectors) to decrease loss of Ca^{2+} in the urine, so more is retained in the blood, and stimulates formation of **calcitriol,** a hormone that promotes absorption of calcium from the gastrointestinal tract. Both of these actions also help elevate blood Ca^{2+} level.

Another hormone makes a potential contribution to the homeostasis of blood Ca^{2+} by its influence on bone. When blood Ca^{2+} rises above normal, *parafollicular cells* in the thyroid gland secrete **calcitonin (CT).** CT inhibits activity of osteoclasts, speeds blood Ca^{2+} uptake by bone, and accelerates Ca^{2+} deposition into bones. The net result is that CT promotes bone formation and decreases blood Ca^{2+} level. Despite these effects, the role of CT in normal calcium homeostasis is uncertain because it can be completely absent without causing symptoms. Nevertheless, calcitonin harvested from salmon (Miacalcin) is an effective drug for treating osteoporosis because it slows bone resorption.

Figure 18.14 on page 608 summarizes the roles of parathyroid hormone, calcitriol, and calcitonin in regulation of blood Ca^{2+} level.

► CHECKPOINT

13. Define remodeling, and describe the roles of osteoblasts and osteoclasts in the process.
14. Define a fracture and outline the four steps involved in fracture repair.
15. How do hormones act on bone to regulate calcium homeostasis?

Figure 6.12 Negative feedback system for the regulation of blood calcium (Ca^{2+}) concentration. PTH = parathyroid hormone.

Release of calcium from bone matrix and retention of calcium by the kidneys are the two main ways that blood calcium level can be increased.

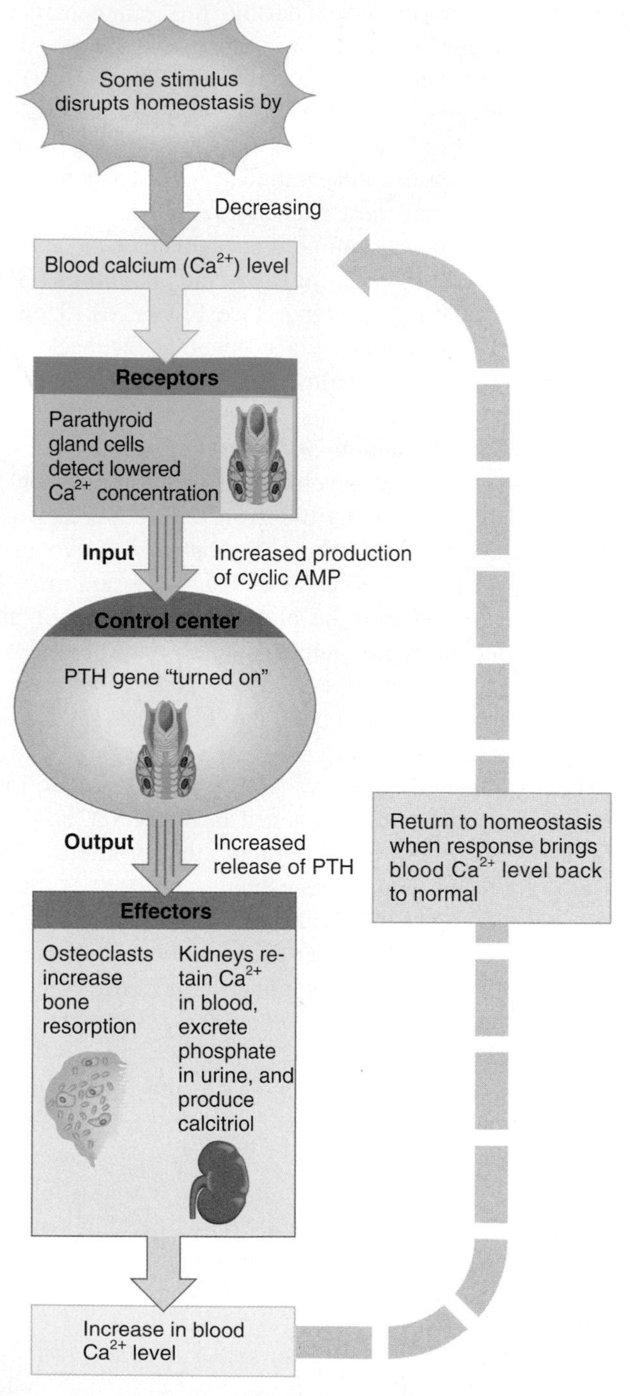

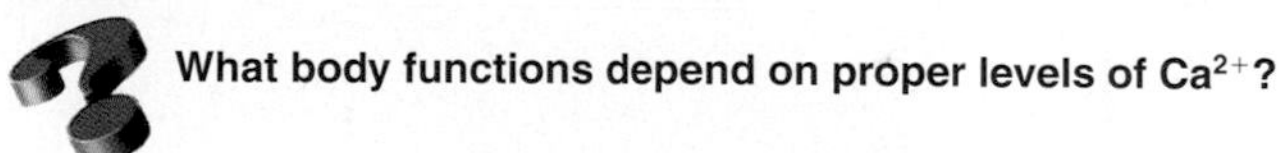

What body functions depend on proper levels of Ca^{2+}?

EXERCISE AND BONE TISSUE

► OBJECTIVE

- **Describe how exercise and mechanical stress affect bone tissue.**

Within limits, bone has the ability to alter its strength in response to changes in mechanical stress. When placed under stress, bone tissue adapts by becoming stronger through increased deposition of mineral salts and production of collagen fibers. Another effect of stress is to increase the production of calcitonin, a hormone that inhibits bone resorption. Without mechanical stress, bone does not remodel normally because bone resorption outstrips bone formation. Removal of mechanical stress weakens bone through demineralization (loss of bone minerals) and decreased numbers of collagen fibers.

The main mechanical stresses on bone are those that result from the pull of skeletal muscles and the pull of gravity. If a person is bedridden or has a fractured bone in a cast, the strength of the unstressed bones diminishes. Astronauts subjected to the microgravity of space also lose bone mass. In both cases, bone loss can be dramatic—as much as 1% per week. Bones of athletes, which are repetitively and highly stressed, become notably thicker and stronger than those of nonathletes. Weight-bearing activities, such as walking or moderate weight lifting, help build and retain bone mass. Adolescents and young adults should engage in regular weight-bearing exercise prior to the closure of the epiphyseal plates to help build total mass prior to its inevitable reduction with aging. Even elderly people can strengthen their bones by engaging in weight-bearing exercise.

► CHECKPOINT

16. What types of mechanical stresses may be used to strengthen bone tissue?

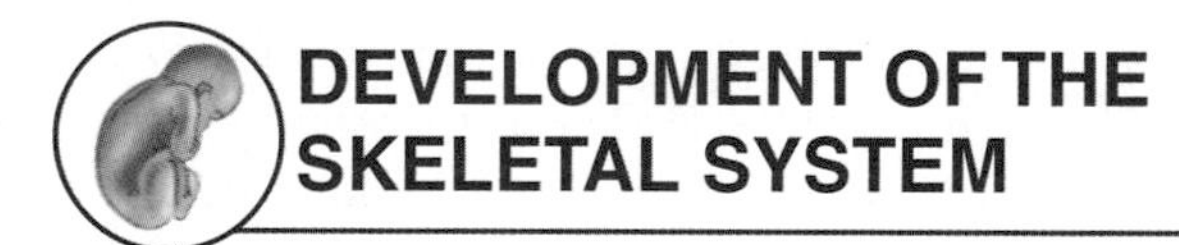

DEVELOPMENT OF THE SKELETAL SYSTEM

► OBJECTIVE

- **Describe the development of the skeletal system and the limbs.**

Both intramembranous and endochondral ossification begin when *mesenchymal cells,* connective tissue cells derived from **mesoderm,** migrate into the area where bone formation will occur. In some skeletal structures, mesenchymal cells develop into chondroblasts that form cartilage. In other skeletal structures, mesenchymal cells develop into osteoblasts that form *bone tissue* by intramembranous or endochondral ossification.

Discussion of the development of the skeletal system provides an excellent opportunity to trace limb development. During the middle of the fourth week, the upper limbs appear as small elevations at the sides of the trunk called **upper limb buds** (Figure 6.13a). By the end of the fourth week, the **lower**

Figure 6.13 Features of a human embryo during weeks four through eight of development. The liver and heart prominences are surface bulges resulting from the underlying growth of these organs. The embryo receives oxygen and nutrients from its mother through blood vessels in the umbilical cord.

After the limb buds develop, endochondral ossification of the limb bones begins during the seventh embryonic week.

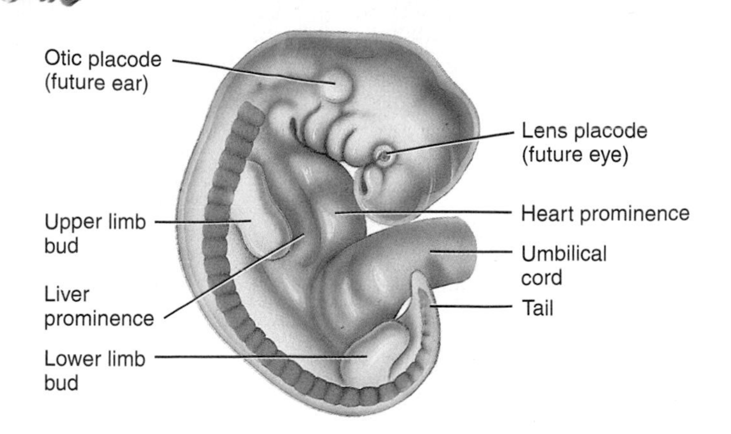

(a) Four-week embryo showing development of limb buds

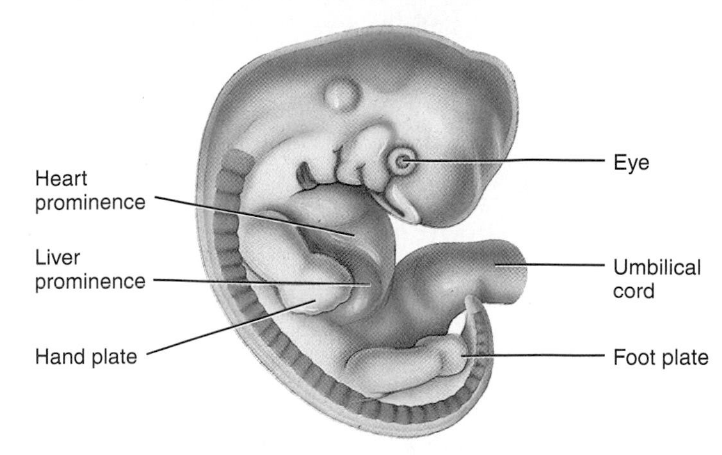

(b) Six-week embryo showing development of hand and foot plates

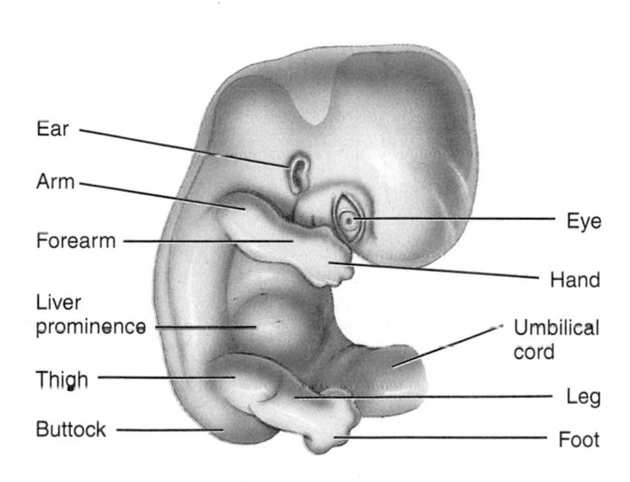

(c) Seven-week embryo showing development of arm, forearm, and hand in upper limb bud and thigh, leg, and foot in lower limb bud

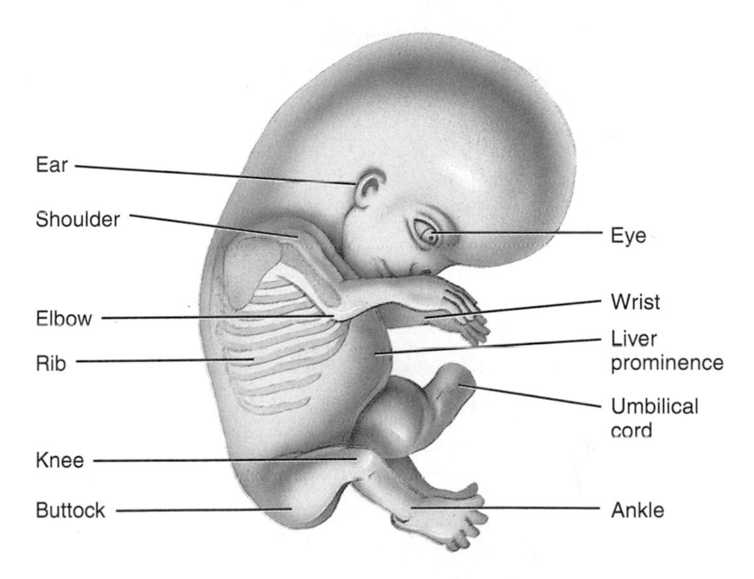

(d) Eight-week embryo in which limb buds have developed into upper and lower limbs

Which of the three basic embryonic tissues—ectoderm, mesoderm, and endoderm—gives rise to the skeletal system?

limb buds appear. The limb buds consist of masses of general **mesoderm** covered by **ectoderm.** At this point, a mesenchymal skeleton exists in the limbs; some of the masses of mesoderm surrounding the developing bones will become the skeletal muscles of the limbs.

By the sixth week, the limb buds develop a constriction around the middle portion. The constriction produces distal segments of the upper buds called **hand plates** and distal segments of the lower buds called **foot plates** (Figure 6.13b). These plates represent the beginnings of the hands and feet, respectively. At this stage of limb development, a cartilaginous skeleton is present. By the seventh week (Figure 6.13c), the arm, forearm, and hand are evident in the upper limb bud, and the thigh, leg, and foot appear in the lower limb bud. Endochondral ossification has begun. By the eighth week (Figure 6.13d), as the shoulder, elbow, and wrist areas become apparent, the upper limb bud is appropriately called the upper limb, and the lower limb bud is now the lower limb.

About 22 to 24 days after fertilization, mesodermal cells form a solid cylinder of cells called the **notochord** (nō-tō-KORD; *noto-* = the back; *chord* = a string). See Figure 29.9a on page 1074. This structure plays an extremely important role in inducing, through chemical signals, unspecialized embryonic cells to develop into specialized tissues and organs. For example, the notochord induces certain mesodermal cells to develop into the vertebral bodies and forms the nucleus pulposus of the intervertebral discs (see Figure 7.16 on pages 203–204).

▶ CHECKPOINT

17. When and how do the limbs develop?

AGING AND BONE TISSUE

▶ OBJECTIVE

- **Describe the effects of aging on bone tissue.**

From birth through adolescence, more bone tissue is produced than is lost during bone remodeling. In young adults the rates of bone deposition and resorption are about the same. As the level of sex steroids diminishes during middle age, especially in women after menopause, a decrease in bone mass occurs because bone resorption outpaces bone deposition. In old age, loss of bone through resorption occurs more rapidly than bone gain. Because women's bones generally are smaller and less massive than men's bones to begin with, loss of bone mass in old age typically has a greater adverse effect in women.

There are two principal effects of aging on bone tissue: loss of bone mass and brittleness. Loss of bone mass results from **demineralization,** the loss of calcium and other minerals from bone matrix. This loss usually begins after age 30 in females, accelerates greatly around age 45 as levels of estrogens decrease, and continues until as much as 30% of the calcium in bones is lost by age 70. Once bone loss begins in females, about 8% of bone mass is lost every ten years. In males, calcium loss typically does not begin until after age 60, and about 3% of bone mass is lost every ten years. The loss of calcium from bones is one of the problems in osteoporosis (described shortly).

The second principal effect of aging on the skeletal system, brittleness, results from a decreased rate of protein synthesis. Recall that the organic part of bone matrix, mainly collagen fibers, gives bone its tensile strength. The loss of tensile strength causes the bones to become very brittle and susceptible to fracture. In some elderly people, collagen fiber synthesis slows, in part, due to diminished production of human growth hormone. In addition to increasing the susceptibility to fractures, loss of bone mass also leads to deformity, pain, stiffness, loss of height, and loss of teeth.

▶ CHECKPOINT

18. What is demineralization, and how does it affect the functioning of bone?
19. What changes occur in the organic part of bone matrix with aging?

DISORDERS: HOMEOSTATIC IMBALANCES

Osteoporosis

Osteoporosis (os′-tē-ō-pō-RŌ-sis; *por-* = passageway; *-osis* = condition) is literally a condition of porous bones (Figure 6.14). The basic problem is that bone resorption outpaces bone deposition. In large part this is due to depletion of calcium from the body—more calcium is lost in urine, feces, and sweat than is absorbed from the diet. Bone mass becomes so depleted that bones fracture, often spontaneously, under the mechanical stresses of everyday living. For example, a hip fracture might result from simply sitting down too quickly. In the United States, osteoporosis causes more than a million fractures a year, mainly in the hip, wrist, and vertebrae. Osteoporosis afflicts the entire skeletal system. In addition to fractures, osteoporosis causes shrinkage of vertebrae, height loss, hunched backs, and bone pain.

Figure 6.14 Comparison of spongy bone tissue from (a) a normal young adult and (b) a person with osteoporosis. Notice the weakened trabeculae in (b). Compact bone tissue is similarly affected by osteoporosis.

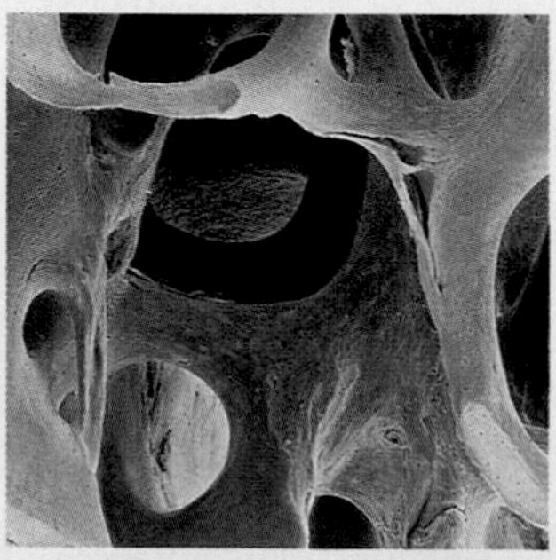

SEM 30x

(a) Normal bone

SEM 30x

(b) Osteoporotic bone

Thirty million people in the United States suffer from osteoporosis. The disorder primarily affects middle-aged and elderly people, 80% of them women. Older women suffer from osteoporosis more often than men for two reasons: Women's bones are less massive than men's bones, and production of estrogens in women declines dramatically at menopause, whereas production of the main androgen, testosterone in older men wanes gradually and only slightly. Estrogens and testosterone stimulate osteoblast activity and synthesis of bone matrix. Besides gender, risk factors for developing osteoporosis include a family history of the disease, European or Asian ancestry, thin or small body build, an inactive lifestyle, cigarette smoking, a diet low in calcium and vitamin D, more than two alcoholic drinks a day, and the use of certain medications.

In postmenopausal women, treatment of osteoporosis may include estrogen replacement therapy (ERT; low doses of estrogens) or hormone replacement therapy (HRT; a combination of estrogens and progesterone, another sex steroid). Although such treatments help combat osteoporosis, they increase a woman's risk of breast cancer. The drug Raloxifene (Evista) mimics the beneficial effects of estrogens on bone without increasing the risk of breast cancer. Another drug that may be used is the nonhormone drug Alendronate (Fosamax), which blocks resorption of bone by osteoclasts.

Perhaps more important than treatment is prevention. Adequate calcium intake and weight-bearing exercise in her early years may be more beneficial to a woman than drugs and calcium supplements when she is older.

Rickets and Osteomalacia

Rickets and **osteomalacia** (os′-tē-ō-ma-LĀ-shē-a; *-malacia* = softness) are disorders in which bones fail to calcify. Although the organic matrix is still produced, calcium salts are not deposited, and the bones become "soft" or rubbery and easily deformed. Rickets affects the growing bones of children. Because new bone formed at the epiphyseal plates fails to ossify, bowed legs and deformities of the skull, rib cage, and pelvis are common. In osteomalacia, sometimes called "adult rickets," new bone formed during remodeling fails to calcify. This disorder causes varying degrees of pain and tenderness in bones, especially in the hip and leg. Bone fractures also result from minor trauma.

MEDICAL TERMINOLOGY

Osteoarthritis (os′-tē-ō-ar-THRĪ-tis; *arthr* = joint) The degeneration of articular cartilage such that the bony ends touch; the resulting friction of bone against bone worsens the condition. Usually associated with the elderly.

Osteogenic sarcoma (os′-tē-Ō-JEN-ik sar-KŌ-ma; *sarcoma* = connective tissue tumor) Bone cancer that primarily affects osteoblasts and occurs most often in teenagers during their growth spurt; the most common sites are the metaphyses of the thigh bone (femur), shin bone (tibia), and arm bone (humerus). Metastases occur most often in lungs; treatment consists of multidrug chemotherapy and removal of the malignant growth, or amputation of the limb.

Osteomyelitis (os′-tē-ō-mī-e-LĪ-tis) An infection of bone characterized by high fever, sweating, chills, pain, and nausea, pus formation, edema, and warmth over the affected bone and rigid overlying muscles. Bacteria, usually *Staphylococcus aureus,* often cause it. The bacteria may reach the bone from outside the body (through open fractures, penetrating wounds, or orthopedic surgical procedures); from other sites of infection in the body (abscessed teeth, burn infections, urinary tract infections, or upper respiratory infections) via the blood; and from adjacent soft tissue infections (as occurs in diabetes mellitus).

Osteopenia (os′-tē-ō-PĒ-nē-a; *penia* = poverty) Reduced bone mass due to a decrease in the rate of bone synthesis to a level insufficient to compensate for normal bone resorption; any decrease in bone mass below normal. An example is osteoporosis.

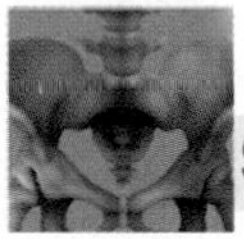

STUDY OUTLINE

INTRODUCTION (p. 162)

1. A bone is made up of several different tissues: bone or osseous tissue, cartilage, dense connective tissues, epithelium, various blood-forming tissues, adipose tissue, and nervous tissue.
2. The entire framework of bones and their cartilages constitute the skeletal system.

FUNCTIONS OF BONE AND THE SKELETAL SYSTEM (p. 162)

1. The skeletal system functions in support, protection, movement, mineral homeostasis, blood cell production, and triglyceride storage.

STRUCTURE OF BONE (p. 162)

1. Parts of a typical long bone are the diaphysis (shaft), proximal and distal epiphyses (ends), metaphyses, articular cartilage, periosteum, medullary (marrow) cavity, and endosteum.

HISTOLOGY OF BONE TISSUE (p. 164)

1. Bone tissue consists of widely separated cells surrounded by large amounts of matrix.
2. The four principal types of cells in bone tissue are osteogenic cells, osteoblasts, osteocytes, and osteoclasts.
3. The matrix of bone contains abundant mineral salts (mostly hydroxyapatite) and collagen fibers.
4. Compact bone tissue consists of osteons (Haversian systems) with little space between them.
5. Compact bone tissue lies over spongy bone tissue in the epiphyses and makes up most of the bone tissue of the diaphysis. Functionally, compact bone tissue protects, supports, and resists stress.
6. Spongy bone tissue does not contain osteons. It consists of trabeculae surrounding many red bone marrow–filled spaces.
7. Spongy bone tissue forms most of the structure of short, flat, and irregular bones, and the epiphyses of long bones. Functionally, spongy bone tissue trabeculae offer resistance along lines of stress, support and protect red bone marrow, and make bones lighter for easier movement.

BLOOD AND NERVE SUPPLY OF BONE (p. 167)

1. Long bones are supplied by periosteal, nutrient, and epiphyseal arteries; veins accompany the arteries.
2. Nerves accompany blood vessels in bone; the periosteum is rich in sensory neurons.

BONE FORMATION (p. 168)

1. Bone forms by a process called ossification (osteogenesis), which begins when mesenchymal cells become transformed into osteogenic cells. These undergo cell division and give rise to cells that differentiate into osteoblasts and osteoclasts.
2. Ossification begins during the sixth or seventh week of embryonic life. The two types of ossification, intramembranous and endochondral, involve the replacement of a preexisting connective tissue with bone.
3. Intramembranous ossification occurs within fibrous connective tissue membranes.
4. Endochondral ossification occurs within a hyaline cartilage model. The primary ossification center of a long bone is in the diaphysis. Cartilage degenerates, leaving cavities that merge to form the medullary cavity. Osteoblasts lay down bone. Next, ossification occurs in the epiphyses, where bone replaces cartilage, except for the epiphyseal plate.

BONE GROWTH (p. 170)

1. The epiphyseal plate consists of four zones: zones of resting cartilage, proliferating cartilage, hypertrophic cartilage, and calcified cartilage.
2. Because of the activity of the epiphyseal plate, the diaphysis of a bone increases in length.
3. Bone grows in thickness or diameter due to the addition of new bone tissue by periosteal osteoblasts around the outer surface of the bone (appositional growth).
4. Dietary minerals (especially calcium and phosphorus) and vitamins (C, K, and B_{12}) are needed for bone growth and maintenance. Insulinlike growth factors (IGFs), human growth hormone, thyroid hormones, insulin, estrogens, and androgens stimulate bone growth. The sex steroids, especially estrogens, also shut down growth at the epiphyseal plates.

BONES AND HOMEOSTASIS (p. 174)

1. Remodeling is an ongoing process whereby osteoclasts carve out small tunnels in old bone tissue and then osteoblasts rebuild it anew.
2. In bone resorption, osteoclasts release enzymes and acids that degrade collagen fibers and dissolve mineral salts.
3. Sex steroids slow resorption of old bone and promote new bone deposition. Estrogens slow resorption by promoting apoptosis of osteoclasts.
4. A fracture is any break in a bone.
5. Fracture repair involves formation of a fracture hematoma, of fibrocartilaginous callus, of bony callus, and bone remodeling.
6. Types of fractures include closed (simple), open (compound), comminuted, greenstick, impacted, stress, Pott's, and Colles'.
7. Bone is the major reservoir for calcium in the body.
8. Parathyroid hormone (PTH) secreted by the parathyroid gland increases blood Ca^{2+} level, whereas calcitonin (CT) from the thyroid gland has the potential to decrease blood Ca^{2+} level.

EXERCISE AND BONE TISSUE (p. 178)

1. Mechanical stress increases bone strength by increasing deposition of mineral salts and production of collagen fibers.
2. Removal of mechanical stress weakens bone through demineralization and collagen fiber reduction.

DEVELOPMENT OF THE SKELETAL SYSTEM (p. 178)

1. Bone forms from mesoderm by intramembranous or endochondral ossification.
2. Limbs develop from limb buds, which consist of mesoderm and ectoderm.

AGING AND BONE TISSUE (p. 180)

1. The principal effect of aging is demineralization, a loss of calcium from bones, which may result in osteoporosis.
2. Another effect is a decreased production of matrix proteins (mostly collagen fibers), which makes bones more brittle and thus more susceptible to fracture.

Q SELF-QUIZ QUESTIONS

Fill in the blanks in the following statements.

1. The crystallized mineral salts in bone contribute to bone's ________, while the collagen fibers and other organic molecules provide bone with ________.
2. Compact bone tissue is composed of ________; spongy bone tissue is composed of ________.
3. Endochondral ossification refers to the formation of bone within ________; intramembranous ossification refers to the formation of bone directly from ________.

Indicate whether the following statements are true or false.

4. The activity of the epiphyseal plate is the only mechanism by which the diaphysis can increase in length.
5. Homeostasis in bone occurs when there is a balance between formation of bone by osteocytes and resorption of bone by osteoclasts.

Choose the one best answer to the following questions.

6. Chrondocytes are actively dividing in the (a) zone of hypertrophic cartilage, (b) zone of proliferative cartilage, (c) epiphyseal line, (d) zone of resting cartilage, (e) zone of calcified cartilage.

7. Which of the following statements are *true*? (1) Osteogenic cells develop directly into osteoclasts. (2) Osteogenic cells are unspecialized stem cells. (3) Osteoblasts form bone. (4) Osteocytes are the principal cells of bone tissue. (5) Osteoclasts function in bone resorption. (6) Osteoblasts maintain daily cellular activities of bone tissue. (a) 1, 2, 4, 5, and 6, (b) 2, 3, 4, and 5, (c) 1, 3, 5, and 6, (d) 2, 4, and 6, (e) 1, 2, 4, 5, and 6.

8. Which of the following statements is *true*? (a) The osteons in compact bone are arranged parallel to the lines of stress. (b) Spongy bone tissue composes the shaft of the long bones of the limbs. (c) Yellow bone marrow is found only in embryonic bone. (d) The sex steroids are the most important hormones regulating blood calcium levels. (e) In adult bones, hemopoiesis occurs in compact bone tissue only.

9. Place in order the steps involved in intramembranous ossification. (1) Fusion of bony matrices to form trabeculae. (2) Clusters of osteoblasts form a center of ossification that secretes the organic matric. (3) Replacement of spongy bone with compact bone on the bone's surface. (4) Development of periosteum on the bone's periphery. (5) Hardening of the matrix by deposition of calcium and mineral salts. (a) 2, 4, 5, 1, 3, (b) 4, 3, 5, 1, 2, (c) 1, 2, 5, 4, 3, (d) 2, 5, 1, 4, 3, (e) 5, 1, 3, 4, 2.

10. Which of the following statements are *true*? (1) Weight-bearing activities help build and retain bone mass. (2) Removal of mechanical stress weakens bone. (3) In the absence of mechanical stress, resorption outstrips bone formation. (4) Mechanical stress decreases the production of calcitonin, a hormone that inhibits bone resorption. (5) The main mechanical stresses on bone are those resulting from the contraction of skeletal muscles and the pull of gravity. (a) 1, 2, 3, and 5, (b) 2, 3, 4, and 5, (c) 1, 3, and 5, (d) 2, 4, and 5, (e) 2, 3, and 4.

11. Which of the following are functions of bone tissue and the skeletal system? (1) support, (2) excretion, (3) assistance in movement, (4) mineral homeostasis, (5) blood cell production. (a) 1, 2, and 3, (b) 2, 3, and 4, (c) 3, 4, and 5, (d) 1, 3, 4, and 5, (e) 2, 3, 4, and 5.

12. Match the following.

____ (a) space within the shaft of a bone that contains red or yellow bone marrow
____ (b) triglyceride storage tissue
____ (c) hemopoietic tissue
____ (d) thin layer of hyaline cartilage covering the ends of bones where they form a joint
____ (e) distal and proximal ends of bones
____ (f) the long, cylindrical main portion of the bone; the shaft
____ (g) in a growing bone, the region that contains the growth plate
____ (h) the tough membrane that surrounds the bone surface wherever cartilage is not present
____ (i) membrane lining the medullary cavity
____ (j) a remnant of the active epiphyseal plate; a sign that the bone has stopped growing in length

(1) articular cartilage
(2) endosteum
(3) medullary cavity
(4) diaphysis
(5) epiphyses
(6) metaphysis
(7) periosteum
(8) red bone marrow
(9) yellow bone marrow
(10) epiphyseal line

13. Match the following.

____ (a) small spaces between lamellae that contain osteocytes
____ (b) perforating canals that penetrate compact bone; carry blood vessels, lymphatic vessels, and nerves from the periosteum
____ (c) areas between osteons
____ (d) microscopic unit of compact bone tissue
____ (e) tiny canals filled with extracellular fluid; connect lacunae to each other and to the central canal
____ (f) canals that extend longitudinally through the bone and connect blood vessels and nerves to the osteocytes
____ (g) irregular lattice work of thin columns of bone found in spongy bone tissue
____ (h) rings of hard calcified matrix found around the central canals
____ (i) an opening in the shaft of the bone allowing an artery to pass into the bone
____ (j) a layer of hyaline cartilage in the area between the shaft and end of a growing bone

(1) osteon
(2) Volkmann's canals
(3) Haversian canals
(4) concentric lamellae
(5) epiphyseal plate
(6) trabeculae
(7) interstitial lamellae
(8) canaliculi
(9) nutrient foramen
(10) lacunae

14. Match the following.

____ (a) blood clot that forms around the fracture site
____ (b) a condition of porous bones characterized by decreased bone mass and increased susceptibility to fractures
____ (c) splintered bone, with smaller fragments lying between main fragments
____ (d) a broken bone that does not break through the skin
____ (e) a partial break in a bone in which one side of the bone is broken and the other side bends
____ (f) a broken bone that protrudes through the skin
____ (g) microscopic bone breaks resulting from inability to withstand repeated stressful impact
____ (h) connective tissue mass that forms a bridge between the broken ends of bones
____ (i) condition in which new bone formed by remodeling fails to calcify in adults
____ (j) an infection of bone

(1) closed (simple) fracture
(2) open (compound) fracture
(3) fibrocartiagenous callus
(4) greenstick fracture
(5) stress fracture
(6) comminuted fracture
(7) osteoporosis
(8) osteomalacia
(9) fracture hematoma
(10) osteomyelitis

15. Match the following.

____ (a) decreases blood calcium levels by accelerating calcium deposition in bones	(1) PTH
____ (b) required for collagen synthesis	(2) CT
____ (c) during childhood, it promotes growth at epiphyseal plate; production stimulated by human growth hormone	(3) insulinlike growth factors
____ (d) involved in bone growth by increasing osteoblast activity; causes long bones to stop growing in length	(4) sex steroids
____ (e) raises blood calcium level by increasing bone resorption	(5) vitamin C

CRITICAL THINKING QUESTIONS

1. A father brings his young daughter Lynne Marie to the emergency room after she fell off her bicycle. The attending physician tells the father that his daughter has suffered a greenstick fracture in her forearm. The father is confused—there weren't any sticks where she fell in the street. What should the physician explain to the father?
HINT ***Only children suffer this type of fracture.***

2. Aunt Edith is 95 years old today. She claims that she's been getting shorter every year, and that soon she'll disappear altogether. What's happening to Aunt Edith?
HINT ***Her mind is as sharp as ever; think about her bones.***

3. Astronauts in space exercise as part of their daily routine, yet they still have problems with bone weakness after prolonged stays in space. Why does this happen?
HINT ***What is missing in space that you have to work against on Earth?***

ANSWERS TO FIGURE QUESTIONS

6.1 The periosteum is essential for growth in bone diameter, bone repair, and bone nutrition. It also serves as a point of attachment for ligaments and tendons.

6.2 Bone resorption is necessary for the development, growth, maintenance, and repair of bone.

6.3 Because the central (Haversian) canals are the main blood supply to the osteocytes of an osteon, their blockage would lead to death of the osteocytes.

6.4 Spongy bone makes up most of short, flat, and irregularly shaped bones; most of the epiphyses of long bones; and a narrow rim around the medullary cavity of the diaphysis of long bones.

6.5 Periosteal arteries enter bone tissue through perforations (Volkmann's canals).

6.6 Flat bones of the skull and mandible (lower jawbone) develop by intramembranous ossification.

6.7 Yes, she probably will grow taller. Epiphyseal lines are indications of growth zones that have ceased to function. The absence of epiphyseal lines indicates that the bone is still lengthening.

6.8 The lengthwise growth of the diaphysis is caused by cell divisions in the zone of proliferating cartilage and maturation of the cells in the zone of hypertrophic cartilage.

6.9 The medullary cavity enlarges by activity of the osteoclasts in the endosteum.

6.10 An open fracture breaks through the skin whereas a closed fracture does not.

6.11 Healing of bone fractures can take months because calcium and phosphorus deposition is a slow process, and bone cells generally grow and reproduce slowly.

6.12 Heartbeat, respiration, nerve cell functioning, enzyme functioning, and blood clotting all depend on proper levels of calcium.

6.13 The skeletal system develops from mesoderm.

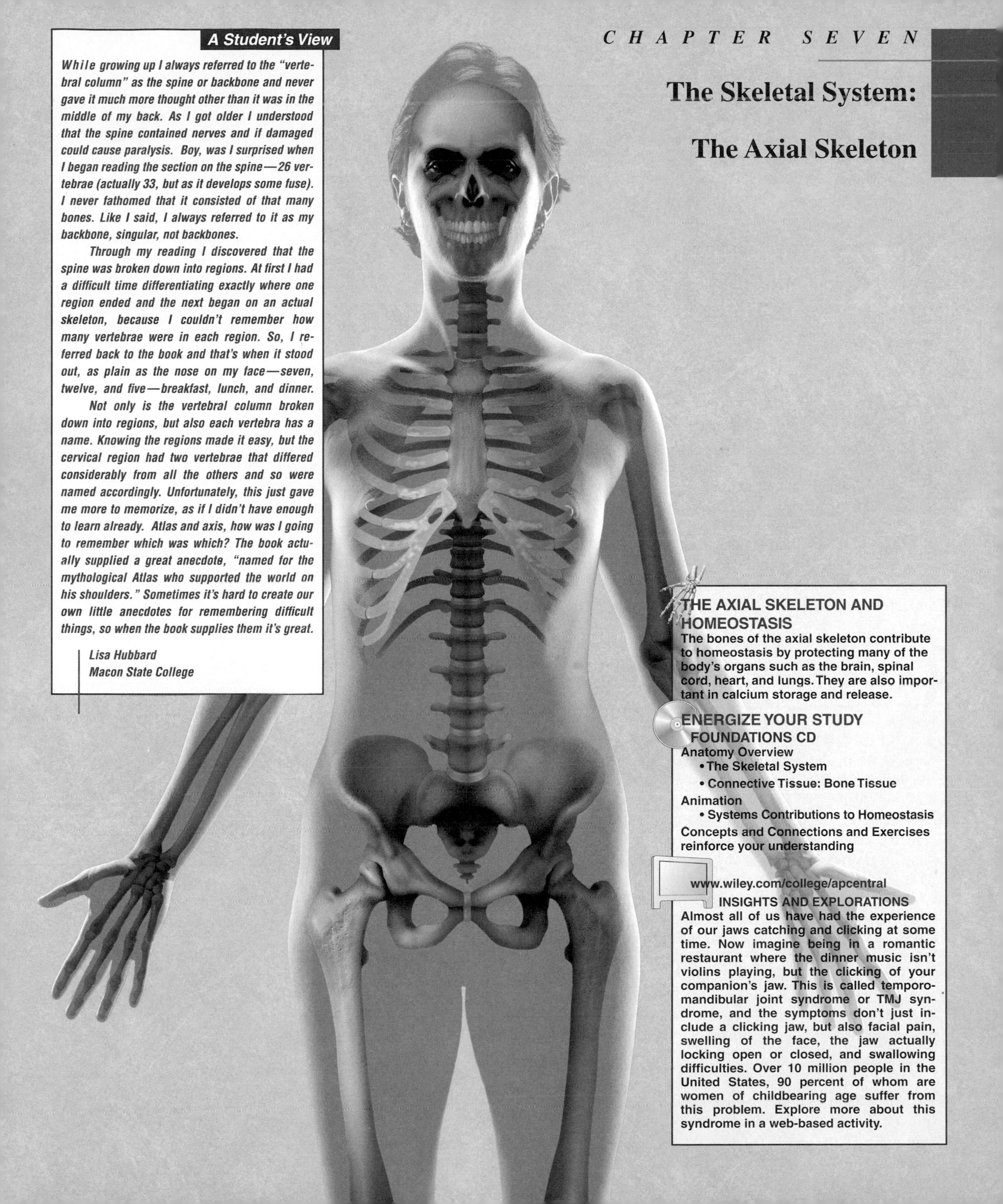

The Skeletal System: The Axial Skeleton

A Student's View

While growing up I always referred to the "vertebral column" as the spine or backbone and never gave it much more thought other than it was in the middle of my back. As I got older I understood that the spine contained nerves and if damaged could cause paralysis. Boy, was I surprised when I began reading the section on the spine—26 vertebrae (actually 33, but as it develops some fuse). I never fathomed that it consisted of that many bones. Like I said, I always referred to it as my backbone, singular, not backbones.

Through my reading I discovered that the spine was broken down into regions. At first I had a difficult time differentiating exactly where one region ended and the next began on an actual skeleton, because I couldn't remember how many vertebrae were in each region. So, I referred back to the book and that's when it stood out, as plain as the nose on my face—seven, twelve, and five—breakfast, lunch, and dinner.

Not only is the vertebral column broken down into regions, but also each vertebra has a name. Knowing the regions made it easy, but the cervical region had two vertebrae that differed considerably from all the others and so were named accordingly. Unfortunately, this just gave me more to memorize, as if I didn't have enough to learn already. Atlas and axis, how was I going to remember which was which? The book actually supplied a great anecdote, "named for the mythological Atlas who supported the world on his shoulders." Sometimes it's hard to create our own little anecdotes for remembering difficult things, so when the book supplies them it's great.

Lisa Hubbard
Macon State College

THE AXIAL SKELETON AND HOMEOSTASIS

The bones of the axial skeleton contribute to homeostasis by protecting many of the body's organs such as the brain, spinal cord, heart, and lungs. They are also important in calcium storage and release.

ENERGIZE YOUR STUDY FOUNDATIONS CD

Anatomy Overview
- The Skeletal System
- Connective Tissue: Bone Tissue

Animation
- Systems Contributions to Homeostasis

Concepts and Connections and Exercises reinforce your understanding

www.wiley.com/college/apcentral

INSIGHTS AND EXPLORATIONS

Almost all of us have had the experience of our jaws catching and clicking at some time. Now imagine being in a romantic restaurant where the dinner music isn't violins playing, but the clicking of your companion's jaw. This is called temporomandibular joint syndrome or TMJ syndrome, and the symptoms don't just include a clicking jaw, but also facial pain, swelling of the face, the jaw actually locking open or closed, and swallowing difficulties. Over 10 million people in the United States, 90 percent of whom are women of childbearing age suffer from this problem. Explore more about this syndrome in a web-based activity.

Because the skeletal system forms the framework of the body, a familiarity with the names, shapes, and positions of individual bones will help you locate other organs. For example, the radial artery, the site where the pulse is usually taken, is named for its closeness to the radius, the lateral bone of the forearm. The ulnar nerve is named for its closeness to the ulna, the medial bone of the forearm. The frontal lobe of the brain lies deep to the frontal (forehead) bone. The tibialis anterior muscle lies along the anterior surface of the tibia (shin bone).

Movements such as throwing a ball, biking, and walking require an interaction between bones and muscles. To understand how muscles produce different movements, you will learn where the muscles attach on individual bones and the types of joints acted on by the contracting muscles. The bones, muscles, and joints together form an integrated system called the **musculoskeletal system.** The branch of medical science concerned with the prevention or correction of disorders of the musculoskeletal system is called **orthopedics** (or′-thō-PĒ-diks; *ortho-* = correct; *pedi* = child).

DIVISIONS OF THE SKELETAL SYSTEM

OBJECTIVE

- **Describe how the skeleton is divided into axial and appendicular divisions.**

The adult human skeleton consists of 206 named bones, most of which are paired on the right and left sides of the body. The skeletons of infants and children have more than 206 bones because some of their bones fuse later in life. Examples are the hip bones and some bones of the backbone.

Bones of the adult skeleton are grouped into two principal divisions: the 80 bones of the **axial skeleton** and the 126 bones of the **appendicular skeleton** (*appendic-* = to hang onto). Table 7.1 presents the 80 bones of the axial skeleton and the 126 bones of the appendicular skeleton. Figure 7.1 shows how both divisions join to form the complete skeleton. The longitudinal **axis,** or center, of the human body is a vertical line that runs through the body's center of gravity, extending down through the head to the space between the feet. The axial skeleton consists of the bones arranged along the axis: skull bones, auditory ossicles (ear bones), hyoid bone (see Figure 7.4), ribs, breastbone, and bones of the backbone. The appendicular skeleton consists of the bones of the **upper** and **lower limbs (extremities),** plus the bones forming the **girdles** that connect the limbs to the axial skeleton. Functionally, the auditory ossicles in the middle ear are not part of either the axial or appendicular skeleton, but they are grouped with the axial skeleton for convenience. The auditory ossicles vibrate in response to sound waves that strike the eardrum and have a key role in hearing (see Chapter 16).

We will organize our study of the skeletal system around the two divisions of the skeleton, with emphasis on how the many bones of the body are interrelated. In this chapter we focus on the axial skeleton, looking first at the skull and then at the bones of the backbone and the chest. In Chapter 8 we explore the appendicular skeleton, examining in turn the bones of the pectoral (shoulder) girdle and upper limbs, and then the pelvic (hip) girdle and the lower limbs. But before we examine the skull, we direct our attention to some general characteristics of bones.

► CHECKPOINT

1. Which bones make up the axial and appendicular divisions of the skeleton?

Table 7.1 The Bones of the Adult Skeletal System

Division of the Skeleton	Structure	Number of Bones
Axial Skeleton	**Skull**	
	Cranium	8
	Face	14
	Hyoid	1
	Auditory ossicles	6
	Vertebral column	26
	Thorax	
	Sternum	1
	Ribs	24
	Subtotal =	80
Appendicular Skeleton	**Pectoral (shoulder) girdles**	
	Clavicle	2
	Scapula	2
	Upper limbs (extremities)	
	Humerus	2
	Ulna	2
	Radius	2
	Carpals	16
	Metacarpals	10
	Phalanges	28
	Pelvic (hip) girdle	
	Hip, pelvic, or coxal bone	2
	Lower limbs (extremities)	
	Femur	2
	Fibula	2
	Tibia	2
	Patella	2
	Tarsals	14
	Metatarsals	10
	Phalanges	28
	Subtotal =	126
	Total =	206

Figure 7.1 Divisions of the skeletal system. The axial skeleton is colored blue, the appendicular skeleton tan. (See Tortora, *A Photographic Atlas of the Human Body,* Figure 3.1.)

The adult human skeleton consists of 206 bones grouped into axial and appendicular divisions.

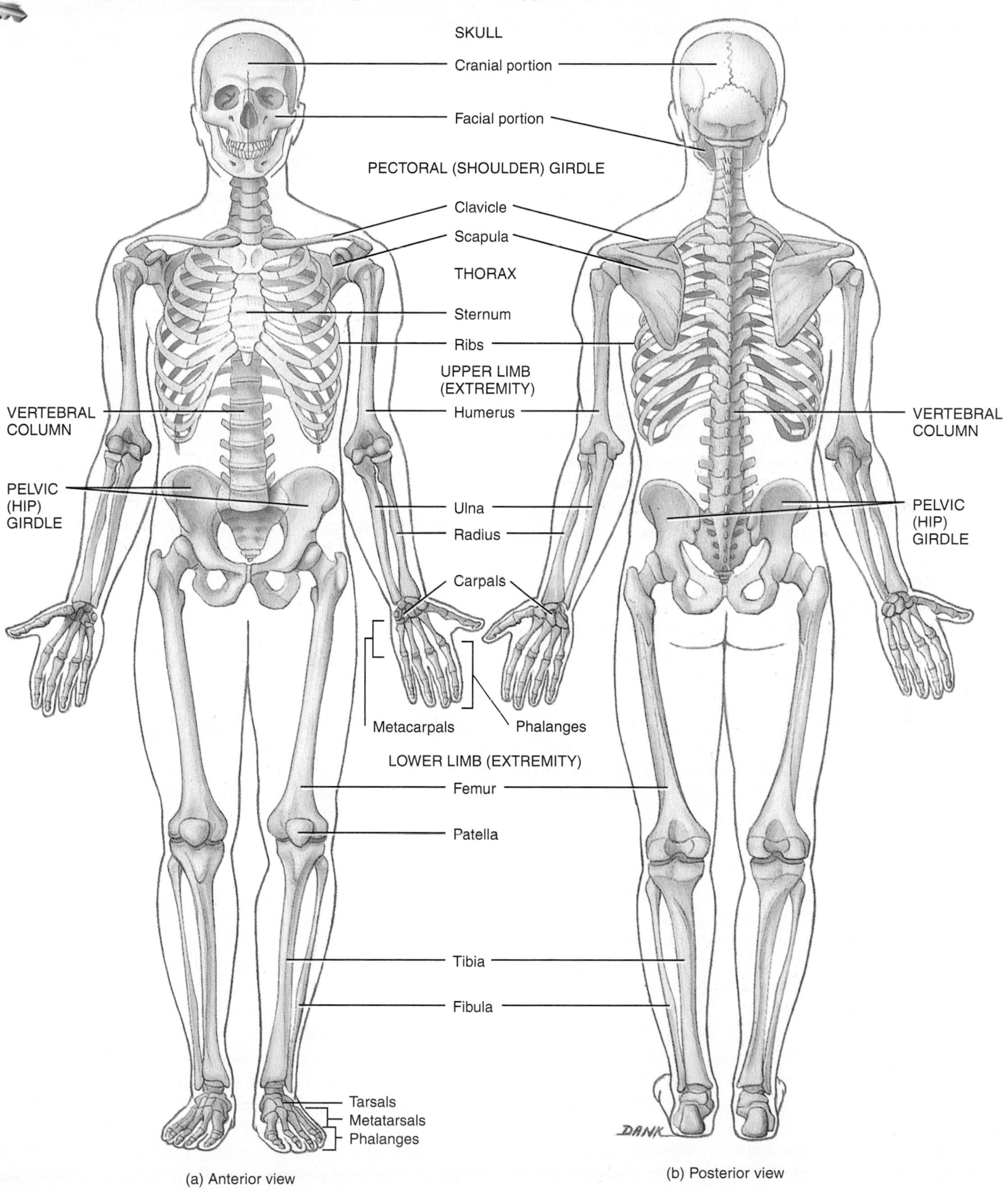

SKULL
Cranial portion
Facial portion
PECTORAL (SHOULDER) GIRDLE
Clavicle
Scapula
THORAX
Sternum
Ribs
UPPER LIMB (EXTREMITY)
Humerus
VERTEBRAL COLUMN
PELVIC (HIP) GIRDLE
Ulna
Radius
Carpals
Metacarpals
Phalanges
LOWER LIMB (EXTREMITY)
Femur
Patella
Tibia
Fibula
Tarsals
Metatarsals
Phalanges
DANK

(a) Anterior view

(b) Posterior view

Which of the following structures are part of the axial skeleton, and which are part of the appendicular skeleton? Skull, clavicle, vertebral column, shoulder girdle, humerus, pelvic girdle, and femur.

TYPES OF BONES

OBJECTIVE

- **Classify bones based on their shape and location.**

Most bones of the body can be classified into five main types based on shape: long, short, flat, irregular, and sesamoid (Figure 7.2). **Long bones** have greater length than width and consist of a shaft and a variable number of extremities (ends). They are usually somewhat curved for strength. A slightly curved bone absorbs the stress of the body's weight at several different points, so that it is evenly distributed. If such bones were straight, the weight of the body would be unevenly distributed, and the bone would fracture more easily. Long bones consist mostly of compact bone tissue in their diaphyses but they also contain considerable amounts of spongy bone tissue in their epiphyses. Long bones include those in the thigh (femur), leg (tibia and fibula), arm (humerus), forearm (ulna and radius), and fingers and toes (phalanges).

Short bones are somewhat cube-shaped because they are nearly equal in length and width. They consist of spongy bone tissue except at the surface, where there is a thin layer of compact bone tissue. Examples of short bones are the wrist or carpal bones (except for the pisiform, which is a sesamoid bone) and the ankle or tarsal bones (except for the calcaneus, which is an irregular bone).

Flat bones are generally thin and composed of two nearly parallel plates of compact bone tissue enclosing a layer of spongy bone tissue. Flat bones afford considerable protection and provide extensive areas for muscle attachment. Flat bones include the cranial bones, which protect the brain; the breastbone (sternum) and ribs, which protect organs in the thorax; and the shoulder blades (scapulae).

Irregular bones have complex shapes and cannot be grouped into any of the previous categories. They vary in the amount of spongy and compact bone present. Such bones include the vertebrae of the backbone and some facial bones.

Sesamoid bones (= shaped like a sesame seed) develop in certain tendons where there is considerable friction, tension, and physical stress, such as the palms and soles. They may vary in number from person to person, are not always completely ossified, and typically measure only a few millimeters in diameter. Notable exceptions are the two patellae (kneecaps), which are large sesamoid bones that are normally present in everyone. Functionally, sesamoid bones protect tendons from excessive wear and tear, and they often change the direction of pull of a tendon, which improves the mechanical advantage at a joint.

An additional type of bone is not included in this classification by shape, but instead is classified by location. **Sutural bones** (SOO-chur-al; *sutur-* = seam) are small bones located within joints, called sutures, between certain cranial bones (see Figure 7.6). Their number varies greatly from person to person.

Figure 7.2 Types of bones based on shape. The bones are not drawn to scale.

The shapes of bones largely determine their functions.

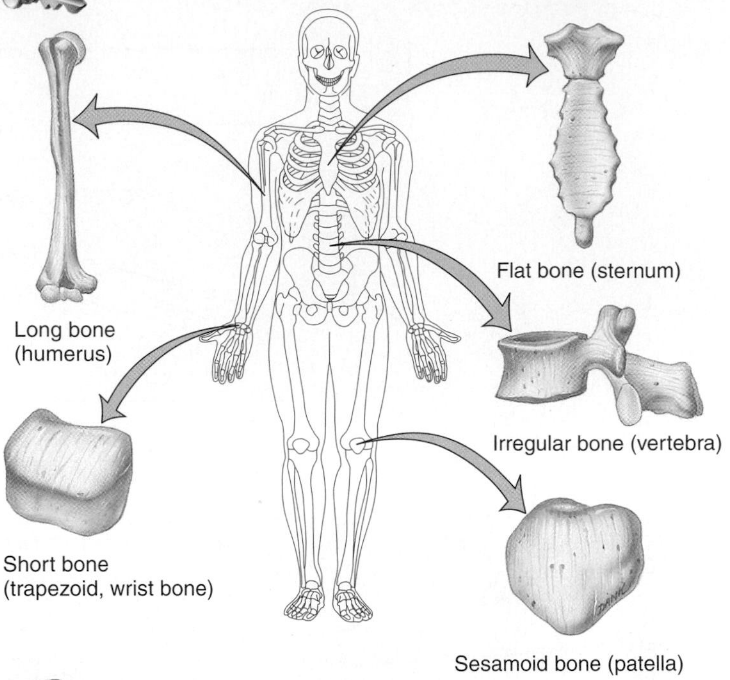

Long bone (humerus)

Flat bone (sternum)

Irregular bone (vertebra)

Short bone (trapezoid, wrist bone)

Sesamoid bone (patella)

Which type of bone primarily protects and provides a large surface area for muscle attachment?

CHECKPOINT

2. Give examples of long, short, flat, and irregular bones.

BONE SURFACE MARKINGS

OBJECTIVE

- **Describe the principal surface markings on bones and the functions of each.**

Bones have characteristic **surface markings,** structural features adapted for specific functions. Sometimes they are not present at birth but develop later in response to certain forces and are most prominent during adult life. In response to tension on a bone surface where tendons, ligaments, aponeuroses, and fasciae pull on the periosteum of bone, new bone is deposited, resulting in raised or roughened areas. Conversely, compression on a bone surface results in a depression.

There are two major types of surface markings: (1) *depressions and openings*, which form joints or allow the passage of soft tissues (such as blood vessels and nerves), and (2) *processes*, which are projections or outgrowths that either help form joints or serve as attachment points for connective tissue (such as ligaments and tendons). Table 7.2 describes the various surface markings and provides examples of each.

Table 7.2 Bone Surface Markings

Marking	Description	Example
Depressions and Openings: Sites allowing the passage of soft tissue (nerves, blood vessels, ligaments, tendons) or formation of joints		
Fissure (FISH-ur)	Narrow slit between adjacent parts of bones through which blood vessels or nerves pass.	Superior orbital fissure of the sphenoid bone (Figure 7.13).
Foramen (fō-RĀ-men); plural is **foramina**	Opening (*foramen* = hole) through which blood vessels, nerves, or ligaments pass.	Optic foramen of the sphenoid bone (Figure 7.13).
Fossa (FOS-a)	Shallow depression (*fossa* = trench).	Coronoid fossa of the humerus (Figure 8.5d).
Sulcus (SUL-kus)	Furrow (*sulcus* = groove) along a bone surface that accommodates a blood vessel, nerve, or tendon.	Intertubercular sulcus of the humerus (Figure 8.5d).
Meatus (mē-Ā-tus)	Tubelike opening (*meatus* = passageway).	External auditory meatus of the temporal bone (Figure 7.4).
Processes: Projections or outgrowths on bone that form joints or attachment points for connective tissue, such as ligaments and tendons.		
Processes that form joints:		
Condyle (KON-dīl)	Large, round protuberance (*condylus* = knuckle) at the end of a bone.	Lateral condyle of the femur (Figure 8.12a).
Facet	Smooth flat articular surface.	Superior articular facet of a vertebra (Figure 7.18d).
Head	Rounded articular projection supported on the neck (constricted portion) of a bone.	Head of the femur (Figure 8.12a).
Processes that form attachment points for connective tissue:		
Crest	Prominent ridge or elongated projection.	Iliac crest of the hip bone (Figure 8.9b).
Epicondyle	Projection above (*epi-* = above) a condyle.	Medial epicondyle of the femur (Figure 8.12a).
Line	Long, narrow ridge or border (less prominent than a crest).	Linea aspera of the femur (Figure 8.12b).
Spinous process	Sharp, slender projection.	Spinous process of a vertebra (Figure 7.17).
Trochanter (trō-KAN-ter)	Very large projection.	Greater trochanter of the femur (Figure 8.12b).
Tubercle (TOO-ber-kul)	Small, rounded projection (*tuber-* = knob).	Greater tubercle of the humerus (Figure 8.5a).
Tuberosity	Large, rounded, usually roughened projection.	Ischial tuberosity of the hip bone (Figure 8.9b).

▶ CHECKPOINT

3. List and describe several bone surface markings, and give an example of each. Check your list against Table 7.2.

SKULL

▶ OBJECTIVES

- **Name the cranial and facial bones and indicate the number of each.**
- **Describe the following special features of the skull: sutures, paranasal sinuses, and fontanels.**

The **skull,** which contains 22 bones, rests on the superior end of the vertebral column. It includes two sets of bones: cranial bones and facial bones. The **cranial bones** (*crani-* = brain case) form the cranial cavity, which encloses and protects the brain. The eight cranial bones are the frontal bone, two parietal bones, two temporal bones, the occipital bone, the sphenoid bone, and the ethmoid bone. Fourteen **facial bones** form the face: two nasal bones, two maxillae (or maxillas), two zygomatic bones, the mandible, two lacrimal bones, two palatine bones, two inferior nasal conchae, and the vomer. Figures 7.3 through 7.8 illustrate these bones from different viewing directions.

General Features

Besides forming the large cranial cavity, the skull also forms several smaller cavities, including the nasal cavity and orbits (eye sockets), which open to the exterior. Certain skull bones also contain cavities that are lined with mucous membranes and are called paranasal sinuses. The sinuses open into the nasal cavity. Also within the skull are small cavities that house the structures involved in hearing and equilibrium.

Other than the auditory ossicles, which are involved in hearing and are located within the temporal bones, the mandible is the only movable bone of the skull. Immovable joints called sutures, which are especially noticeable on the outer surface of the skull, hold most of the skull bones together.

The skull has many surface markings, such as foramina and fissures through which blood vessels and nerves pass. You will learn the names of important skull bone surface markings as we describe each bone.

In addition to protecting the brain, the cranial bones also have other functions. Their inner surfaces attach to membranes (meninges) that stabilize the positions of the brain, blood vessels, and nerves. The outer surfaces of cranial bones provide large areas of attachment for muscles that move various parts of the head. The bones also provide attachment for some muscles that produce facial expressions. Besides forming the framework of the face, the facial bones protect and provide support for the entrances to the digestive and respiratory systems. Together, the cranial and facial bones protect and support the delicate special sense organs for vision, taste, smell, hearing, and equilibrium.

Cranial Bones

Frontal Bone

The **frontal bone** forms the forehead (the anterior part of the cranium), the roofs of the orbits, and most of the anterior part of the cranial floor (Figure 7.3). Soon after birth, the left and right sides of the frontal bone are united by the *metopic suture*, which usually disappears by age six to eight.

If you examine the anterior view of the skull in Figure 7.3, you will note the *frontal squama,* a scalelike plate of bone that forms the forehead. It gradually slopes inferiorly from the coronal suture, on the top of the skull, then angles abruptly and becomes almost vertical. Superior to the orbits the frontal bone thickens, forming the *supraorbital margin* (*supra* = above; *-orbi* = circle). From this margin, the frontal bone extends posteriorly to form the roof of the orbit, which is part of the floor of the cranial cavity. Within the supraorbital margin, slightly medial to its midpoint, is a hole called the *supraorbital foramen.* As you read about each foramen associated with a cranial bone, refer to Table 7.3 to note which structures pass through it. The *frontal sinuses* lie deep to the frontal squama. Sinuses, or more technically paranasal sinuses, are mucous membrane–lined cavities in certain skull bones (discussed later).

Figure 7.3 Skull. (See Tortora, *A Photographic Atlas of the Human Body,* Figure 3.2.)

The skull consists of 8 cranial bones and 14 facial bones.

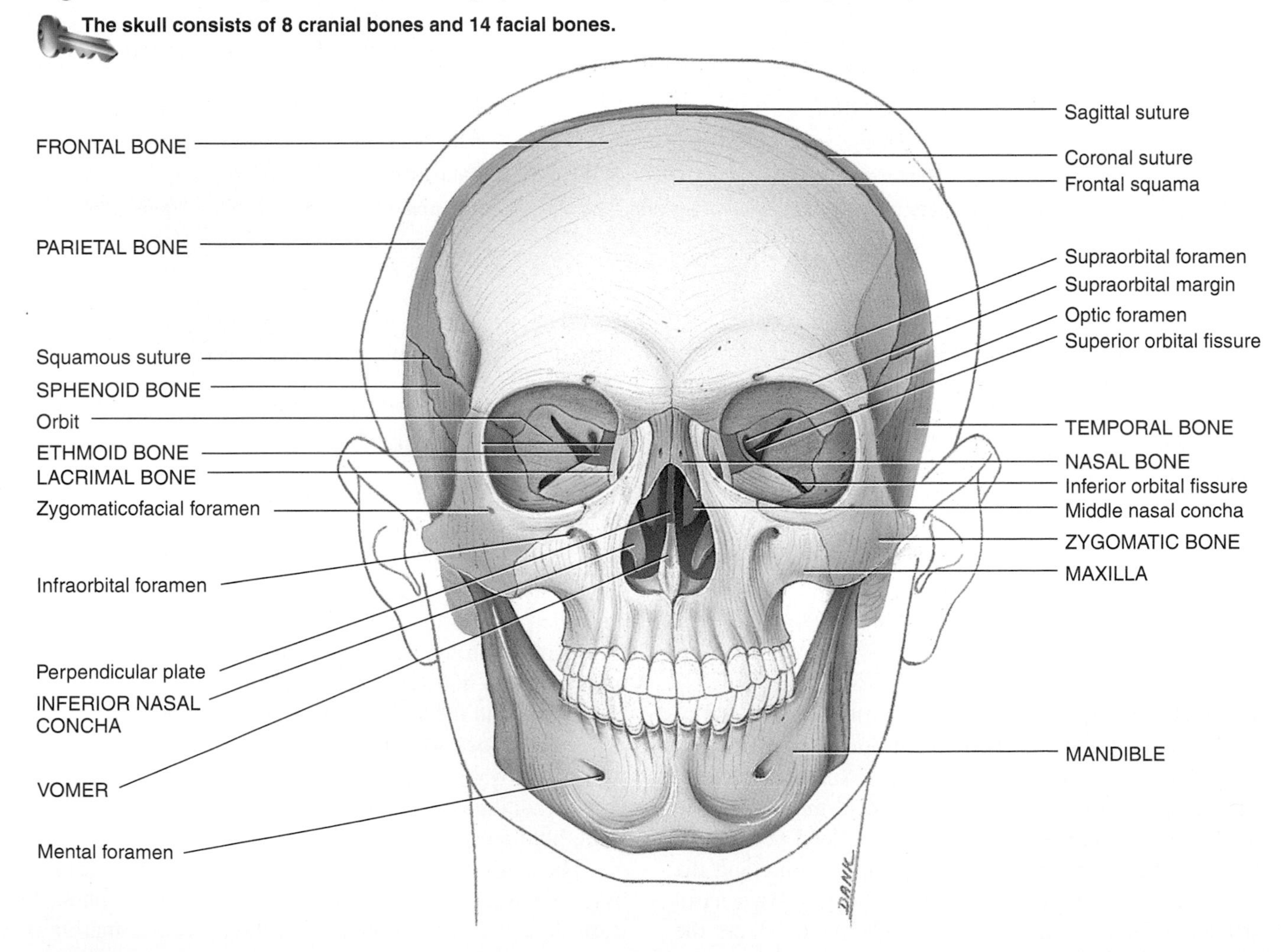

Anterior view

Which of the bones shown here are cranial bones?

Black Eyes

Just superior to the supraorbital margin is a sharp ridge. A blow to the ridge often fractures the bone or lacerates the skin over it, resulting in bleeding. Bruising of the skin over the ridge causes tissue fluid and blood to accumulate in the surrounding connective tissue. The resulting swelling and discoloration is called a black eye. ■

Parietal Bones

The two **parietal bones** (pa-RĪ-e-tal; *pariet-* = wall) form the greater portion of the sides and roof of the cranial cavity (Figure 7.4). The internal surfaces of the parietal bones contain many protrusions and depressions that accommodate the blood vessels supplying the dura mater, the superficial connective tissue covering of the brain. There are no foramina in the parietal bones.

Temporal Bones

The two **temporal bones** (*tempor-* = temple) form the inferior lateral aspects of the cranium and part of the cranial floor. In the lateral view of the skull (Figure 7.4), note the *temporal squama*, the thin, flat part of the temporal bone that forms the anterior and superior part of the temple. Projecting from the inferior portion of the temporal squama is the *zygomatic process,* which articulates (forms a joint) with the temporal process of the zygomatic (cheek) bone. Together, the zygomatic process of the temporal bone and the temporal process of the zygomatic bone form the *zygomatic arch.*

A socket called the *mandibular fossa* is located on the inferior posterior surface of the zygomatic process of the temporal bone. Anterior to the mandibular fossa is a rounded elevation, the *articular tubercle* (Figure 7.4). The mandibular fossa and

Figure 7.4 Skull. Although the hyoid bone is not part of the skull, it is included in the illustration for reference. (See Tortora, *A Photographic Atlas of the Human Body,* Figure 3.3.)

Together the zygomatic process of the temporal bone and the temporal process of the zygomatic bone form the zygomatic arch.

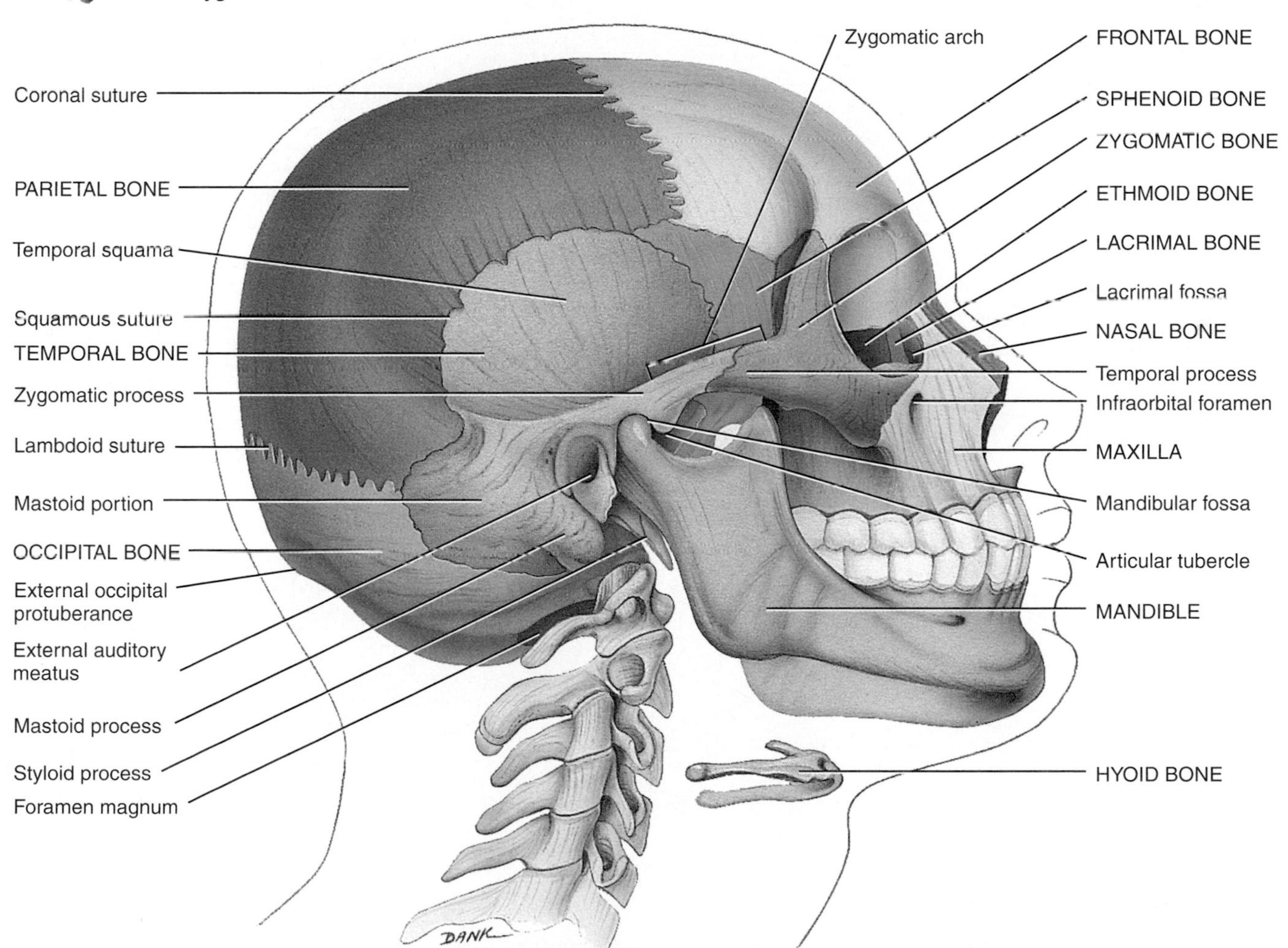

Right lateral view

What are the major bones on either side of the squamous suture, the lambdoid suture, and the coronal suture?

articular tubercle articulate with the mandible (lower jawbone) to form the *temporomandibular joint (TMJ).*

Located posteriorly on the temporal bone is the *mastoid portion* (*mastoid* = breast-shaped). See Figure 7.4. It is located posterior and inferior to the *external auditory meatus* (*meatus* = passageway), or ear canal, which directs sound waves into the ear. In an adult, this portion of the bone contains several *mastoid "air cells."* These tiny air-filled compartments are separated from the brain by thin bony partitions. In cases of **mastoiditis** (inflammation of the mastoid air cells, for example, from a middle-ear infection), the infection may spread to the brain.

The *mastoid process* is a rounded projection of the mastoid portion of the temporal bone posterior to the external auditory meatus. It is the point of attachment for several neck muscles. The *internal auditory meatus* (Figure 7.5) is the opening through which the facial nerve (cranial nerve VII) and vestibulocochlear nerve (cranial nerve VIII) pass. The *styloid process* (*styl-* = stake or pole) projects inferiorly from the inferior surface of the temporal bone and serves as a point of attachment for muscles and ligaments of the tongue and neck (see Figure 7.4). Between the styloid process and the mastoid process is the *stylomastoid foramen* (see Figure 7.7).

At the floor of the cranial cavity (see Figure 7.8a) is the *petrous portion* (*petrous* = rock) of the temporal bone. This part is triangular and is located at the base of the skull between the sphenoid and occipital bones. The petrous portion houses the internal ear and the middle ear, structures involved in hearing and equilibrium. It also contains the *carotid foramen,* through which

Figure 7.5 Skull. (See Tortora, *A Photographic Atlas of the Human Body,* Figure 3.4.)

The cranial bones are the frontal, parietal, temporal, occipital, sphenoid, and ethmoid bones. The facial bones are the nasal bone, maxillae, zygomatic bones, lacrimal bones, palatine bones, mandible, and vomer.

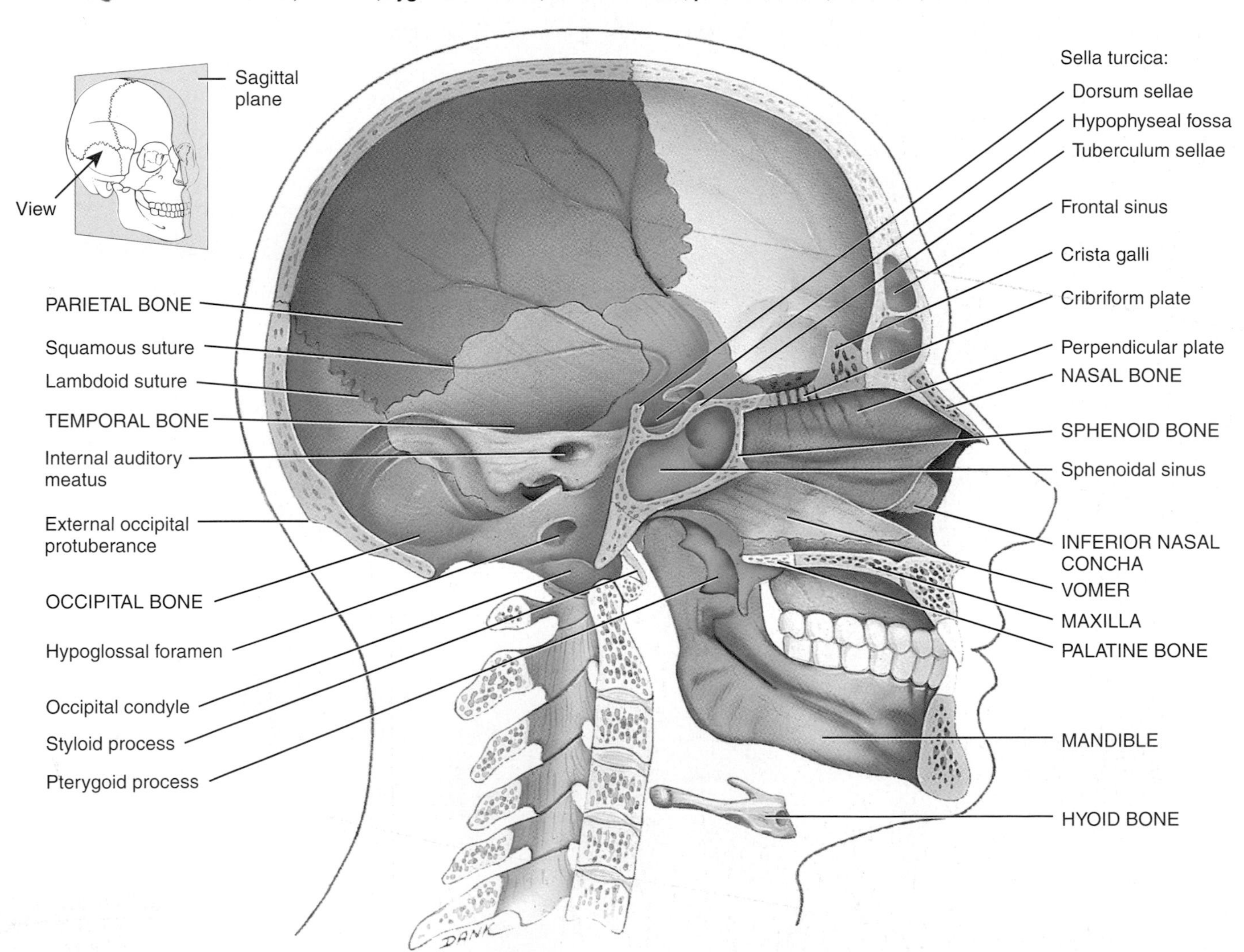

Medial view of sagittal section

With which bones does the temporal bone articulate?

the carotid artery passes (see Figure 7.7). Posterior to the carotid foramen and anterior to the occipital bone is the *jugular foramen,* a passageway for the jugular vein.

Occipital Bone

The **occipital bone** (ok-SIP-i-tal; *occipit-* = back of head) forms the posterior part and most of the base of the cranium (Figure 7.6; also see Figure 7.4). Also view the occipital bone and surrounding structures in the inferior view of the skull in Figure 7.7. The *foramen magnum* (= large hole) is in the inferior part of the bone. Within this foramen, the medulla oblongata (inferior part of the brain) connects with the spinal cord. The vertebral and spinal arteries also pass through this foramen. The *occipital condyles* are oval processes with convex surfaces, one on either side of the foramen magnum (Figure 7.7 on page 195). They articulate with depressions on the first cervical vertebra (atlas) to form the *atlanto-occipital joint*. Superior to each occipital condyle on the inferior surface of the skull is the *hypoglossal foramen* (*hypo-* = under; *-glossal* = tongue). (See Figure 7.5.)

The *external occipital protuberance* is a prominent midline projection on the posterior surface of the bone just above the foramen magnum. You may be able to feel this structure as a bump on the back of your head, just above your neck. (See Figure 7.4.) A large fibrous, elastic ligament, the *ligamentum nuchae* (*nucha-* = nape of neck), which helps support the head, extends from the external occipital protuberance to the seventh cervical vertebra. Extending laterally from the protuberance are two curved ridges, the *superior nuchal lines,* and below these are two *inferior nuchal lines,* which are areas of muscle attachment (Figure 7.7).

Sphenoid Bone

The **sphenoid bone** (SFĒ-noyd = wedge-shaped) lies at the middle part of the base of the skull (Figures 7.7 and 7.8). This

Figure 7.6 Skull. The sutures are exaggerated for emphasis. (See Tortora, *A Photographic Atlas of the Human Body,* Figure 3.5.)

The occipital bone forms most of the posterior and inferior portions of the cranium.

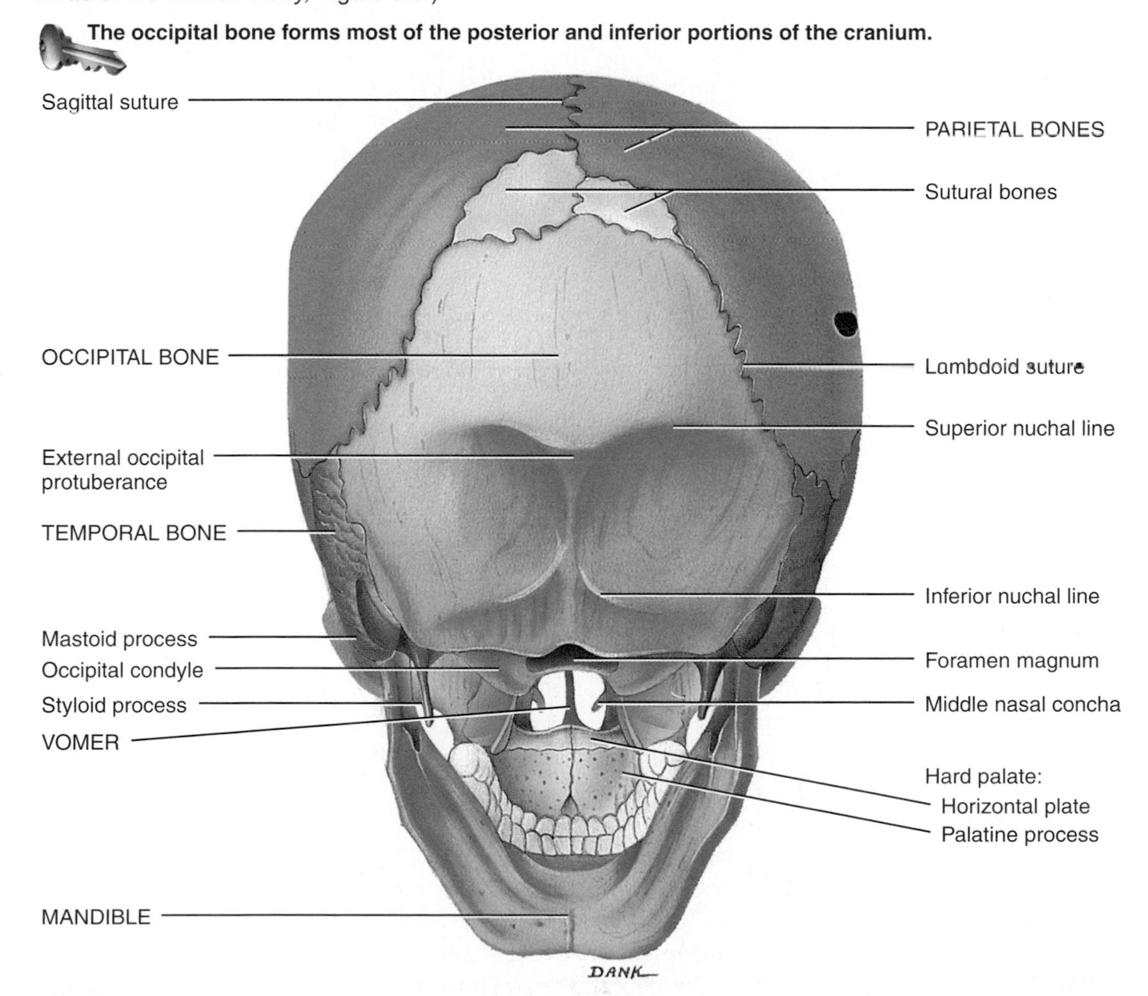

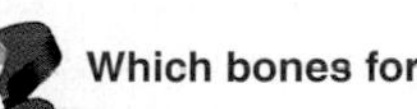

Which bones form most of the lateral aspect of the cranium posterior to the ears?

Figure 7.7 Skull. The mandible (lower jawbone) has been removed. (See Tortora, *A Photographic Atlas of the Human Body,* Figure 3.7.)

The occipital condyles of the occipital bone articulate with the first cervical vertebra to form the atlanto-occipital joints.

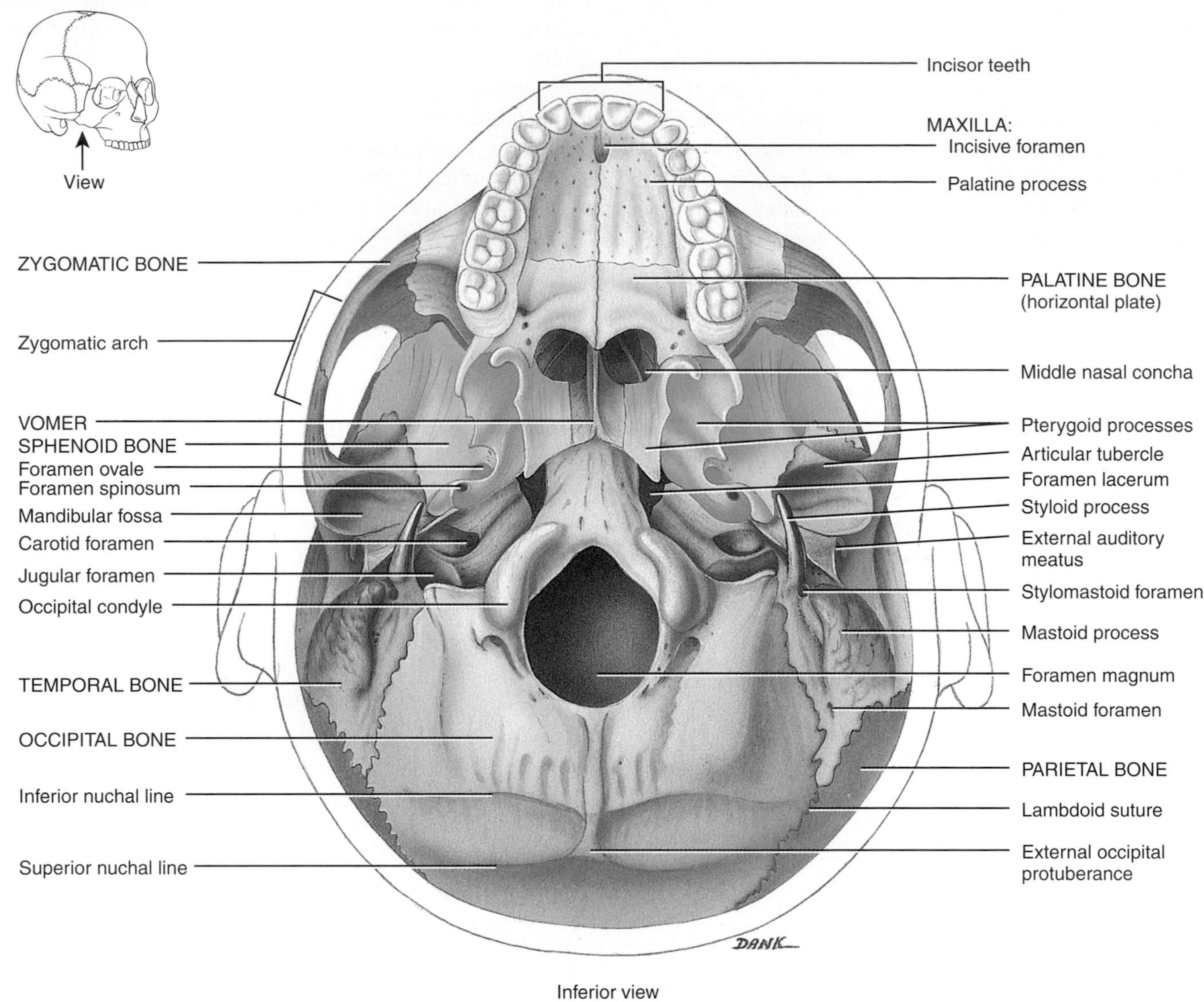

Inferior view

What parts of the nervous system join within the foramen magnum?

bone is the keystone of the cranial floor because it articulates with all the other cranial bones, holding them together. Viewing the floor of the cranium from a superior perspective (Figure 7.8a), note the sphenoid articulations. It joins anteriorly with the frontal bone, laterally with the temporal bones, and posteriorly with the occipital bone. The sphenoid lies posterior and slightly superior to the nasal cavity and forms part of the floor, sidewalls, and rear wall of the orbit (see Figure 7.12).

The shape of the sphenoid resembles a bat with outstretched wings (see Figure 7.8b). The *body* of the sphenoid is the cube-like medial portion between the ethmoid and occipital bones. It contains the *sphenoidal sinuses,* which drain into the nasal cavity (see Figure 7.13). The *sella turcica* (SEL-a TUR-si-ka; *sella* = saddle; *turcica* = Turkish) is a bony saddle-shaped structure on the superior surface of the body of the sphenoid (Figure 7.8a). The anterior part of the sella turcica, which forms the horn of the saddle, is a ridge called the **tuberculum sellae.** The seat of the saddle is a depression, the **hypophyseal fossa** (hī-pō-FIZ-ē-al), which contains the pituitary gland. The posterior part of the sella turcica, which forms the back of the saddle, is another ridge called the **dorsum sellae.**

The *greater wings* of the sphenoid project laterally from the body and form the anterolateral floor of the cranium. The greater wings also form part of the lateral wall of the skull just anterior

Figure 7.8 Sphenoid bone. (See Tortora, *A Photographic Atlas of the Human Body,* Figures 3.8. and 3.9.)

The sphenoid bone is called the keystone of the cranial floor because it articulates with all other cranial bones, holding them together.

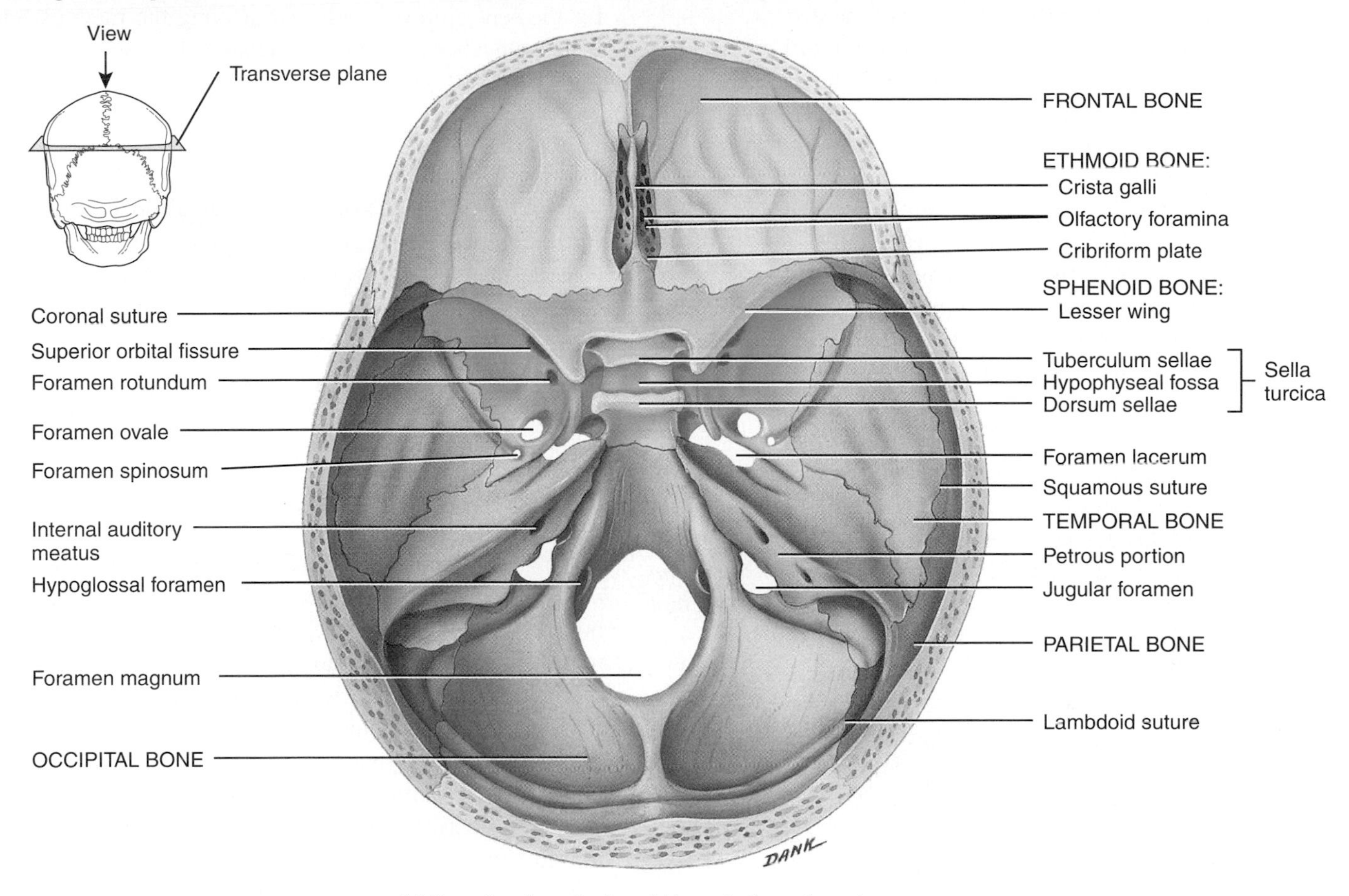

(a) Superior view of sphenoid bone in floor of cranium

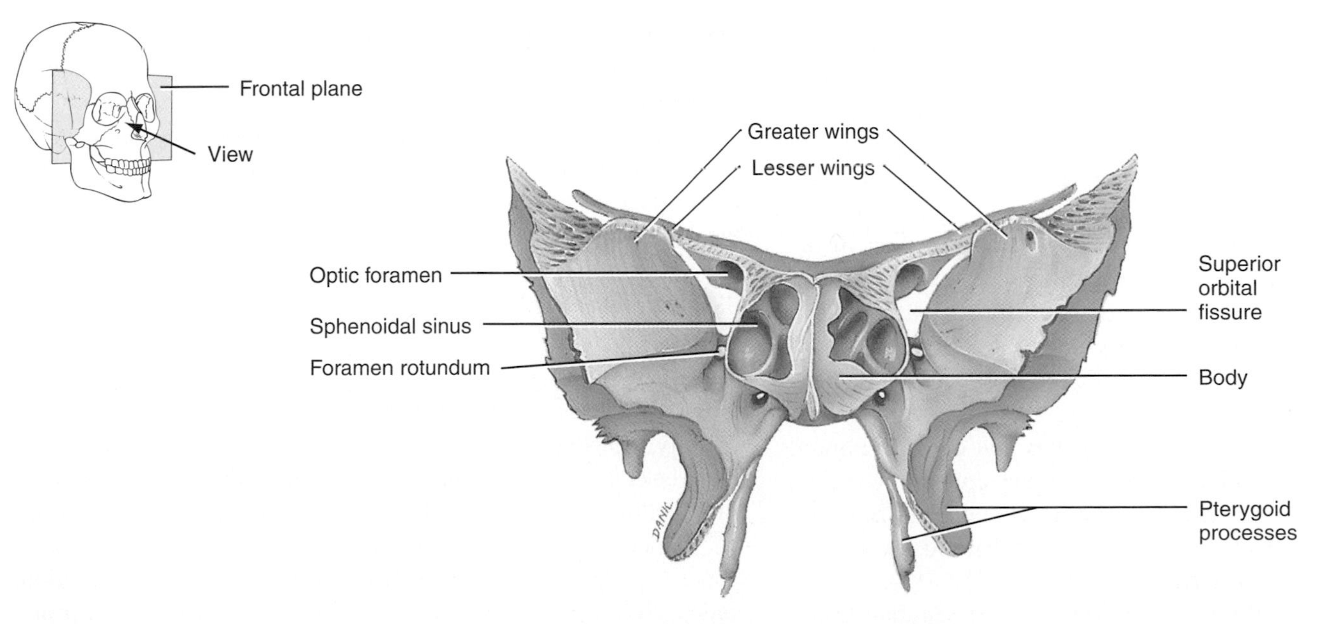

(b) Anterior view of sphenoid bone

Starting at the crista galli of the ethmoid bone and going in a clockwise direction, what are the names of the bones that articulate with the sphenoid bone?

to the temporal bone and can be viewed externally. The *lesser wings,* which are smaller, form a ridge of bone anterior and superior to the greater wings. They form part of the floor of the cranium and the posterior part of the orbit of the eye.

Between the body and lesser wing just anterior to the sella turcica is the *optic foramen* (*optic* = eye). Lateral to the body between the greater and lesser wings is a triangular slit called the *superior orbital fissure.* This fissure may also be seen in the anterior view of the orbit in Figure 7.12.

In Figures 7.7 and 7.8b, you can see the *pterygoid processes* (TER-i-goyd = winglike) by looking at the inferior part of the sphenoid bone. These structures project inferiorly from the points where the body and greater wings unite and form the lateral posterior region of the nasal cavity. Some of the muscles that move the mandible attach to the pterygoid processes. At the base of the lateral pterygoid process in the greater wing is the *foramen ovale* (= oval hole). The *foramen lacerum* (= lacerated) is bounded anteriorly by the sphenoid bone and medially by the sphenoid and occipital bones. The foramen is covered in part by a layer of fibrocartilage in living subjects. Another foramen associated with the sphenoid bone is the *foramen rotundum* (= round hole) located at the junction of the anterior and medial parts of the sphenoid bone.

Ethmoid Bone

The **ethmoid bone** (ETH-moyd = like a sieve) is a light, sponge-like bone located on the midline in the anterior part of the cranial floor medial to the orbits (Figure 7.9). It is anterior to the sphenoid and posterior to the nasal bones. The ethmoid bone forms (1) part of the anterior portion of the cranial floor; (2) the medial wall of the orbits; (3) the superior portions of the nasal septum, a partition that divides the nasal cavity into right and left sides; and (4) most of the superior sidewalls of the nasal cavity.

The *lateral masses* of the ethmoid bone compose most of the wall between the nasal cavity and the orbits. They contain 3 to 18 air spaces, or "cells." The ethmoidal cells together form the *ethmoidal sinuses* (see Figure 7.13). The *perpendicular plate* forms the superior portion of the nasal septum (Figure 7.9). The *cribriform plate* (*cribri-* = sieve) lies in the anterior floor of the cranium and forms the roof of the nasal cavity. The cribriform plate contains the *olfactory foramina* (*olfact-* = to smell) through which axons of the olfactory nerve pass. Projecting upward from the cribriform plate is a triangular process called the *crista galli* (*crista* = crest; *galli* = cock). This structure serves as a point of attachment for the membranes that cover the brain.

The lateral masses of the ethmoid bone contain two thin, scroll-shaped projections lateral to the nasal septum. These are the *superior nasal concha* (KONG-ka = shell) and the *middle nasal concha.* The plural term is *conchae* (KONG-kē). A third pair of conchae, the inferior nasal conchae, are separate bones (discussed shortly). The conchae cause turbulence in inhaled air, which results in many inhaled particles striking and becoming trapped in the mucus that lines the nasal passageways. This turbulence thus cleanses the inhaled air before it passes into the rest of the respiratory tract. Turbulent airflow around the superior nasal conchae also aids in the distribution of olfactory stimulants for the sensation of smell. Air striking the mucous lining of the conchae is also warmed and moistened.

Facial Bones

The shape of the face changes dramatically during the first two years after birth. The brain and cranial bones expand, the teeth form and erupt (emerge), and the paranasal sinuses increase in size. Growth of the face ceases at about 16 years of age.

Nasal Bones

The paired **nasal bones** meet at the midline (see Figure 7.3) and form part of the bridge of the nose. The rest of the supporting tissue of the nose consists of cartilage.

Maxillae

The paired **maxillae** (mak-SIL-ē = jawbones; singular is *maxilla*) unite to form the upper jawbone. They articulate with every bone of the face except the mandible, or lower jawbone (see Figures 7.4 and 7.7). The maxillae form part of the floors of the orbits, part of the lateral walls and floor of the nasal cavity, and most of the hard palate. The hard palate is a bony partition formed by the palatine processes of the maxillae and horizontal plates of the palatine bones that forms the roof of the mouth.

Each maxilla contains a large *maxillary sinus* that empties into the nasal cavity (see Figure 7.13). The *alveolar process* (al-VĒ-ō-lar; *alveol-* = small cavity) of the maxilla is an arch that contains the *alveoli* (sockets) for the maxillary (upper) teeth. The *palatine process* is a horizontal projection of the maxilla that forms the anterior three-quarters of the hard palate. The union and fusion of the maxillary bones normally is completed before birth.

The *infraorbital foramen* (*infra-* = below), which can be seen in the anterior view of the skull in Figure 7.3, is an opening in the maxilla below the orbit. A final structure associated with the maxilla and sphenoid bone is the *inferior orbital fissure.* It is located between the greater wing of the sphenoid and the maxilla (see Figure 7.12).

Cleft Palate and Cleft Lip

Usually the palatine processes of the maxillary bones unite during weeks 10 to 12 of embryonic development. Failure to do so can result in one type of **cleft palate.** The condition may also involve incomplete fusion of the horizontal plates of the palatine bones (see Figures 7.6 and 7.7). Another form of this condition, called **cleft lip,** involves a split in the upper lip. Cleft lip and cleft palate often occur together. Depending on the extent and position of the cleft, speech and swallowing may be affected. In addition, children with cleft palate tend to have many ear infec-

Figure 7.9 Ethmoid bone. (See Tortora, *A Photographic Atlas of the Human Body,* Figure 3.10.)

The ethmoid bone forms part of the anterior portion of the cranial floor, the medial wall of the orbits, the superior portions of the nasal septum, and most of the sidewalls of the nasal cavity.

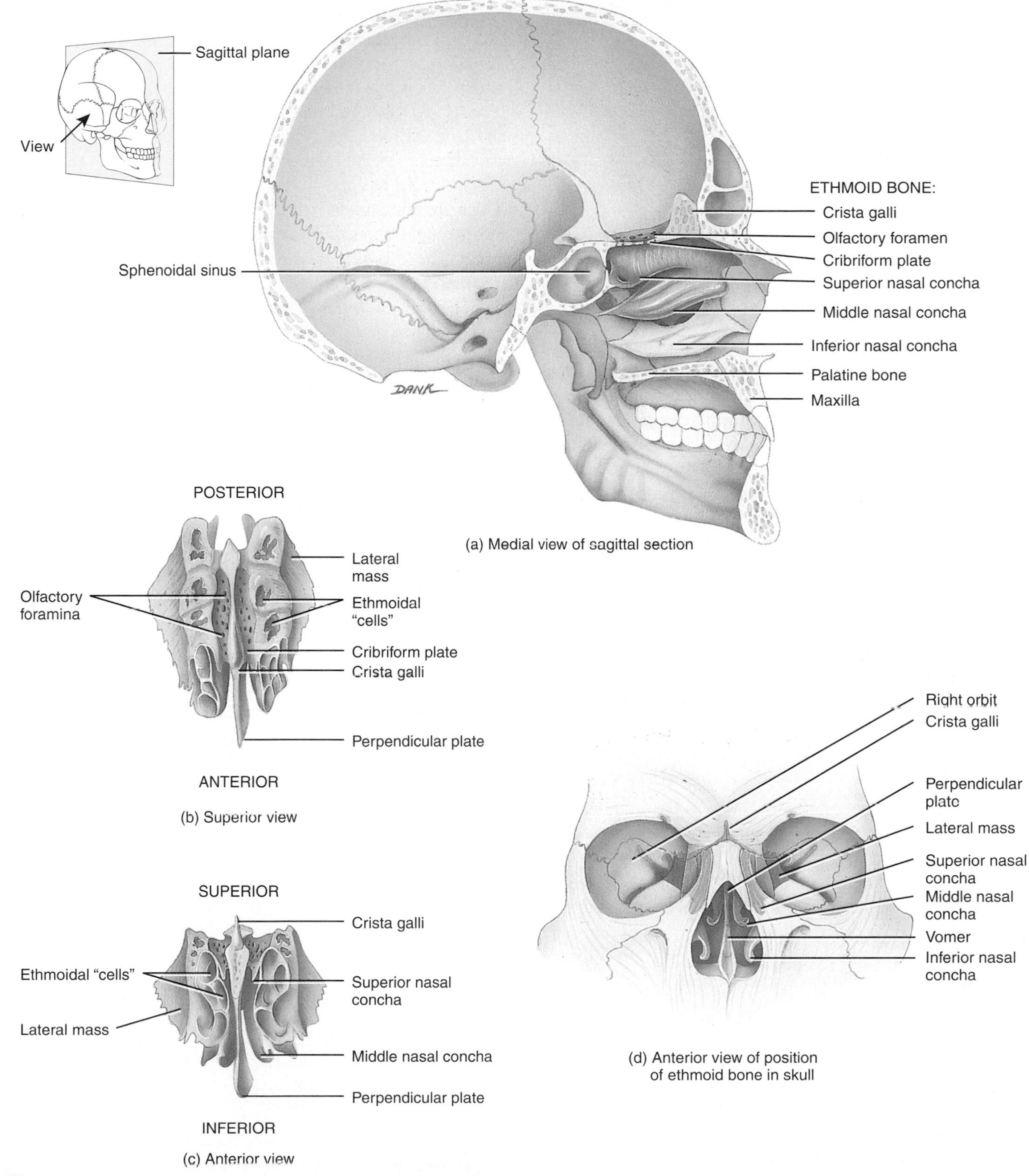

(a) Medial view of sagittal section

(b) Superior view

(c) Anterior view

(d) Anterior view of position of ethmoid bone in skull

What part of the ethmoid bone forms the superior part of the nasal septum? The medial walls of the orbits?

tions that can lead to hearing loss. Facial and oral surgeons recommend closure of cleft lip during the first few weeks following birth, and surgical results are excellent. Repair of cleft palate typically is done between 12 and 18 months of age, ideally before the child begins to talk. Speech therapy may be needed, because the palate is important for pronouncing consonants, and orthodontic therapy may be needed to align the teeth. Again, results are usually excellent. ■

Zygomatic Bones

The two **zygomatic bones** (*zygo-* = like a yoke), commonly called cheekbones, form the prominences of the cheeks and part of the lateral wall and floor of each orbit (see Figure 7.12). They articulate with the maxilla and the frontal, sphenoid, and temporal bones.

Lacrimal Bones

The paired **lacrimal bones** (LAK-ri-mal; *lacrim-* = teardrops), the smallest bones of the face, are thin and resemble a fingernail in size and shape (see Figures 7.3 and 7.4). These bones are posterior and lateral to the nasal bones and form a part of the medial wall of each orbit. The lacrimal bones each contain a *lacrimal fossa,* a vertical tunnel formed with the maxilla, that houses the lacrimal sac, a structure that gathers tears and passes them into the nasal cavity (see Figure 7.12).

Palatine Bones

The two **palatine bones** (PAL-a-tīn) are L-shaped (see Figure 7.7). They form the posterior portion of the hard palate, part of the floor and lateral wall of the nasal cavity, and a small portion of the floors of the orbits. The *horizontal plates* of the palatine bones form the posterior portion of the hard palate, which separates the nasal cavity from the oral cavity (see Figures 7.6 and 7.7).

Inferior Nasal Conchae

The two **inferior nasal concha** are inferior to the middle nasal conchae of the ethmoid bone (see Figures 7.3 and 7.9a). They are scroll-like bones that form a part of the inferior lateral wall of the nasal cavity and project into the nasal cavity. The inferior nasal conchae are separate bones; they are not part of the ethmoid bone. All three pairs of nasal conchae help swirl and filter air before it passes into the lungs. However, only the superior nasal conchae are involved in the sense of smell.

Vomer

The **vomer** (VŌ-mer = plowshare) is a roughly triangular bone on the floor of the nasal cavity that articulates superiorly with the perpendicular plate of the ethmoid bone and inferiorly with both the maxillae and palatine bones along the midline (see Figures 7.3 and 7.7). It is part of the nasal septum, the partition that divides the nasal cavity into right and left sides.

Mandible

The **mandible** (*mand-* = to chew), or lower jawbone, is the largest, strongest facial bone (Figure 7.10). Other than the auditory ossicles, it is the only movable skull bone. Notice that the mandible consists of a curved, horizontal portion, the *body,* and two perpendicular portions, the *rami* (RĀ-mī = branches; singular is *ramus*). The *angle* of the mandible is the area where each ramus meets the body. Each ramus has a posterior *condylar process* (KON-di-lar) that articulates with the mandibular fossa and articular tubercle of the temporal bone (see Figure 7.4). This articulation is the **temporomandibular joint (TMJ).** The mandible also has an anterior *coronoid process* (KOR-ō-noyd) to which the temporalis muscle attaches. The depression between the coronoid and condylar processes is called the *mandibular notch.* The *alveolar process* is an arch containing the *alveoli* (sockets) for the mandibular (lower) teeth.

The *mental foramen* (*ment-* = chin) is located below the mandibular second premolar tooth. Another foramen in the mandible is the *mandibular foramen* on the medial surface of each ramus. Dentists use both foramina to reach nerves while injecting anesthetics. The mandibular foramen is the beginning of the *mandibular canal,* which runs obliquely in the ramus and anteriorly to the body deep to the roots of the teeth. The inferior alveolar nerves and blood vessels, which are distributed to the mandibular teeth, pass through the canal.

Temporomandibular Joint Syndrome

One problem associated with the temporomandibular joint (TMJ) is **temporomandibular joint (TMJ) syndrome.** It is

Figure 7.10 Mandible.

The mandible is the largest and strongest facial bone.

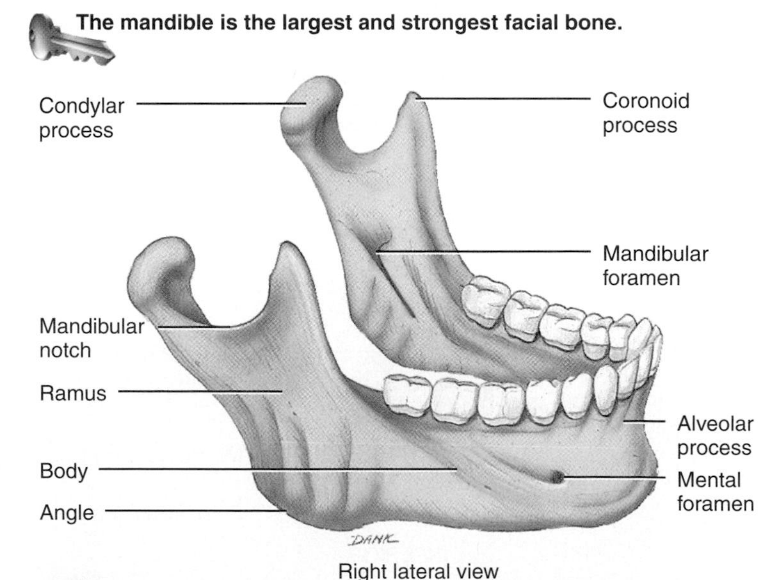

Right lateral view

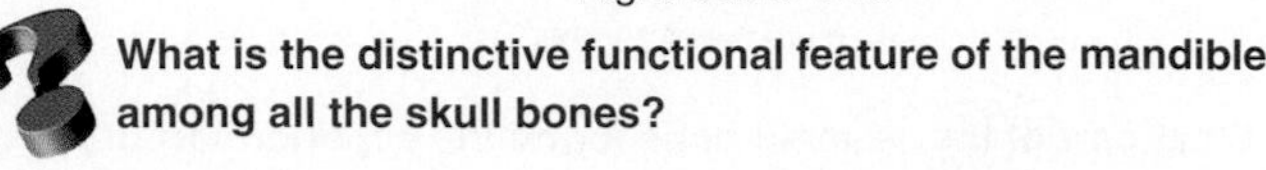
What is the distinctive functional feature of the mandible among all the skull bones?

characterized by dull pain around the ear, tenderness of the jaw muscles, a clicking or popping noise when opening or closing the mouth, limited or abnormal opening of the mouth, headache, tooth sensitivity, and abnormal wearing of the teeth. TMJ syndrome can be caused by improperly aligned teeth, grinding or clenching the teeth, trauma to the head and neck, or arthritis. Treatments include applying moist heat or ice, eating a soft diet, taking pain relievers such as aspirin, muscle retraining, adjusting or reshaping the teeth, orthodontic treatment, or surgery. ■

Nasal Septum

The inside of the nose, called the nasal cavity, is divided into right and left sides by a vertical partition called the **nasal septum.** The three components of the nasal septum are the vomer, septal cartilage, and the perpendicular plate of the ethmoid bone (Figure 7.11). The anterior border of the vomer articulates with the septal cartilage, which is hyaline cartilage, to form the anterior portion of the septum. The superior border of the vomer articulates with the perpendicular plate of the ethmoid bone to form the remainder of the nasal septum.

A **deviated nasal septum** is one that is deflected laterally from the midline of the nose. The deviation usually occurs at the junction of the vomer bone with the septal cartilage. Septal deviations may occur due to a developmental abnormality or trauma. If the deviation is severe, it may entirely block the nasal passageway. Even a partial blockage may lead to infection. If inflammation occurs, it may cause nasal congestion, blockage of the paranasal sinus openings, chronic sinusitis, headache, and nosebleeds. The condition usually can be corrected surgically. ■

Orbits

Seven bones of the skull join to form each **orbit** (eye socket), which contains the eyeball and associated structures (Figure 7.12). The three cranial bones of the orbit are the frontal, sphenoid, and ethmoid; the four facial bones are the palatine, zygomatic, lacrimal, and maxilla. Each pyramid-shaped orbit has four regions that converge posteriorly:

1. Parts of the frontal and sphenoid bones comprise the roof of the orbit.
2. Parts of the zygomatic and sphenoid bones form the lateral wall of the orbit.
3. Parts of the maxilla, zygomatic, and palatine bones make up the floor of the orbit.
4. Parts of the maxilla, lacrimal, ethmoid, and sphenoid bones form the medial wall of the orbit.

Associated with each orbit are five openings:

1. The *optic foramen* is at the junction of the roof and medial wall.
2. The *superior orbital fissure* is at the superior lateral angle of the apex.
3. The *inferior orbital fissure* is at the junction of the lateral wall and floor.
4. The *supraorbital foramen* is on the medial side of the supraorbital margin of the frontal bone.
5. The *lacrimal fossa* is in the lacrimal bone.

Foramina

We mentioned most of the **foramina** of the skull in the descriptions of the cranial and facial bones that they penetrate.

Figure 7.11 Nasal septum. (See Tortora, *A Photographic Atlas of the Human Body,* Figure 3.4.)

The structures that form the nasal septum are the perpendicular plate of the ethmoid bone, the vomer, and septal cartilage.

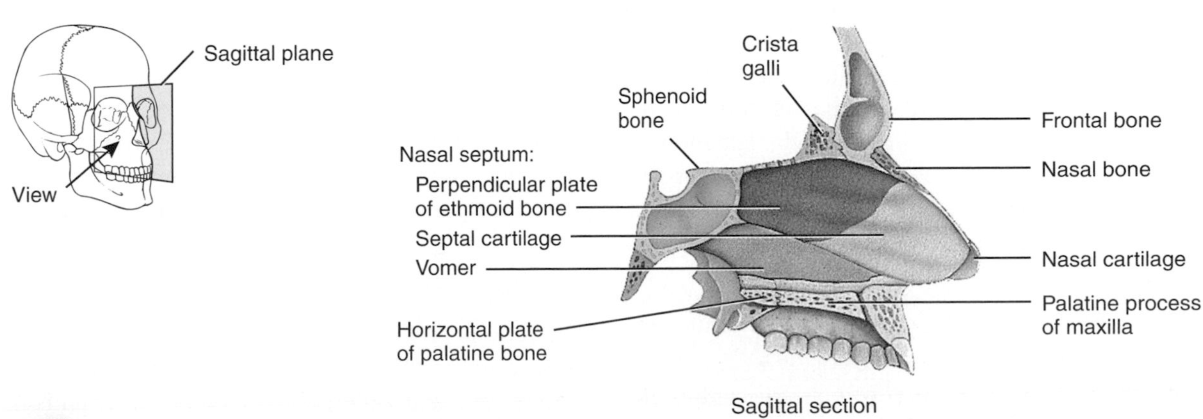

What is the function of the nasal septum?

Figure 7.12 Details of the orbit (eye socket). (See Tortora, *A Photographic Atlas of the Human Body,* Figure 3.11.)

The orbit is a pyramid-shaped structure that contains the eyeball and associated structures.

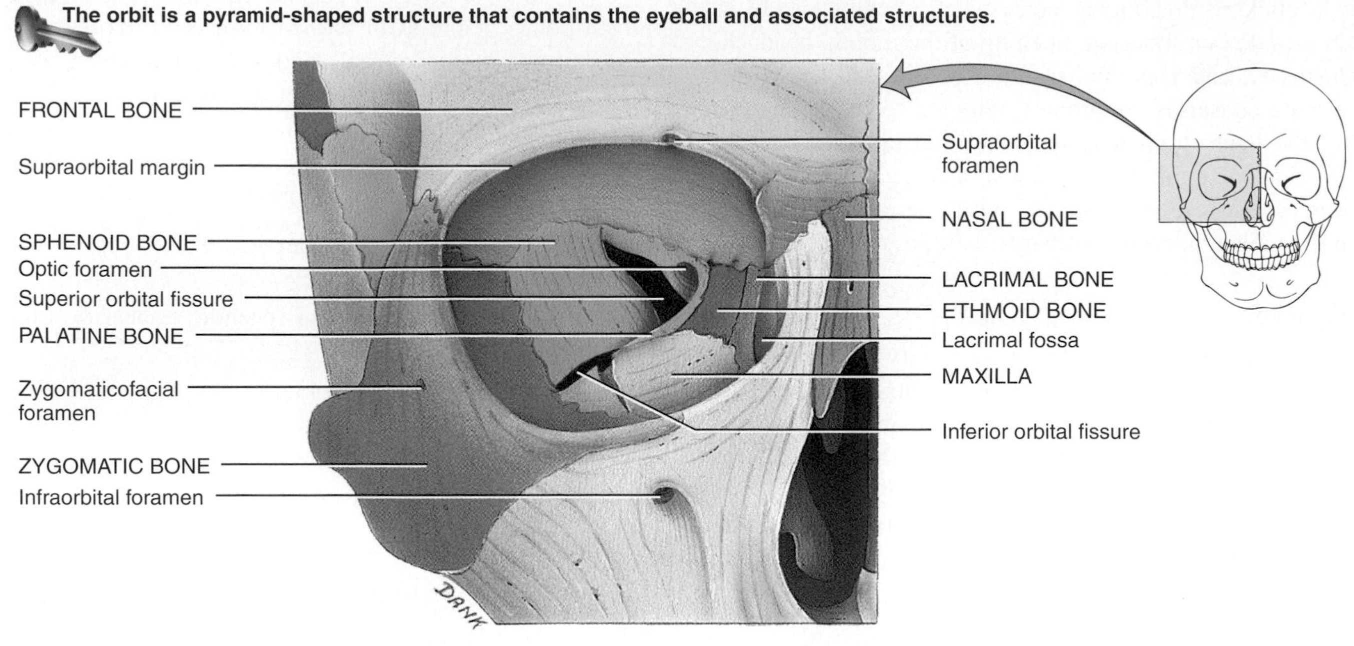

Anterior view showing the bones of the right orbit

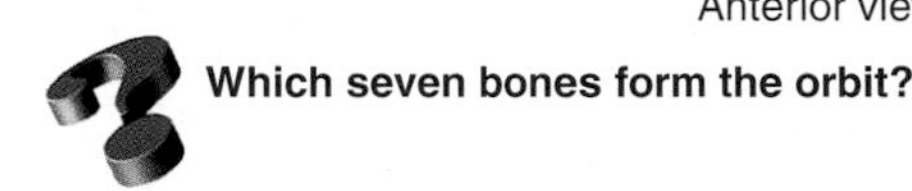

Which seven bones form the orbit?

As preparation for studying other systems of the body, especially the nervous and cardiovascular systems, these foramina and the structures passing through them are listed in Table 7.3. For your convenience and for future reference, the foramina are listed alphabetically.

Unique Features of the Skull

The skull exhibits several unique features not seen in other bones of the body. These include sutures, paranasal sinuses, and fontanels.

Sutures

A **suture** (SOO-chur = seam) is an immovable joint in an adult that is found only between skull bones and that holds most skull bones together. Sutures in the skulls of infants and children often are movable. The names of many sutures reflect the bones they unite. For example, the frontozygomatic suture is between the frontal bone and the zygomatic bone. Similarly, the sphenoparietal suture is between the sphenoid bone and the parietal bone. In other cases, however, the names of sutures are not so obvious. Of the many sutures found in the skull, we will identify only four prominent ones:

1. The **coronal suture** (kō-RŌ-nal; *coron-* = crown) unites the frontal bone and both parietal bones (see Figure 7.4).
2. The **sagittal suture** (SAJ-i-tal; *sagitt-* = arrow) unites the two parietal bones on the superior midline of the skull (see Figure 7.6). The sagittal suture is so named because in the infant, before the bones of the skull are firmly united, the suture and the fontanels (soft spots) associated with it resemble an arrow.
3. The **lambdoid suture** (LAM-doyd) unites the two parietal bones to the occipital bone. This suture is so named because of its resemblance to the Greek letter lambda (Λ), as can be seen in Figure 7.6. Sutural bones may occur within the sagittal and lambdoid sutures.
4. The **squamous sutures** (SKWĀ-mus; *squam-* = flat) unite the parietal and temporal bones on the lateral aspects of the skull (see Figure 7.4).

Paranasal Sinuses

The **paranasal sinuses** (*para-* = beside) are paired cavities in certain cranial and facial bones near the nasal cavity. They are most evident in a sagittal section of the skull (Figure 7.13 on page 202). The paranasal sinuses are lined with mucous membranes that are continuous with the lining of the nasal cavity. Skull bones containing the paranasal sinuses are the frontal, sphenoid, ethmoid, and maxillary. Besides producing mucus, the paranasal sinuses serve as resonating chambers for sound as we speak or sing.

Sinusitis

Secretions produced by the mucous membranes of the paranasal sinuses drain into the nasal cavity. An inflammation of the membranes due to an allergic reaction or infection is called **sinusitis.** If the membranes swell enough to block drainage into the nasal cavity, fluid pressure builds up in the paranasal sinuses, and a si-

Table 7.3 Principal Foramina of the Skull

Foramen	Location	Structures Passing Through*
Carotid (relating to carotid artery in neck)	Petrous portion of temporal bone (Figure 7.7).	Internal carotid artery and sympathetic nerves for eyes.
Hypoglossal (*hypo-* = under; *glossus* = tongue)	Superior to base of occipital condyles (Figure 7.8).	Cranial nerve XII (hypoglossal) and branch of ascending pharyngeal artery.
Infraorbital (*infra-* = below)	Inferior to orbit in maxilla (Figure 7.13).	Infraorbital nerve and blood vessels and a branch of the maxillary division of cranial nerve V (trigeminal).
Jugular (*jugul-* = the throat)	Posterior to carotid canal between petrous portion of temporal bone and occipital bone (Figure 7.8).	Internal jugular vein, cranial nerves IX (glossopharyngeal), X (vagus), and XI (accessory).
Lacerum (*lacerum* = lacerated)	Bounded anteriorly by sphenoid bone, posteriorly by petrous portion of temporal bone, and medially by sphenoid and occipital bones (Figure 7.8a).	Branch of ascending pharyngeal artery.
Magnum (= large)	Occipital bone (Figure 7.7).	Medulla oblongata and its membranes (meninges), cranial nerve XI (accessory), and vertebral and spinal arteries.
Mandibular (*mand-* = to chew)	Medial surface of ramus of mandible (Figure 7.10).	Inferior alveolar nerve and blood vessels.
Mastoid (= breast-shaped)	Posterior border of mastoid process of temporal bone (Figure 7.7).	Emissary vein to transverse sinus and branch of occipital artery to dura mater.
Mental (*ment-* = chin)	Inferior to second premolar tooth in mandible (Figure 7.10).	Mental nerve and vessels.
Olfactory (*olfact* = to smell)	Cribriform plate of ethmoid bone (Figure 7.8).	Cranial nerve I (olfactory).
Optic (= eye)	Between superior and inferior portions of small wing of sphenoid bone (Figure 7.13).	Cranial nerve II (optic) and ophthalmic artery.
Ovale (= oval)	Greater wing of sphenoid bone (Figure 7.8).	Mandibular branch of cranial nerve V (trigeminal).
Rotundum (= round)	Junction of anterior and medial parts of sphenoid bone (Figure 7.8).	Maxillary branch of cranial nerve V (trigeminal).
Stylomastoid (*stylo-* = stake or pole)	Between styloid and mastoid processes of temporal bone (Figure 7.7).	Cranial nerve VII (facial) and stylomastoid artery.
Supraorbital (*supra-* = above)	Supraorbital margin of orbit in frontal bone (Figure 7.13).	Supraorbital nerve and artery

* The cranial nerves listed here are described in Table 14.3 on pages 485–489.

nus headache results. A severely deviated nasal septum or nasal polyps, growths that can be removed surgically, may also cause chronic sinusitis. ■

Fontanels

The skeleton of a newly formed embryo consists of cartilage and fibrous connective tissue membrane structures shaped like bones. Gradually, ossification occurs—bone replaces the cartilage and fibrous connective tissue membranes. At birth, membrane-filled spaces called **fontanels** (fon-ta-NELZ = little fountains) are present between the cranial bones (Figure 7.14). Commonly called "soft spots," fontanels are areas of fibrous connective tissue membranes. Eventually, they will be replaced with bone by intramembranous ossification and become sutures. Functionally, the fontanels provide some flexibility to the fetal skull. They allow the skull to change shape as it passes through the birth canal and permit rapid growth of the brain during infancy. Although an infant may have many fontanels at birth, the form and location of six are fairly constant:

- The unpaired **anterior fontanel,** located at the midline between the two parietal bones and the frontal bone, is roughly diamond-shaped and is the largest fontanel. It usually closes 18 to 24 months after birth.
- The unpaired **posterior fontanel** is located at the midline between the two parietal bones and the occipital bone. Because it is much smaller than the anterior fontanel, it generally closes about 2 months after birth.
- The paired **anterolateral fontanels,** located laterally between the frontal, parietal, temporal, and sphenoid bones, are small and irregular in shape. Normally, they close about 3 months after birth.

Figure 7.13 Paranasal sinuses. (See Tortora, *A Photographic Atlas of the Human Body,* Figure 3.4.)

Paranasal sinuses are mucous membrane-lined spaces in the frontal, sphenoid, ethmoid, and maxillary bones that connect to the nasal cavity.

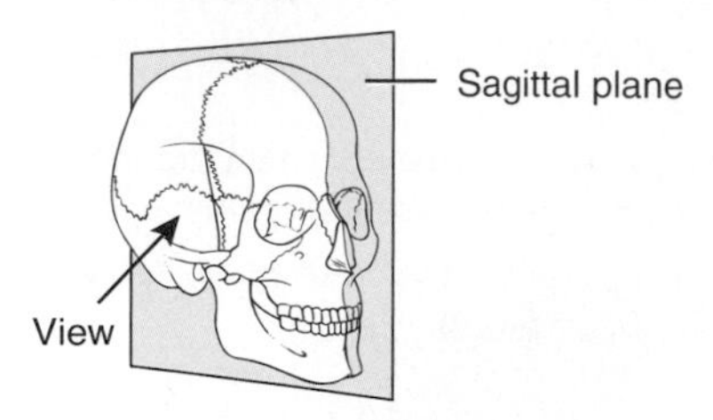

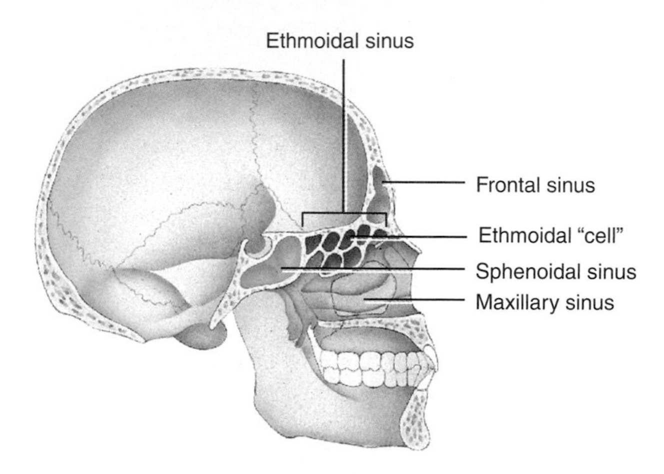

Sagittal section

What are the functions of the paranasal sinuses?

Figure 7.14 Fontanels at birth. (See Tortora, *A Photographic Atlas of the Human Body,* Figure 3.12.)

Fontanels are fibrous connective tissue membrane-filled spaces between cranial bones that are present at birth.

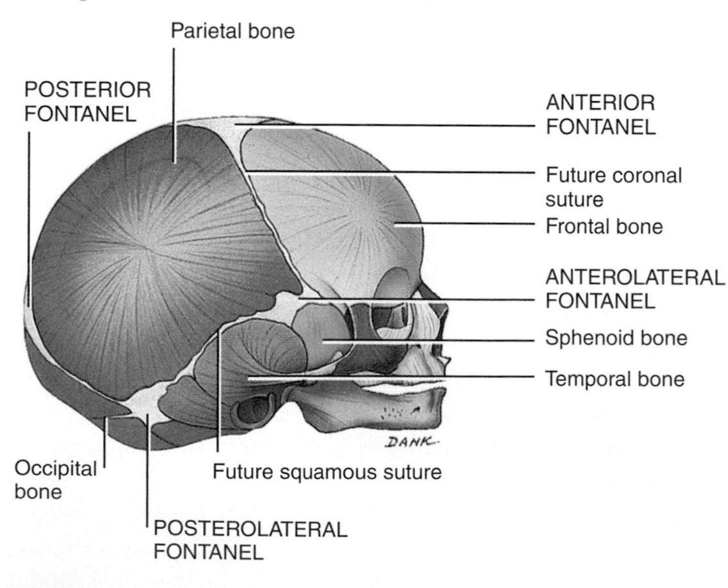

Right lateral view

Which fontanel do four different skull bones border?

- The paired **posterolateral fontanels,** located laterally between the parietal, occipital, and temporal bones, are irregularly shaped. They begin to close 1 to 2 months after birth, but closure is generally not complete until 12 months.

The amount of closure in fontanels helps a physician gauge the degree of brain development. In addition, the anterior fontanel serves as a landmark for withdrawal of blood for analysis from the superior sagittal sinus (a large vein on the midline surface of the brain).

HYOID BONE

OBJECTIVE

- **Describe the relationship of the hyoid bone to the skull.**

The single **hyoid bone** (HĪ-oyd = U-shaped) is a unique component of the axial skeleton because it does not articulate with any other bone (see Figure 7.4). Rather, it is suspended from the styloid processes of the temporal bones by ligaments and muscles. Located in the anterior neck between the mandible and larynx, the hyoid bone supports the tongue, providing attachment sites for some tongue muscles and for muscles of the neck and pharynx. The hyoid bone consists of a horizontal *body* and paired projections called the *lesser horns* and the *greater horns* (Figure 7.15). Muscles and ligaments attach to these paired projections.

The hyoid bone and the cartilages of the larynx and trachea are often fractured during strangulation. As a result, they are carefully examined at autopsy when strangulation is suspected.

CHECKPOINT

4. Describe the general features of the skull.
5. What bones constitute the orbit?
6. What structures make up the nasal septum?
7. Define the following: foramen, suture, paranasal sinus, and fontanel.
8. What are the functions of the hyoid bone?

Figure 7.15 Hyoid bone. (See Tortora, *A Photographic Atlas of the Human Body,* Figure 3.13.)

The hyoid bone supports the tongue, providing attachment sites for muscles of the tongue, neck, and pharynx.

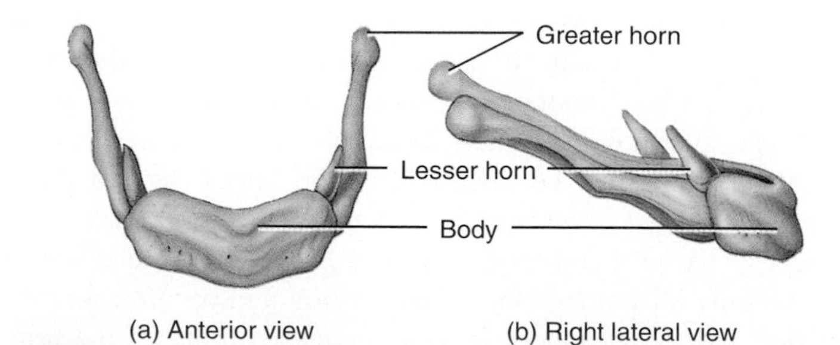

(a) Anterior view (b) Right lateral view

In what way is the hyoid bone different from all the other bones of the axial skeleton?

VERTEBRAL COLUMN

OBJECTIVE

- **Identify the regions and normal curves of the vertebral column and describe its structural and functional features.**

The **vertebral column,** also called the *spine* or *backbone,* makes up about two-fifths of the total height of the body and is composed of a series of bones called **vertebrae** (VER-te-brē; singular is *vertebra*). Together with the sternum and ribs, the vertebral column forms the skeleton of the trunk of the body. Whereas the vertebral column consists of bone and connective tissue, the spinal cord consists of nervous tissue. The length of the column is about 71 cm (28 in.) in an average adult male and about 61 cm (24 in.) in an average adult female. The vertebral column functions as a strong, flexible rod that can move forward, backward, sideways, and rotate. It encloses and protects the spinal cord, supports the head, and serves as a point of attachment for the ribs, pelvic girdle, and muscles of the back.

The total number of vertebrae during early development is 33. Then, several vertebrae in the sacral and coccygeal regions fuse. As a result, the adult vertebral column, also called the spinal column, typically contains 26 vertebrae (Figure 7.16). These are distributed as follows:

- 7 **cervical vertebrae** (*cervic-* = neck) are in the neck region.
- 12 **thoracic vertebrae** (*thorax* = chest) are posterior to the thoracic cavity.
- 5 **lumbar vertebrae** (*lumb-* = loin) support the lower back.
- 1 **sacrum** (SĀ-krum = sacred bone) consists of five fused *sacral vertebrae.*
- 1 **coccyx** (KOK-siks = cuckoo, because the shape resembles the bill of a cuckoo bird), consists of four fused *coccygeal vertebrae* (kok-SIJ-ē-al).

Whereas the cervical, thoracic, and lumbar vertebrae are movable, the sacrum and coccyx are not. We will discuss each of these regions in detail shortly.

Figure 7.16 Vertebral column. The numbers in parentheses in (a) indicate the number of vertebrae in each region. In (d), the relative size of the disc has been enlarged for emphasis. A "window" has been cut in the annulus fibrosus to view the nucleus pulposus. (See Tortora, *A Photographic Atlas of the Human Body,* Figure 3.15.)

The adult vertebral column typically contains 26 vertebrae.

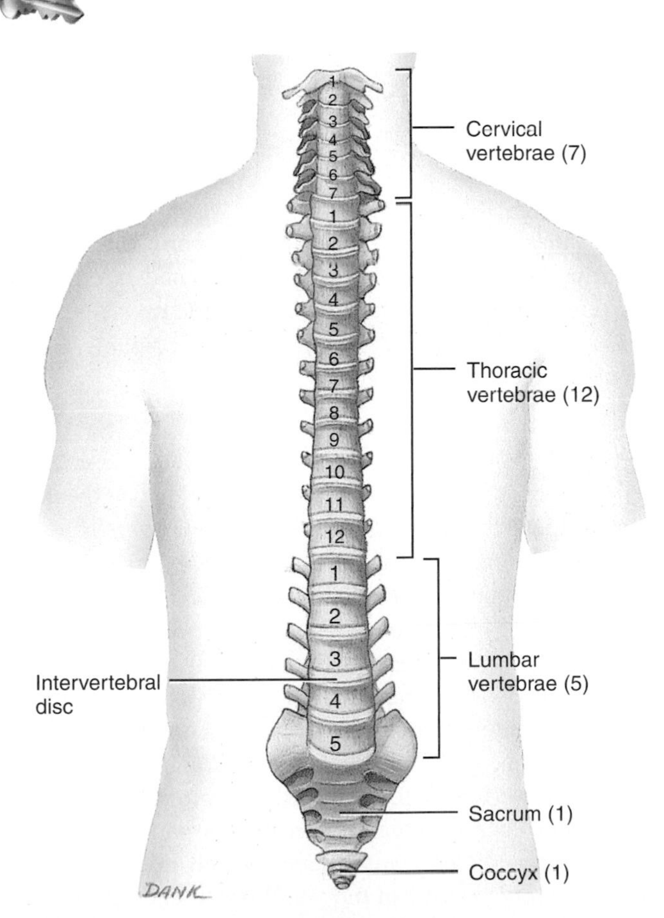

(a) Anterior view showing regions of the vertebral column

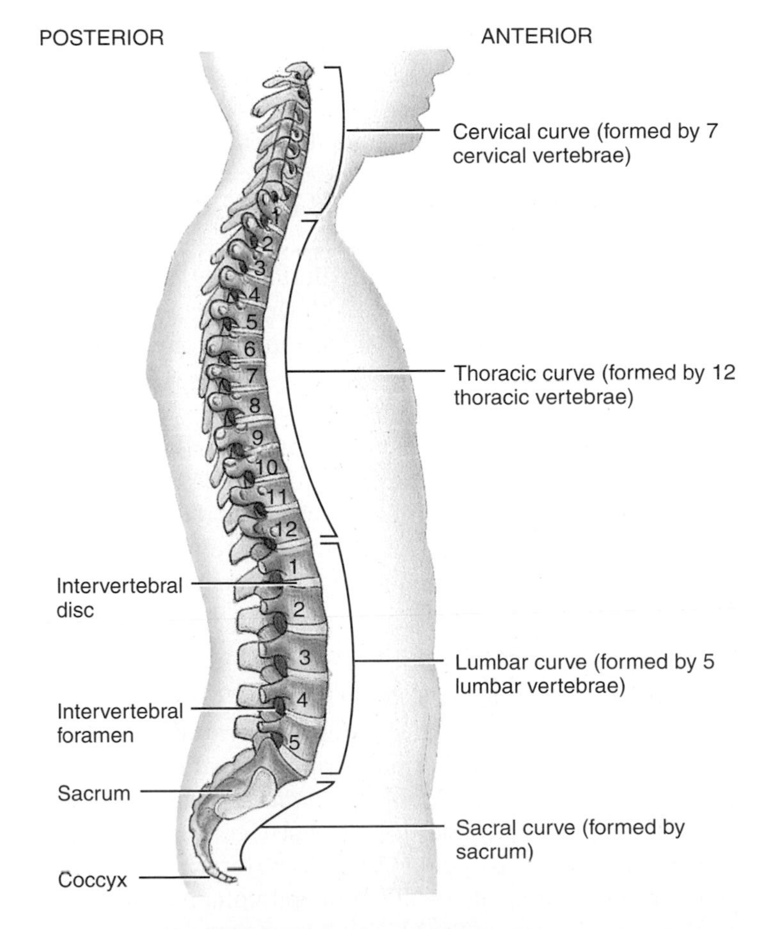

(b) Right lateral view showing four normal curves

(continues)

Figure 7.16 *(continued)*

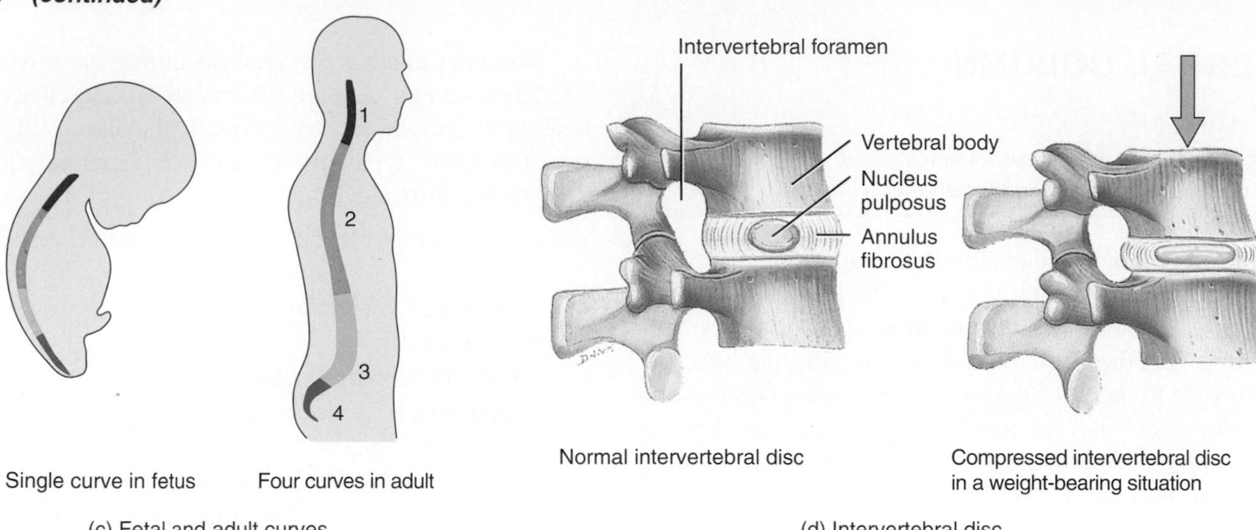

(c) Fetal and adult curves

(d) Intervertebral disc

Which curves of the adult vertebral column are concave (relative to the anterior side of the body)?

Normal Curves of the Vertebral Column

When viewed from the side, the vertebral column shows four slight bends called **normal curves** (Figure 7.16b). Relative to the front of the body, the *cervical* and *lumbar curves* are convex (bulging out), whereas the *thoracic* and *sacral curves* are concave (cupping in). The curves of the vertebral column increase its strength, help maintain balance in the upright position, absorb shocks during walking, and help protect the vertebrae from fracture.

In the fetus, there is only a single anteriorly concave curve (Figure 7.16c). At about the third month after birth, when an infant begins to hold its head erect, the cervical curve develops. Later, when the child sits up, stands, and walks, the lumbar curve develops. The thoracic and sacral curves are called *primary curves* because they form first during fetal development. The cervical and lumbar curves are known as *secondary curves* because they begin to form later, several months after birth. All curves are fully developed by age 10. However, secondary curves may be progressively lost in old age.

Various conditions may exaggerate the normal curves of the vertebral column, or the column may acquire a lateral bend, resulting in **abnormal curves** of the vertebral column. Three such abnormal curves—kyphosis, lordosis, and scoliosis—are described in the Medical Terminology section on page 214.

Intervertebral Discs

Between the bodies of adjacent vertebrae from the second cervical vertebra to the sacrum are **intervertebral discs** (Figure 7.16d). Each disc has an outer fibrous ring consisting of fibrocartilage called the *annulus fibrosus* (*annulus* = ringlike) and an inner soft, pulpy, highly elastic substance called the *nucleus pulposus* (*pulposus* = pulplike). The discs form strong joints, permit various movements of the vertebral column, and absorb vertical shock. Under compression, they flatten, broaden, and bulge from their intervertebral spaces. Superior to the sacrum, the intervertebral discs constitute about one-fourth the length of the vertebral column.

Parts of a Typical Vertebra

Vertebrae in different regions of the spinal column vary in size, shape, and detail, but they are similar enough that we can discuss the structure and functions of a typical vertebra (Figure 7.17). Vertebrae typically consist of a body, a vertebral arch, and several processes.

Body

The **body** is the thick, disk-shaped anterior portion that is the weight-bearing part of a vertebra. Its superior and inferior surfaces are roughened for the attachment of cartilaginous intervertebral discs. The anterior and lateral surfaces contain nutrient foramina for blood vessels.

Vertebral Arch

The **vertebral arch** extends posteriorly from the body of the vertebra and together with the body of the vertebra surrounds the spinal cord. Two short, thick processes, the *pedicles* (PED-i-kuls = little feet), form the vertebral arch. The pedicles project posteriorly from the body to unite with the laminae. The *laminae* (LAM-i-nē = thin layers) are the flat parts that join to form the posterior portion of the vertebral arch. The *vertebral foramen* lies between the vertebral arch and body and contains the spinal cord, adipose tissue, areolar connective tissue, and blood vessels. Collectively, the vertebral foramina of all vertebrae form the **vertebral (spinal) canal,** which is the inferior part of the dorsal body cavity. The pedicles exhibit superior and inferior notches called *vertebral notches.* When the vertebral notches are stacked on top of one another, they form an opening between adjoining

Figure 7.17 Structure of a typical vertebra, as illustrated by a thoracic vertebra. In (b), only one spinal nerve has been included, and it has been extended beyond the intervertebral foramen for clarity. The sympathetic chain is part of the autonomic nervous system (see Figure 17.2 on page 569). (See Tortora, *A Photographic Atlas of the Human Body,* Figure 3.16.)

A vertebra consists of a body, a vertebral arch, and several processes.

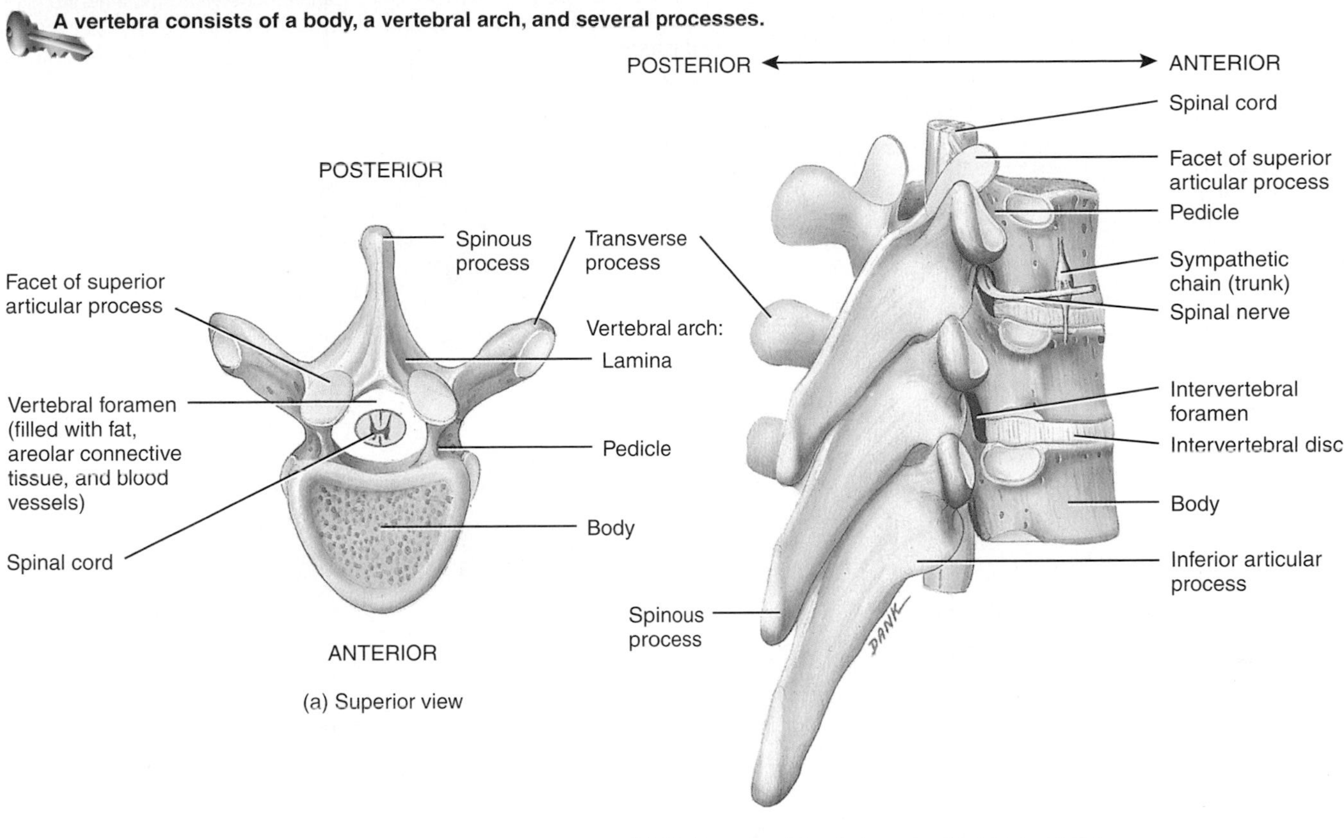

(a) Superior view

(b) Right posterolateral view of articulated vertebrae

What are the functions of the vertebral and intervertebral foramina?

vertebrae on both sides of the column. Each opening, called an *intervertebral foramen,* permits the passage of a single spinal nerve. Thoracic and lumbar spinal nerves pass through the intervertebral foramina.

Processes

Seven *processes* arise from the vertebral arch. At the point where a lamina and pedicle join, a *transverse process* extends laterally on each side. A single *spinous process* (*spine*) projects posteriorly from the junction of the laminae. These three processes serve as points of attachment for muscles. The remaining four processes form joints with other vertebrae above or below. The two *superior articular processes* of a vertebra articulate with the two inferior *articular processes* of the vertebra immediately above them. The two *inferior articular processes* of a vertebra articulate with the two superior articular processes of the vertebra immediately below them. The articulating surfaces of the articular processes are called *facets* (= little faces). The articulations formed between the bodies and articular facets of successive vertebrae are termed *intervertebral joints.*

Regions of the Vertebral Column

We turn now to the five regions of the vertebral column, beginning at the superior end. Note that vertebrae in each region are numbered in sequence, from superior to inferior.

Cervical Region

The bodies of **cervical vertebrae** (C1–C7) are smaller than those of thoracic vertebrae (Figure 7.18a). Their vertebral arches, however, are larger. All cervical vertebrae have three foramina: one vertebral foramen and two transverse foramina (Figure 7.18b). The vertebral foramina of cervical vertebrae are the largest in the spinal column because they house the cervical enlargement of the spinal cord. Each cervical transverse process contains a *transverse foramen* through which the vertebral artery and its accompanying vein and nerve pass. The spinous processes of C2 through C6 are often *bifid*—that is, split into two parts (Figure 7.18c, d).

The first two cervical vertebrae differ considerably from the others. The first cervical vertebra (C1), the **atlas,** supports

Figure 7.18 Cervical vertebrae. (See Tortora, *A Photographic Atlas of the Human Body,* Figure 3.17.)

The cervical vertebrae are found in the neck region.

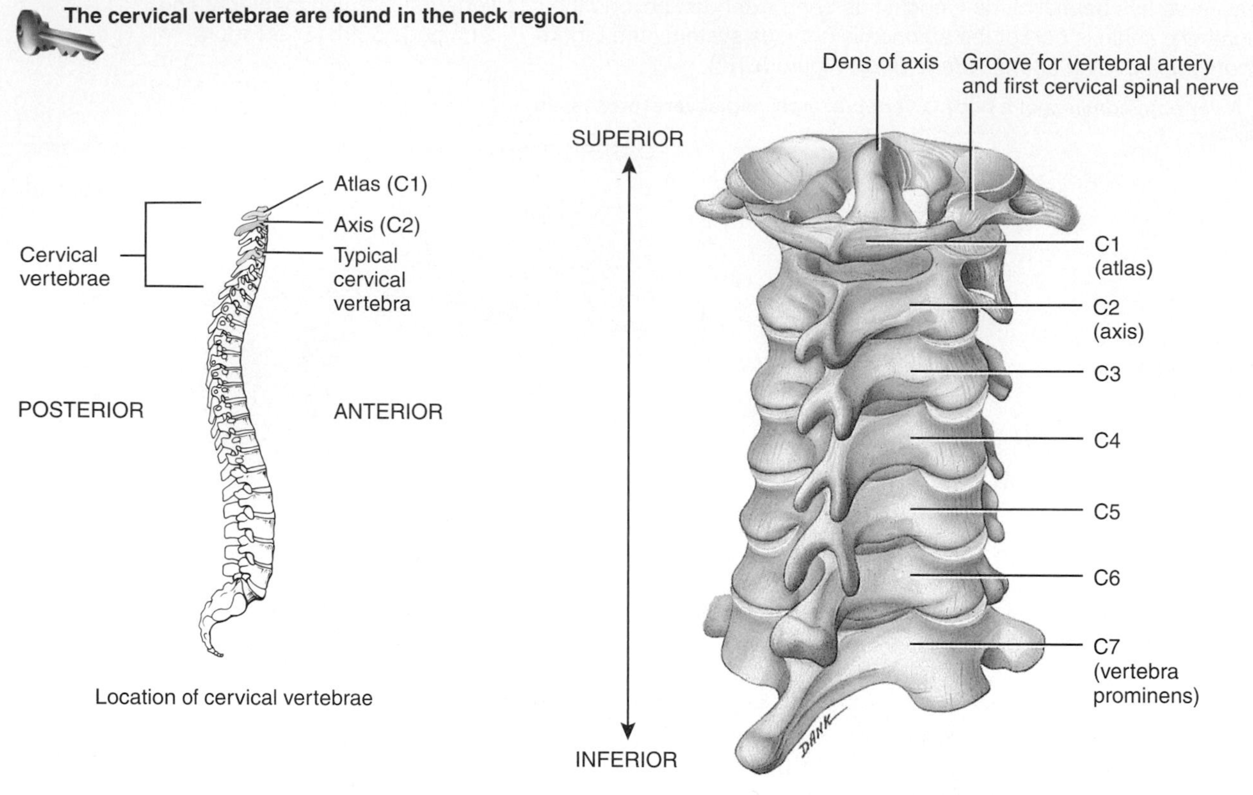

(a) Posterior view of articulated cervical vertebrae

the head and is named for the mythological Atlas who supported the world on his shoulders (Figure 7.18a, b). The atlas is a ring of bone with *anterior* and *posterior arches* and large *lateral masses.* It lacks a body and a spinous process. The superior surfaces of the lateral masses, called *superior articular facets,* are concave. They articulate with the occipital condyles of the occipital bone to form the *atlanto-occipital joints.* These articulations permit the movement seen when moving the head to signify "yes." The inferior surfaces of the lateral masses, the *inferior articular facets,* articulate with the second cervical vertebra. The transverse processes and transverse foramina of the atlas are quite large.

The second cervical vertebra (C2), the **axis** (see Figure 7.18a, c), does have a body. A peglike process called the *dens* (= tooth) or *odontoid process* projects up through the anterior portion of the vertebral foramen of the atlas. The dens makes a pivot on which the atlas and head rotate, as in moving the head to signify "no." This arrangement permits side-to-side rotation of the head. The articulation formed between the anterior arch of the atlas and dens of the axis, and between their articular facets, is called the *atlanto-axial joint.* In some instances of trauma, the dens of the axis may be driven into the medulla oblongata of the brain. When whiplash injuries result in death, this type of injury is the usual cause.

The third through sixth cervical vertebrae (C3–C6), represented by the vertebra in Figure 7.18d, correspond to the structural pattern of the typical cervical vertebra previously described. The seventh cervical vertebra (C7), called the *vertebra prominens,* is somewhat different (see Figure 7.18a). It has a single large spinous process that can be seen and felt at the base of the neck.

Thoracic Region

Thoracic vertebrae (T1–T12; Figure 7.19 on page 208) are considerably larger and stronger than cervical vertebrae. In addition, the spinous processes on T1 and T2 are long, laterally flattened, and directed inferiorly. In contrast, the spinous processes on T11 and T12 are shorter, broader, and directed more posteriorly. Compared to cervical vertebrae, thoracic vertebrae also have longer and larger transverse processes.

The most distinctive feature of the thoracic vertebrae is that they articulate with the ribs. The articulating surfaces of the vertebrae are called *facets* and *demifacets* (= half-facets). Except for T11 and T12, the transverse processes have facets for articulating with the *tubercles* of the ribs. The bodies of thoracic vertebrae also have facets or demifacets for articulation with the *heads* of the ribs. The articulations between the thoracic vertebrae and ribs are called *vertebrocostal joints.* As you can see in Figure 7.19a, T1 has a superior facet and an inferior demifacet, one on each side of the vertebral body. T2 through T8 have a superior and inferior demifacet, one on each side of the vertebral

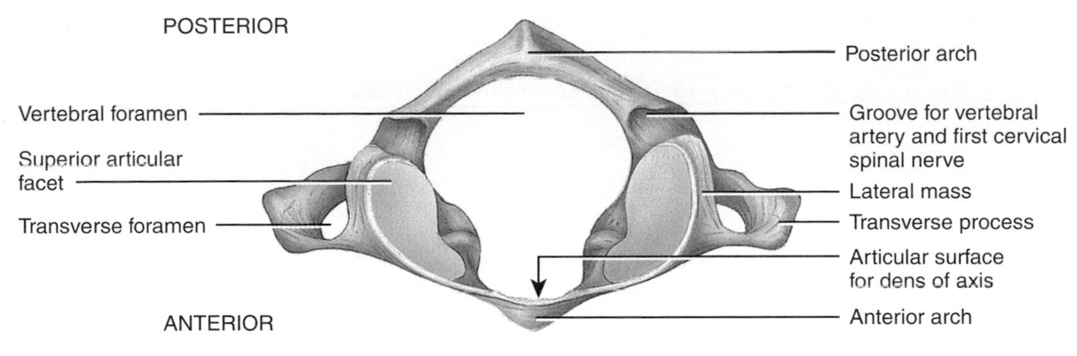

(b) Superior view of the atlas (C1)

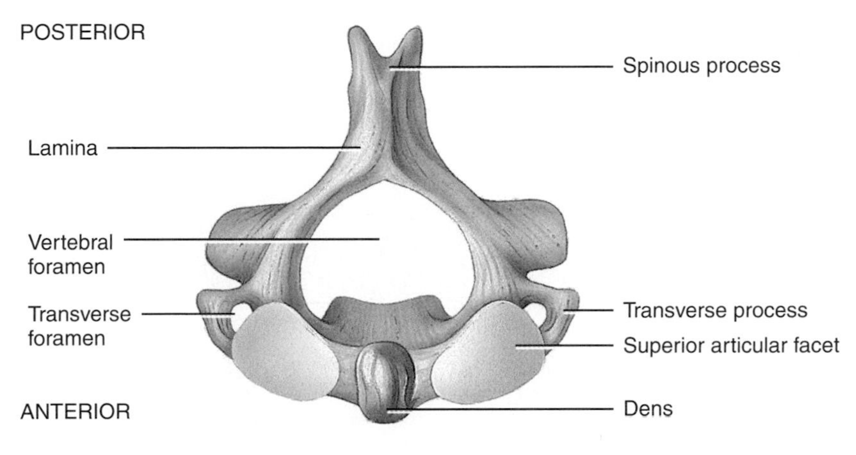

(c) Superior view of the axis (C2)

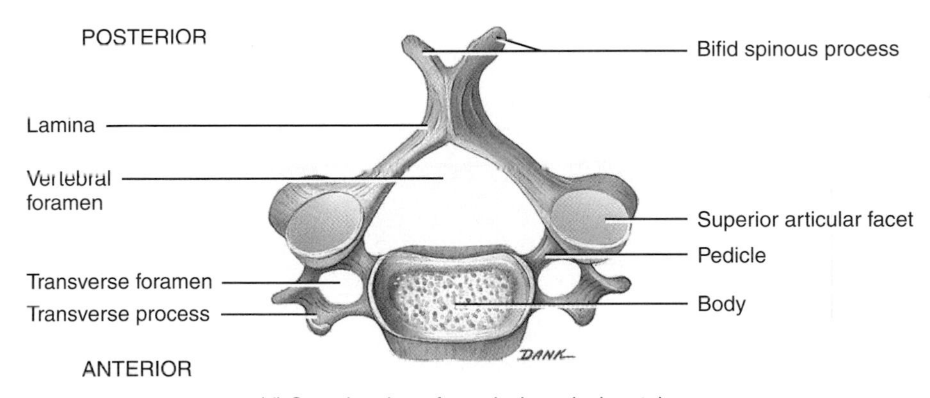

(d) Superior view of a typical cervical vertebra

Which bones permit the movement of the head to signify no?

body. T9 has a superior demifacet on each side of the vertebral body, and T10 through T12 have a superior facet on each side of the vertebral body. Movements of the thoracic region are limited by thin intervertebral discs and by the attachment of the ribs to the sternum.

Lumbar Region

The **lumbar vertebrae** (L1–L5) are the largest and strongest in the vertebral column (Figure 7.20 on page 209) because the amount of body weight supported by the vertebrae increases toward the inferior end of the backbone. Their processes are short and thick. The superior articular processes are directed medially instead of superiorly and the inferior articular processes are directed laterally instead of inferiorly. The spinous processes are thick and broad and project posteriorly. The spinous processes are well adapted for the attachment of the large back muscles.

Table 7.4 on page 210 summarizes the major structural differences among cervical, thoracic, and lumbar vertebrae.

Figure 7.19 Thoracic vertebrae. (See Tortora, *A Photographic Atlas of the Human Body,* Figure 3.16.)

The thoracic vertebrae are found in the chest region and articulate with the ribs.

POSTERIOR
ANTERIOR
Location of thoracic vertebrae

Transverse process
Pedicle
Superior articular facet
Facet for articular part of tubercle of rib
Spinous process
Inferior articular process
Intervertebral foramen
POSTERIOR
Superior vertebral notch
Superior facet
Inferior vertebral notch
Inferior demifacet
Superior demifacet
Inferior demifacet
Superior demifacet
Body
Superior facet
Superior articular process
ANTERIOR
T1
T2-8
T9
T10
T11
T12
DANK

(a) Right lateral view of several articulated thoracic vertebrae

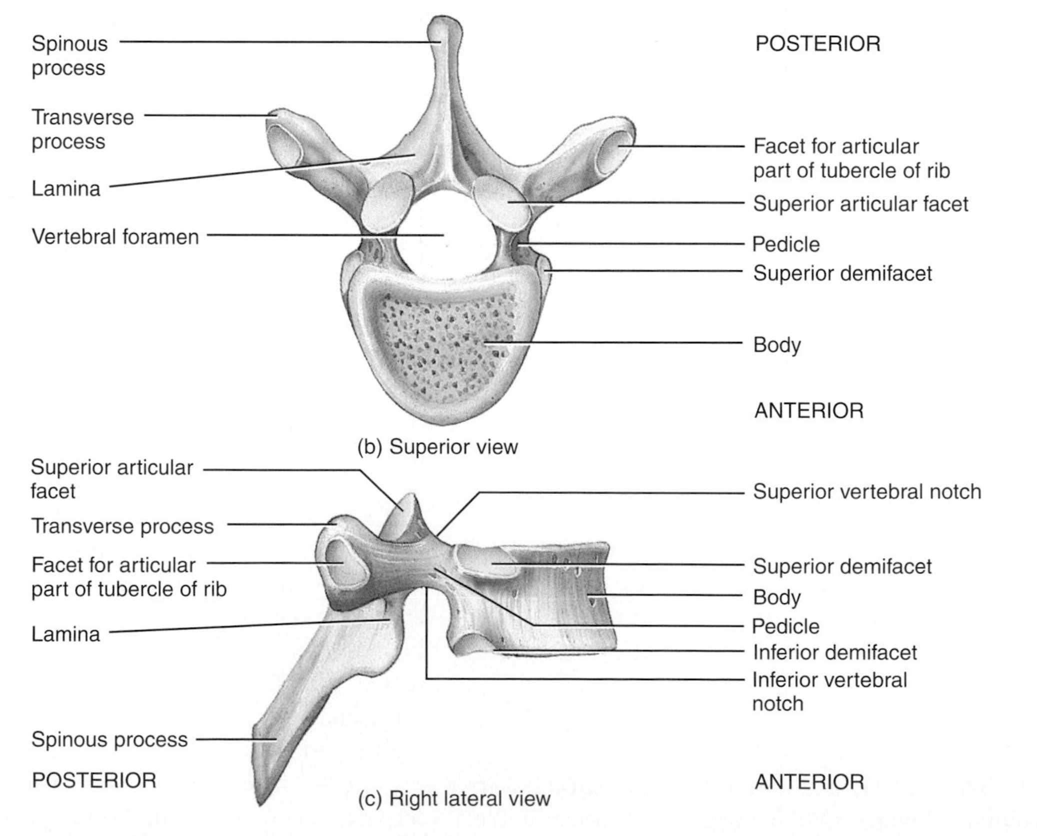

(b) Superior view

(c) Right lateral view

Which parts of thoracic vertebrae articulate with the ribs?

Figure 7.20 Lumbar vertebrae. (See Tortora, *A Photographic Atlas of the Human Body,* Figure 3.18.)

Lumbar vertebrae are found in the lower back.

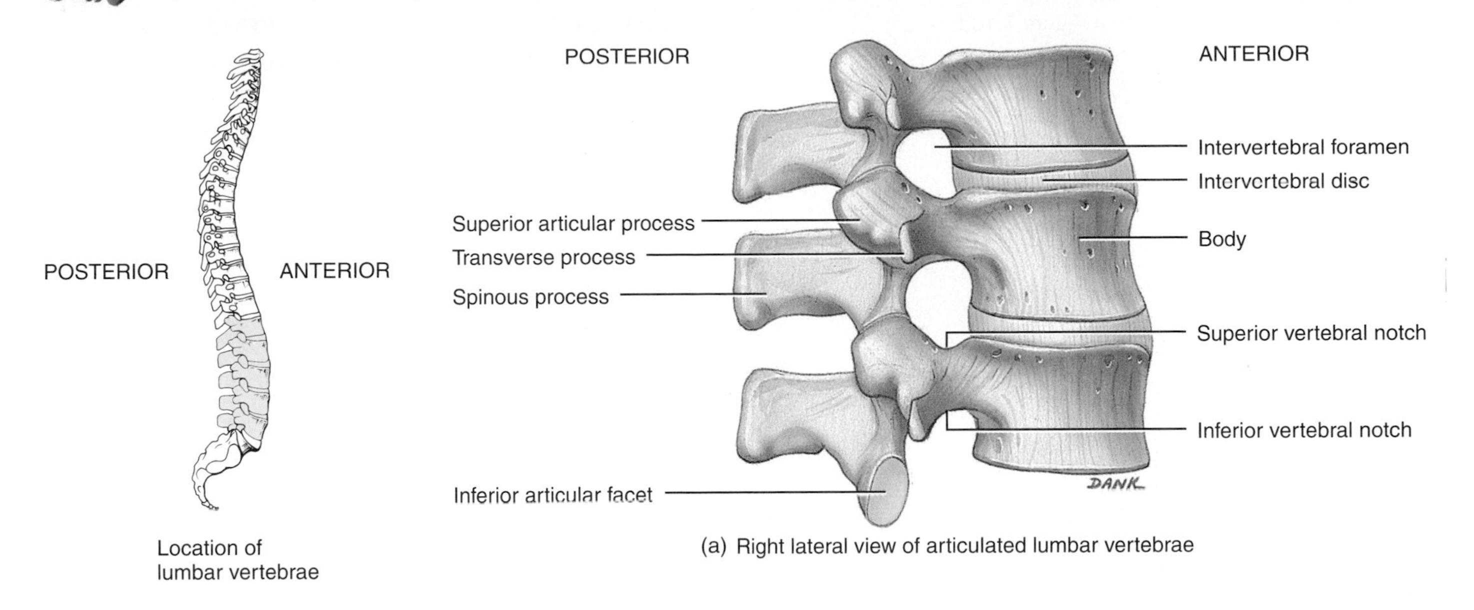

(a) Right lateral view of articulated lumbar vertebrae

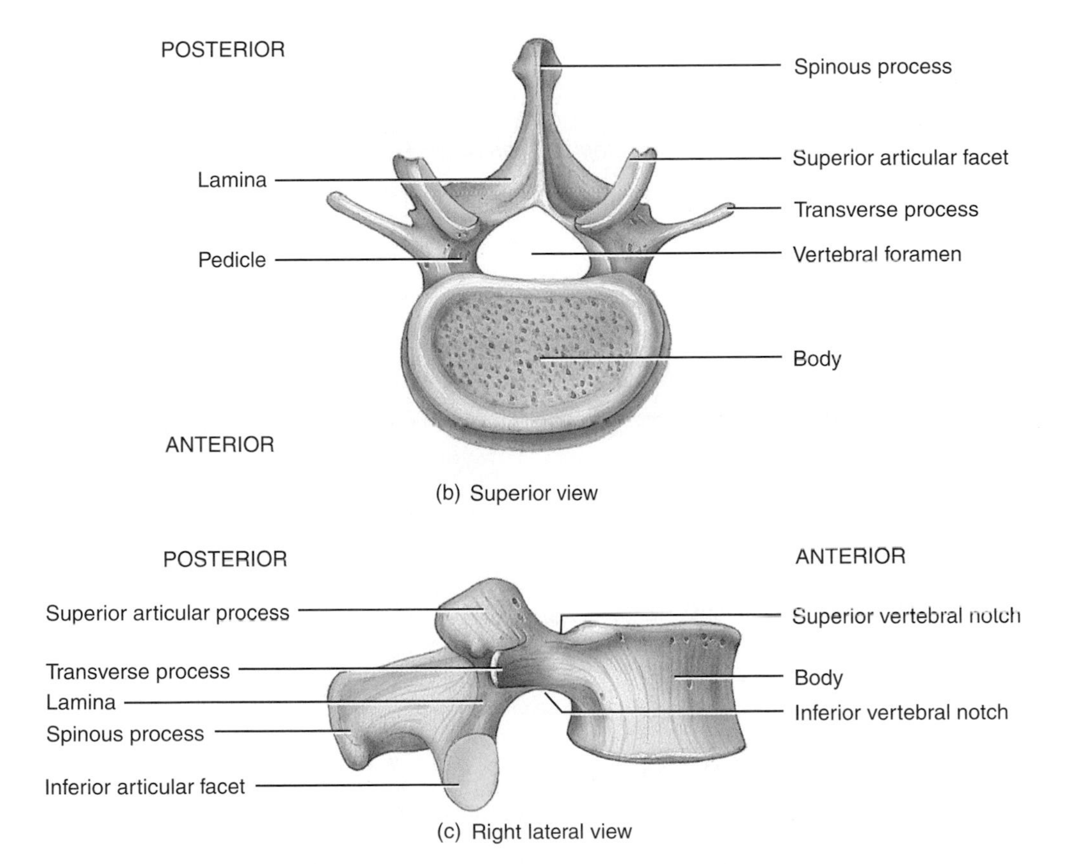

(b) Superior view

(c) Right lateral view

Why do the lumbar vertebrae need to be the largest and strongest in the vertebral column?

Sacrum

The **sacrum** is a triangular bone formed by the union of five sacral vertebrae, indicated in Figure 7.21a as S1–S5. The fusion of the sacral vertebrae begins between ages 16 and 18 and is usually completed by age 30. The sacrum serves as a strong foundation for the pelvic girdle. It is positioned at the posterior portion of the pelvic cavity medial to the two hip bones. The female sacrum is shorter, wider, and more curved

Table 7.4 Comparison of Major Structural Features of Cervical, Thoracic, and Lumbar Vetebrae

Characteristic	Cervical	Thoracic	Lumbar
Overall structure	See Figure 7.18d	See Figure 7.19b	See Figure 7.20b
Size	Small	Larger	Largest
Foramina	One vertebral and two transverse	One vertebral	One vertebral
Spinous processes	Slender and often bifid (C2–C6)	Long and fairly thick (most project inferiorly)	Short and blunt (project posteriorly rather than inferiorly)
Transverse processes	Small	Fairly large	Large and blunt
Articular facets for ribs	Absent	Present	Absent
Direction of articular facets			
Superior	Posterosuperior	Posterolateral	Medial
Inferior	Anteroinferior	Anteromedial	Lateral
Size of intervertebral discs	Thick relative to size of vertebral bodies	Thin relative to vertebral bodies	Massive

between S2 and S3 than the male sacrum (see Table 8.1 on page 231).

The concave anterior side of the sacrum faces the pelvic cavity. It is smooth and contains four *transverse lines (ridges)* that mark the joining of the sacral vertebral bodies (Figure 7.21a). At the ends of these lines are four pairs of *anterior sacral foramina.* The lateral portion of the superior surface contains a smooth surface called the *sacral ala* (= wing), which is formed by the fused transverse processes of the first sacral vertebra (S1).

The convex, posterior surface of the sacrum contains a *median sacral crest,* which is the fused spinous processes of the upper sacral vertebrae, and a *lateral sacral crest,* which is the

Figure 7.21 Sacrum and coccyx. (See Tortora, *A Photographic Atlas of the Human Body,* Figure 3.19.)

The union of five sacral vertebrae forms the sacrum, and the union of usually four coccygeal vertebrae forms the coccyx.

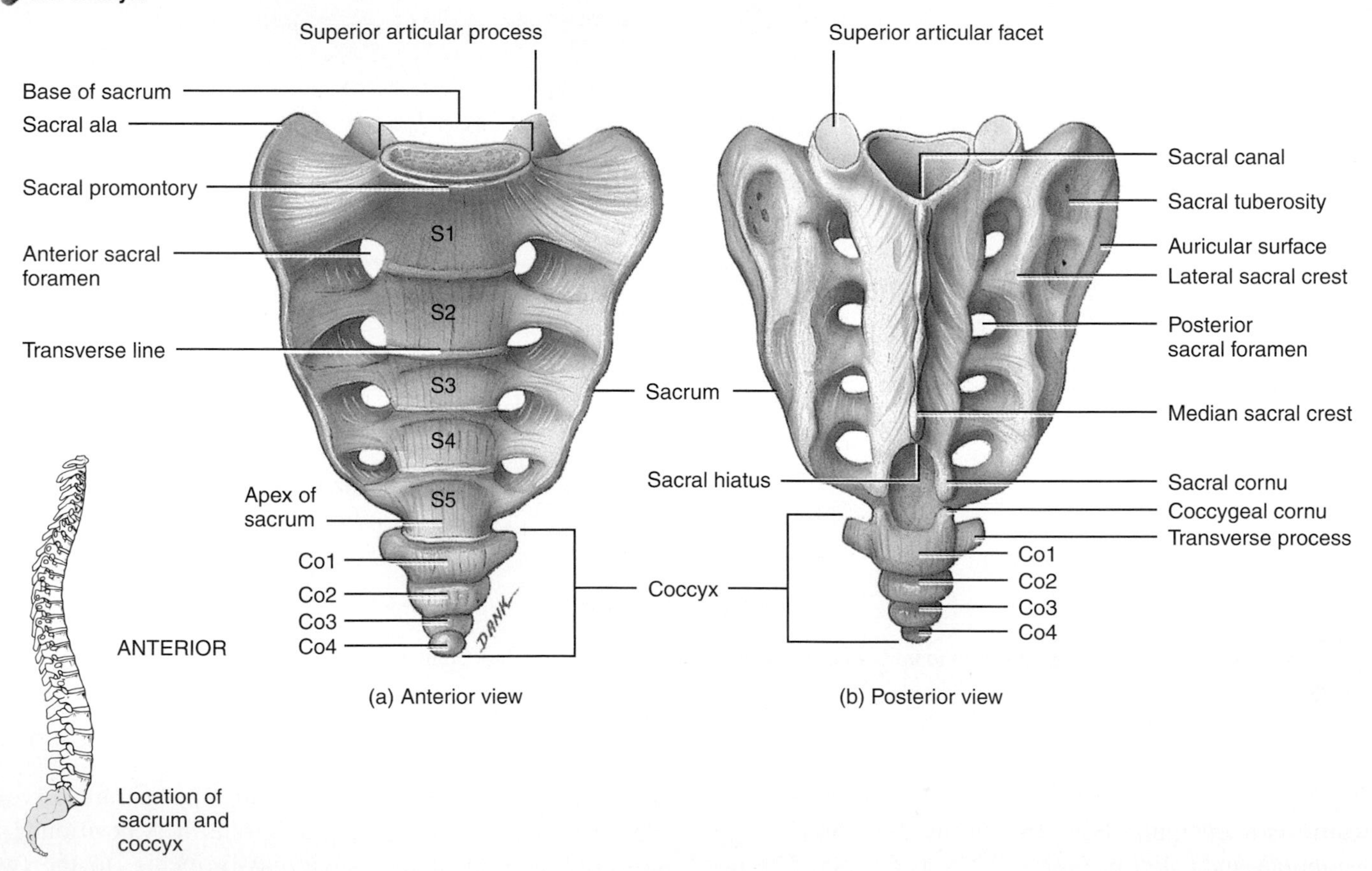

How many foramina pierce the sacrum, and what is their function?

fused transverse processes of the sacral vertebrae. Four pairs of *posterior sacral foramina* (Figure 7.21b) communicate with the anterior sacral foramina through which nerves and blood vessels pass. The *sacral canal* is a continuation of the vertebral canal. The laminae of the fifth sacral vertebra, and sometimes the fourth, fail to meet. This leaves an inferior entrance to the vertebral canal called the *sacral hiatus* (hī-Ā-tus = opening). On either side of the sacral hiatus are the *sacral cornua,* the inferior articular processes of the fifth sacral vertebra. Ligaments connect them to the coccyx.

The narrow inferior portion of the sacrum is known as the *apex.* The broad superior portion of the sacrum is the *base.* The anteriorly projecting border of the base, called the *sacral promontory* (PROM-on-tō-rē), is one of the points used for measurements of the pelvis. On both lateral surfaces, the sacrum has a large *auricular surface* (*auricular* = ear) that articulates with the ilium of each hipbone to form the *sacroiliac joint.* Posterior to the auricular surface is a roughened surface, the *sacral tuberosity,* that contains depressions for the attachment of ligaments. The sacral tuberosity is another surface of the sacrum that unites with the hip bones to form the sacroiliac joints. The *superior articular processes* of the sacrum articulate with the fifth lumbar vertebra, and the base of the sacrum articulates with the body of the fifth lumbar vertebra, to form the *lumbosacral joint.*

Coccyx

The **coccyx** is also triangular and is formed by the fusion of usually four coccygeal vertebrae, indicated in Figure 7.21 as Co1–Co4. The coccygeal vertebrae fuse when a person is between 20 and 30 years of age. The dorsal surface of the body of the coccyx contains two long *coccygeal cornua* that are connected by ligaments to the sacral cornua. The coccygeal cornua are the pedicles and superior articular processes of the first coccygeal vertebra. On the lateral surfaces of the coccyx are a series of *transverse processes,* the first pair being the largest. The coccyx articulates superiorly with the apex of the sacrum. In females, the coccyx points inferiorly; in males, it points anteriorly (see Table 8.1 on page 231).

Caudal Anesthesia

Anesthetic agents that act on the sacral and coccygeal nerves are sometimes injected through the sacral hiatus, a procedure called **caudal anesthesia** or **epidural block.** The procedure is used most often to relieve pain during labor and to provide anesthesia to the perineal area. Because the sacral hiatus is between the sacral cornua, the cornua are important bony landmarks for locating the hiatus. Anesthetic agents may also be injected through the posterior sacral foramina. ■

► CHECKPOINT

9. What are the functions of the vertebral column?
10. When do the secondary vertebral curves develop?
11. What are the main distinguishing characteristics of the bones of the various regions of the vertebral column?

THORAX

► OBJECTIVE

- **Identify the bones of the thorax.**

The term **thorax** refers to the entire chest. The skeletal part of the thorax, the **thoracic cage,** is a bony enclosure formed by the sternum, costal cartilages, ribs, and the bodies of the thoracic vertebrae (Figure 7.22). The thoracic cage is narrower at its superior end and broader at its inferior end and is flattened from front to back. It encloses and protects the organs in the thoracic and superior abdominal cavities and provides support for the bones of the shoulder girdle and upper limbs.

Sternum

The **sternum,** or breastbone, is a flat, narrow bone located in the center of the anterior thoracic wall that measures about 15 cm (6 in.) in length and consists of three parts (Figure 7.22). The superior part is the **manubrium** (ma-NOO-brē-um = handlelike); the middle and largest part is the **body;** and the inferior, smallest part is the **xiphoid process** (ZĪ-foyd = sword-shaped).

The junction of the manubrium and body forms the *sternal angle.* The manubrium has a depression on its superior surface, the *suprasternal notch.* Lateral to the suprasternal notch are *clavicular notches* that articulate with the medial ends of the clavicles to form the *sternoclavicular joints.* The manubrium also articulates with the costal cartilages of the first and second ribs to form the *sternocostal joints.*

The body of the sternum articulates directly or indirectly with the costal cartilages of the second through tenth ribs. The xiphoid process consists of hyaline cartilage during infancy and childhood and does not ossify completely until about age 40. No ribs are attached to it, but the xiphoid process provides attachment for some abdominal muscles. Incorrect positioning of the hands of a rescuer during cardiopulmonary resuscitation (CPR) may fracture the xiphoid process, driving it into internal organs. During thoracic surgery, the sternum may be split along the midline to allow surgeons access to structures in the thoracic cavity such as the thymus, heart, and great vessels of the heart.

Ribs

Twelve pairs of **ribs** give structural support to the sides of the thoracic cavity (see Figure 7.22). The ribs increase in length from the first through seventh, and then decrease in length to the twelfth rib. Each rib articulates posteriorly with its corresponding thoracic vertebra.

The first through seventh pairs of ribs have a direct anterior attachment to the sternum by a strip of hyaline cartilage called *costal cartilage* (*cost-* = rib). These ribs are called *true (vertebrosternal) ribs.* The remaining five pairs of ribs are termed *false ribs* because their costal cartilages either attach indirectly to the sternum or do not attach to the sternum at all. The cartilages of

Figure 7.22 Skeleton of the thorax. (See Tortora, *A Photographic Atlas of the Human Body,* Figure 3.20.)

The bones of the thorax enclose and protect organs in the thoracic cavity and in the superior abdominal cavity.

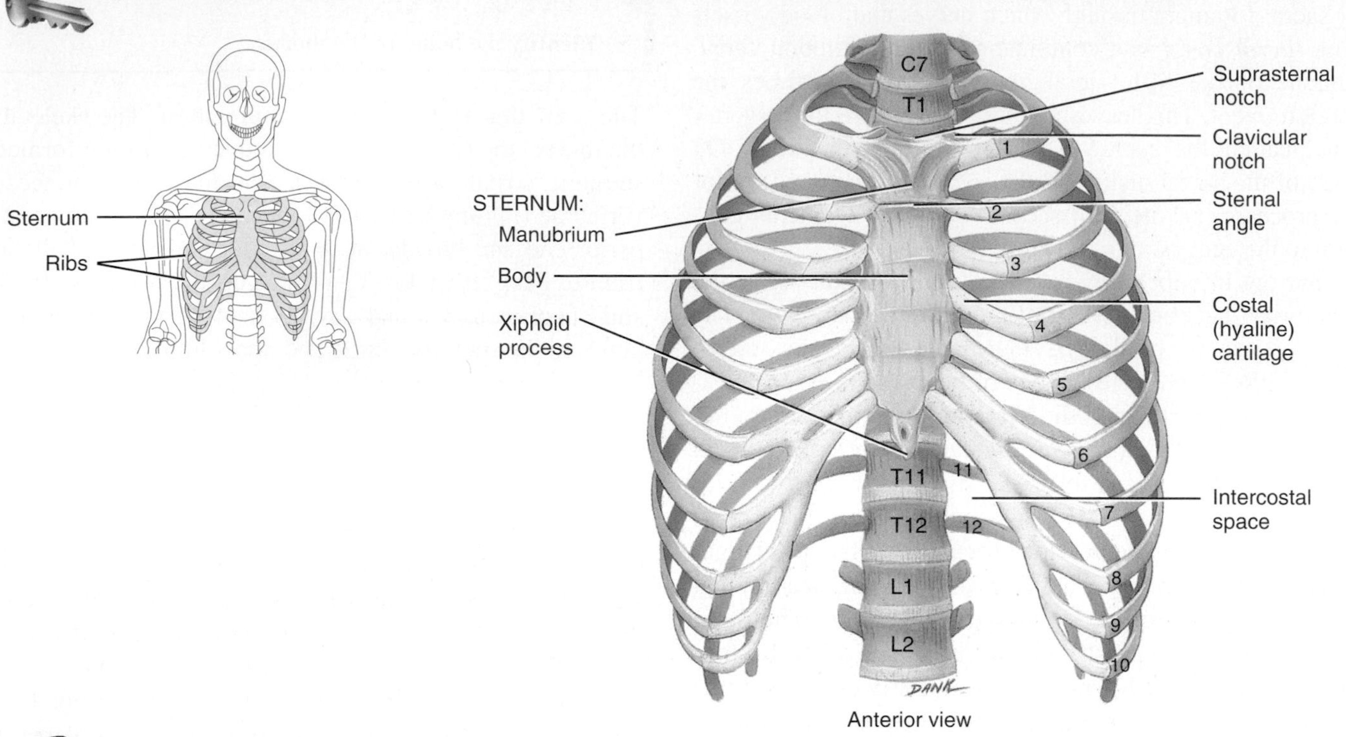

Which ribs are true ribs? Which are false ribs?

the eighth, ninth, and tenth pairs of ribs attach to each other and then to the cartilages of the seventh pair of ribs. These false ribs are called *vertebrochondral ribs.* The eleventh and twelfth false ribs are also known as *floating ribs* because the costal cartilage at their anterior ends does not attach to the sternum at all. Floating ribs attach only posteriorly to the thoracic vertebrae. Inflammation of one or more costal cartilages, called *costochondritis,* is characterized by local tenderness and pain in the anterior chest wall that may radiate. The symptoms mimic the chest pain associated with a heart attack (angina pectoris).

Figure 7.23a shows the parts of a typical (third through ninth) rib. The *head* is a projection at the posterior end of the rib. It consists of one or two *facets* that articulate with facets on the bodies of adjacent thoracic vertebrae to form *vertebrocostal joints.* The *neck* is a constricted portion just lateral to the head. A knoblike structure on the posterior surface where the neck joins the body is called a *tubercle* (TOO- ber-kul). The *nonarticular part* of the tubercle affords attachment to the ligament of the tubercle. The *articular part* of the tubercle articulates with the facet of a transverse process of the inferior of the two vertebrae to which the head of the rib is connected. These articulations also form vertebrocostal joints. The *body* is the main part of the rib. A short distance beyond the tubercle, an abrupt change in the curvature of the body occurs. This point is called the *costal angle.* The inner surface of the rib has a *costal groove* that protects blood vessels and a small nerve.

In summary, the posterior portion of the rib connects to a thoracic vertebra by its head and the articular part of a tubercle. The facet of the head fits into a facet on the body of one vertebra or into the demifacets of two adjoining vertebrae. The articular part of the tubercle articulates with the facet of the transverse process of the vertebra.

Intercostal muscles, blood vessels, and nerves occupy spaces between ribs, called *intercostal spaces.* Surgical access to the lungs or other structures in the thoracic cavity is commonly obtained through an intercostal space. Special rib retractors are used to create a wide separation between ribs. The costal cartilages are sufficiently elastic in younger persons to permit considerable bending without breaking.

Rib Fractures

Rib fractures are the most common chest injuries, and they usually result from direct blows, most often from impact with a steering wheel, falls, and crushing injuries to the chest. Ribs tend to break at the point where the greatest force is applied, but they may also break at their weakest point—the site of greatest curvature, which is just anterior to the costal angle. In some cases, fractured ribs may puncture the heart, great vessels of the heart, lungs, trachea, bronchi, esophagus, spleen, liver, and kidneys. Rib fractures are usually quite painful. ■

▶ CHECKPOINT

12. What are the functions of the bones of the thorax?
13. What are the parts of the sternum?
14. How are ribs classified based on their attachment to the sternum?

Figure 7.23 The structure of ribs. Each rib has a head, a neck, and a body. The facets and the articular part of the tubercle are where the rib articulates with a vertebra. (See Tortora, *A Photographic Atlas of the Human Body,* Figure 3.21.)

Each rib articulates posteriorly with its corresponding thoracic vertebra.

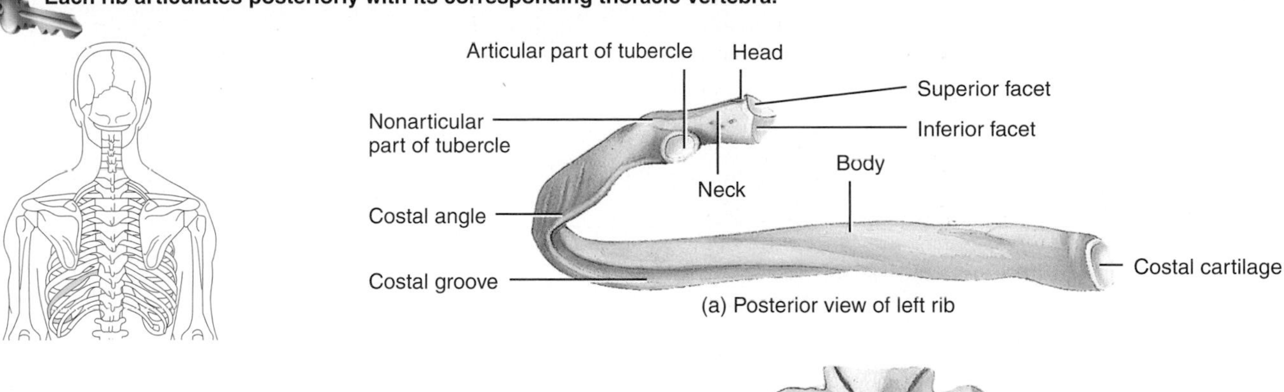

(a) Posterior view of left rib

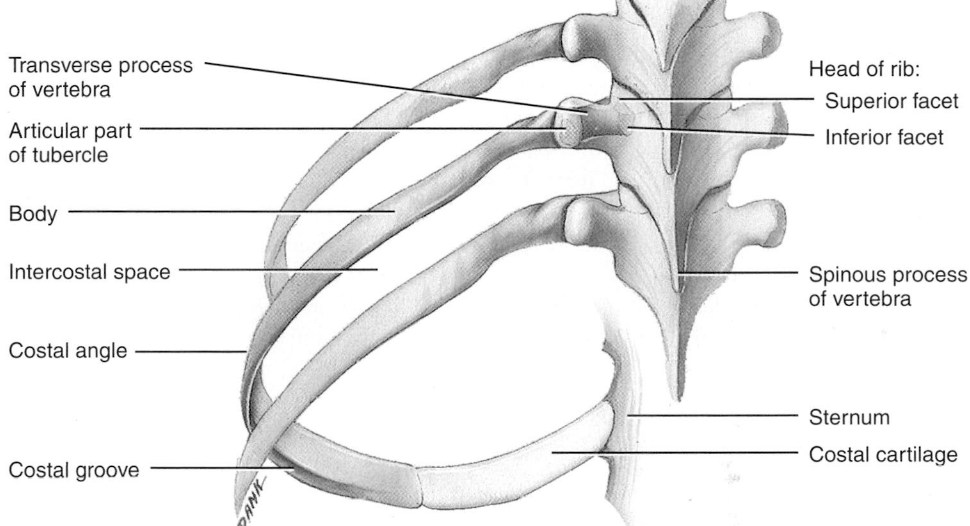

(b) Posterior view of left ribs articulated with thoracic vertebrae and sternum

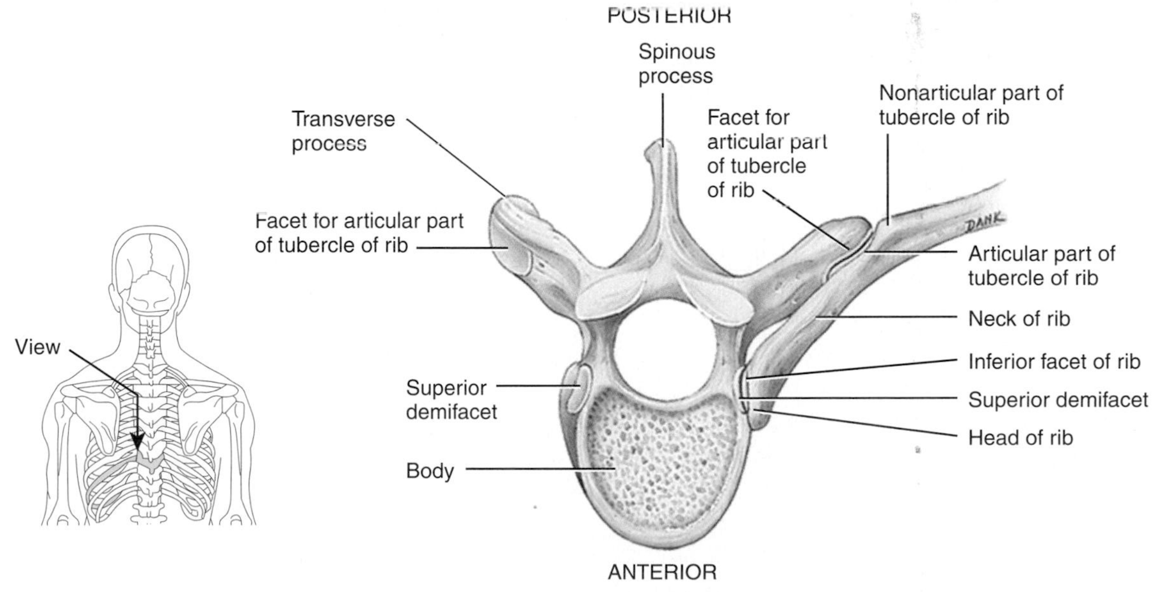

(c) Superior view of left rib articulated with thoracic vertebra

How does a rib articulate with a thoracic vertebra?

DISORDERS: HOMEOSTATIC IMBALANCES

Herniated (Slipped) Disc

In their function as shock absorbers, intervertebral discs are constantly compressed. If the anterior and posterior ligaments of the discs become injured or weakened, the pressure developed in the nucleus pulposus may be great enough to rupture the surrounding fibrocartilage (annulus fibrosus). Then, the nucleus pulposus may herniate (protrude) posteriorly or into one of the adjacent vertebral bodies. This condition is called a **herniated (slipped) disc.** It occurs most often in the lumbar region because that part of the vertebral column bears much of the weight of the body, and it is the region of the most flexing and bending.

Most often, the nucleus pulposus slips posteriorly toward the spinal cord and spinal nerves (Figure 7.24). This movement exerts pressure on the spinal nerves, causing acute pain. If the roots of the sciatic nerve, which passes from the spinal cord to the foot, are compressed, the pain radiates down the posterior thigh, through the calf, and occasionally into the foot. If pressure is exerted on the spinal cord itself, some of its neurons may be destroyed. Treatment of a herniated disc may involve a *laminectomy,* a procedure in which parts of the laminae of the vertebra and intervertebral disc are removed to relieve pressure on the nerves.

Figure 7.24 Herniated (slipped) disc.

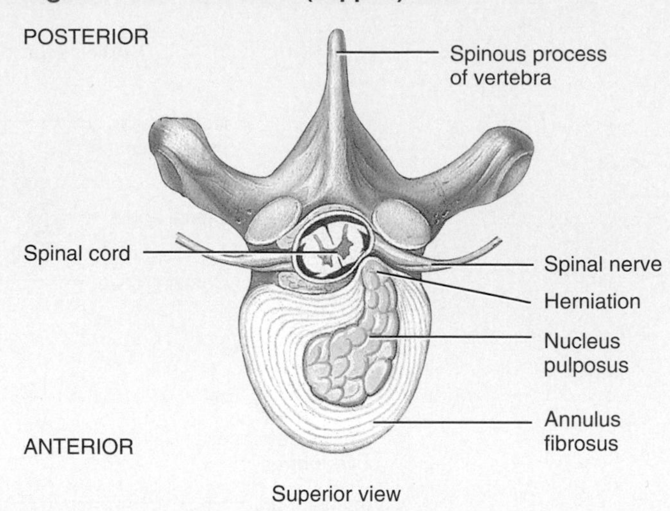

Spina Bifida

Spina bifida (SPĪ-na BIF-i-da) is a congenital defect of the vertebral column in which laminae fail to unite at the midline. In serious cases, the membranes (meninges) around the spinal cord or the spinal cord itself protrude through the opening and produce serious problems, such as partial or complete paralysis, partial or complete loss of urinary bladder control, and the absence of reflexes. An increased risk of spina bifida is associated with low levels of folic acid, one of the B vitamins, during pregnancy. Spina bifida may be diagnosed before birth by a test of the mother's blood for alphafetoprotein, a protein produced by the fetus; by sonography; or by amniocentesis (withdrawal of amniotic fluid for analysis).

MEDICAL TERMINOLOGY

Kyphosis (kī-FŌ-sis; *kyphos-* = hump; *-osis* = condition). An exaggeration of the thoracic curve of the vertebral column. In tuberculosis of the spine, vertebral bodies may partially collapse, causing an acute angular bending of the vertebral column. In the elderly, degeneration of the intervertebral discs leads to kyphosis. Rickets or poor posture may also cause kyphosis. It is also common in females with advanced osteoporosis. The term *round-shouldered* refers to mild kyphosis.

Lordosis (lor-DŌ-sis; *lord-* = bent backward), sometimes called *swayback.* An exaggeration of the lumbar curve of the vertebral column. It may result from increased weight of the abdomen as in pregnancy or extreme obesity, poor posture, rickets, or tuberculosis of the spine.

Lumbar spine stenosis (*sten-* = narrowed). Narrowing of the spinal canal in the lumbar part of the vertebral column, due to hypertrophy of surrounding bone or soft tissues. It may be caused by arthritic changes in the intervertebral discs and is a common cause of back and leg pain.

Scoliosis (skō-lē-Ō-sis; *scolio* = crooked). Lateral bending of the vertebral column, usually in the thoracic region. It may result from congenitally (present at birth) malformed vertebrae, chronic sciatica, paralysis of muscles on one side of the vertebral column, poor posture, or one leg being shorter than the other.

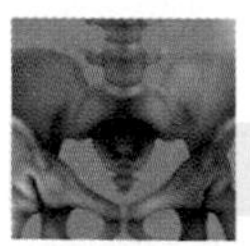

STUDY OUTLINE

INTRODUCTION (p. 186)

1. Bones, joints, and muscles working together form the musculoskeletal system.

DIVISIONS OF THE SKELETAL SYSTEM (p. 186)

1. The axial skeleton consists of bones arranged along the longitudinal axis. The parts of the axial skeleton are the skull, auditory os-

sicles (ear bones), hyoid bone, vertebral column, sternum, and ribs.
2. The appendicular skeleton consists of the bones of the pectoral (shoulder) girdles, bones of the upper limbs, pelvic (hip) girdles, and bones of the lower limbs.

TYPES OF BONES (p. 188)

1. On the basis of shape, bones are classified as long, short, flat, irregular, or sesamoid. Sesamoid bones develop in tendons or ligaments.
2. Sutural bones are found within the sutures of some cranial bones.

BONE SURFACE MARKINGS (p. 188)

1. Surface markings are structural features visible on the surfaces of bones.
2. Each marking—whether a depression, an opening, or a process—is structured for a specific function, such as joint formation, muscle attachment, or passage of nerves and blood vessels (Table 7.2 on page 189).

SKULL (p. 189)

1. The skull consists of 22 cranial and facial bones.
2. The eight cranial bones are the frontal, parietal (2), temporal (2), occipital, sphenoid, and ethmoid.
3. The 14 facial bones are the nasal (2), maxillae (2), zygomatic (2), lacrimal (2), palatine (2), inferior nasal conchae (2), vomer, and mandible.
4. The nasal septum consists of the vomer, perpendicular plate of the ethmoid, and septal cartilage. It divides the nasal cavity into left and right sides.
5. Seven skull bones form each of the orbits (eye sockets).
6. The foramina of the skull bones provide passages for nerves and blood vessels (Table 7.3 on page 201).
7. Sutures are immovable joints that connect most bones of the skull. Examples are the coronal, sagittal, lambdoid, and squamous sutures.
8. Paranasal sinuses are cavities in bones of the skull that communicate with the nasal cavity. They are lined by mucous membranes. The cranial bones containing paranasal sinuses are the frontal, sphenoid, ethmoid, and maxillae.
9. Fontanels are fibrous, connective tissue membrane-filled spaces between the cranial bones of fetuses and infants. The major fontanels are the anterior, posterior, anterolaterals, and posterolaterals. After birth, they fill in with bone and become sutures.

HYOID BONE (p. 202)

1. The hyoid bone is a U-shaped bone that does not articulate with any other bone.
2. It supports the tongue and provides attachment for some tongue muscles and for some muscles of the pharynx and neck.

VERTEBRAL COLUMN (p. 203)

1. The vertebral column, sternum, and ribs constitute the skeleton of the trunk of the body.
2. The 26 bones of the adult vertebral column are the cervical vertebrae (7), the thoracic vertebrae (12), the lumbar vertebrae (5), the sacrum (5 fused vertebrae), and the coccyx (usually 4 fused vertebrae).
3. The vertebral column contains normal curves (cervical, thoracic, lumbar, and sacral) that give strength, support, and balance.
4. The vertebrae are similar in structure, each usually consisting of a body, vertebral arch, and seven processes. Vertebrae in the different regions of the column vary in size, shape, and detail.

THORAX (p. 211)

1. The thoracic skeleton consists of the sternum, ribs, costal cartilages, and thoracic vertebrae.
2. The thoracic cage protects vital organs in the chest area and upper abdomen.

SELF-QUIZ QUESTIONS

Fill in the blanks in the following statements.

1. Membrane-filled spaces between cranial bones that enable the fetal skull to change size and shape during passage through the birth canal are called ________.
2. The ribs articulate posteriorly with the ________ vertebrae.
3. ________ separate adjacent vertebrae, from the second cervical vertebra to the sacrum.
4. The hypophyseal fossa of the sella turcica of the sphenoid bone contains the ________.

Indicate whether the following statements are true or false.

5. The axial skeleton consists of the bones that lie along the axis: skull bones, ear bones, hyoid bone, ribs, breastbone, and pelvic and pectoral girdles.
6. The first cervical vertebra is the axis, and the second is the atlas.

Choose the one best answer to the following questions.

7. Which of the following are true for the cranial and/or facial bones? (1) Their inner surfaces attach to membranes that stabilize positions of the brain, blood vessels, and nerves. (2) Their outer surfaces provide areas of attachment for muscles that move various parts of the head. (3) They protect and provide support for the entrances to the digestive, respiratory, and integumentary systems. (4) They provide attachment for the muscles of facial expression. (5) They protect and support the delicate special sense organs. (a) 1, 2, 4, and 5, (b) 2, 3, 4, and 5, (c) 1, 3, 4, and 5, (d) 1, 2, 3, and 5, (e) 1, 2, 3, 4, and 5.
8. In which of the following bones are paranasal sinuses *not* found? (a) frontal bone, (b) sphenoid bone, (c) lacrimal bones, (d) ethmoid bone, (e) maxillae.
9. Which of the following are functions of the vertebral curves? (1) protection of the heart and lungs, (2) increase in strength, (3) maintenance of balance in the upright position, (4) absorption of shock during walking, (5) protection of vertebrae against fracture. (a) 1, 2, 3, and 4, (b) 2, 3, 4, and 5, (c) 1, 3, 4, and 5, (d) 1, 2, 3, and 5, (e) 1, 2, 3, 4, and 5.

10. The partition that divides the nasal cavity into right and left sides in the (a) pterygoid process, (b) petrous portion of the temporal bone, (c) hypophyseal fossa, (d) hard palate, (e) nasal septum.

11. The suture that unites the two parietal bones is the (a) coronal suture, (b) lambdoid suture, (c) squamous suture, (d) sagittal suture, (e) frontal suture.

12. Match the following.

____ (a) supraorbital foramen
____ (b) temporomandibular joint
____ (c) external auditory meatus
____ (d) foramen magnum
____ (e) optic foramen
____ (f) cribriform plate
____ (g) palatine process
____ (h) ramus, body, and condylar process
____ (i) transverse foramen, bifid processes
____ (j) dens
____ (k) promontory
____ (l) costal cartilages
____ (m) xiphoid process

(1) temporal bone
(2) sphenoid bone
(3) cervical vertebrae
(4) ethmoid bone
(5) articulation of mandibular fossa and articular tubercle of the temporal bone to the mandible
(6) occipital bone
(7) frontal bone
(8) maxillae
(9) mandible
(10) axis
(11) sacrum
(12) sternum
(13) ribs

13. Match the following (answers may be used more than once).

____ (a) bones that have greater length than width and consist of a shaft and a variable number of extremities
____ (b) cube-shaped bones that are nearly equal in length and width
____ (c) bones that develop in certain tendons where there is considerable friction, tension, and physical stress
____ (d) small bones located within joints between certain cranial bones
____ (e) thin bones composed of two nearly parallel plates of compact bone enclosing a layer of spongy bone
____ (f) bones with complex shapes, including the vertebrae and some facial bones
____ (g) patellae are examples
____ (h) bones that provide considerable protection and extensive areas for muscle attachment
____ (i) include femur, tibia, fibula, humerus, ulna, and radius
____ (j) include cranial bones, sternum, and ribs
____ (k) include almost all of the carpal (wrist) and tarsal (ankle) bones

(1) irregular bones
(2) long bones
(3) short bones
(4) flat bones
(5) sesamoid bones
(6) sutural bones

14. Match the following.

____ (a) prominent ridge or elongated projection
____ (b) tubelike opening
____ (c) large round protuberance used as attachment point for tendons or ligaments
____ (d) smooth, flat articular surface
____ (e) sharp, slender projection
____ (f) hole for passage of blood vessels, nerves or ligaments
____ (g) large, rounded, rough projection

(1) foramen
(2) tuberosity
(3) spinous process
(4) crest
(5) facet
(6) condyle
(7) meatus

15. Match the following.

____ (a) forms the forehead
____ (b) form the inferior lateral aspects of the cranium and part of the cranial floor, contain zygomatic process and mastoid process
____ (c) forms part of the anterior portion of the cranial floor, medial wall of the orbits, superior portions of nasal septum, and most of the sidewalls of the nasal cavity
____ (d) form the prominence of the cheek and part of the lateral wall and floor of each orbit
____ (e) the largest, strongest facial bone; is the only movable skull bone
____ (f) a roughly triangular bone on the floor of the nasal cavity; one of the components of the nasal septum
____ (g) form greater portion of the sides and roof of the cranial cavity
____ (h) forms the posterior part and most of the base of the cranium; contains the foramen magnum
____ (i) called the keystone of the cranial floor; contains the sella turcica, optic foramen, and pterygoid processes
____ (j) form the bridge of the nose
____ (k) the smallest bones of the face; contain a vertical groove that houses a structure that gathers tears and passes them into the nasal cavity
____ (l) does not articulate with any other bone
____ (m) unite to form the upper jawbone and articulate with every bone of the face except the lower jawbone
____ (n) form the posterior part of the hard palate, part of the floor and lateral wall of the nasal cavity, and a small portion of the floors of the orbits
____ (o) scroll-like bones that form a part of the lateral walls of the nasal cavity; functions in the turbulent circulation and filtration of air

(1) temporal bones
(2) parietal bones
(3) frontal bone
(4) occipital bone
(5) sphenoid bone
(6) ethmoid bone
(7) nasal bones
(8) maxillae
(9) zygomatic bones
(10) lacrimal bones
(11) palatine bones
(12) vomer
(13) mandible
(14) inferior nasal conchae
(15) hyoid bone

CRITICAL THINKING QUESTIONS

1. While investigating her new baby brother, 4-year-old Latisha found a soft spot on the baby's skull and announced that the baby needed to go back because "it's not finished yet." Explain the presence of soft spots in the infant's skull.
   HINT ***An infant should be "soft in the head," even if an adult should not.***
2. Thirty-five-year old Barbara (old enough to know better) was bouncing down the stairs in her stocking feet when she slipped and landed hard on her buttocks. She felt a pain sharp enough to bring tears to her eyes, and she needed to sit very gingerly on the way to and at the emergency room. The attending physician said that she had broken a bone but that she wouldn't be given a cast. What bone do you think she broke?
   HINT ***Your buttocks usually keep this bone from being banged when you sit.***
3. The advertisement reads "New Posture Perfect Mattress! Keeps spine perfectly straight—just like when you were born! A straight spine equals a great sleep!" Would you buy a mattress from this company? Explain.
   HINT ***Take a sideways view of the vertebral column.***

ANSWERS TO FIGURE QUESTIONS

7.1 Axial skeleton: skull and vertebral column. Appendicular skeleton: clavicle, shoulder girdle, humerus, pelvic girdle, and femur.

7.2 Flat bones protect and provide a large surface area for muscle attachment.

7.3 The frontal, parietal, sphenoid, ethmoid, and temporal bones are cranial bones.

7.4 Squamous suture: parietal and temporal bones. Lambdoid suture: parietal and occipital bones. Coronal suture: parietal and frontal bones.

7.5 The temporal bone articulates with the parietal, sphenoid, zygomatic, and occipital bones.

7.6 The parietal bones form the posterior, lateral portion of the cranium.

7.7 The medulla oblongata of the brain connects with the spinal cord in the foramen magnum.

7.8 Crista galli of ethmoid bone, frontal, parietal, temporal, occipital, temporal, parietal, frontal, crista galli of ethmoid bone.

7.9 The perpendicular plate of the ethmoid bone forms the superior part of the nasal septum, and the lateral masses compose most of the medial walls of the orbits.

7.10 The mandible is the only movable skull bone, other than the auditory ossicles.

7.11 The nasal septum divides the nasal cavity into right and left sides.

7.12 Bones forming the orbit are the frontal, sphenoid, zygomatic, maxilla, lacrimal, ethmoid, and palatine.

7.13 The paranasal sinuses produce mucus and serve as resonating chambers for vocalization.

7.14 Four different skull bones border the anterolateral fontanel.

7.15 The hyoid bone does not articulate with any other bone.

7.16 The thoracic and sacral curves are concave.

7.17 The vertebral foramina enclose the spinal cord, whereas the intervertebral foramina provide spaces for spinal nerves to exit the vertebral column.

7.18 The atlas moving on the axis permits movement of the head to signify "no."

7.19 The facets and demifacets on the bodies of the thoracic vertebrae articulate with the facets on the head of the ribs, and the facets on the transverse processes of these vertebrae articulate with the tubercles of the ribs.

7.20 The lumbar vertebrae are stout because the amount of body weight supported by vertebrae increases toward the inferior end of the vertebral column.

7.21 There are four pairs of sacral foramina, for a total of eight. Each anterior sacral foramen joins a posterior sacral foramen at the intervertebral foramen. Nerves and blood vessels pass through these tunnels in the bone.

7.22 True ribs: pairs 1–7; false ribs: pairs 8–12.

7.23 The facet on the head of a rib fits into a facet on the body of a vertebra, and the articular part of the tubercle of a rib articulates with the facet of the transverse process of a vertebra.

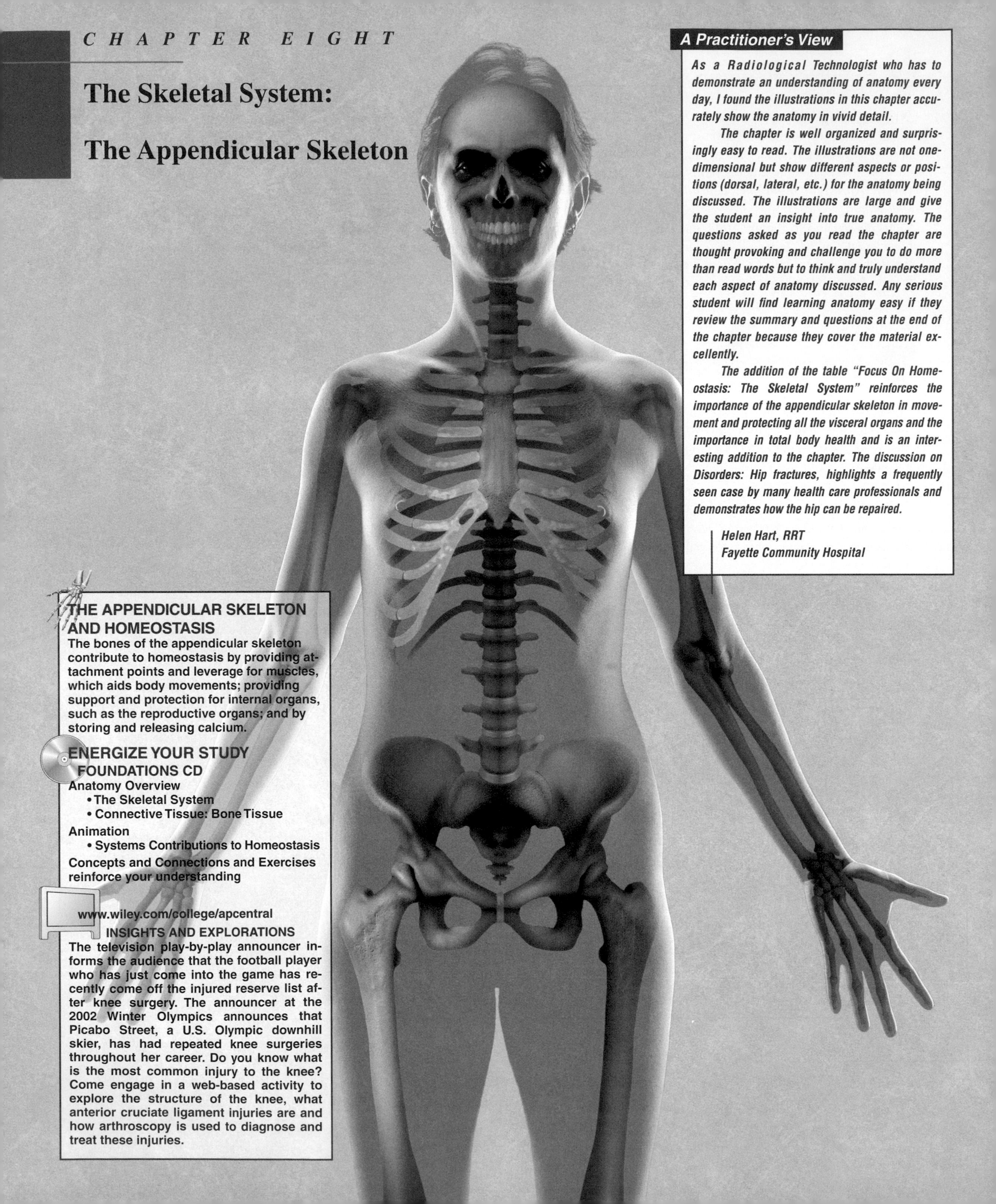

CHAPTER EIGHT

The Skeletal System:

The Appendicular Skeleton

A Practitioner's View

As a Radiological Technologist who has to demonstrate an understanding of anatomy every day, I found the illustrations in this chapter accurately show the anatomy in vivid detail.

The chapter is well organized and surprisingly easy to read. The illustrations are not one-dimensional but show different aspects or positions (dorsal, lateral, etc.) for the anatomy being discussed. The illustrations are large and give the student an insight into true anatomy. The questions asked as you read the chapter are thought provoking and challenge you to do more than read words but to think and truly understand each aspect of anatomy discussed. Any serious student will find learning anatomy easy if they review the summary and questions at the end of the chapter because they cover the material excellently.

The addition of the table "Focus On Homeostasis: The Skeletal System" reinforces the importance of the appendicular skeleton in movement and protecting all the visceral organs and the importance in total body health and is an interesting addition to the chapter. The discussion on Disorders: Hip fractures, highlights a frequently seen case by many health care professionals and demonstrates how the hip can be repaired.

Helen Hart, RRT
Fayette Community Hospital

THE APPENDICULAR SKELETON AND HOMEOSTASIS

The bones of the appendicular skeleton contribute to homeostasis by providing attachment points and leverage for muscles, which aids body movements; providing support and protection for internal organs, such as the reproductive organs; and by storing and releasing calcium.

ENERGIZE YOUR STUDY FOUNDATIONS CD

Anatomy Overview
- The Skeletal System
- Connective Tissue: Bone Tissue

Animation
- Systems Contributions to Homeostasis

Concepts and Connections and Exercises reinforce your understanding

www.wiley.com/college/apcentral

INSIGHTS AND EXPLORATIONS

The television play-by-play announcer informs the audience that the football player who has just come into the game has recently come off the injured reserve list after knee surgery. The announcer at the 2002 Winter Olympics announces that Picabo Street, a U.S. Olympic downhill skier, has had repeated knee surgeries throughout her career. Do you know what is the most common injury to the knee? Come engage in a web-based activity to explore the structure of the knee, what anterior cruciate ligament injuries are and how arthroscopy is used to diagnose and treat these injuries.

The two main divisions of the skeletal system are the axial skeleton and the appendicular skeleton. Whereas the axial skeleton serves mainly to protect and support internal organs, the appendicular skeleton, the focus of this chapter, mainly facilitates movement. The appendicular skeleton includes bones that make up the upper and lower limbs as well as bones, called girdles, that attach the limbs to the axial skeleton. As you study this chapter, you will see how the bones of the appendicular skeleton act with each other and with skeletal muscles to bring about a variety of movements.

PECTORAL (SHOULDER) GIRDLE

OBJECTIVE

- **Identify the bones of the pectoral (shoulder) girdle and their principal markings.**

The **pectoral** (PEK-tō-ral) or **shoulder girdles** attach the bones of the upper limbs to the axial skeleton (Figure 8.1). Each of the two pectoral girdles consists of two bones, the *clavicle* on the anterior side and the *scapula* on the posterior side. The clavicle articulates with the manubrium of the sternum at the *sternoclavicular joint.* The scapula articulates with the clavicle at the *acromioclavicular joint* and with the humerus at the *glenohumeral (shoulder) joint.* The pectoral girdles do not articulate with the vertebral column. They are held in position by complex muscle attachments.

Clavicle

Each slender, S-shaped **clavicle** (KLAV-i-kul = key) or *collarbone* lies horizontally in the superior and anterior part of the thorax superior to the first rib (Figure 8.2). The medial half of the clavicle is convex anteriorly, whereas the lateral half is concave anteriorly. The medial end of the clavicle, the *sternal extremity,* is rounded and articulates with the manubrium of the sternum to form the *sternoclavicular joint.* The broad, flat, lateral end, the

Figure 8.1 Right pectoral (shoulder) girdle.

The clavicle is the anterior bone of the pectoral girdle, and the scapula is the posterior bone.

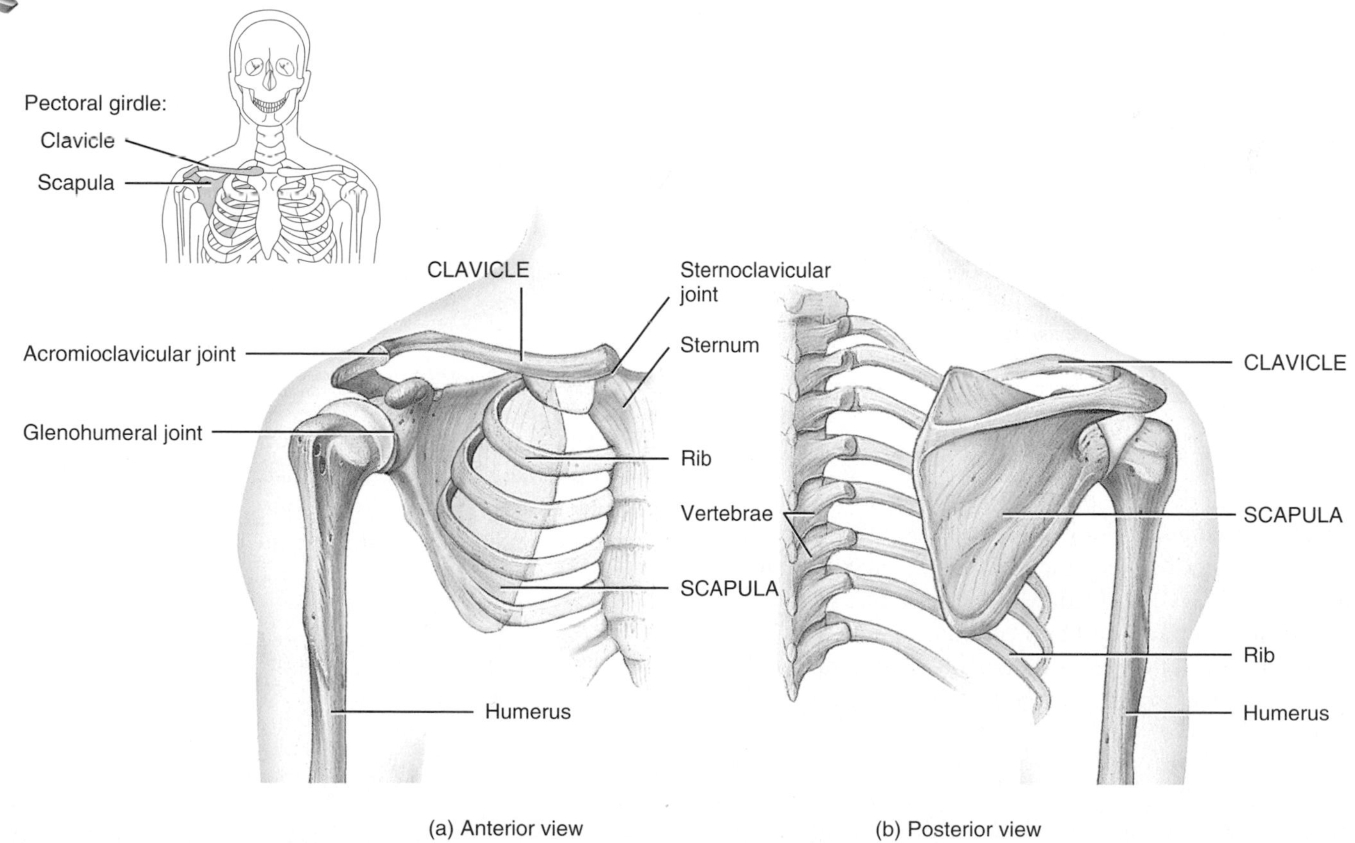

What is the function of the pectoral girdles?

Figure 8.2 Right clavicle.

The clavicle articulates medially with the manubrium of the sternum and laterally with the acromion of the scapula.

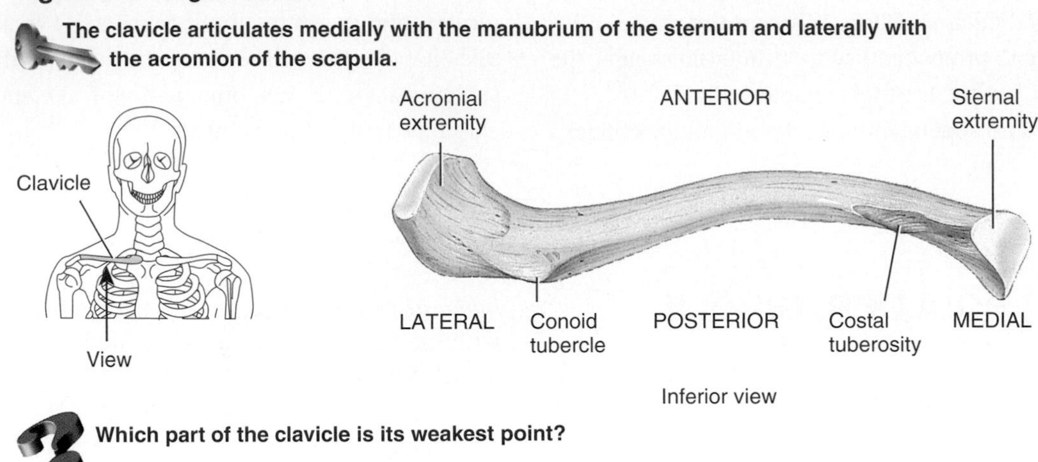

Which part of the clavicle is its weakest point?

acromial extremity (a-KRŌ-mē-al), articulates with the acromion of the scapula. This joint is called the *acromioclavicular joint.* (See Figure 8.1.) The *conoid tubercle* (KŌ-noyd = cone-like) on the inferior surface of the lateral end of the bone is a point of attachment for the conoid ligament. The *costal tuberosity* on the inferior surface of the medial end is a point of attachment for the costoclavicular ligament.

Fractured Clavicle

The clavicle transmits mechanical force from the upper limb to the trunk. If the force transmitted to the clavicle is excessive, as in falling on one's outstretched arm, a **fractured clavicle** may result. The clavicle is one of the most frequently broken bones in the body. Because the junction of the clavicle's two curves is its weakest point, the clavicular midregion is the most frequent fracture site. As a result of automobile accidents involving people who wear shoulder harness seatbelts, compression of the clavicle, even in the absence of fracture, often causes damage to the median nerve, which lies between the clavicle and the second rib. ■

Scapula

Each **scapula** (SCAP-ū-la; plural is *scapulae*), or *shoulder blade,* is a large, triangular, flat bone situated in the superior part of the posterior thorax between the levels of the second and seventh ribs (Figure 8.3). The medial borders of the scapulae lie about 5 cm (2 in.) from the vertebral column.

A sharp ridge, the *spine,* runs diagonally across the posterior surface of the flattened, triangular *body* of the scapula. The lateral end of the spine projects as a flattened, expanded process called the *acromion* (a-KRŌ-mē-on; *acrom-* = topmost), easily felt as the high point of the shoulder. Tailors measure the length of the upper limb from the acromion. The acromion articulates with the acromial extremity of the clavicle to form the *acromioclavicular joint.* Inferior to the acromion is a shallow depression called the *glenoid cavity.* This cavity accepts the head of the humerus (arm bone) to form the *glenohumeral joint* (see Figure 8.1).

The thin edge of the bone near the vertebral column is the *medial (vertebral) border.* The thick edge closer to the arm is the *lateral (axillary) border.* The medial and lateral borders join at the *inferior angle.* The superior edge of the scapula, called the *superior border,* joins the vertebral border at the *superior angle.* The *scapular notch* is a prominent indentation along the superior border through which the suprascapular nerve passes.

At the lateral end of the superior border of the scapula is a projection of the anterior surface called the *coracoid process* (KOR-a-koyd = like a crow's beak), to which the tendons of muscles attach. Superior and inferior to the spine are two fossae: the *supraspinous fossa* (sū-pra-SPĪ-nus) and the *infraspinous fossa* (in-fra-SPĪ-nus), respectively. Both serve as surfaces of attachment for the tendons of shoulder muscles called the supraspinatus and infraspinatus muscles. On the anterior surface is a slightly hollowed-out area called the *subscapular fossa,* also a surface of attachment for the tendons of shoulder muscles.

► CHECKPOINT

1. Which bones or parts of bones of the pectoral girdle form the sternoclavicular, acromioclavicular, and glenohumeral joints?

Figure 8.3 Right scapula (shoulder blade). (See Tortora, *A Photographic Atlas of the Human Body,* Figure 3.22.)

The glenoid cavity of the scapula articulates with the head of the humerus to form the shoulder joint.

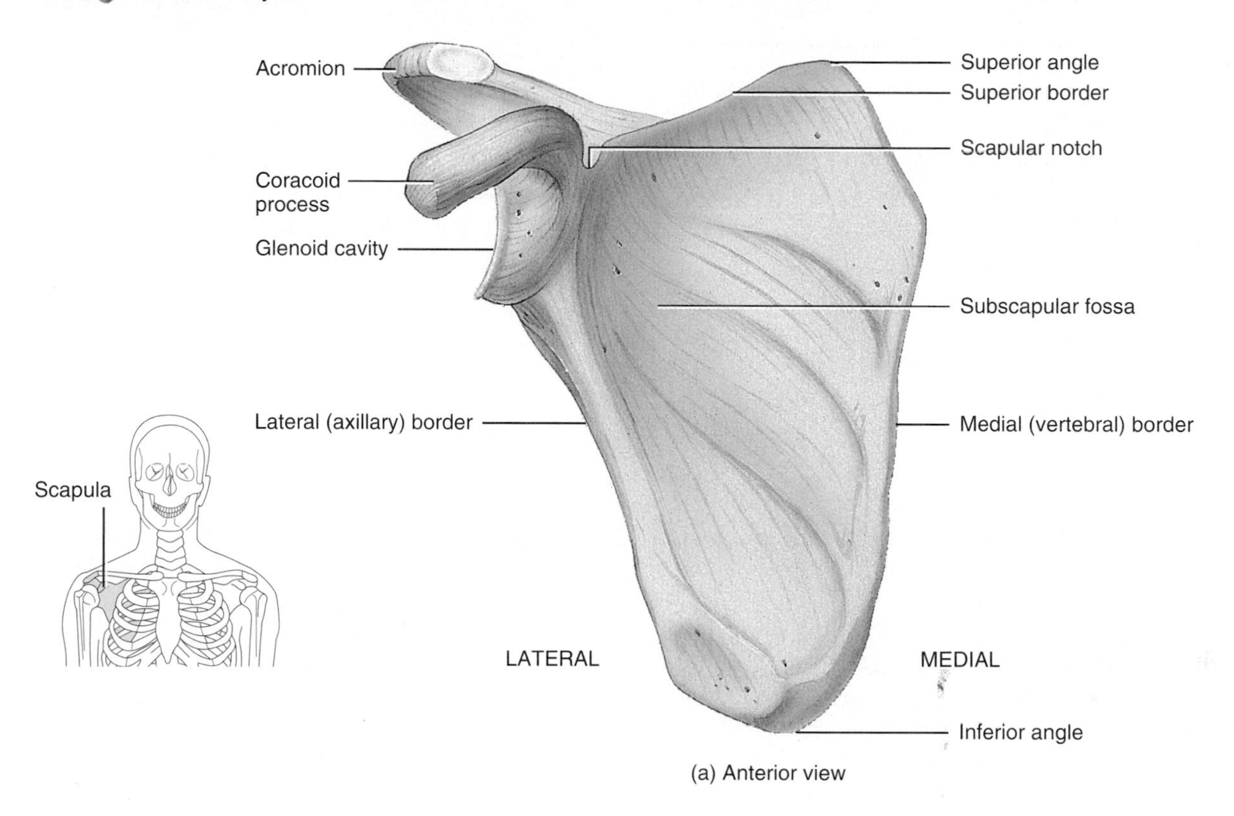

(a) Anterior view

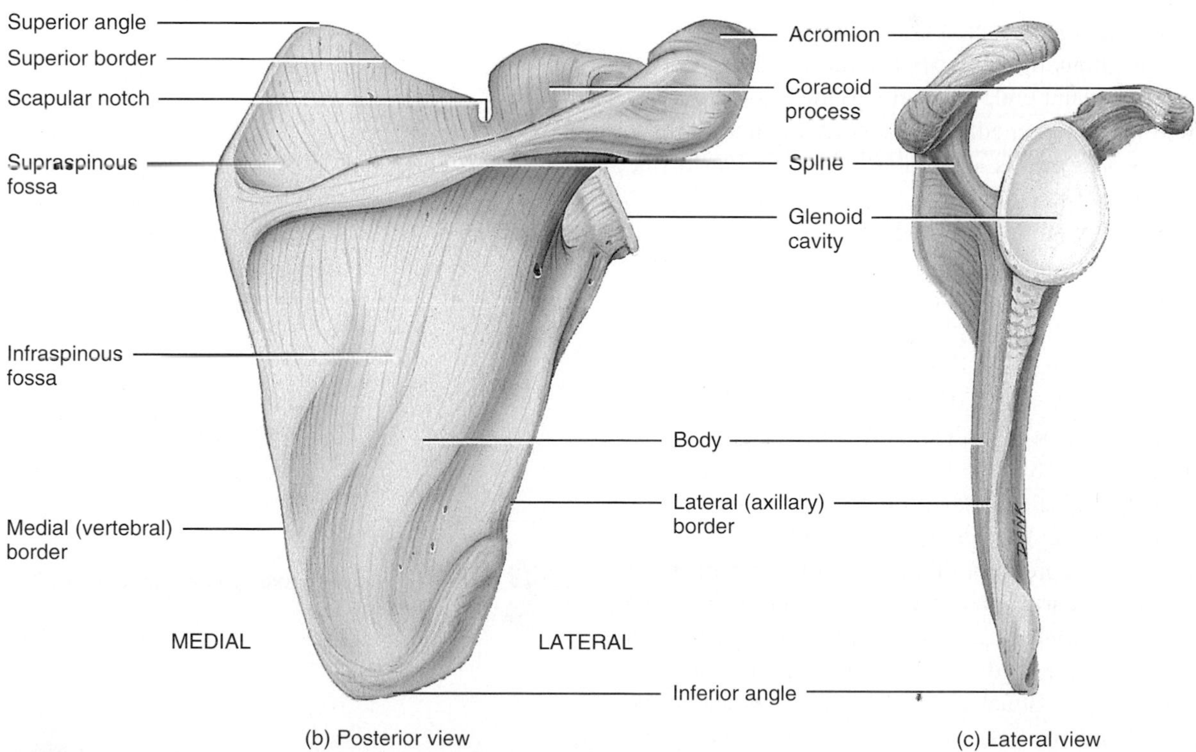

(b) Posterior view

(c) Lateral view

Which part of the scapula forms the high point of the shoulder?

UPPER LIMB

OBJECTIVE

- **Identify the bones of the upper limb and their principal markings.**

Each **upper limb (upper extremity)** has 30 bones in three locations—(1) the humerus in the arm; (2) the ulna and radius in the forearm; and (3) the 8 carpals in the carpus (wrist), the 5 metacarpals in the metacarpus (palm), and the 14 phalanges (bones of the digits) in the hand (Figure 8.4).

Humerus

The **humerus** (HŪ-mer-us), or arm bone, is the longest and largest bone of the upper limb (Figure 8.5). It articulates proximally with the scapula and distally at the elbow with both the ulna and the radius.

The proximal end of the humerus features a rounded *head* that articulates with the glenoid cavity of the scapula to form the *glenohumeral joint.* Distal to the head is the *anatomical neck,* the site of the epiphyseal line, which is visible as an oblique groove. The *greater tubercle* is a lateral projection distal to the anatomical neck. It is the most laterally palpable bony landmark of the shoulder region. The *lesser tubercle* projects anteriorly. Between both tubercles runs an *intertubercular sulcus.* The *surgical neck* is a constriction in the humerus just distal to the tubercles, where the head tapers to the shaft; it is so named because fractures often occur here.

The *body (shaft)* of the humerus is roughly cylindrical at its proximal end, but it gradually becomes triangular until it is flattened and broad at its distal end. Laterally, at the middle portion of the shaft, there is a roughened, V-shaped area called the *deltoid tuberosity.* This area serves as a point of attachment for the tendons of the deltoid muscle.

Several prominent features are evident at the distal end of the humerus. The *capitulum* (ka-PIT-ū-lum; *capit-* = head) is a rounded knob on the lateral aspect of the bone that articulates with the head of the radius. The *radial fossa* is an anterior depression that receives the head of the radius when the forearm is flexed (bent). The *trochlea* (TRŌK-lē-a), located medial to the capitulum, is a spool-shaped surface that articulates with the ulna. The *coronoid fossa* (KŌR-o-noyd = crown-shaped) is an anterior depression that receives the coronoid process of the ulna when the forearm is flexed. The *olecranon fossa* (ō-LEK-ra-non = elbow) is a posterior depression that receives the olecranon of the ulna when the forearm is extended (straightened). The *medial epicondyle* and *lateral epicondyle* are rough projections on either side of the distal end to which the tendons of most muscles of the forearm are attached. The ulnar nerve lies on the posterior surface of the medial epicondyle and may easily be palpated by rolling a finger over the skin surface above the medial epicondyle.

Figure 8.4 Right upper limb.

Each upper limb includes a humerus, ulna, radius, carpals, metacarpals, and phalanges.

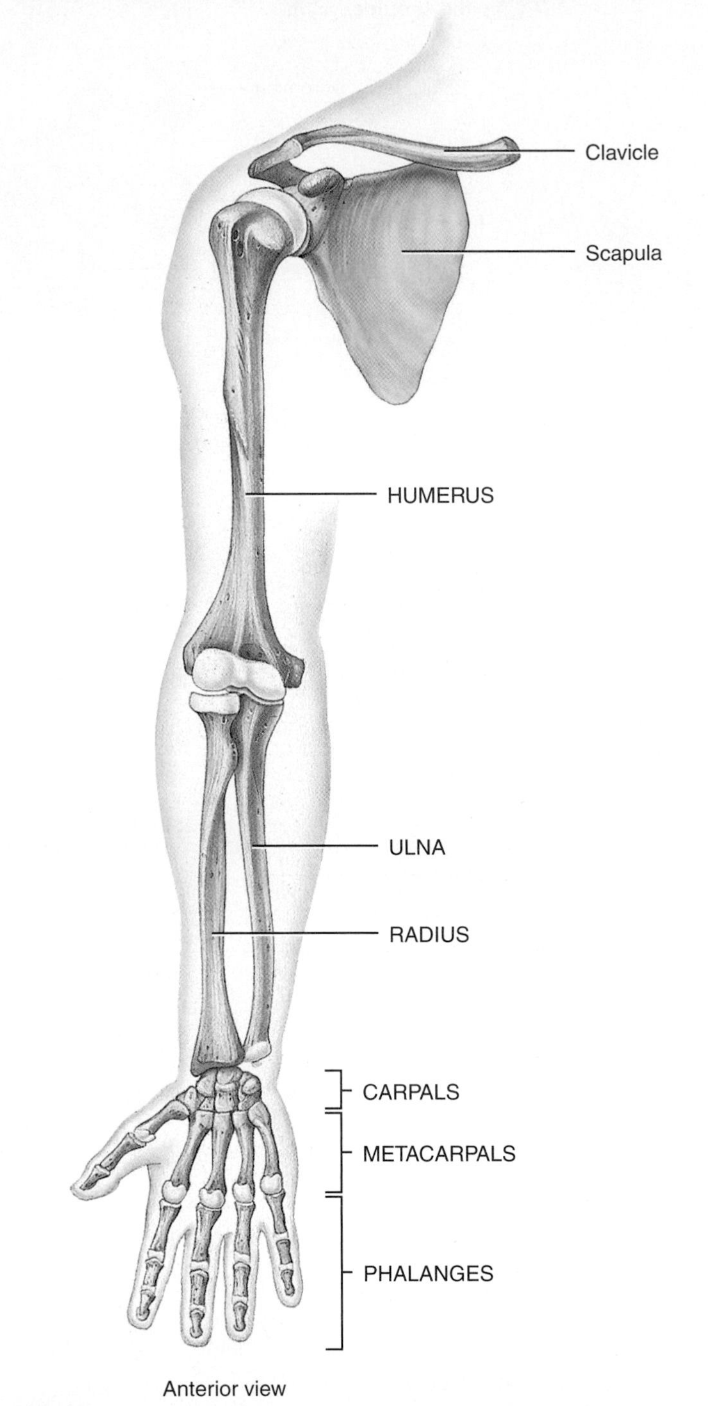

How many bones make up each upper limb?

Figure 8.5 Right humerus in relation to the scapula, ulna, and radius. (See Tortora, *A Photographic Atlas of the Human Body,* Figure 3.23.)

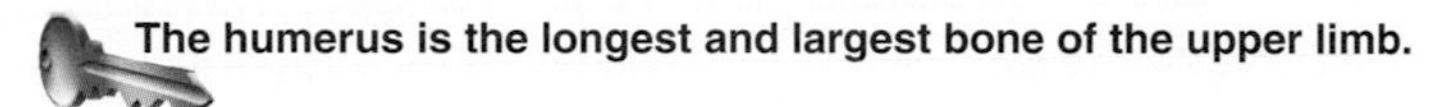
The humerus is the longest and largest bone of the upper limb.

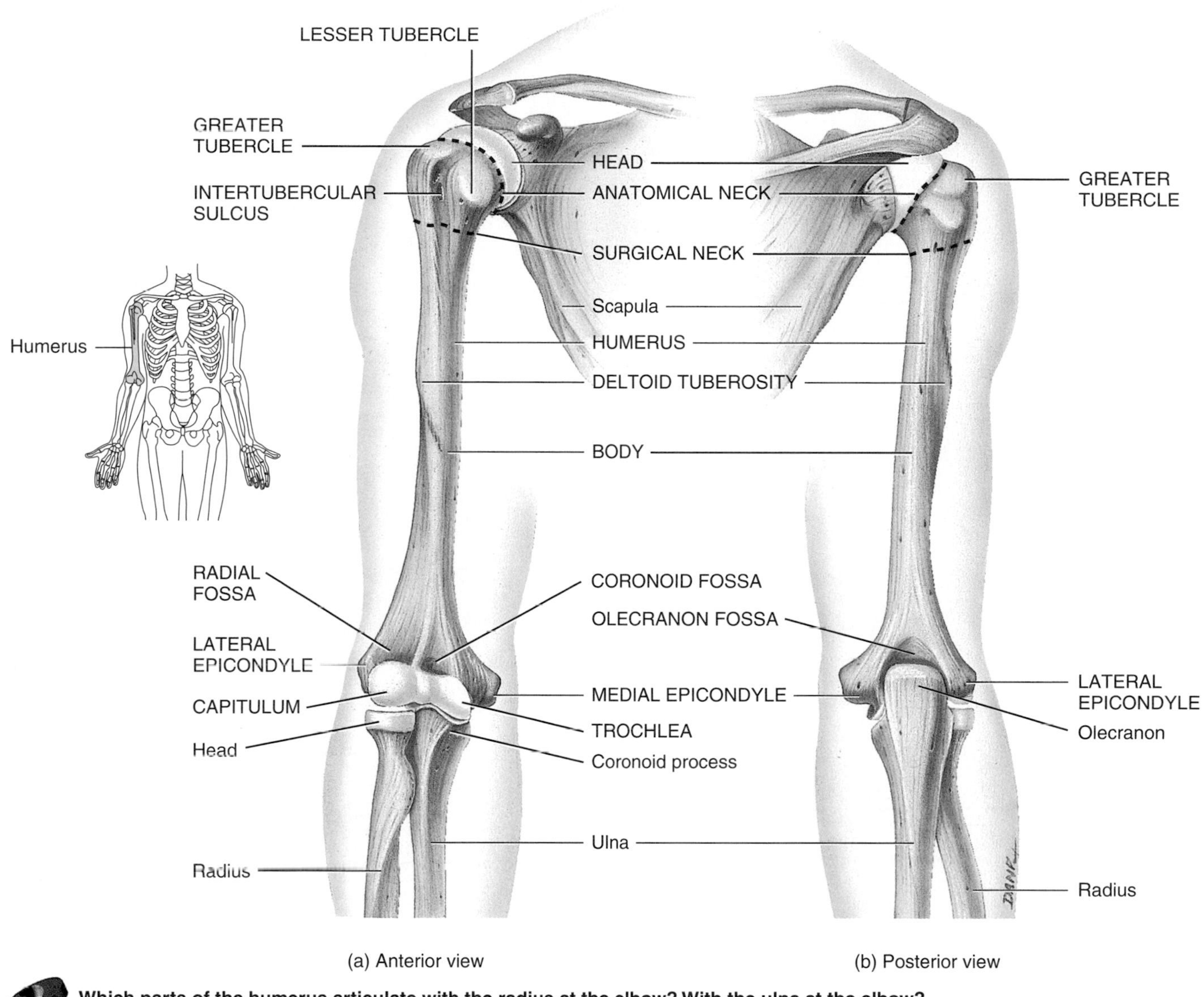

(a) Anterior view

(b) Posterior view

Which parts of the humerus articulate with the radius at the elbow? With the ulna at the elbow?

Ulna and Radius

The **ulna** is located on the medial aspect (the little-finger side) of the forearm and is longer than the radius (Figure 8.6). At the proximal end of the ulna (Figure 8.6b) is the *olecranon,* which forms the prominence of the elbow. The *coronoid process* (Figure 8.6a) is an anterior projection that, together with the olecranon, receives the trochlea of the humerus. The *trochlear notch* is a large curved area between the olecranon and coronoid process that forms part of the elbow joint (see Figure 8.7b). Just inferior to the coronoid process is the *ulnar tuberosity.* The distal end of the ulna consists of a *head* that is separated from the wrist by a fibrocartilage disc. A *styloid process* is on the posterior side of the distal end.

The **radius** is located on the lateral aspect (thumb side) of the forearm (see Figure 8.6). The proximal end of the radius has a disc-shaped *head* that articulates with the capitulum of the humerus and the radial notch of the ulna. Inferior to the head is the constricted *neck.* A roughened area inferior to the neck on the medial side, called the *radial tuberosity,* is a point of attachment for the tendons of the biceps brachii muscle. The shaft of the radius widens distally to form a *styloid process* on the lateral side. Fracture of the distal end of the radius is the most common fracture in adults older than 50 years.

Figure 8.6 Right ulna and radius in relation to the humerus and carpals. (See Tortora, *A Photographic Atlas of the Human Body,* Figure 3.24.)

In the forearm, the longer ulna is on the medial side, whereas the shorter radius is on the lateral side.

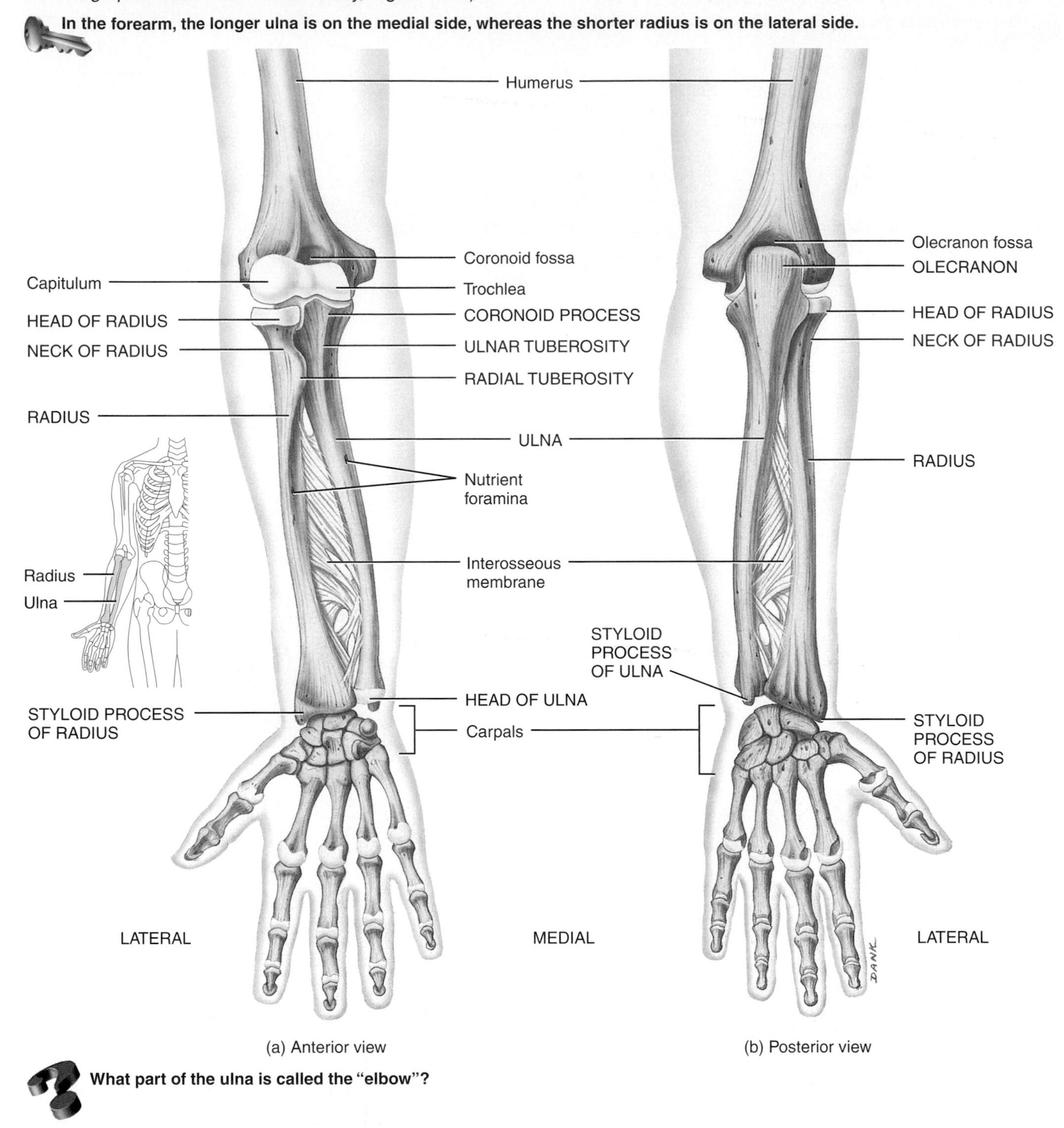

What part of the ulna is called the "elbow"?

The ulna and radius articulate with the humerus at the *elbow joint.* The articulation occurs in two places: where the head of the radius articulates with the capitulum of the humerus (Figure 8.7a), and where the trochlear notch of the ulna receives the trochlea of the humerus (Figure 8.7b).

The ulna and the radius connect with one another at three sites. First, a broad, flat, fibrous connective tissue called the *interosseous membrane* (in′-ter-OS-ē-us; *inter-* = between, *osse-* = bone) joins the shafts of the two bones. This membrane also provides a site of attachment for some tendons of deep

Figure 8.7 Articulations formed by the ulna and radius. (a) Elbow joint. (b) Joint surfaces at proximal end of the ulna. (c) Joint surfaces at distal ends of radius and ulna. The interosseous membrane also connects the ulna and the radius.

The elbow joint is formed by two articulations: (1) the trochlear notch of the ulna with the trochlea of the humerus and (2) the head of the radius with the capitulum of the humerus.

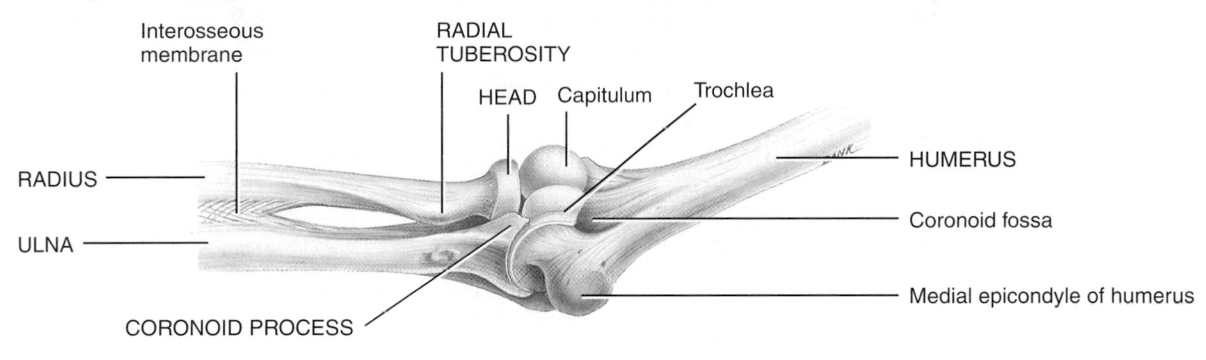

(a) Medial view in relation to humerus

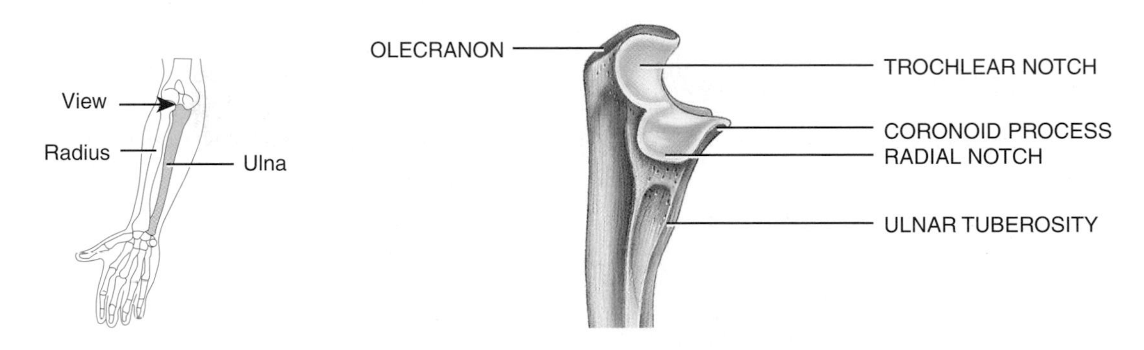

(b) Lateral view of proximal end of ulna

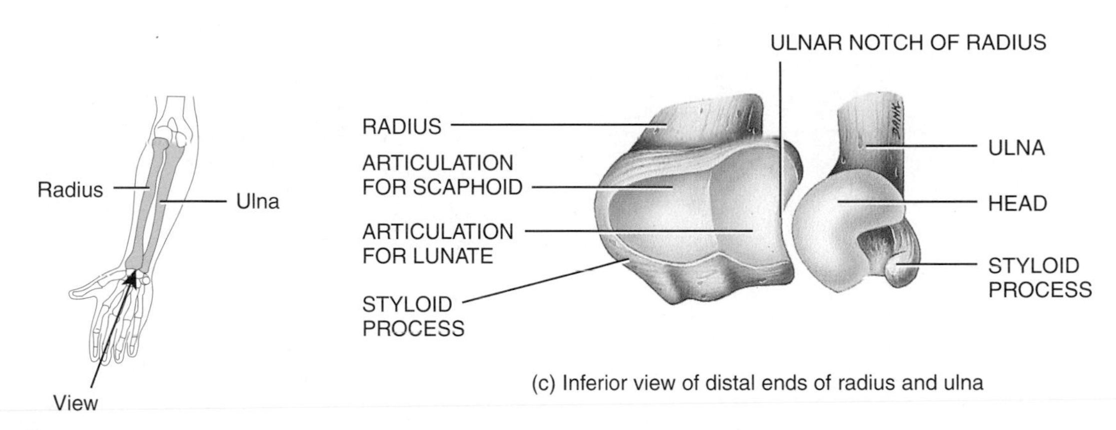

(c) Inferior view of distal ends of radius and ulna

How many joints are formed between the radius and ulna? What are they called?

skeletal muscles of the forearm. The ulna and radius also articulate at their proximal and distal ends. Proximally, the head of the radius articulates with the ulna's *radial notch,* a depression that is lateral and inferior to the trochlear notch (Figure 8.7b). This articulation is the *proximal radioulnar joint.* Distally, the head of the ulna articulates with the *ulnar notch* of the radius (Figure 8.7c). This articulation is the *distal radioulnar joint.* Finally, the distal end of the radius articulates with three bones of the wrist—the lunate, the scaphoid, and the triquetrum—to form the *radiocarpal (wrist) joint.*

Carpals, Metacarpals, and Phalanges

The **carpus** (wrist) is the proximal part of the hand and includes eight small bones, the **carpal bones (carpals),** joined to one another by ligaments (Figure 8.8). Articulations between carpal bones are called *intercarpal joints.* The carpals are arranged in two transverse rows of four bones each. Their names reflect their shapes. The carpals in the proximal row, from lateral to medial, are the **scaphoid** (SKAF-oyd = boatlike), **lunate** (LOO-nāt = moon-shaped), **triquetrum** (trī-KWĒ-trum = three-cornered), and **pisiform** (PIS-i-form = pea-shaped). The carpals in the distal row, from lateral to medial, are the **trapezium** (tra-PĒ-zē-um = four-sided figure with no two sides parallel), **trapezoid** (TRAP-e-zoyd), **capitate** (KAP-i-tāt = head-shaped), and **hamate** (HAM-āt = hooked). The capitate is the largest carpal bone; its rounded projection, the head, articulates with the lunate. The hamate is named for a large hook-shaped projection on its anterior surface. In about 70% of carpal fractures, only the scaphoid is broken. This is because the force of a fall on an outstretched hand is transmitted from the capitate through the scaphoid to the radius.

The concave space formed by the pisiform and hamate (on the ulnar side), and the scaphoid and trapezium (on the radial side), plus the *flexor retinaculum* (deep fascia) is the **carpal tunnel.** The long flexor tendons of the fingers and thumb and the median nerve pass through the carpal tunnel. Narrowing of the carpal tunnel may give rise to a condition called carpal tunnel syndrome (see page 356).

The **metacarpus** (*meta-* = beyond), or palm, is the intermediate region of the hand and consists of five bones called **metacarpal bones (metacarpals).** Each metacarpal bone con-

Figure 8.8 Right wrist and hand in relation to the ulna and radius. (See Tortora, *A Photographic Atlas of the Human Body,* Figure 3.25.)

The skeleton of the hand consists of the proximal carpals, the intermediate metacarpals, and the distal phalanges.

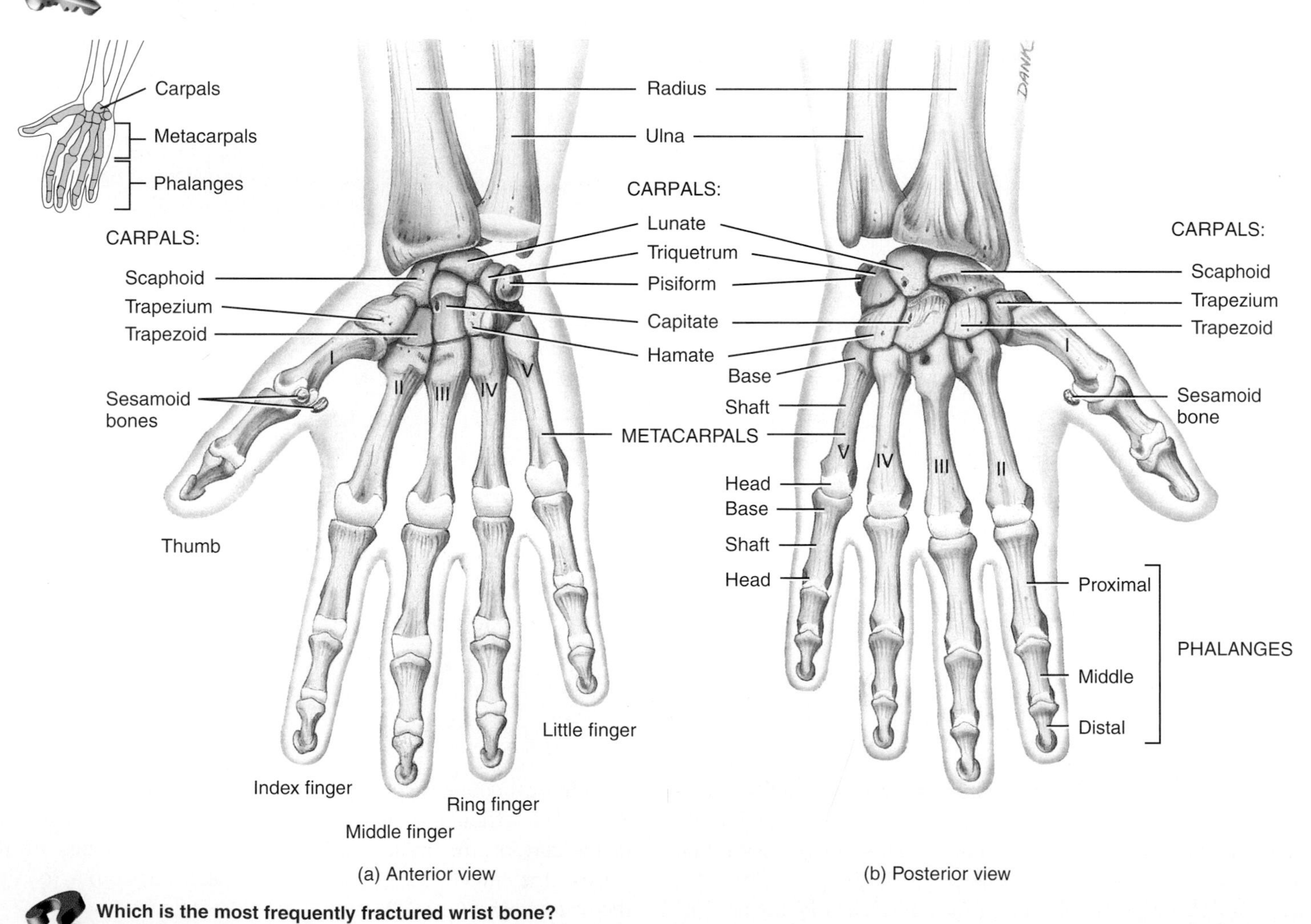

(a) Anterior view

(b) Posterior view

Which is the most frequently fractured wrist bone?

sists of a proximal *base,* an intermediate *shaft,* and a distal *head* (Figure 8.8b). The metacarpals are numbered I to V (or 1–5), starting with the thumb, which is lateral. The bases articulate with the distal row of carpal bones to form the *carpometacarpal joints.* The heads articulate with the proximal phalanges to form the *metacarpophalangeal joints.* The heads of the metacarpals are commonly called "knuckles" and are readily visible in a clenched fist.

The **phalanges** (fa-LAN-jēz; *phalan-* = a battle line), or bones of the digits, make up the distal part of the hand. There are 14 phalanges in each hand and, like the metacarpals, the phalanges are numbered I to V (or 1–5), beginning with the thumb, which is lateral. A single bone of a digit is referred to as a **phalanx** (FĀ-lanks). Each phalanx consists of a proximal *base,* an intermediate *shaft,* and a distal *head.* There are two phalanges in the thumb (*pollex*), and three phalanges in each of the other four digits. In order from the thumb, the other four digits are commonly called the index finger, middle finger, ring finger, and little finger. The first row of phalanges, the *proximal row,* articulates with the metacarpal bones and second row of phalanges. The second row of phalanges, the *middle row,* articulates with the proximal row and the third row. The third row of phalanges, the *distal row,* articulates with the middle row. The thumb has no middle phalanx. Joints between phalanges are called *interphalangeal joints.*

► CHECKPOINT

2. Which bones form the upper limb, from proximal to distal?
3. What joints do the bones of the upper limb form?

PELVIC (HIP) GIRDLE

► OBJECTIVE

- **Identify the bones of the pelvic girdle and their principal markings.**

The **pelvic (hip) girdle** consists of the two **hip bones,** also called **coxal bones** (KOK-sal; *cox-* = hip) (Figure 8.9). The hip bones unite anteriorly at a joint called the **pubic symphysis** (PŪ-bik SIM-fi-sis). They unite posteriorly with the sacrum at the *sacroiliac joints.* The complete ring composed of the hip bones, pubic symphysis, and sacrum forms a deep, basinlike structure called the **bony pelvis** (*pelv-* = basin). The plural is *pelves* (PEL-vēz) or *pelvises.* Functionally, the bony pelvis provides a strong and stable support for the vertebral column and pelvic organs. The pelvic girdle of the bony pelvis also accepts the bones of the lower limbs, connecting them to the axial skeleton.

Each of the two hip bones of a newborn consists of three bones separated by cartilage: a superior *ilium,* an inferior and anterior *pubis,* and an inferior and posterior *ischium* (IS-kē-um). Eventually, the three separate bones fuse together and become one bone (Figure 8.10a). Although the hip bones function as single bones, anatomists commonly discuss them as if they still consisted of three bones.

Ilium

The **ilium** (IL-ē-um = flank) is the largest of the three components of the hip bone (Figure 8.10b, c). A superior *ala* (= wing)

Figure 8.9 Bony pelvis. Shown here is the female bony pelvis. (See Tortora, *A Photographic Atlas of the Human Body,* Figure 3.27.)

The hip bones unite anteriorly at the pubic symphysis and posteriorly at the sacrum to form the bony pelvis.

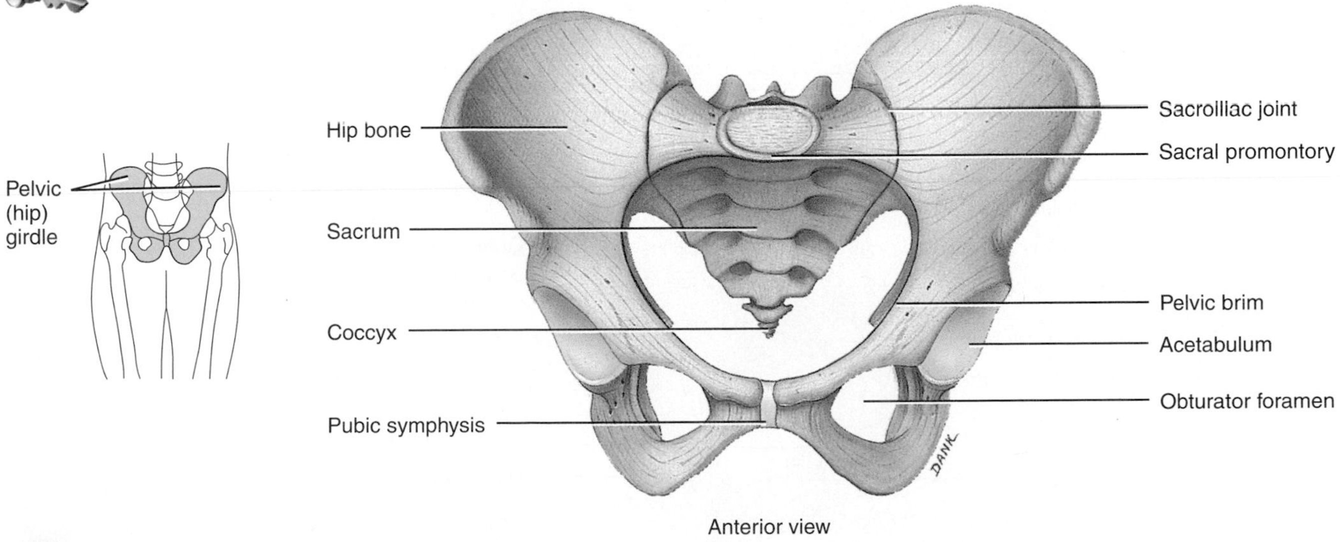

Pelvic (hip) girdle
Hip bone
Sacrum
Coccyx
Pubic symphysis
Sacroiliac joint
Sacral promontory
Pelvic brim
Acetabulum
Obturator foramen
Anterior view

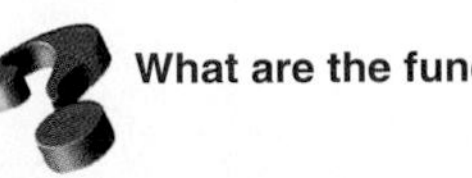

What are the functions of the bony pelvis?

Figure 8.10 Right hip bone. The lines of fusion of the ilium, ischium, and pubis depicted in (a) are not always visible in an adult. (See Tortora, *A Photographic Atlas of the Human Body,* Figure 3.26.)

The acetabulum is the socket formed where the three parts of the hip bone converge.

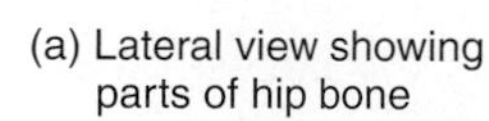

(a) Lateral view showing parts of hip bone

(b) Detailed lateral view

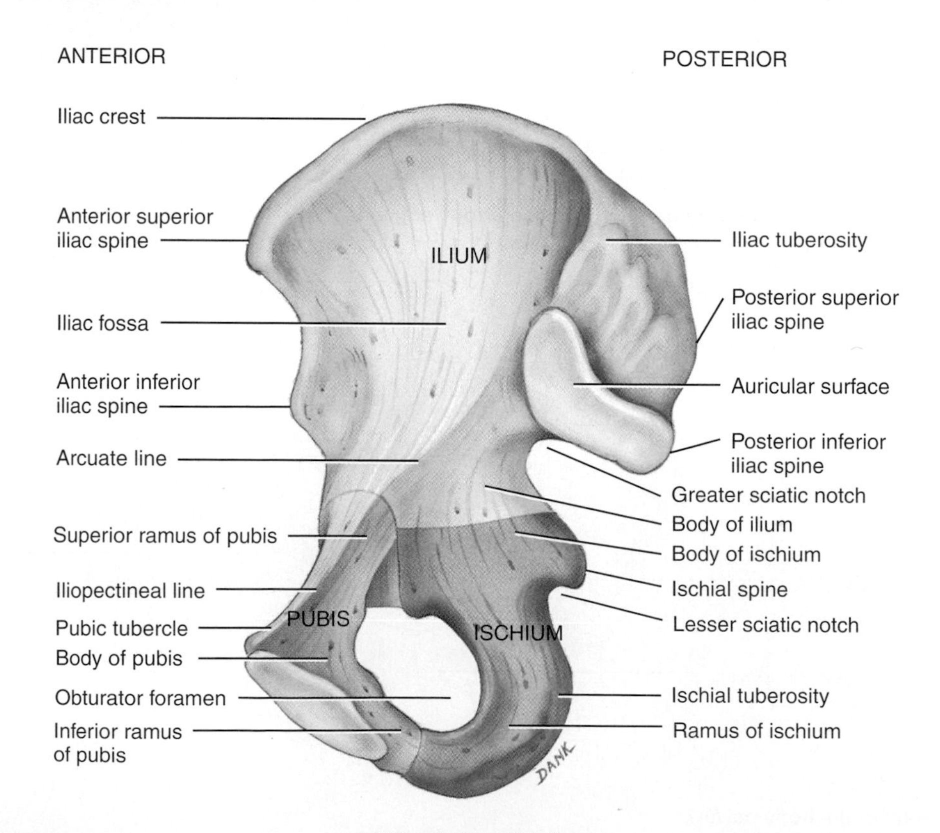

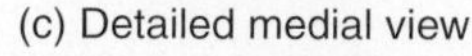

(c) Detailed medial view

Which part of the hip bone articulates with the femur? With the sacrum?

and an inferior *body,* which enters into the formation of the *acetabulum,* the socket for the head of the femur, comprise the ilium. Its superior border, the *iliac crest,* ends anteriorly in a blunt *anterior superior iliac spine.* Below this spine is the *anterior inferior iliac spine.* Posteriorly, the iliac crest ends in a sharp *posterior superior iliac spine.* Below this spine is the *posterior inferior iliac spine.* The spines serve as points of attachment for the tendons of the muscles of the trunk, hip, and thighs. Below the posterior inferior iliac spine is the *greater sciatic notch* (sī-AT-ik), through which the sciatic nerve, the longest nerve in the body, passes.

The medial surface of the ilium contains the *iliac fossa,* a concavity where the tendon of the iliacus muscle attaches. Posterior to this fossa are the *iliac tuberosity,* a point of attachment for the sacroiliac ligament, and the *auricular surface (auric-* = ear-shaped), which articulates with the sacrum to form the *sacroiliac joint* (see Figure 8.9). Projecting anteriorly and inferiorly from the auricular surface is a ridge called the *arcuate line* (AR-kū–at; *arc-* = bow).

The other conspicuous markings of the ilium are three arched lines on its lateral surface called the *posterior gluteal line (glut-* = buttock), the *anterior gluteal line,* and the *inferior gluteal line.* The tendons of the gluteal muscles attach to the ilium between these lines.

Ischium

The **ischium** (IS-kē-um = hip) is the inferior, posterior portion of the hip bone (see Figure 8.10b, c). A superior *body* and an inferior *ramus (ram-* = branch; plural is *rami*), which joins the pubis, comprise the ischium. Its features include the prominent *ischial spine,* a *lesser sciatic notch* below the spine, and a rough and thickened *ischial tuberosity.* This prominent tuberosity may hurt someone's thigh when you sit on their lap. Together, the ramus and the pubis surround the *obturator foramen* (OB-too-rā-tōr; *obtur-* = closed up), the largest foramen in the skeleton. The foramen is so-named because, even though blood vessels and nerves pass through it, it is nearly completely closed by the fibrous *obturator membrane.*

Pubis

The **pubis** or **os pubis,** meaning pubic bone, is the anterior and inferior part of the hip bone (Figure 8.10b, c). A *superior ramus,* an *inferior ramus,* and a *body* between the rami comprise the pubis. The anterior border of the body is the *pubic crest,* and at its lateral end is a projection, the *pubic tubercle.* This tubercle is the beginning of a raised line, the *iliopectineal line* (il-ē-ō-pek-TIN-ē-al), which extends superiorly and laterally along the superior ramus to merge with the arcuate line of the ilium. These lines, as you will see shortly, are important landmarks for distinguishing the superior and inferior portions of the bony pelvis.

The *pubic symphysis* is the joint between the two hip bones (see Figure 8.9). It consists of a disc of fibrocartilage. The inferior rami of the two pubic bones converge to form the pubic arch. In the later stages of pregnancy, the hormone relaxin (produced by the ovaries and placenta) increases the flexibility of the pubic symphysis to ease delivery of the baby. Weakening of the joint, together with an already compromised center of gravity due to an enlarged uterus, also alters the gait during pregnancy.

The *acetabulum* (as-e-TAB-ū-lum = vinegar cup) is a deep fossa formed by the ilium, ischium, and pubis. It functions as the socket that accepts the rounded head of the femur. Together, the acetabulum and the femoral head form the *hip (coxal) joint.* On the inferior side of the acetabulum is a deep indentation, the *acetabular notch.* It forms a foramen through which blood vessels and nerves pass, and it serves as a point of attachment for ligaments of the femur (for example, the ligament of the head of the femur).

True and False Pelves

The bony pelvis is divided into superior and inferior portions by a boundary called the *pelvic brim* (Figure 8.11a). You can trace the pelvic brim by following the landmarks around parts of the hip bones. Beginning posteriorly at the *sacral promontory* of the sacrum, trace laterally and inferiorly along the *arcuate lines* of the ilium of the hip bones. Continue inferiorly along the *iliopectineal lines* of the pubis. Finally, trace anteriorly to the superior portion of the pubic symphysis. Together, these points form an oblique plane that is higher in the back than in the front. The circumference of this plane is the pelvic brim.

The portion of the bony pelvis superior to the pelvic brim is the **false (greater) pelvis** (Figure 8.11b). It is bordered by the lumbar vertebrae posteriorly, the upper portions of the hip bones laterally, and the abdominal wall anteriorly. The space enclosed by the false pelvis is part of the abdomen. It does not contain pelvic organs, except for the urinary bladder (when it is full) and the uterus during pregnancy.

The **true (lesser) pelvis** is the part of the bony pelvis inferior to the pelvic brim (Figure 8.11b). It is bounded by the sacrum and coccyx posteriorly, inferior portions of the ilium and ischium laterally, and the pubic bones anteriorly. The true pelvis surrounds the pelvic cavity (see Figure 1.9 on page 17). The superior opening of the true pelvis is the pelvic brim, also called the *pelvic inlet;* the inferior opening of the true pelvis is the *pelvic outlet.* The *pelvic axis* is an imaginary line that curves through the true pelvis and joins the central points of the planes of the pelvic inlet and outlet. During childbirth, the pelvic axis is the route taken by the baby's head as it descends through the pelvis.

▶ CHECKPOINT

4. Where are the true and false pelves?

Figure 8.11 True and false pelves. Shown here is the female pelvis. For simplicity, the landmarks of the pelvic brim are shown only on the left side of the body, and the outline of the pelvic brim is shown only on the right side. The entire pelvic brim is shown in Figure 8.9.

The pelvic brim separates the true and false pelves.

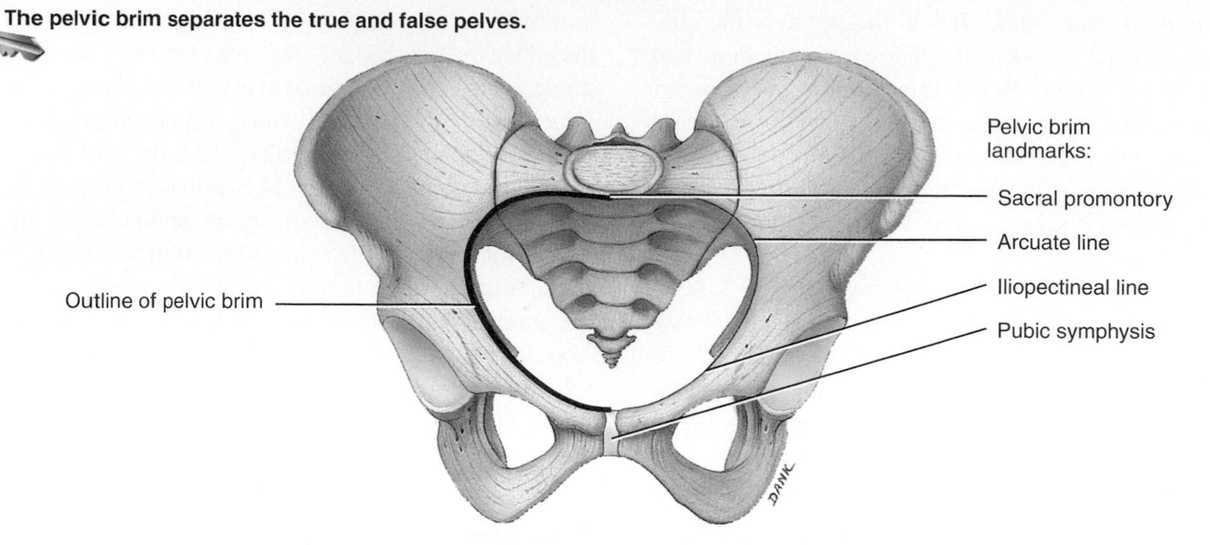

(a) Anterior view of borders of pelvic brim

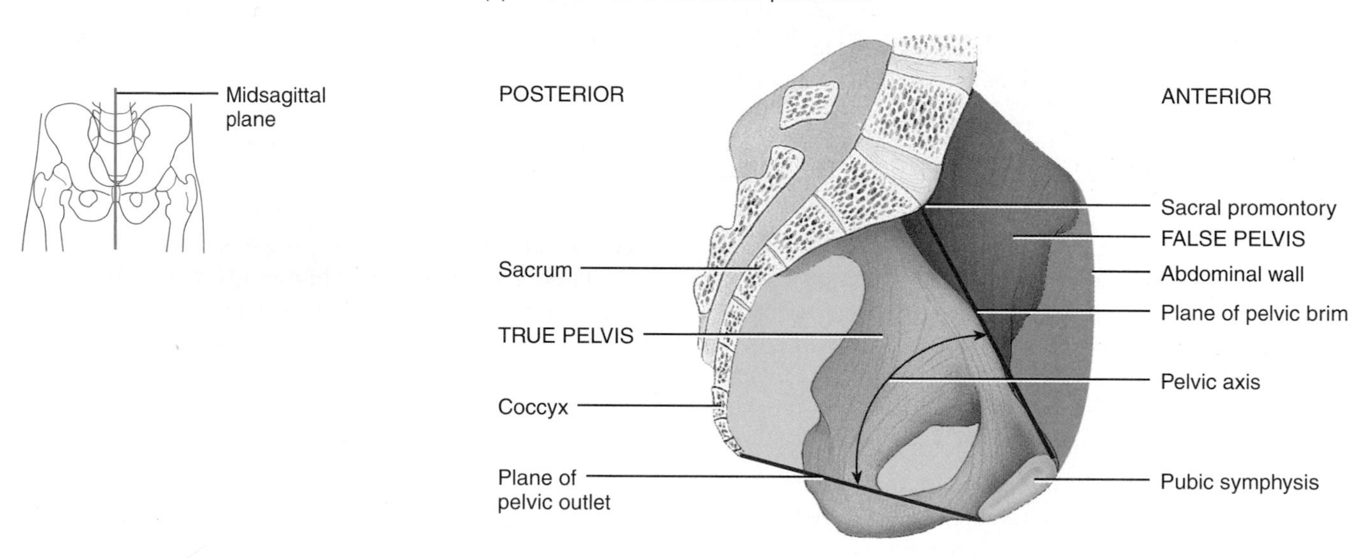

(b) Midsagittal section indicating locations of true and false pelves

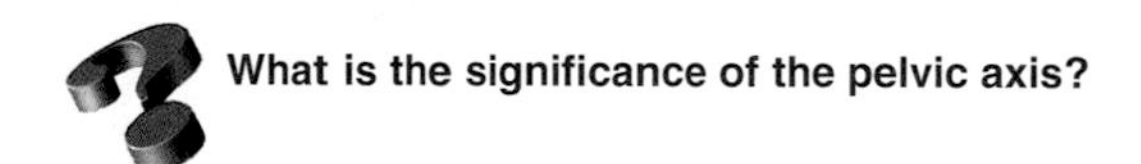

What is the significance of the pelvic axis?

COMPARISON OF FEMALE AND MALE PELVES

OBJECTIVE

- **Compare the principal structural differences between female and male pelves.**

In the following discussion, it is assumed that the male and female are comparable in age and physical stature. Generally, the bones of males are larger and heavier than those of females and possess larger surface markings. Gender-related differences in the features of skeletal bones are readily apparent when comparing the female and male pelves. Most of the structural differences in the pelves are adaptations to the requirements of pregnancy and childbirth. The female's pelvis is wider and shallower than the male's (Table 8.1). As a result, there is more space in the true pelvis of the female, especially in the pelvic inlet and pelvic outlet, which accommodate the passage of the infant's head at birth. Table 8.1 lists and illustrates the structural differences between the female and male pelves.

CHECKPOINT

5. Why are structural differences between female and male pelves important?

Table 8.1 Comparison of Female and Male Pelves

Point of Comparison	Female	Male
General structure	Light and thin.	Heavy and thick.
False (greater) pelvis	Shallow.	Deep.
Pelvic brim (inlet)	Larger and more oval.	Smaller and heart-shaped.
Acetabulum	Small and faces anteriorly.	Large and faces laterally.
Obturator foramen	Oval.	Round.
Pubic arch	Greater than 90° angle.	Less than 90° angle.
Iliac crest	Less curved.	More curved.
Ilium	Less vertical.	More vertical.
Greater sciatic notch	Wide.	Narrow.
Coccyx	More movable and more curved anteriorly.	Less movable and less curved anteriorly.
Sacrum	Shorter, wider (see anterior views), and more curved anteriorly.	Longer, narrower (see anterior views), and less curved anteriorly.

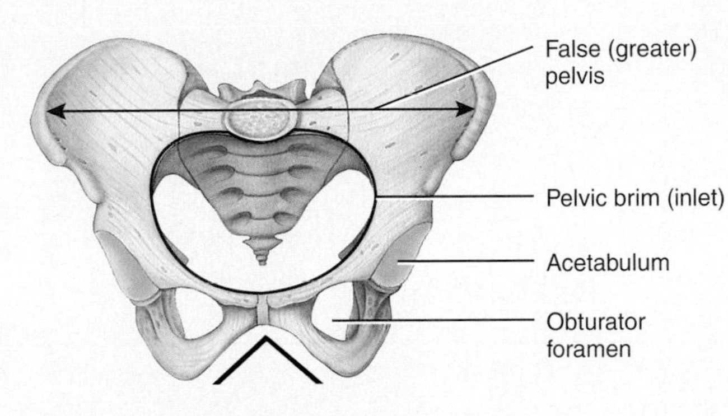

Pubic arch (greater than 90°)

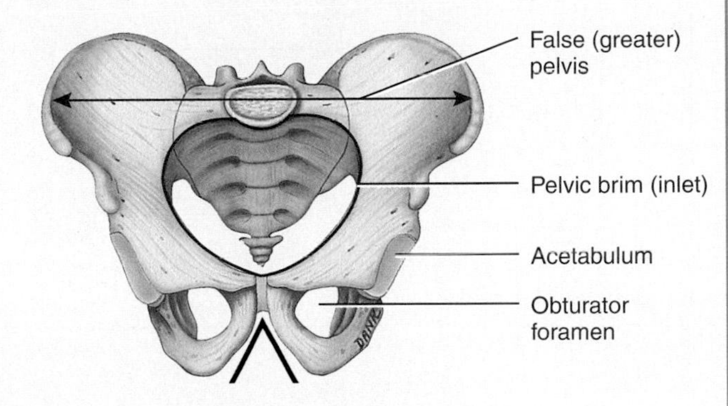

Pubic arch (less than 90°)

Anterior views

Iliac crest
Ilium
Greater sciatic notch
Sacrum
Coccyx

Iliac crest
Ilium
Greater sciatic notch
Sacrum
Coccyx

Right lateral views

(continues)

Table 8.1 Comparison of Female and Male Pelves *(continued)*

Point of Comparison	Female	Male
Pelvic outlet	Wider.	Narrower.
Ischial tuberosity	Shorter, farther apart, and more laterally projecting.	Longer, closer together, and more medially projecting.

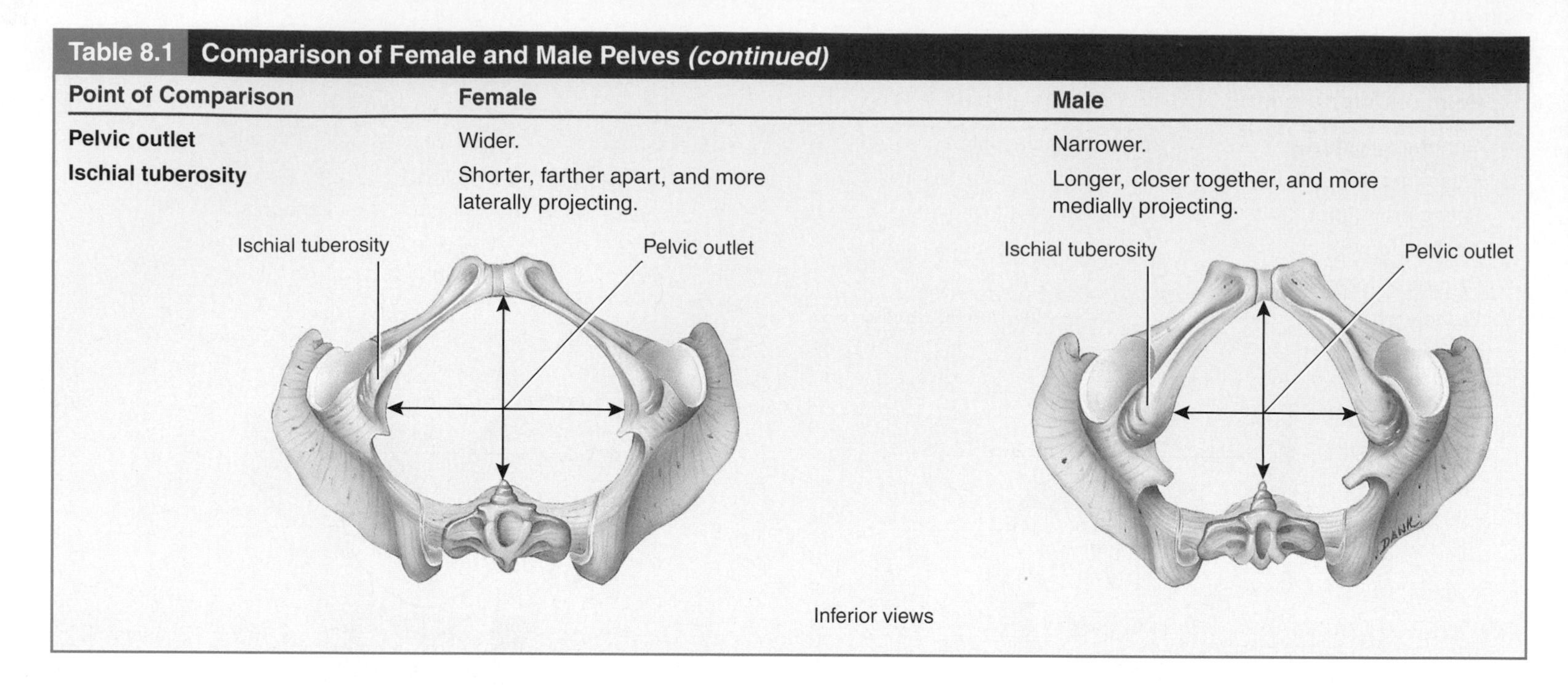

COMPARISON OF PECTORAL AND PELVIC GIRDLES

OBJECTIVE

- **Describe the differences in the pectoral and pelvic girdles.**

Now that we've studied the structures of the pectoral and pelvic girdles, we can note some of their significant differences. The pectoral girdle does not directly articulate with the vertebral column, whereas the pelvic girdle directly articulates with the vertebral column via the sacroiliac joint. The sockets (glenoid fossae) for the upper limbs in the pectoral girdle are shallow and maximize movement, whereas the sockets (acetabula) for the lower limbs in the pelvic girdle are deep and allow less movement. Overall, the structure of the pectoral girdle offers more mobility than strength, whereas that of the pelvic girdle offers more strength and less mobility.

LOWER LIMB

OBJECTIVE

- **Identify the bones of the lower limb and their principal markings.**

Each **lower limb (lower extremity)** has 30 bones in four locations—(1) the femur in the thigh; (2) the patella (kneecap); (3) the tibia and fibula in the leg; (4) and the 7 tarsals in the tarsus (ankle), the 5 metatarsals in the metatarsus, and the 14 phalanges (bones of the digits) in the foot (Figure 8.12).

Femur

The **femur,** or thigh bone, is the longest, heaviest, and strongest bone in the body (Figure 8.13 on page 234). Its proximal end articulates with the acetabulum of the hip bone. Its distal end articulates with the tibia and patella. The *body (shaft)* of the femur angles medially and, as a result, the knee joints are closer to the midline. The angle of convergence is greater in females because the female pelvis is broader.

The proximal end of the femur consists of a rounded *head* that articulates with the acetabulum of the hip bone to form the *hip (coxal) joint.* The head contains a small centered depression (pit) called the *fovea capitis* (FŌ-vē-a CAP-i-tis; *fovea* = pit; *capitis* = of the head). The ligament of the head of the femur connects the fovea capitis of the femur to the acetabulum of the hip bone. The *neck* of the femur is a constricted region distal to the head. The *greater trochanter* (trō-KAN-ter) and *lesser trochanter* are projections that serve as points of attachment for the tendons of some of the thigh and buttock muscles. The greater trochanter is the prominence felt and seen anterior to the hollow on the side of the hip. It is a landmark commonly used to locate the site for intramuscular injections into the lateral surface of the thigh. The lesser trochanter is inferior and medial to the greater trochanter. Between the anterior surface of the trochanters is a narrow *intertrochanteric line* (Figure 8.13a).

Figure 8.12 Right lower limb.

Each lower limb includes a femur, patella (kneecap), tibia, fibula, tarsals (ankle bones), metatarsals, and phalanges (bones of the digits).

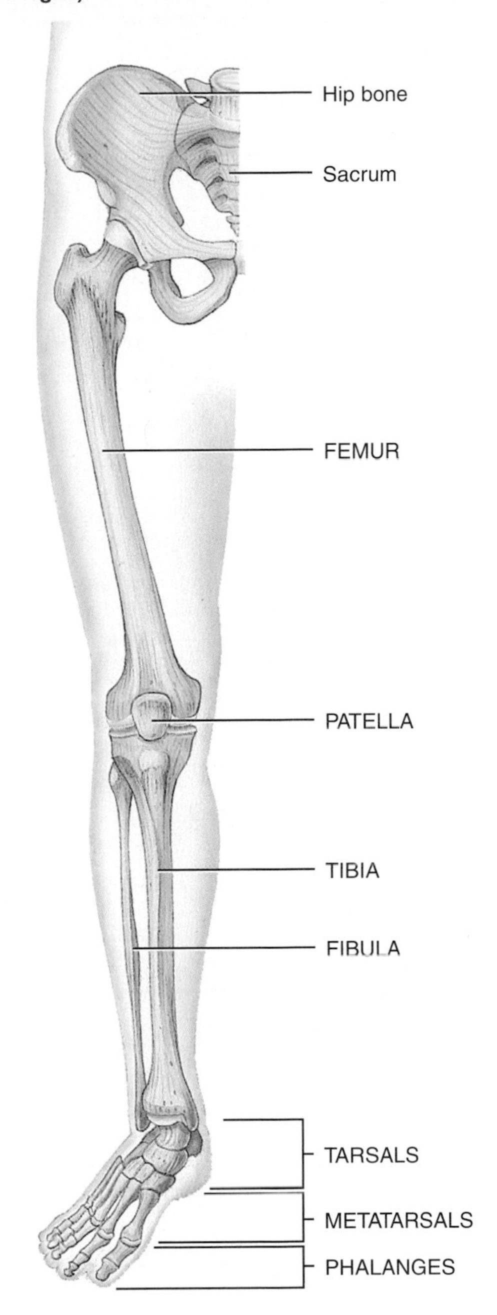

How many bones make up each lower limb?

Between the posterior surface of the trochanters is the *intertrochanteric crest* (Figure 8.13b).

Inferior to the intertrochanteric crest on the posterior surface of the body of the femur is a vertical ridge called the *gluteal tuberosity.* It blends into another vertical ridge called the *linea aspera* (LIN-ē-a AS-per-a; *asper-* = rough). Both ridges serve as attachment points for the tendons of several thigh muscles.

The distal end of the femur is expanded and includes the *medial condyle* and the *lateral condyle.* These articulate with the medial and lateral condyles of the tibia. Superior to the condyles are the *medial epicondyle* and the *lateral epicondyle.* A depressed area between the condyles on the posterior surface is called the *intercondylar fossa* (in-ter-KON-di-lar). The *patellar surface* is located between the condyles on the anterior surface.

Patella

The **patella** (= little dish), or kneecap, is a small, triangular bone located anterior to the knee joint (Figure 8.14 on page 235). It is a sesamoid bone that develops in the tendon of the quadriceps femoris muscle. The broad superior end of the patella is called the *base.* The pointed inferior end is the *apex.* The posterior surface contains two *articular facets,* one for the medial condyle and the other for the lateral condyle of the femur. The patellar ligament attaches the patella to the tibial tuberosity. The *patellofemoral joint,* between the posterior surface of the patella and the patellar surface of the femur, is the intermediate component of the *tibiofemoral (knee) joint.* The patella functions to increase the leverage of the tendon of the quadriceps femoris muscle, to maintain the position of the tendon when the knee is bent (flexed), and to protect the knee joint.

Patellofemoral Stress Syndrome

Patellofemoral stress syndrome ("runner's knee") is one of the most common problems in runners. During normal flexion and extension of the knee, the patella tracks (glides) up and down in the groove between the femoral condyles. In patellofemoral stress syndrome, normal tracking does not occur. Instead, the patella also tracks laterally, and the increased pressure on the joint causes aching or tenderness around or under the patella. The pain typically occurs after a person has been sitting for a while, especially after exercise. Squatting or walking down stairs worsens it. One cause of runner's knee is constantly walking, running, or jogging on the same side of the road. Because roads slope down on the sides, the knee that is closer to the center of the road endures greater mechanical stress because it does not fully extend during a stride. Other predisposing factors include having knock-knees, running on hills, and running long distances. ■

Figure 8.13 Right femur in relation to the hip bone, patella, tibia, and fibula. (See Tortora, *A Photographic Atlas of the Human Body,* Figure 3.28.)

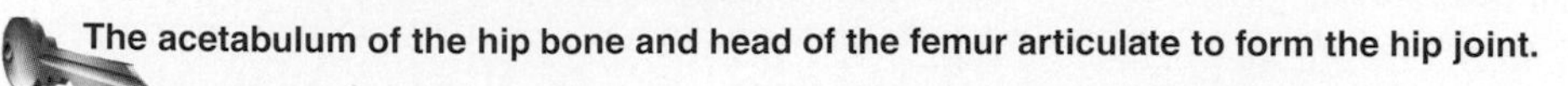

The acetabulum of the hip bone and head of the femur articulate to form the hip joint.

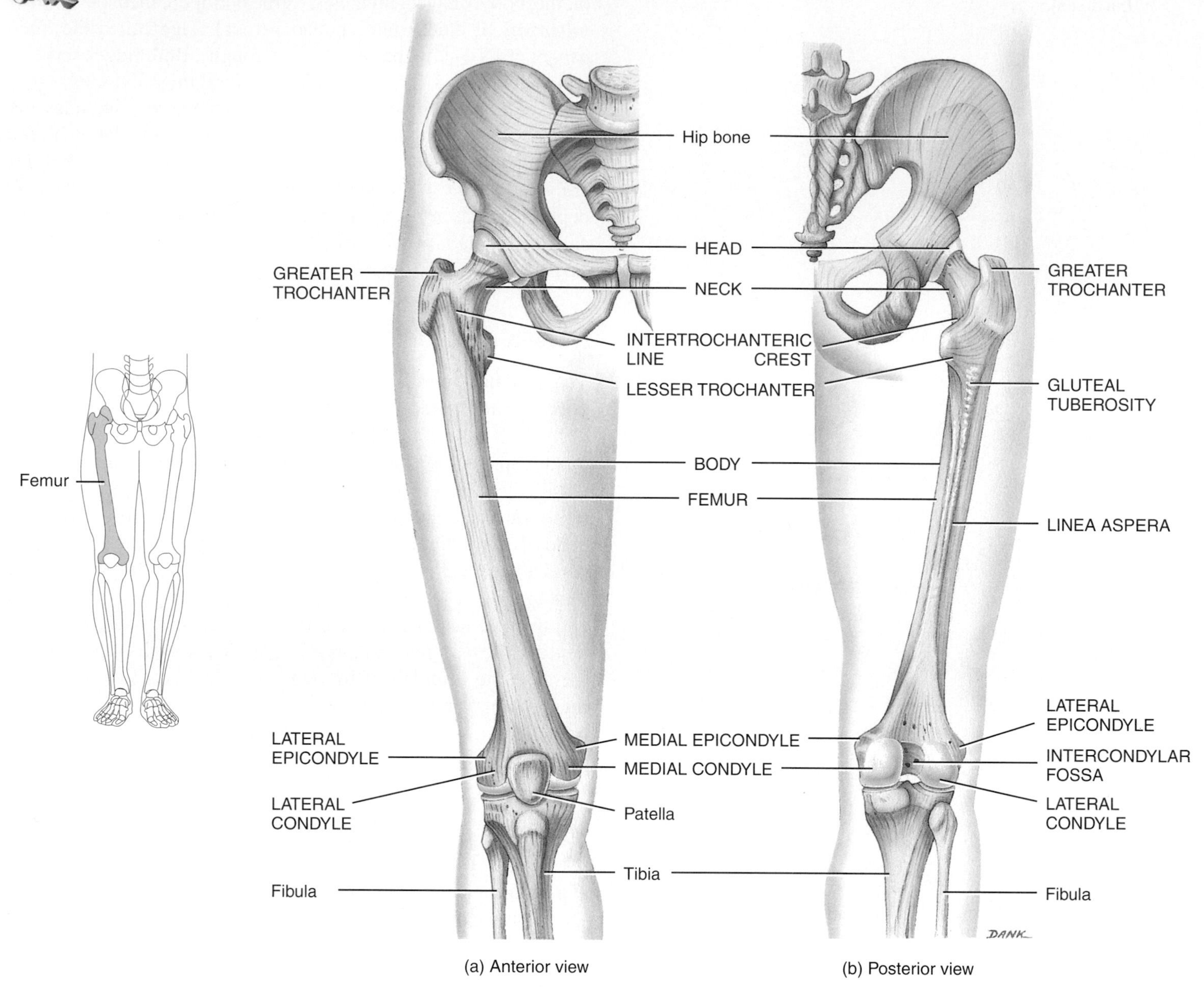

(a) Anterior view

(b) Posterior view

Tibia and Fibula

The **tibia,** or shin bone, is the larger, medial, weight-bearing bone of the leg (Figure 8.15 on page 236). The tibia articulates at its proximal end with the femur and fibula, and at its distal end with the fibula and the talus bone of the ankle. An interosseous membrane connects the tibia and fibula.

The proximal end of the tibia is expanded into a *lateral condyle* and a *medial condyle.* These articulate with the condyles of the femur to form the lateral and medial *tibiofemoral (knee) joints.* The inferior surface of the lateral condyle articulates with the head of the fibula. The slightly concave condyles are separated by an upward projection called the *intercondylar eminence* (Figure 8.15b). The *tibial tuberosity* on the anterior surface is a point of attachment for the patellar ligament. Inferior to and continuous with the tibial tuberosity is a sharp ridge that can be felt below the skin. This ridge is the *anterior border (crest)* or *shin.*

The medial surface of the distal end of the tibia forms the *medial malleolus* (mal-LĒ-ō-lus = hammer). This structure articulates with the talus bone of the ankle and forms the protru-

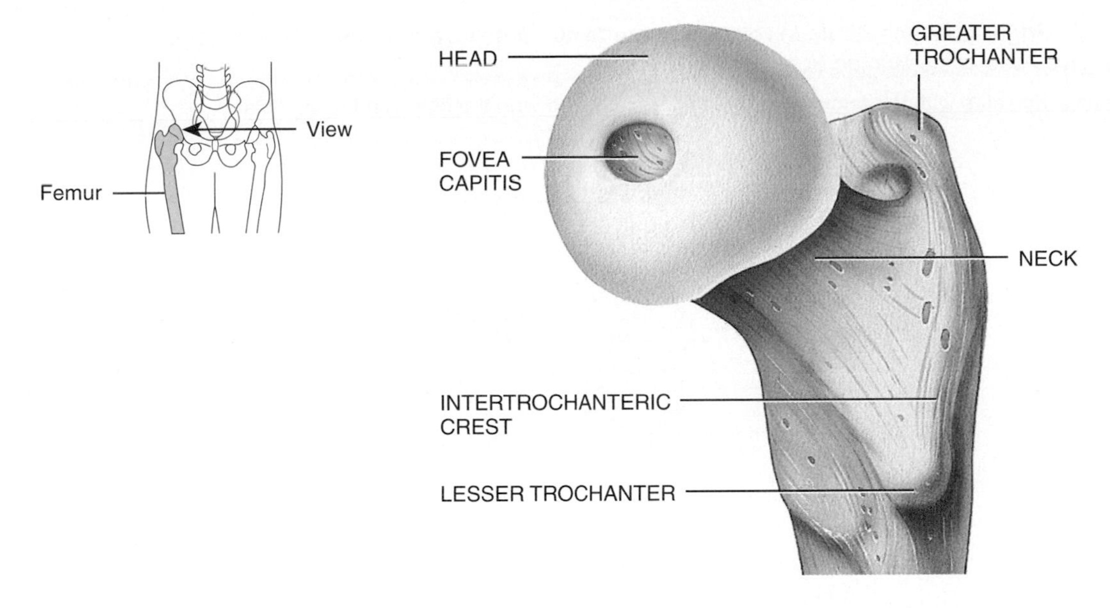

(c) Medial view of proximal end of femur

Why is the angle of convergence of the femurs greater in females than males?

sion on the medial surface of the ankle. The *fibular notch* (Figure 8.15c) articulates with the distal end of the fibula to form the *distal tibiofibular joint.*

The **fibula** is parallel and lateral to the tibia, but it is considerably smaller than the tibia. The proximal end, the *head* of the fibula, articulates with the inferior surface of the lateral condyle of the tibia below the level of the knee joint to form the *proximal tibiofibular joint.* The distal end has a projection called the *lateral malleolus* that articulates with the talus bone of the ankle. The lateral malleolus is the prominence on the lateral surface of the ankle. As noted, the fibula also articulates with the tibia at the fibular notch.

Figure 8.14 Right patella.

The patella articulates with the lateral and medial condyles of the femur.

Patella

Base

Articular facet for medial femoral condyle

Articular facet for lateral femoral condyle

Apex

(a) Anterior view

(b) Posterior view

Because the patella develops in the tendon of a muscle, it is classified as which type of bone?

Figure 8.15 Right tibia and fibula in relation to the femur, patella, and talus. (See Tortora, *A Photographic Atlas of the Human Body,* Figure 3.30.)

The tibia articulates with the femur and fibula proximally, and with the fibula and talus distally.

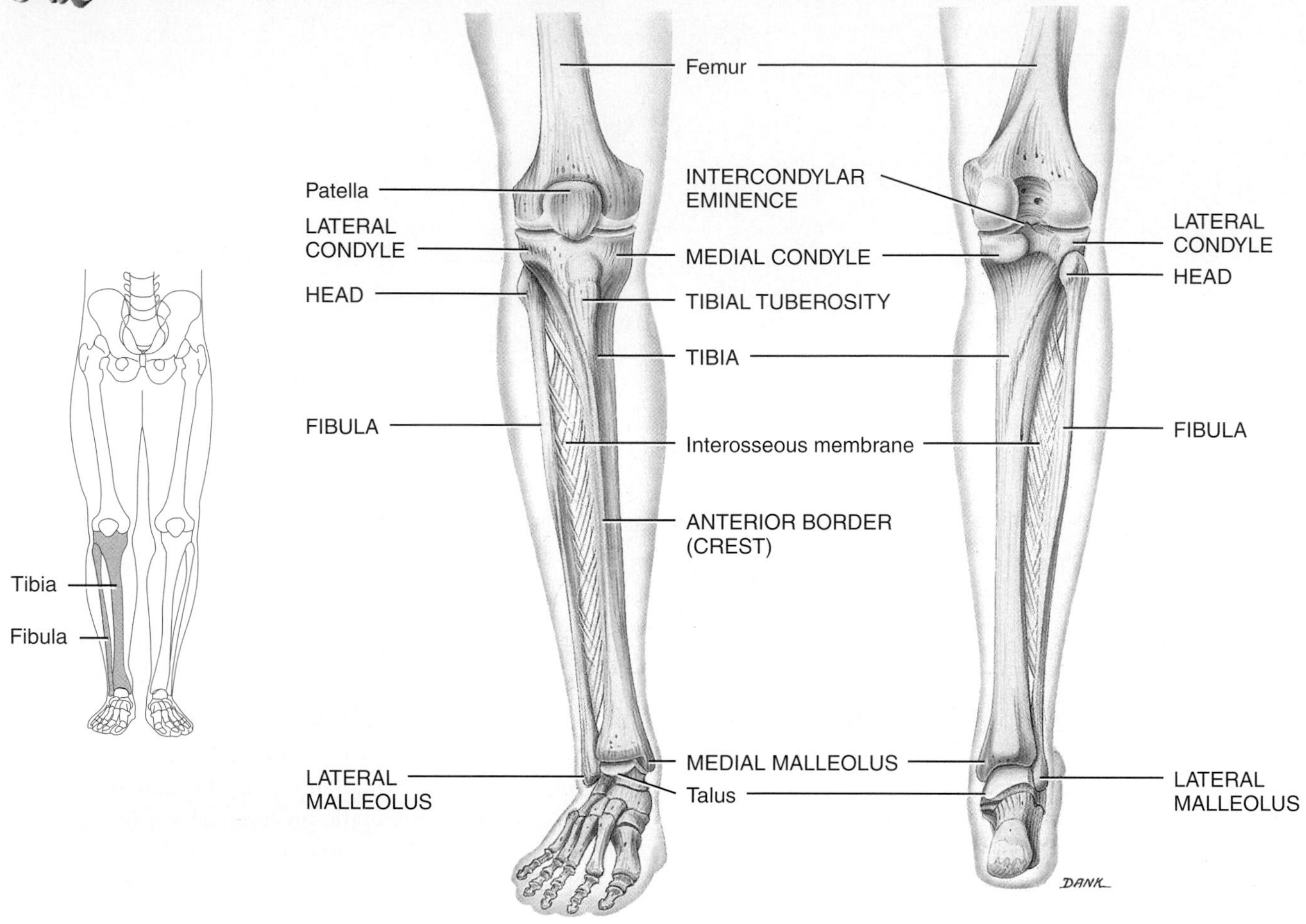

Femur
Patella
LATERAL CONDYLE
HEAD
INTERCONDYLAR EMINENCE
MEDIAL CONDYLE
TIBIAL TUBEROSITY
TIBIA
FIBULA
Interosseous membrane
ANTERIOR BORDER (CREST)
LATERAL MALLEOLUS
MEDIAL MALLEOLUS
Talus
LATERAL CONDYLE
HEAD
FIBULA
LATERAL MALLEOLUS
Tibia
Fibula
DANK

(a) Anterior view

(b) Posterior view

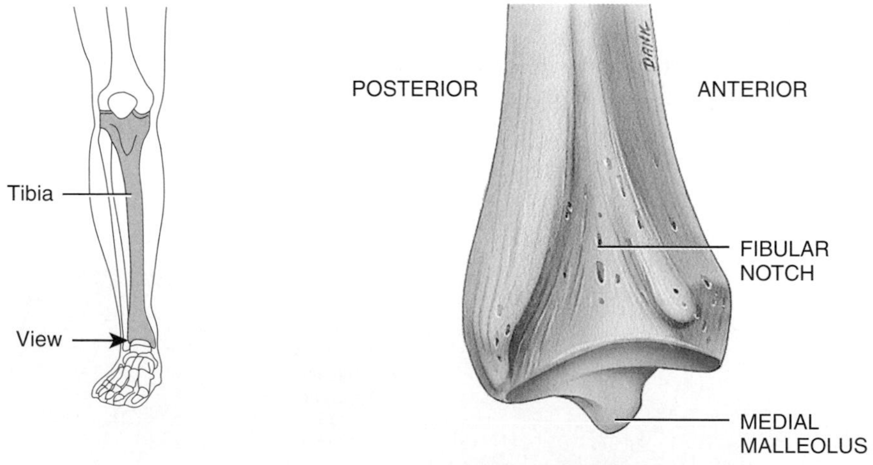

Tibia
View
POSTERIOR
ANTERIOR
FIBULAR NOTCH
MEDIAL MALLEOLUS
DANK

(c) Lateral view of distal end of tibia

Which leg bone bears the weight of the body?

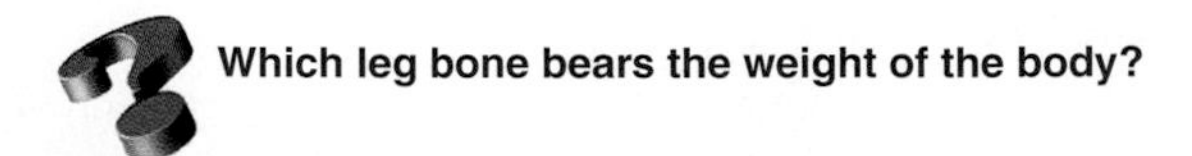

Tarsals, Metatarsals, and Phalanges

The **tarsus** (ankle) is the proximal region of the foot and consists of seven **tarsal bones** (Figure 8.16). They include the **talus** (TĀ-lus = ankle bone) and **calcaneus** (kal-KĀ-nē-us = heel), located in the posterior part of the foot. The calcaneus is the largest and strongest tarsal bone. The anterior tarsal bones are the **cuboid** (= cube-shaped), **navicular** (= like a little boat), and three **cuneiform bones** (= wedge-shaped) called the **first (medial), second (intermediate),** and **third (lateral) cuneiforms.** Joints between tarsal bones are called *intertarsal joints.* The talus, the uppermost tarsal bone, is the only bone of the foot that articulates with the fibula and tibia. It articulates on one side with the medial malleolus of the tibia and on the other side with the lateral malleolus of the fibula. These articulations form the *talocrural (ankle) joint.* During walking, the talus transmits about half the weight of the body to the calcaneus and half to other tarsal bones.

The **metatarsus** is the intermediate region of the foot and consists of five **metatarsal bones** numbered I to V (or 1–5) from the medial to lateral position (Figure 8.16). Like the metacarpals of the palm of the hand, each metatarsal consists of a proximal *base,* an intermediate *shaft,* and a distal *head.* The metatarsals articulate proximally with the first, second, and third cuneiform bones and with the cuboid to form the *tarsometatarsal joints.* Distally, they articulate with the proximal

Figure 8.16 Right foot. (See Tortora, *A Photographic Atlas of the Human Body,* Figure 3.31.)

The skeleton of the foot consists of the proximal tarsals, the intermediate metatarsals, and the distal phalanges.

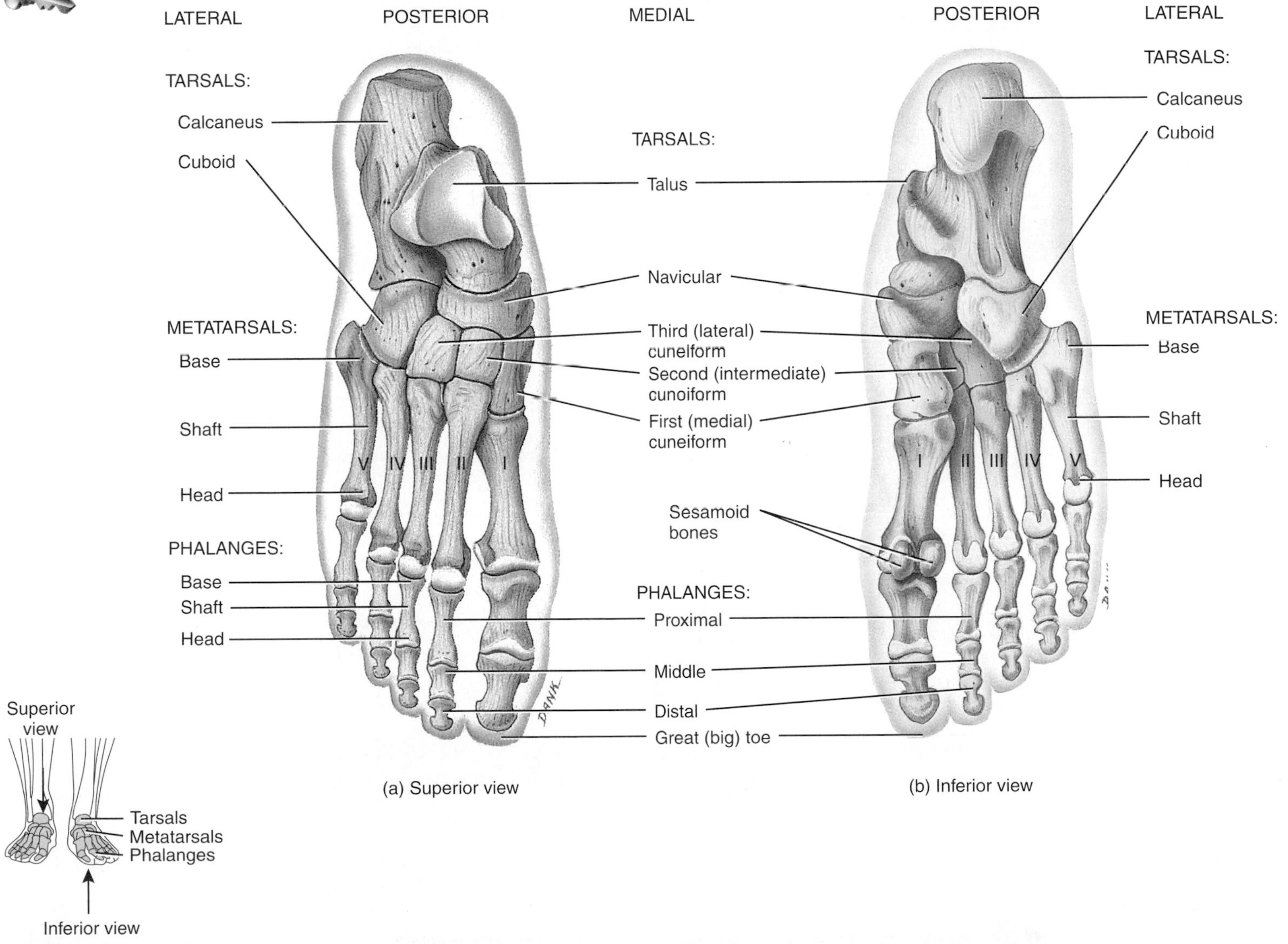

LATERAL
POSTERIOR
MEDIAL
POSTERIOR
LATERAL
TARSALS:
Calcaneus
Cuboid
TARSALS:
Talus
Navicular
Third (lateral) cuneiform
Second (intermediate) cuneiform
First (medial) cuneiform
METATARSALS:
Base
Shaft
Head
V IV III II I
I II III IV V
Sesamoid bones
PHALANGES:
Base
Shaft
Head
PHALANGES:
Proximal
Middle
Distal
Great (big) toe
Superior view
Tarsals
Metatarsals
Phalanges
Inferior view

(a) Superior view

(b) Inferior view

Which tarsal bone articulates with the tibia and fibula?

row of phalanges to form the *metatarsophalangeal joints.* The first metatarsal is thicker than the others because it bears more weight.

The **phalanges** comprise the distal component of the foot and resemble those of the hand both in number and arrangement. They are numbered I to V (or 1–5) beginning with the great toe, which is medial. Each also consists of a proximal *base,* an intermediate *shaft,* and a distal *head.* The great or big toe (*hallux*) has two large, heavy phalanges called proximal and distal phalanges. The other four toes each have three phalanges—proximal, middle, and distal. Joints between phalanges of the foot, like those of the hand, are called *interphalangeal joints.*

Arches of the Foot

The bones of the foot are arranged in two **arches** (Figure 8.17). The arches enable the foot to support the weight of the body, provide an ideal distribution of body weight over the hard and soft tissues of the foot, and provide leverage while walking. The arches are not rigid; they yield as weight is applied and spring back when the weight is lifted, thus helping to absorb shocks. Usually, the arches are fully developed by age 12 or 13.

The *longitudinal arch* has two parts, both of which consist of tarsal and metatarsal bones arranged to form an arch from the anterior to the posterior part of the foot. The *medial part* of the longitudinal arch originates at the calcaneus. It rises to the talus and descends through the navicular, the three cuneiforms, and the heads of the three medial metatarsals. The *lateral part* of the longitudinal arch also begins at the calcaneus. It rises at the cuboid and descends to the heads of the two lateral metatarsals.

The *transverse arch* is found between the medial and lateral aspects of the foot and is formed by the navicular, three cuneiforms, and the bases of the five metatarsals.

Flatfoot and Clawfoot

The bones composing the arches are held in position by ligaments and tendons. If these ligaments and tendons are weakened, the height of the medial longitudinal arch may decrease or "fall." The result is **flatfoot,** the causes of which include excessive weight, postural abnormalities, weakened supporting tissues, and genetic predisposition. A custom-designed arch support often is prescribed to treat flatfoot.

Clawfoot is a condition in which the medial longitudinal arch is abnormally elevated. It is often caused by muscle deformities, such as may occur in diabetics whose neurological lesions lead to atrophy of muscles of the foot. ■

▶ CHECKPOINT

6. Which bones form the lower limb, from proximal to distal?
7. What joints do the bones of the lower limb form with one another?
8. What are the functions of the arches of the foot?

• • •

To appreciate the skeletal system's contributions to homeostasis of other body systems, examine *Focus on Homeostasis: The Skeletal System.* Next, in Chapter 9, we will see how joints both hold the skeleton together and permit it to participate in movements.

Figure 8.17 Arches of the right foot.

Arches help the foot support and distribute the weight of the body and provide leverage during walking.

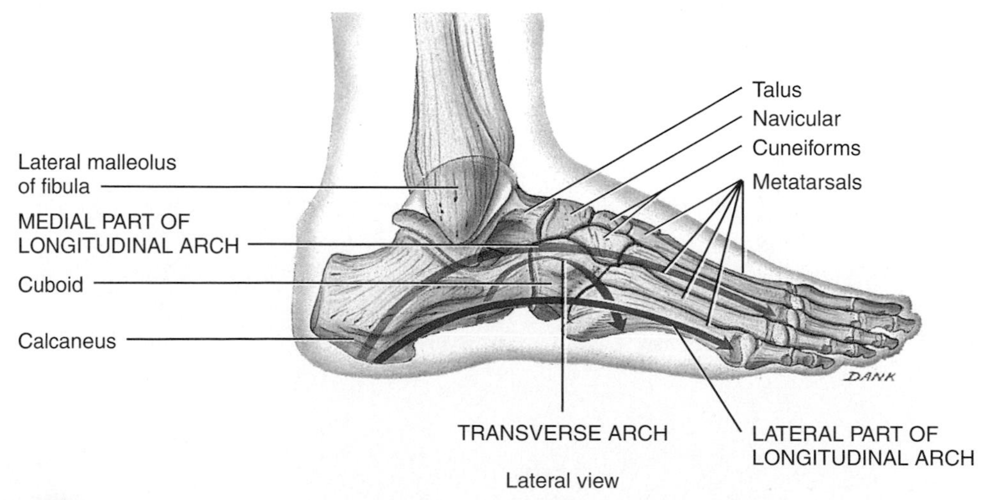

What structural feature of the arches allows them to absorb shocks?

Focus on Homeostasis: The Skeletal System

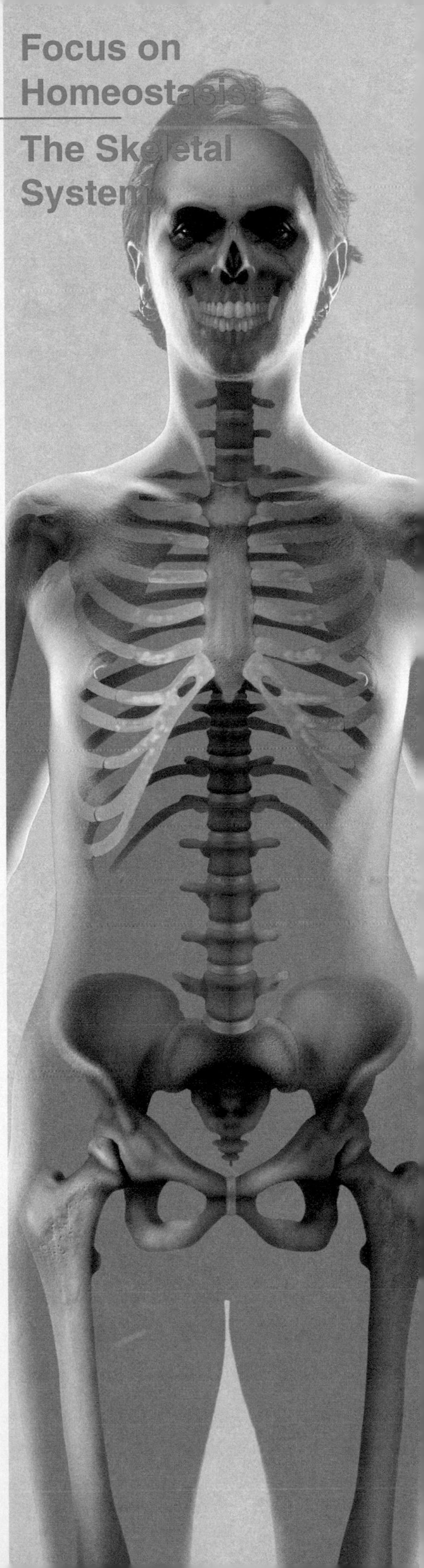

Body System	Contribution of the Skeletal System
For all body systems	Bones provide support and protection for internal organs; bones store and release calcium, which is needed for proper functioning of most body tissues.
Integumentary system	Bones provide strong support for overlying muscles and skin while joints provide flexibility that allows skin to bend.
Muscular system 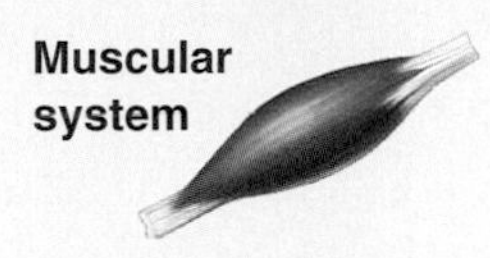	Bones provide attachment points for muscles and leverage for muscles to bring about body movements; contraction of skeletal muscle requires calcium ions.
Nervous system	Skull and vertebrae protect brain and spinal cord; normal blood level of calcium is needed for normal functioning of neurons and neuroglia.
Endocrine system 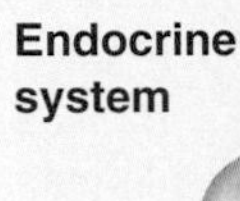	Bones store and release calcium, needed during exocytosis of hormone-filled vesicles and for normal actions of many hormones.
Cardiovascular system 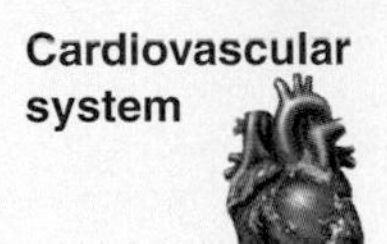	Red bone marrow carries out hemopoiesis (blood cell formation); rhythmical beating of the heart requires calcium ions.
Lymphatic and immune system 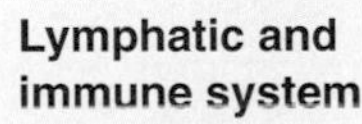 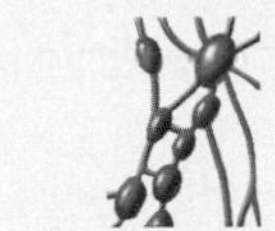	Red bone marrow produces lymphocytes, white blood cells that are involved in immune responses.
Respiratory system 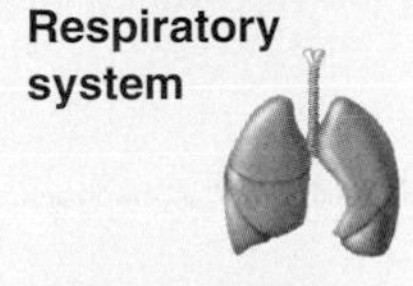	Axial skeleton of thorax protects lungs; rib movements assist breathing; some muscles used for breathing attach to bones via tendons.
Digestive system 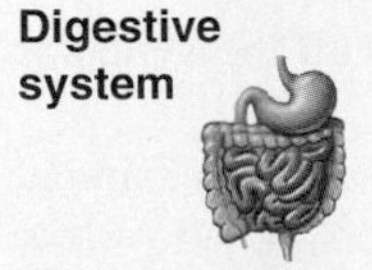	Teeth masticate (chew) food; rib cage protects esophagus, stomach, and liver; pelvis protects portions of the intestines.
Urinary system 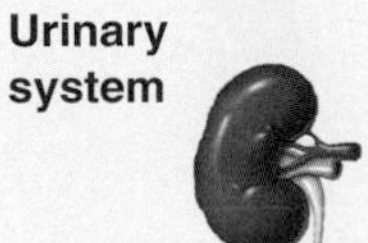	Ribs partially protect kidneys and pelvis protects urinary bladder and urethra.
Reproductive systems 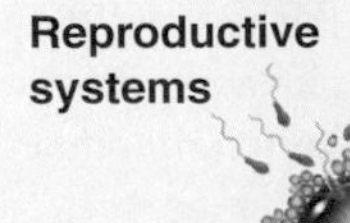	Pelvis protects ovaries, uterine (Fallopian) tubes, and uterus in females and part of ductus (vas) deferens and accessory glands in males; bones are an important source of calcium needed for milk synthesis during lactation.

DISORDERS: HOMEOSTATIC IMBALANCES

Hip Fracture

Although any region of the hip girdle may fracture, the term **hip fracture** most commonly applies to a break in the bones associated with the hip joint—the head, neck, or trochanteric regions of the femur, or the bones that form the acetabulum. In the United States, 300,000 to 500,000 people sustain hip fractures each year. The incidence of hip fractures is increasing, in part due to longer life spans. Decreases in bone mass due to osteoporosis and an increased tendency to fall predispose elderly people to hip fractures.

Hip fractures often require surgical treatment, the goal of which is to repair and stabilize the fracture, increase mobility, and decrease pain. Sometimes the repair is accomplished by using surgical pins, screws, nails, and plates to secure the head of the femur. In severe hip fractures, the femoral head or the acetabulum of the hip bone may be replaced by prostheses (artificial devices). The procedure of replacing either the femoral head or the acetabulum is *hemiarthroplasty* (hem-ē-AR-thrō-plas-tē; *hemi-* = one half; *-arthro-* = joint; *-plasty* = molding). Replacement of both the femoral head and acetabulum is *total hip arthroplasty.* The acetabular prosthesis is made of plastic, whereas the femoral prosthesis is metal; both are designed to withstand a high degree of stress. The prostheses are attached to healthy portions of bone with acrylic cement and screws.

MEDICAL TERMINOLOGY

Clubfoot or ***talipes equinovarus*** (*-pes* = foot; *equino-* = horse) An inherited deformity in which the foot is twisted inferiorly and medially, and the angle of the arch is increased; occurs in 1 of every 1000 births. Treatment consists of manipulating the arch to a normal curvature by casts or adhesive tape, usually soon after birth. Corrective shoes or surgery may also be required.

Genu valgum (JĒ-noo VAL-gum; *genu-* = knee; *valgum* = bent outward) A decreased space between the knees and increased space between the ankles due to a medial bending of the legs in relation to the thigh. Also called **knock-knee.**

Genu varum (JĒ-noo VAR-um; *varum* = bent toward the midline) An increased space between the knees due to a lateral bending of the legs relative to the thigh. Also called **bowleg.**

Hallux valgus (HAL-uks VAL-gus; *hallux* = great toe) Angulation of the great toe away from the midline of the body, typically caused by wearing tightly fitting shoes. Involves lateral deviation of the proximal phalanx of the great toe and medial displacement of metatarsal I. Also called a **bunion.**

Pelvimetry (pel-VIM-e-trē) Measurement of the size of the inlet and outlet of the birth canal; may be done by ultrasound or physical examination.

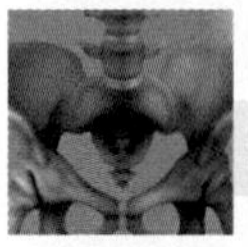

STUDY OUTLINE

PECTORAL (SHOULDER) GIRDLE (p. 219)

1. Each pectoral (shoulder) girdle consists of a clavicle and scapula.
2. Each pectoral girdle attaches an upper limb to the axial skeleton.

UPPER LIMB (p. 222)

1. There are 60 bones in the two upper limbs (extremities).
2. The bones of each upper limb include the humerus, the ulna, the radius, the carpals, the metacarpals, and the phalanges.

PELVIC (HIP) GIRDLE (p. 227)

1. The pelvic (hip) girdle consists of two hip bones.
2. Each hip bone consists of three fused components: ilium, pubis, and ischium.
3. The hip bones, sacrum, and pubic symphysis form the bony pelvis. It supports the vertebral column and pelvic viscera and attaches the lower limbs to the axial skeleton.
4. The true pelvis is separated from the false pelvis by the pelvic brim.

COMPARISON OF FEMALE AND MALE PELVES (p. 230)

1. Bones of males are generally larger and heavier than bones of females. They also have more prominent markings for muscle attachments.
2. The female pelvis is adapted for pregnancy and childbirth. Gender-related differences in pelvic structure are listed and illustrated in Table 8.1 on pages 231–232.

COMPARISON OF PECTORAL AND PELVIC GIRDLES (p. 232)

1. The pectoral girdle does not directly articulate with the vertebral column; the pelvic girdle does.
2. The glenoid fossae of the scapulae are shallow and maximize movement; the acetabula of the hip bones are deep and allow less movement.

LOWER LIMB (p. 232)

1. There are 60 bones in the two lower limbs (extremities).
2. The bones of each lower limb include the femur, the patella, the tibia, the fibula, the tarsals, the metatarsals, and the phalanges.
3. The bones of the foot are arranged in two arches, the longitudinal arch and the transverse arch, to provide support and leverage.

SELF-QUIZ QUESTIONS

Fill in the blanks in the following statements.

1. Functions of the axial skeleton are related mainly to ________ of internal organs, whereas functions of the appendicular skeleton are related more to ________.
2. The bones that comprise the palm are called the ________.

Indicate whether the following statements are true or false.

3. The arches of the foot enable the foot to support the weight of the body, provide an ideal distribution of body weight over the hard and soft tissues of the foot, and provide leverage for walking.
4. Functionally, the pelvis provides a strong and stable support for the vertebral column and pelvic viscera.

Choose the one best answer to the following questions.

5. Which of the following statements are *true*? (1) The pectoral girdle consists of the scapula, the clavicle, and the sternum. (2) Although the joints of the pectoral girdle are not very stable, they allow free movement in many directions. (3) The anterior component of the pectoral girdle is the scapula. (4) The pectoral girdle articulates directly with the vertebral column. (5) The posterior component of the pectoral girdle is the sternum. (a) 1, 2, and 3, (b) only 2, (c) only 4, (d) 2, 3, and 5, (e) 3, 4, and 5.
6. Which of the following bones are carpal bones? (1) phalanges, (2) scaphoid, lunate, and pisiform, (3) trapezium, trapezoid, capitate, (4) hamate and triquetrum, (5) calcaneus and talus. (a) 1, 2, and 3, (b) 2, 3, and 4, (c) 1, 2, 3, and 4, (d) 2, 3, 4, and 5, (e) 1, 2, 3, 4, and 5.
7. Which of the following statements is *true*? (a) The ilium is the smallest component of the hip bone. (b) The greater sciatic notch is found on the ischium. (c) The ischium is the anterior, superior portion of the hip bone. (d) The pubic symphysis is the joint between the two hip bones. (e) The obturator foramen is the smallest foramen in the body.
8. Which of the following articulates with the distal end of the tibia? (a) calcaneous, (b) navicular, (c) cuboid, (d) cuneiform, (e) talus.
9. Which of the following statements are *true*? (1) Bones of males are generally larger and heavier than those of females. (2) The female pelvis is wider than the male pelvis. (3) The true pelvis of males contains more space than that of the females. (4) Skeletal bones of males generally possess larger surface features than those of females. (5) Most structural differences between male and female pelves are related to pregnancy and childbirth. (a) 1, 2, and 3, (b) 2, 3, and 4, (c) 3, 4, and 5, (d) 1, 2, 3, and 5, (e) 1, 2, 4, and 5.
10. The acetabulum is part of the (a) fibula, (b) humerus, (c) hip bone, (d) femur, (e) tibia.
11. Which of the following is *not* a tarsal bone? (a) patella, (b) navicular, (c) calcaneus, (d) talus, (e) cuneiform.
12. Which of the following are *true* concerning the elbow joint? (1) When the forearm is extended, the olecranon fossa receives the olecranon. (2) When the forearm is flexed, the radial fossa receives the coronoid process. (3) The head of the radius articulates with the capitulum. (4) The trochlea articulates with the trochlear notch. (5) The head of the ulna atriculates with the ulnar notch of the radius. (a) 1, 2, 3, 4, and 5, (b) 1, 3, and 4, (c) 1, 3, 4, and 5, (d) 1, 2, 3, and 4, (e) 2, 3, and 4.
13. Which of the following is mismatched? (a) narrowing of carpal tunnel: carpal tunnel syndrome, (b) abnormal elevation of medial longitudinal arch: flat foot, (c) inherited foot deformity causing the foot to twist inferiorly and medially: clubfoot, (d) medial bending of legs resulting in decreased space between the knees and increased space between the ankle: genu valgum, (e) condition where great toes angle away from the midline of the body; often caused by tight-fitting shoes: bunion.
14. Match the following:

____ (a) olecranon process	(1) clavicle
____ (b) olecranon fossa	(2) scapula
____ (c) trochlea	(3) humerus
____ (d) greater trochanter	(4) ulna
____ (e) medial malleolus	(5) radius
____ (f) acromial extremity	(6) femur
____ (g) capitulum	(7) tibia
____ (h) acromion	(8) fibula
____ (i) radial tuberosity	(9) coxal bone
____ (j) acetabulum	
____ (k) lateral malleolus	
____ (l) glenoid cavity	
____ (m) coronoid process	
____ (n) linea aspera	
____ (o) anterior border	

15. Match the following:

____ (a) a large, triangular, flat bone found in the posterior part of the thorax	(1) calcaneus
____ (b) an S-shaped bone lying horizontally in the superior and anterior part of the thorax	(2) scapula
____ (c) articulates proximally with the scapula and distally with the radius and ulna	(3) scaphoid
____ (d) located on the medial aspect of the forearm	(4) radius
____ (e) located on the lateral aspect of the forearm	(5) femur
____ (f) the longest, heaviest, and strongest bone of the body	(6) clavicle
____ (g) the larger, medial bone of the leg	(7) ulna
____ (h) the smaller, lateral bone of the leg	(8) tibia
____ (i) heel bone	(9) humerus
____ (j) carpal bone	(10) fibula

CRITICAL THINKING QUESTIONS

1. Young Tommy watches way too much television, especially cartoons and monster movies. So when he heard that his great uncle had "clawfoot," he couldn't decide if he should be thrilled or terrified! Explain "clawfoot" to Tommy.
   HINT ***His great uncle is not changing his species—it's a human condition.***
2. The local newspaper reported that Farmer White caught his hand in a piece of machinery last Tuesday. He lost the lateral two fingers of his left hand. Science reporter Kent Clark, really a high school junior, reports that Farmer White has three remaining phalanges. Is Kent correct, or does he need a refresher course in anatomy?
   HINT ***How many phalanges are in each finger?***
3. Given that the hips and shoulders are located in the trunk of the body, why aren't they part of the axial skeleton like the rest of the bones that make up the trunk?
   HINT ***Have you ever seen a snake with shoulders?***

ANSWERS TO FIGURE QUESTIONS

8.1 The pectoral girdles attach the upper limbs to the axial skeleton.

8.2 The weakest part of the clavicle is its midregion at the junction of the two curves.

8.3 The highest point of the shoulder is the acromion.

8.4 Each upper limb has 30 bones.

8.5 The radius articulates at the elbow with the capitulum and radial fossa of the humerus. The ulna articulates at the elbow with the trochlea, coronoid fossa, and olecranon fossa of the humerus. The interosseous membrane connects their shafts.

8.6 The "elbow" part of the ulna is the olecranon.

8.7 The radius and ulna form three joints, the connection at the interosseus membrane and the proximal and distal radioulnar joints.

8.8 The most frequently fractured wrist bone is the scaphoid.

8.9 The bony pelvis attaches the lower limbs to the axial skeleton and supports the backbone and pelvic viscera.

8.10 The femur articulates with the acetabulum of the hip bone; the sacrum articulates with the auricular surface of the hip bone.

8.11 The pelvic axis is the course taken by a baby's head as it descends through the pelvis during childbirth.

8.12 Each lower limb has 30 bones.

8.13 The angle of convergence of the femurs is greater in females than males because the female pelvis is broader.

8.14 The patella is a sesamoid bone, the largest one in the body.

8.15 The tibia is the weight-bearing bone of the leg.

8.16 The talus articulates with the tibia and the fibula.

8.17 The arches are not rigid; they yield when weight is applied and spring back when weight is lifted.

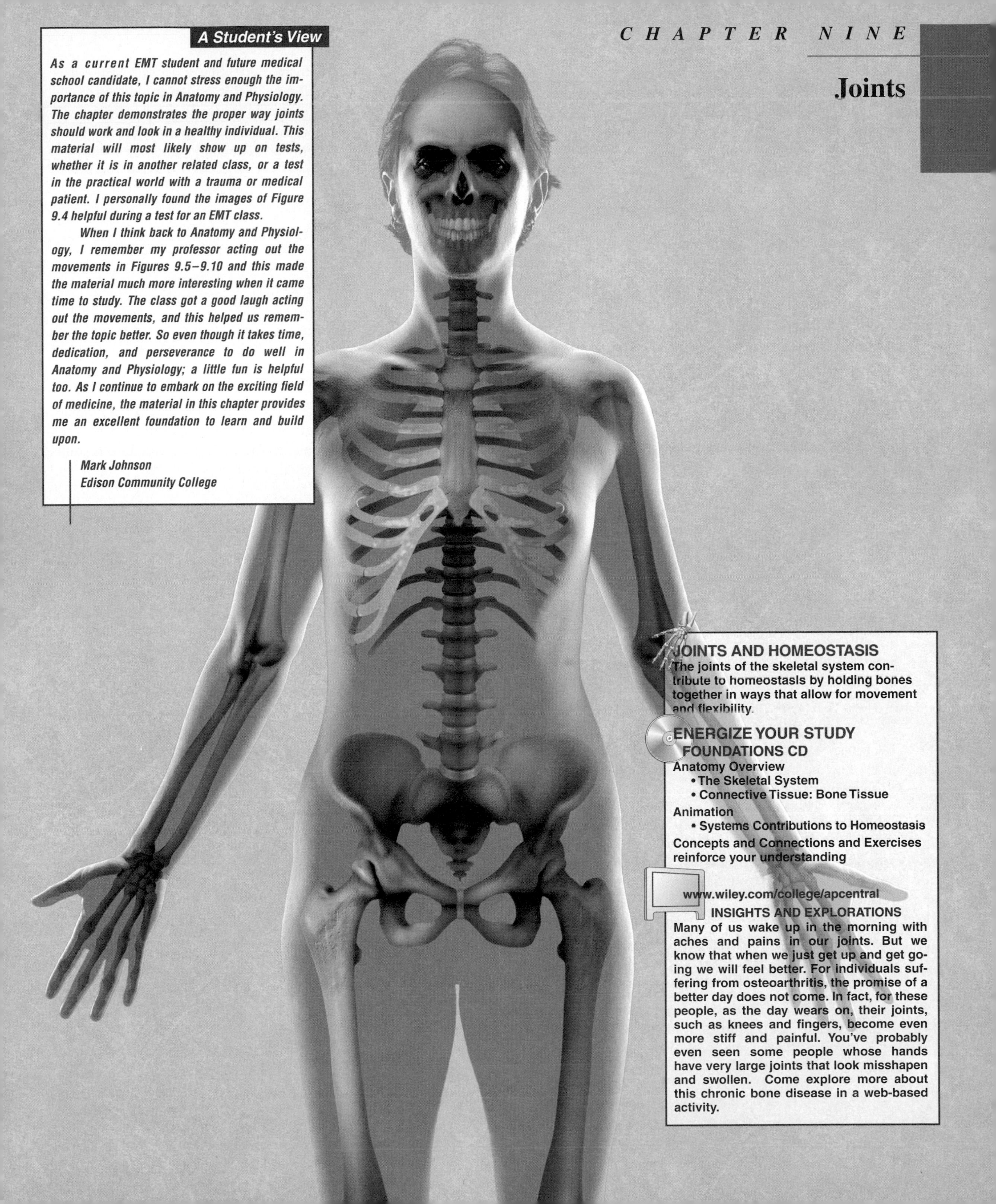

CHAPTER NINE

Joints

A Student's View

As a current EMT student and future medical school candidate, I cannot stress enough the importance of this topic in Anatomy and Physiology. The chapter demonstrates the proper way joints should work and look in a healthy individual. This material will most likely show up on tests, whether it is in another related class, or a test in the practical world with a trauma or medical patient. I personally found the images of Figure 9.4 helpful during a test for an EMT class.

When I think back to Anatomy and Physiology, I remember my professor acting out the movements in Figures 9.5–9.10 and this made the material much more interesting when it came time to study. The class got a good laugh acting out the movements, and this helped us remember the topic better. So even though it takes time, dedication, and perseverance to do well in Anatomy and Physiology; a little fun is helpful too. As I continue to embark on the exciting field of medicine, the material in this chapter provides me an excellent foundation to learn and build upon.

Mark Johnson
Edison Community College

JOINTS AND HOMEOSTASIS

The joints of the skeletal system contribute to homeostasis by holding bones together in ways that allow for movement and flexibility.

ENERGIZE YOUR STUDY
FOUNDATIONS CD

Anatomy Overview
- The Skeletal System
- Connective Tissue: Bone Tissue

Animation
- Systems Contributions to Homeostasis

Concepts and Connections and Exercises reinforce your understanding

www.wiley.com/college/apcentral

INSIGHTS AND EXPLORATIONS

Many of us wake up in the morning with aches and pains in our joints. But we know that when we just get up and get going we will feel better. For individuals suffering from osteoarthritis, the promise of a better day does not come. In fact, for these people, as the day wears on, their joints, such as knees and fingers, become even more stiff and painful. You've probably even seen some people whose hands have very large joints that look misshapen and swollen. Come explore more about this chronic bone disease in a web-based activity.

Bones are too rigid to bend without being damaged. Fortunately, flexible connective tissues form joints that hold bones together while still permitting, in most cases, some degree of movement. A **joint,** also called an **articulation** or **arthrosis,** is a point of contact between two bones, between bone and cartilage, or between bone and teeth. When we say one bone *articulates* with another bone, we mean that the bones form a joint. Because most movements of the body occur at joints, you can appreciate their importance if you imagine how a cast over your knee joint makes walking difficult, or how a splint on a finger limits your ability to manipulate small objects. The scientific study of joints is termed **arthrology** (ar-THROL-ō-jē; *arthr-* = joint; *-logy* = study of). The study of motion of the human body is called **kinesiology** (ki-nē-sē′-OL-ō-jē; *kinesi* = movement).

JOINT CLASSIFICATIONS

OBJECTIVE

- **Describe the structural and functional classifications of joints.**

Joints are classified structurally, based on their anatomical characteristics, and functionally, based on the type of movement they permit.

The structural classification of joints is based on two criteria: (1) the presence or absence of a space between the articulating bones, called a synovial cavity, and (2) the type of connective tissue that binds the bones together. Structurally, joints are classified as one of the following types:

- **Fibrous joints:** The bones are held together by fibrous connective tissue that is rich in collagen fibers; they lack a synovial cavity.
- **Cartilaginous joints:** The bones are held together by cartilage; they lack a synovial cavity.
- **Synovial joints:** The bones forming the joint have a synovial cavity and are united by the dense irregular connective tissue of an articular capsule, and often by accessory ligaments.

The functional classification of joints relates to the degree of movement they permit. Functionally, joints are classified as one of the following types:

- **Synarthrosis** (sin′-ar-THRŌ-sis; *syn-* = together): An immovable joint. The plural is *synarthroses.*
- **Amphiarthrosis** (am′-fē-ar-THRŌ-sis; *amphi-* = on both sides): A slightly movable joint. The plural is *amphiarthroses.*
- **Diarthrosis** (dī-ar-THRŌ-sis = movable joint): A freely movable joint. The plural is *diarthroses.* All diarthroses are synovial joints. They have a variety of shapes and permit several different types of movements.

The following sections present the joints of the body according to their structural classification. As we examine the structure of each type of joint, we will also explore its functional attributes.

FIBROUS JOINTS

OBJECTIVE

- **Describe the structure and functions of the three types of fibrous joints.**

Fibrous joints lack a synovial cavity, and the articulating bones are held very closely together by fibrous connective tissue. They permit little or no movement. The three types of fibrous joints are sutures, syndesmoses, and gomphoses.

Sutures

A **suture** (SOO-chur; *sutur-* = seam) is a fibrous joint composed of a thin layer of dense fibrous connective tissue that unites only bones of the skull. An example is the coronal suture between the parietal and frontal bones (Figure 9.1a). The irregular, interlocking edges of sutures give them added strength and decrease their chance of fracturing. Because a suture is immovable, it is classified functionally as a synarthrosis.

Some sutures, although present during childhood, are replaced by bone in the adult. Such a suture is called a **synostosis** (sin′-os-TŌ-sis; *os-* = bone), or bony joint—a joint in which there is a complete fusion of bone across the suture line. An example is the *metopic suture* between the left and right sides of the frontal bone that begins to fuse during infancy.

Syndesmoses

A **syndesmosis** (sin′-dez-MŌ-sis; *syndesmo-* = band or ligament) is a fibrous joint in which, compared to a suture, there is a greater distance between the articulating bones and more fibrous connective tissue. The fibrous connective tissue is arranged either as a bundle (ligament) or as a sheet (interosseous membrane) (Figure 9.1b). One example of a syndesmosis is the distal tibiofibular joint, where the anterior tibiofibular ligament connects the tibia and fibula. Another example is the interosseous membrane between the parallel borders of the tibia and fibula. Because it permits slight movement, a syndesmosis is classified functionally as an amphiarthrosis.

Figure 9.1 Fibrous joints.

At a fibrous joint the bones are held together by fibrous connective tissue.

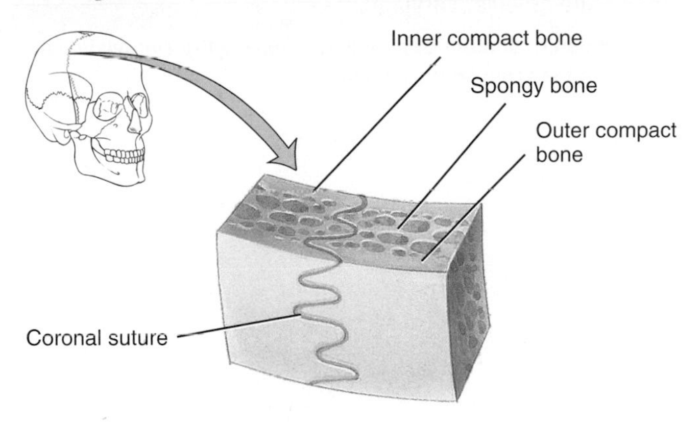

(a) Suture between skull bones

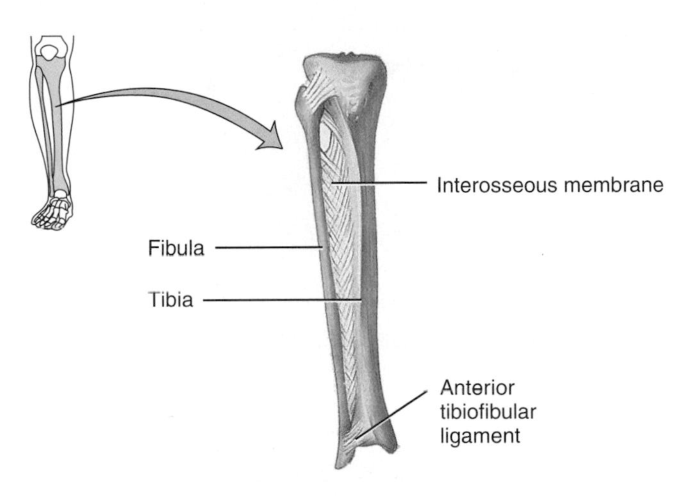

(b) Syndesmoses between tibia and fibula

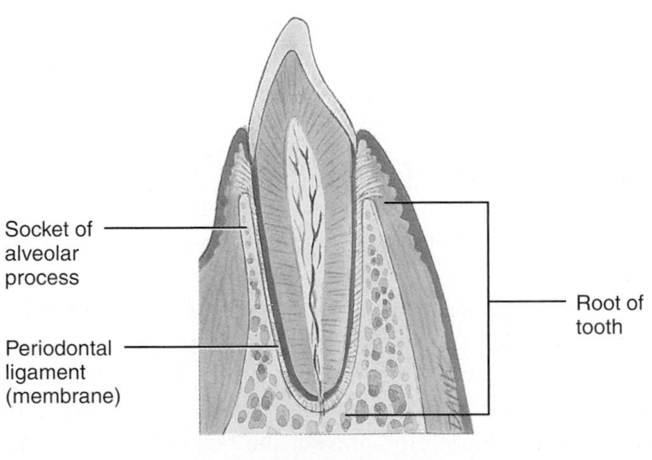

(c) Gomphosis between tooth and socket of alveolar process

Functionally, why are sutures classified as synarthroses, and syndesmoses as amphiarthroses?

Gomphoses

A **gomphosis** (gom-FŌ-sis; *gompho-* = a bolt or nail) or *dento-alveolar joint* is a type of fibrous joint in which a cone-shaped peg fits into a socket. The only examples of gomphoses are the articulations of the roots of the teeth with the sockets (alveoli) of the alveolar processes of the maxillae and mandible (Figure 9.1c). The dense fibrous connective tissue between a tooth and its socket is the periodontal ligament (membrane). A gomphosis is classified functionally as a synarthrosis, an immovable joint.

CARTILAGINOUS JOINTS

OBJECTIVE

- **Describe the structure and functions of the two types of cartilaginous joints.**

Like a fibrous joint, a **cartilaginous joint** lacks a synovial cavity and allows little or no movement. Here the articulating bones are tightly connected by either hyaline cartilage or fibrocartilage. The two types of cartilaginous joints are synchondroses and symphyses.

Synchondroses

A **synchondrosis** (sin′-kon-DRŌ-sis; *chondro-* = cartilage) is a cartilaginous joint in which the connecting material is hyaline cartilage. An example of a synchondrosis is the epiphyseal plate that connects the epiphysis and diaphysis of a growing bone (Figure 9.2a). A photomicrograph of the epiphyseal plate is shown in Figure 6.8 on page 174. Functionally, a synchondrosis is a synarthrosis. When bone elongation ceases, bone replaces the hyaline cartilage, and the synchondrosis becomes a synostosis, a bony joint. Another example of a synchondrosis is the joint between the first rib and the manubrium of the sternum, which also ossifies during adult life and becomes an immovable synostosis.

Symphyses

A **symphysis** (SIM-fi-sis = growing together) is a cartilaginous joint in which the ends of the articulating bones are covered with hyaline cartilage, but a broad, flat disc of fibrocartilage connects the bones. All symphyses occur in the midline of the body. The pubic symphysis between the anterior surfaces of the hip bones is one example of a symphysis (Figure 9.2b). This type of joint is also found at the intervertebral joints between the bodies of vertebrae (see Figure 7.17 on page 205). A portion of the intervertebral disc is fibrocartilage. A symphysis is an amphiarthrosis, a slightly movable joint.

Figure 9.2 Cartilaginous joints.

At a cartilaginous joint the bones are held together by cartilage.

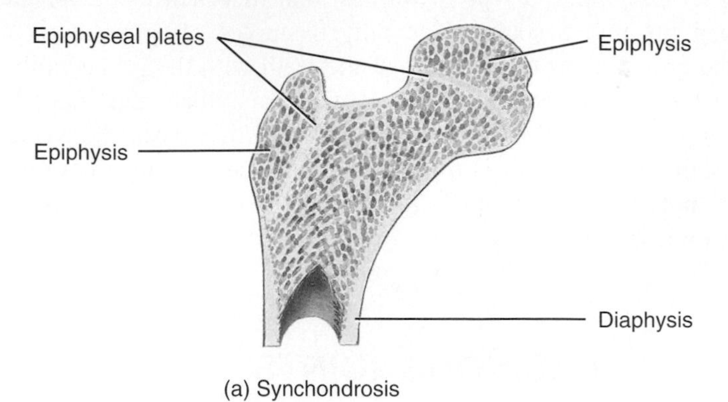

(a) Synchondrosis

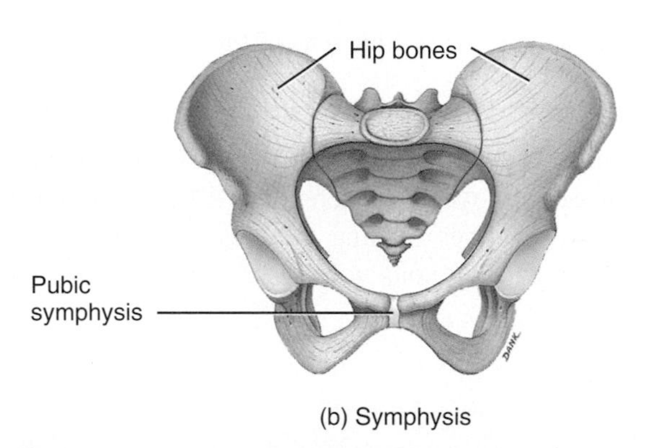

(b) Symphysis

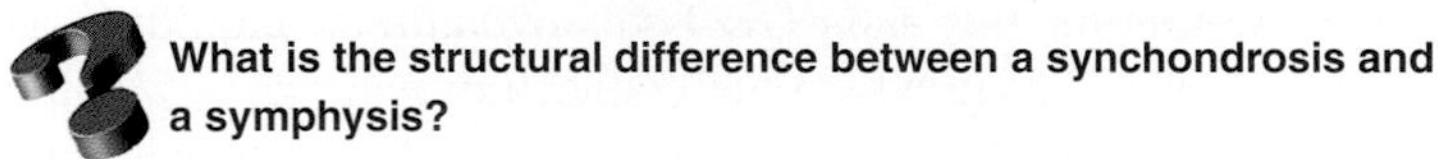

What is the structural difference between a synchondrosis and a symphysis?

Figure 9.3 Structure of a typical synovial joint. Note the two layers of the articular capsule—the fibrous capsule and the synovial membrane. Synovial fluid fills the joint cavity between the synovial membrane and the articular cartilage.

The distinguishing feature of a synovial joint is the synovial cavity between the articulating bones.

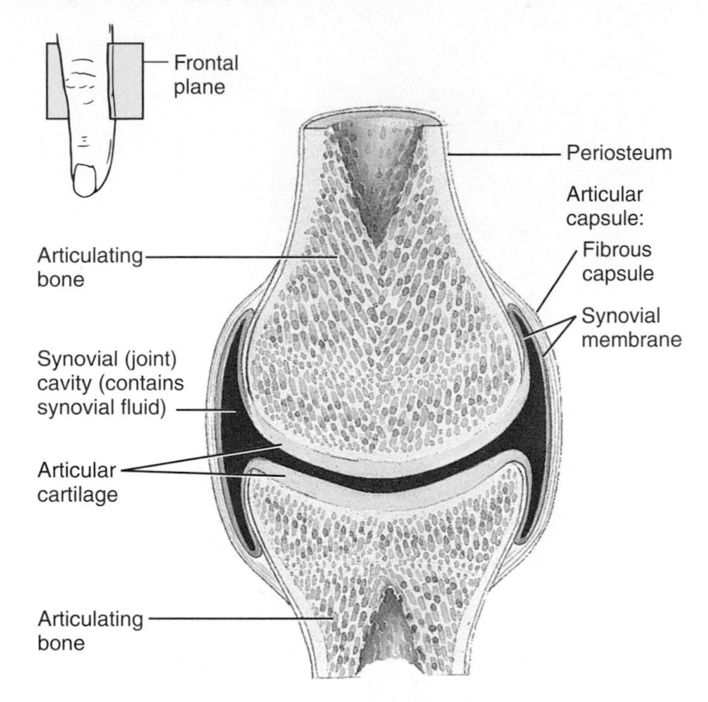

Frontal section

What is the functional classification of synovial joints?

▶ **CHECKPOINT**

1. Where in the body can each type of fibrous and cartilaginous joint be found?
2. What type of tissue holds the articulating bones together in fibrous and cartilaginous joints?
3. Which of the fibrous and cartilaginous joints are classified as amphiarthroses?

SYNOVIAL JOINTS

▶ **OBJECTIVES**

- **Describe the structure of synovial joints.**
- **Describe the six subtypes of synovial joints.**
- **Describe the structure and function of bursae and tendon sheaths.**

Structure of Synovial Joints

Synovial joints (si-NŌ-vē-al) have certain characteristics that distinguish them from other joints. The unique characteristic of a synovial joint is the presence of a space called a **synovial (joint) cavity** between the articulating bones (Figure 9.3). The synovial cavity allows a joint to be freely moveable. Hence, all synovial joints are classified functionally as diarthroses. The bones at a synovial joint are covered by **articular cartilage,** which is hyaline cartilage. The cartilage covers the articulating surface of the bones with a smooth, slippery surface but does not bind them together. Articular cartilage reduces friction between bones in the joint during movement and helps to absorb shock.

Articular Capsule

A sleevelike **articular capsule** surrounds a synovial joint, encloses the synovial cavity, and unites the articulating bones. The articular capsule is composed of two layers, an outer fibrous capsule and an inner synovial membrane (Figure 9.3). The outer layer, the **fibrous capsule,** usually consists of dense irregular connective tissue that attaches to the periosteum of the articulating bones. The flexibility of the fibrous capsule permits considerable movement at a joint while its great tensile strength (resistance to stretching) helps prevent the bones from dislocating. The fibers of some fibrous capsules are arranged in parallel bundles that are highly adapted for resisting strains. Such fiber bundles are called **ligaments** (*liga-* = bound or tied) and are often designated by individual names. The strength of ligaments is one of the main mechanical factors that hold bones close together in a synovial joint. The inner layer of the articular capsule, the **synovial membrane,** is composed of areolar connective tissue with elastic fibers. At many synovial joints the synovial membrane includes accumulations of adipose tissue, called **articular fat pads.** An example is the infrapatellar fat pad in the knee (see Figure 9.14c).

Synovial Fluid

The synovial membrane secretes **synovial fluid** (*ov-* = egg), which forms a thin film over the surfaces within the articular capsule. This viscous, clear or pale yellow fluid was named for its similarity in appearance and consistency to uncooked egg white (albumin). Synovial fluid consists of hyaluronic acid, secreted by fibroblast-like cells in the synovial membrane, and interstitial fluid filtered from blood plasma. Its several functions include reducing friction by lubricating the joint, and supplying nutrients to and removing metabolic wastes from the chondrocytes within articular cartilage. (Recall that cartilage is an avascular tissue.) Synovial fluid also contains phagocytic cells that remove microbes and the debris that results from normal wear and tear in the joint. When a synovial joint is immobile for a time, the fluid is quite viscous (gel-like), but as joint movement increases, the fluid becomes less viscous. One of the benefits of a warm-up before exercise is that it stimulates the production and secretion of synovial fluid.

Accessory Ligaments and Articular Discs

Many synovial joints also contain **accessory ligaments** called extracapsular ligaments and intracapsular ligaments. *Extracapsular ligaments* lie outside the articular capsule. Examples are the fibular and tibial collateral ligaments of the knee joint (see Figure 9.14d). *Intracapsular ligaments* occur within the articular capsule but are excluded from the synovial cavity by folds of the synovial membrane. Examples are the anterior and posterior cruciate ligaments of the knee joint (see Figure 9.14d).

Inside some synovial joints, such as the knee, are pads of fibrocartilage that lie between the articular surfaces of the bones and are attached to the fibrous capsule. These pads are called **articular discs** or **menisci** (me-NIS-sī; singular is *meniscus*). Figure 9.14d depicts the lateral and medial menisci in the knee joint. The discs usually subdivide the synovial cavity into two separate spaces as in the temporomandibular joint (TMJ), thus permitting separate movements in the respective spaces. By modifying the shape of the joint surfaces of the articulating bones, articular discs allow two bones of different shapes to fit more tightly. Articular discs also help to maintain the stability of the joint and direct the flow of synovial fluid to the areas of greatest friction.

Torn Cartilage and Arthroscopy

The tearing of articular discs (menisci) in the knee, commonly called **torn cartilage,** occurs often among athletes. Such damaged cartilage will begin to wear and may precipitate arthritis unless surgically removed (meniscectomy). Surgical repair of the torn cartilage may be assisted by **arthroscopy** (ar-THROS-kō-pē; *-scopy* = observation). This procedure involves examination of the interior of a joint, usually the knee, with an arthroscope, a lighted, pencil-thin instrument used for visualization. Arthroscopy is used to determine the nature and extent of damage following knee injury; to help remove torn cartilage and repair damaged cruciate ligaments in the knee; to help obtain tissue samples for analysis; to monitor the progression of disease and the effects of therapy; and to help perform surgery on other joints, such as the shoulder, elbow, ankle, and wrist. ■

Nerve and Blood Supply

The nerves that supply a joint are the same as those that supply the skeletal muscles that move the joint. Synovial joints contain many nerve endings that are distributed to the articular capsule and associated ligaments. Some of the nerve endings convey information about pain from the joint to the spinal cord and brain for processing. Other nerve endings are responsive to the degree of movement and stretch at a joint. This information, too, is relayed to the spinal cord and brain, which may respond by sending impulses through different nerves to the muscles to adjust body movements.

Arteries in the vicinity of a synovial joint send out many branches that penetrate the ligaments and articular capsule to deliver oxygen and nutrients. Veins remove carbon dioxide and wastes from the joints. The arterial branches from several different arteries typically join together around a joint before penetrating the articular capsule. The articulating portions of a synovial joint receive nourishment from synovial fluid, whereas all other joint tissues are supplied by blood capillaries.

Sprain and Strain

A **sprain** is the forcible wrenching or twisting of a joint that stretches or tears its ligaments but does not dislocate the bones. It occurs when the ligaments are stressed beyond their normal capacity. Sprains also may damage surrounding blood vessels, muscles, tendons, or nerves. Severe sprains may be so painful that the joint cannot be moved. There is considerable swelling, which results from hemorrhage of ruptured blood vessels. The ankle joint is most often sprained; the lower back is another frequent location for sprains. A **strain** is a stretched or partially torn muscle. It often occurs when a muscle contracts suddenly and powerfully—for example, in sprinters when they accelerate too quickly. ■

Bursae and Tendon Sheaths

The various movements of the body create friction between moving parts. Saclike structures called **bursae** (BER-sē = purses; singular is *bursa*) are strategically situated to reduce friction in some synovial joints, such as the shoulder and knee joints (see Figures 9.11 and 9.14c). Bursae are not strictly part of synovial joints, but do resemble joint capsules because their walls consist of connective tissue lined by a synovial membrane. They are also filled with a fluid similar to synovial fluid. Bursae are located between the skin and bone in places where skin rubs over bone. They are also found between tendons and bones, muscles and bones, and ligaments and bones. The fluid-filled bursal sacs cushion the movement of one body part over another.

In addition to bursae, structures called tendon sheaths also reduce friction at joints. **Tendon sheaths** are tubelike bursae that wrap around tendons where there is considerable friction. This occurs where tendons pass through synovial cavities, such as the tendon of the biceps brachii muscle at the shoulder joint (see Figure 9.11c). Other areas of considerable friction where tendon sheaths are found are at the wrist and ankle, where many tendons come together in a confined space (see Figure 11.22 on page 374), and in the fingers and toes, where there is a great deal of movement (see Figure 11.18a on page 358).

Bursitis

An acute or chronic inflammation of a bursa is called **bursitis.** The condition may be caused by trauma, by an acute or chronic infection (including syphilis and tuberculosis), or by rheumatoid arthritis (described on page 267). Repeated, excessive exertion of a joint often results in bursitis, with local inflammation and the accumulation of fluid. Symptoms include pain, swelling, tenderness, and limited movement. Treatment may include oral anti-inflammatory agents and injections of cortisol-like steroids. ■

Types of Synovial Joints

Although all synovial joints have a generally similar structure, the shapes of the articulating surfaces vary and thus various types of movements are possible. Accordingly, synovial joints are divided into six subtypes: planar, hinge, pivot, condyloid, saddle, and ball-and-socket joints.

Planar Joints

The articulating surfaces of bones in a **planar joint** are flat or slightly curved (Figure 9.4a). Planar joints primarily permit side-to-side and back-and-forth gliding movements (described shortly). These joints are said to be *nonaxial* because the motion they allow does not occur around an axis. However, x-ray films made during wrist and ankle movements reveal some rotation of the small carpal and tarsal bones in addition to their predominant gliding movements. Examples of planar joints are the intercarpal joints (between carpal bones at the wrist), intertarsal joints (between tarsal bones at the ankle), sternoclavicular joints (between the manubrium of the sternum and the clavicle), acromioclavicular joints (between the acromion of the scapula and the clavicle), sternocostal joints (between the sternum and ends of the costal cartilages at the tips of the second through seventh pairs of ribs), and vertebrocostal joints (between the heads and tubercles of ribs and transverse processes of thoracic vertebrae).

Hinge Joints

In a **hinge joint,** the convex surface of one bone fits into the concave surface of another bone (Figure 9.4b). As the name implies, hinge joints produce an angular, opening-and-closing motion like that of a hinged door. Hinge joints are *monaxial (uniaxial)* because they typically allow motion around a single axis. Examples of hinge joints are the knee, elbow, ankle, and interphalangeal joints.

Pivot Joints

In a **pivot joint,** the rounded or pointed surface of one bone articulates with a ring formed partly by another bone and partly by a ligament (Figure 9.4c). A pivot joint is monaxial because it allows rotation around its own longitudinal axis only. Examples of pivot joints are the atlanto-axial joint, in which the atlas rotates around the axis and permits the head to turn from side to side as in signifying "no" (see Figure 9.9a), and the radioulnar joints that enable the palms to turn anteriorly and posteriorly (see Figure 9.10h).

Condyloid Joints

In a **condyloid joint** (KON-di-loyd; *condyl-* = knuckle) or *ellipsoidal* joint, the convex oval-shaped projection of one bone fits into the oval-shaped depression of another bone (Figure 9.4d). A condyloid joint is *biaxial* because the movement it permits is around two axes. Notice that you can move your index finger both up-and-down and from side-to-side. Examples are the wrist and metacarpophalangeal joints for the second through fifth digits.

Saddle Joints

In a **saddle joint,** the articular surface of one bone is saddle-shaped, and the articular surface of the other bone fits into the "saddle" as a sitting rider would sit (Figure 9.4e). A saddle joint is a modified condyloid joint in which the movement is somewhat freer. Saddle joints are *biaxial,* producing side-to-side and up-and-down movements. An example of a saddle joint is the carpometacarpal joint between the trapezium of the carpus and metacarpal of the thumb.

Ball-and-Socket Joints

A **ball-and-socket joint** consists of the ball-like surface of one bone fitting into a cuplike depression of another bone (Figure 9.4f). Such joints are *multiaxial (polyaxial)* because they permit movement around three axes plus all directions in between. Examples of functional ball-and-socket joints are the shoulder

Figure 9.4 Subtypes of synovial joints. For each subtype, a drawing of the actual joint and a simplified diagram are shown.

Synovial joints are classified into subtypes based on the shapes of the articulating bone surfaces.

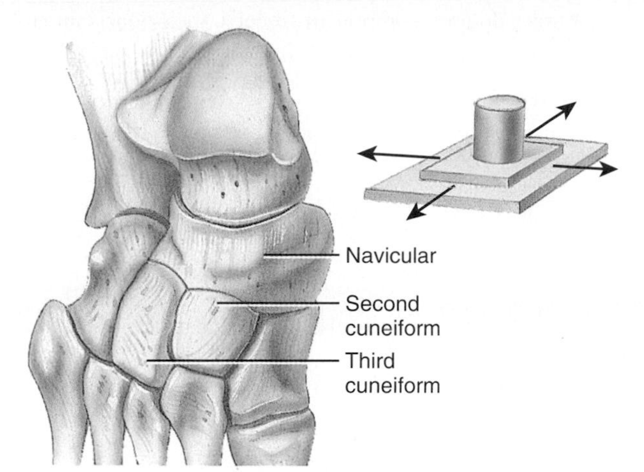

(a) Planar joint between the navicular and second and third cuneiforms of the tarsus in the foot

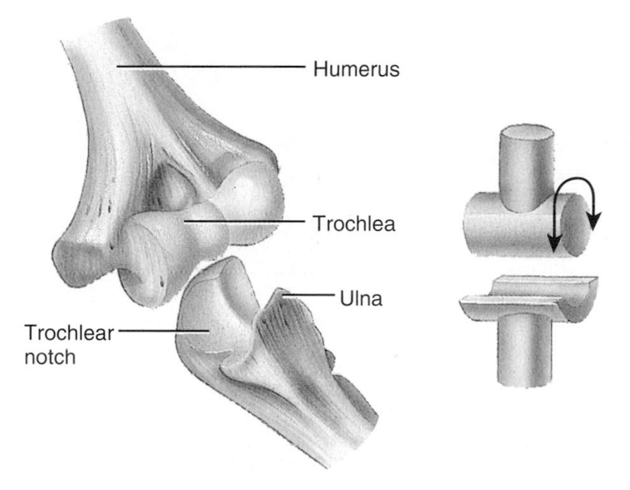

(b) Hinge joint between trochlea of humerus and trochlear notch of ulna at the elbow

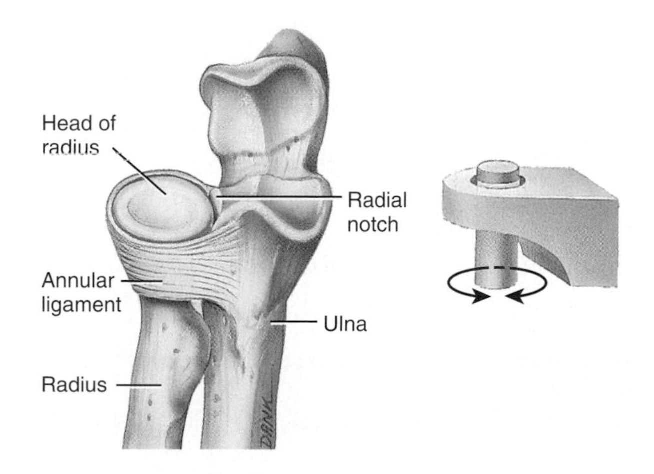

(c) Pivot joint between head of radius and radial notch of ulna

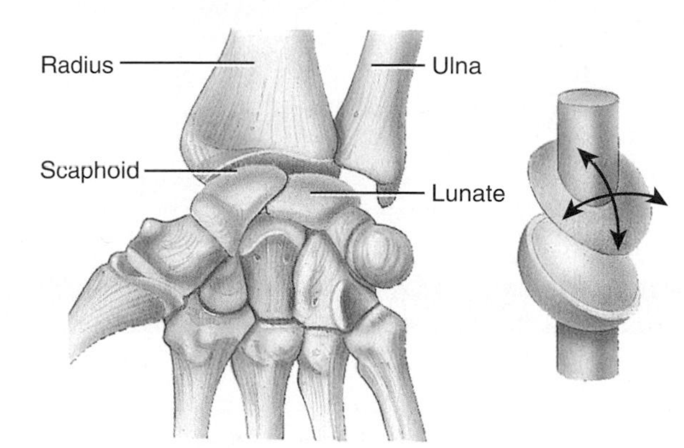

(d) Condyloid joint between radius and scaphoid and lunate bones of the carpus (wrist)

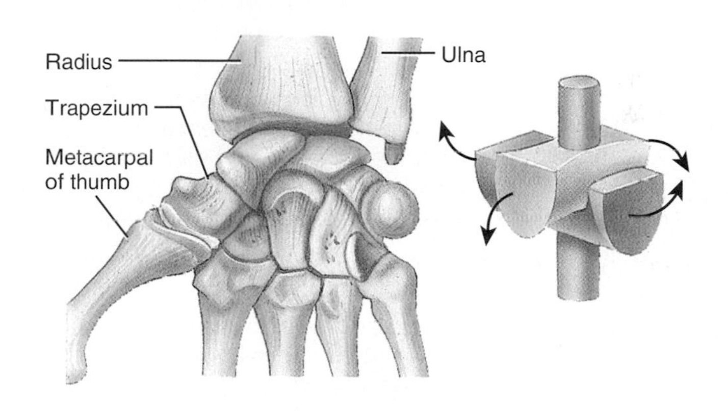

(e) Saddle joint between trapezium of carpus (wrist) and metacarpal of thumb

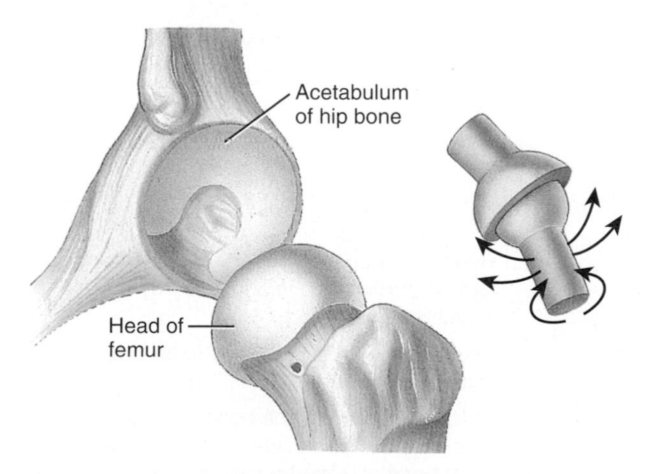

(f) Ball-and-socket joint between head of the femur and acetabulum of the hip bone

Which of the joints shown here are biaxial?

and hip joints. At the shoulder joint, the head of the humerus fits into the glenoid cavity of the scapula. At the hip joint, the head of the femur fits into the acetabulum of the hip bone.

Table 9.1 summarizes the structural and functional categories of joints.

▶ CHECKPOINT

4. How does the structure of synovial joints classify them as diarthroses?
5. What are the functions of articular cartilage, synovial fluid, and articular discs?
6. What types of sensations are perceived at joints, and from what sources do joint components receive their nourishment?
7. In what ways are bursae similar to joint capsules? How do they differ?
8. Which types of joints are nonaxial, monaxial, biaxial, and multiaxial?

Table 9.1 Summary of Structural and Functional Classifications of Joints

Structural Classification	Description	Functional Classification	Example
Fibrous	Articulating bones held together by fibrous connective tissue; no synovial cavity.		
Suture	Articulating bones united by a thin layer of dense fibrous connective tissue, found between bones of the skull. With age, some sutures are replaced by a synostosis, in which the bones fuse across the former suture.	Synarthrosis (immovable).	Frontal suture.
Syndesmosis	Articulating bones united by dense fibrous connective tissue, either a ligament or an interosseous membrane.	Amphiarthrosis (slightly movable).	Distal tibiofibular joint.
Gomphosis	Articulating bones united by periodontal ligament; cone-shaped peg fits into a socket.	Synarthrosis.	At roots of teeth in alveoli (sockets) of maxillae and mandible.
Cartilaginous	Articulating bones united by cartilage; no synovial cavity.		
Synchondrosis	Connecting material is hyaline cartilage; becomes a synostosis when bone elongation ceases.	Synarthrosis.	Epiphyseal plate at between the joint diaphysis and epiphysis of a long bone.
Symphysis	Connecting material is a broad, flat disc of fibrocartilage.	Amphiarthrosis.	Intervertebral joints and pubic symphysis.
Synovial	Characterized by a synovial cavity, articular cartilage, and an articular capsule; may contain accessory ligaments, articular discs, and bursae.		
Planar	Articulated surfaces are flat or slightly curved.	Nonaxial diarthrosis (freely movable); gliding motion.	Intercarpal, intertarsal, sternocostal (between sternum and the 2nd–7th pairs of ribs), and vertebrocostal joints.
Hinge	Convex surface fits into a concave surface.	Monaxial diarthrosis; angular motion.	Elbow, ankle, and interphalangeal joints.
Pivot	Rounded or pointed surface fits into a ring formed partly by bone and partly by a ligament.	Monaxial diarthrosis; rotation.	Atlanto-axial and radioulnar joints.
Condyloid	Oval-shaped projection fits into an oval-shaped depression.	Biaxial diarthrosis; angular motion.	Radiocarpal and metacarpophalangeal joints.
Saddle	Articular surface of one bone is saddle-shaped, and the articular surface of the other bone "sits" in the saddle.	Biaxial diarthrosis; angular motion.	Carpometarcarpal joint between trapezium and thumb.
Ball-and-socket	Ball-like surface fits into a cuplike depression.	Multiaxial diarthrosis; angular motion and rotation.	Shoulder and hip joints.

TYPES OF MOVEMENTS AT SYNOVIAL JOINTS

OBJECTIVE

- **Describe the types of movements that can occur at synovial joints.**

Anatomists, physical therapists, and kinesiologists use specific terminology to designate the movements that can occur at synovial joints. These precise terms may indicate the form of motion, the direction of movement, or the relationship of one body part to another during movement. Movements at synovial joints are grouped into four main categories: (1) gliding, (2) angular movements, (3) rotation, and (4) special movements. This last category includes movements that occur only at certain joints.

Gliding

Gliding is a simple movement in which relatively flat bone surfaces move back-and-forth and from side-to-side with respect to one another (Figure 9.5). There is no significant alteration of the angle between the bones. Gliding movements are limited in range due to the structure of the articular capsule and associated ligaments and bones. Gliding occurs at planar joints.

Angular Movements

In **angular movements,** there is an increase or a decrease in the angle between articulating bones. The major angular movements are flexion, extension, lateral extension, hyperextension, abduction, adduction, and circumduction. These movements are discussed with respect to the body in the anatomical position.

Figure 9.5 Gliding movements at synovial joints.

Gliding motion consists of side-to-side and back-and-forth movements.

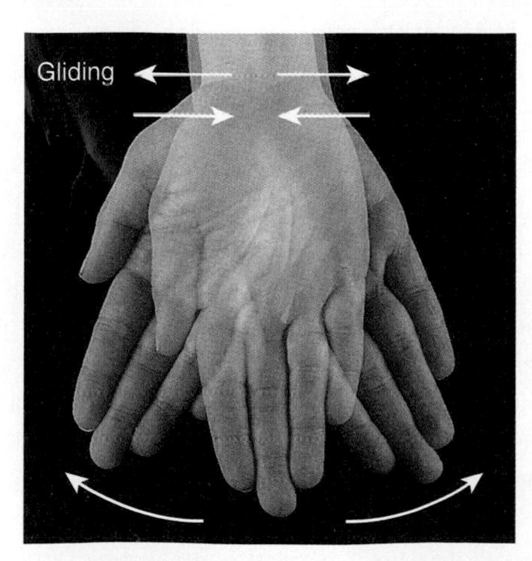

Intercarpal joints

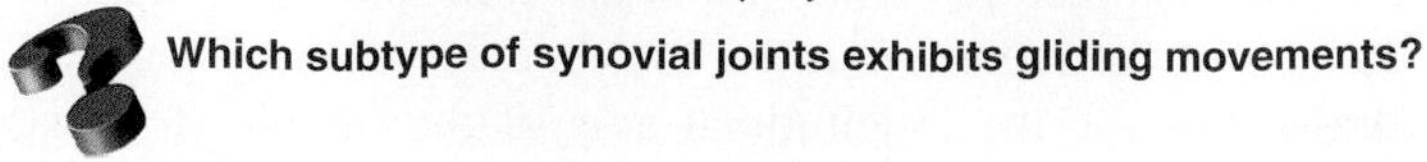

Which subtype of synovial joints exhibits gliding movements?

Flexion, Extension, Lateral Flexion, and Hyperextension

Flexion and extension are opposite movements. In **flexion** (FLEK-shun; *flex-* = to bend) there is a decrease in the angle between articulating bones; in **extension** (eks-TEN-shun; *exten-* = to stretch out) there is an increase in the angle between articulating bones, often to restore a part of the body to the anatomical position after it has been flexed (Figure 9.6). Both movements usually occur in the sagittal plane. Hinge, pivot, condyloid, saddle, and ball-and-socket joints all permit flexion and extension. All of the following are examples of flexion:

- Bending the head toward the chest at the atlanto-occipital joint between the atlas (the first vertebra) and the occipital bone of the skull, and at the cervical intervertebral joints between the cervical vertebrae (Figure 9.6a)
- Bending the trunk forward at the intervertebral joints
- Moving the humerus forward at the shoulder joint, as in swinging the arms forward while walking (Figure 9.6b)
- Moving the forearm toward the arm at the elbow joint between the humerus, ulna, and radius (Figure 9.6c)
- Moving the palm toward the forearm at the wrist or radiocarpal joint between the radius and carpals (Figure 9.6d)
- Bending the digits of the hand or feet at the interphalangeal joints between phalanges
- Moving the femur forward at the hip joint between the femur and hip bone, as in walking (Figure 9.6e)
- Moving the leg toward the thigh at the knee or tibiofemoral joint between the tibia, femur, and patella, as occurs when bending the knee (Figure 9.6f)

Although flexion and extension usually occur in the sagittal plane, there are a few exceptions. For example, flexion of the thumb involves movement of the thumb medially across the palm at the carpometacarpal joint between the trapezium and metacarpal of the thumb, as in touching the thumb to the opposite side of the palm. Another example is movement of the trunk sideways to the right or left at the waist. This movement, which occurs in the frontal plane and involves the intervertebral joints, is called **lateral flexion** (Figure 9.6g).

Continuation of extension beyond the anatomical position is called **hyperextension** (*hyper-* = beyond or excessive). Examples of hyperextension include:

- Bending the head backward at the atlanto-occipital and cervical intervertebral joints (Figure 9.6a)
- Bending the trunk backward at the intervertebral joints
- Moving the humerus backward at the shoulder joint, as in swinging the arms backward while walking (Figure 9.6b)
- Moving the palm backward at the wrist joint (Figure 9.6d)
- Moving the femur backward at the hip joint, as in walking (Figure 9.6e)

Figure 9.6 Angular movements at synovial joints—flexion, extension, hyperextension, and lateral flexion.

In angular movements, there is an increase or decrease in the angle between articulating bones.

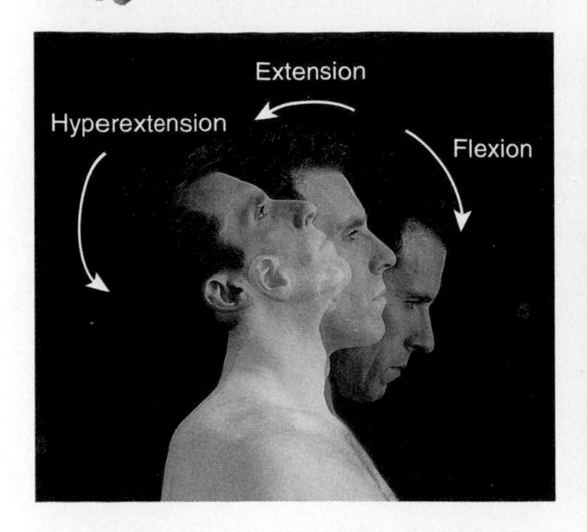

(a) Atlanto-occipital and cervical intervertebral joints

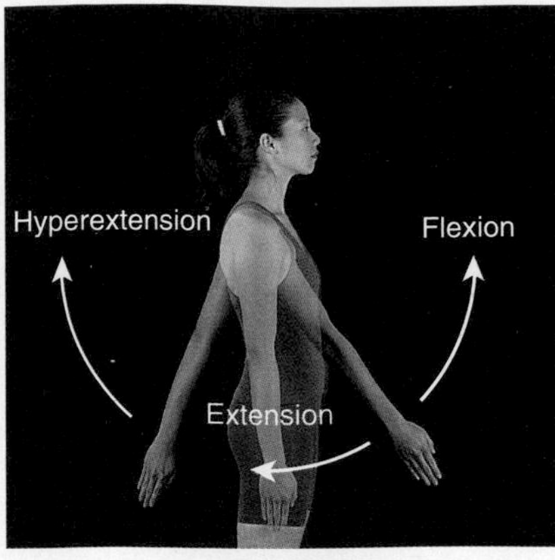

(b) Shoulder joint

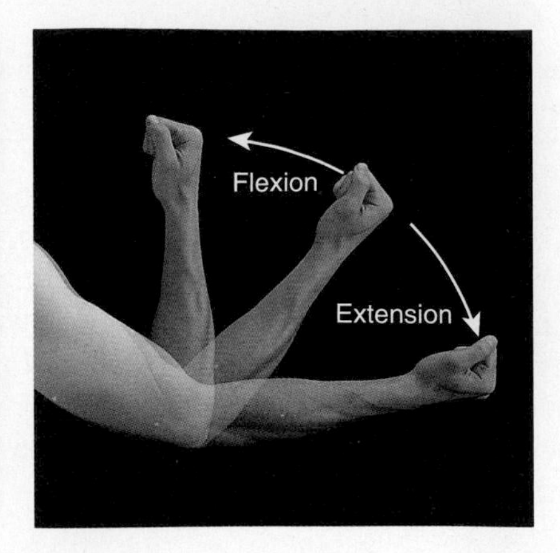

(c) Elbow joint

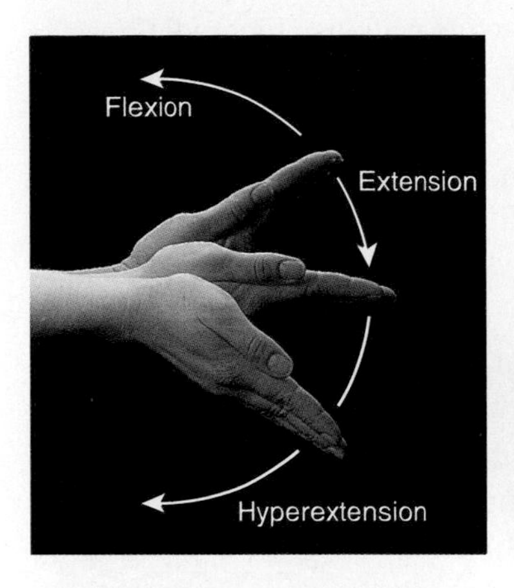

(d) Wrist joint

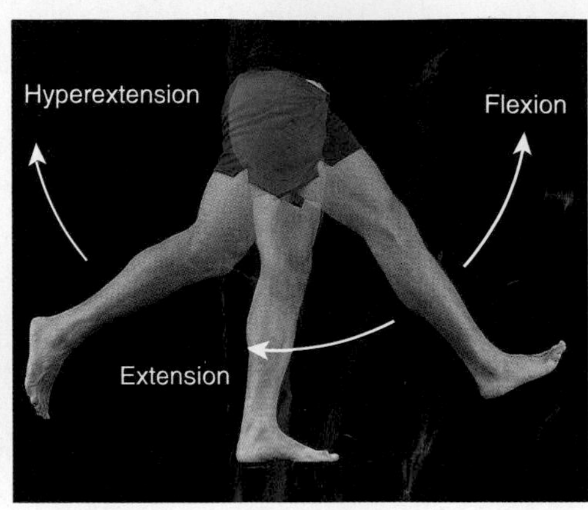

(e) Hip joint

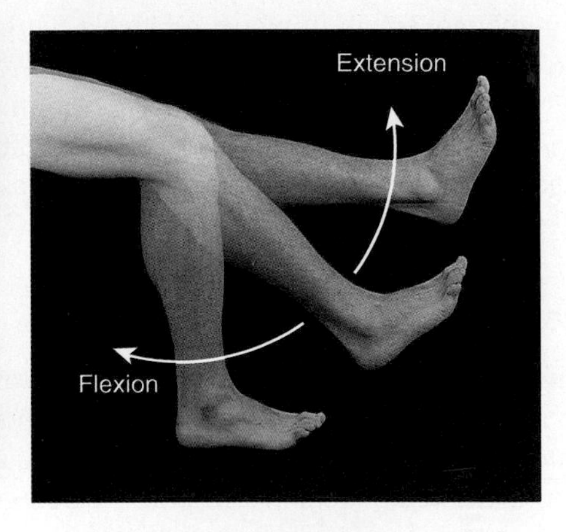

(f) Knee joint

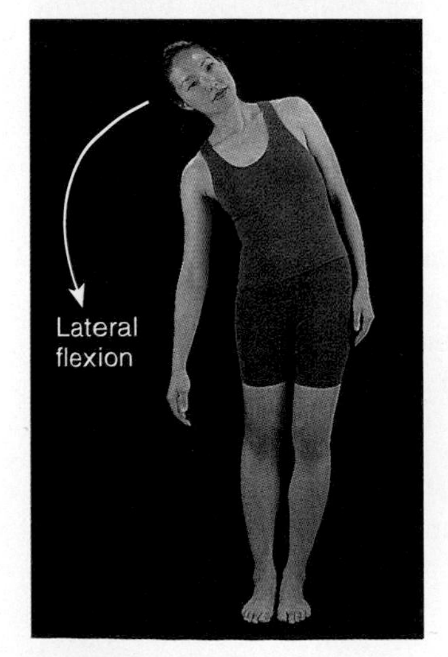

(g) Intervertebral joints

What are two examples of flexion that do not occur in the sagittal plane?

Hyperextension of hinge joints, such as the elbow, interphalangeal, and knee joints, is usually prevented by the arrangement of ligaments and the anatomical alignment of the bones.

Abduction, Adduction, and Circumduction

Abduction (ab-DUK-shun; *ab-* = away; *-duct* = to lead) is the movement of a bone away from the midline, whereas **adduction** (ad-DUK-shun; *ad-* = toward) is the movement of a bone toward the midline. Both movements usually occur in the frontal plane. Condyloid, saddle, and ball-and-socket joints permit abduction and adduction. Examples of abduction include moving the humerus laterally at the shoulder joint, moving the palm laterally at the wrist joint, and moving the femur laterally at the hip joint (Figure 9.7a–c). The movement that returns each of these body parts to the anatomical position is adduction (Figure 9.7a–c).

With respect to the digits, the midline of the body is not used as a point of reference for abduction and adduction. In abduction of the fingers (but not the thumb), an imaginary line is drawn through the longitudinal axis of the middle (longest)

Figure 9.7 Angular movements at synovial joints—abduction and adduction.

Abduction and adduction usually occur in the frontal plane.

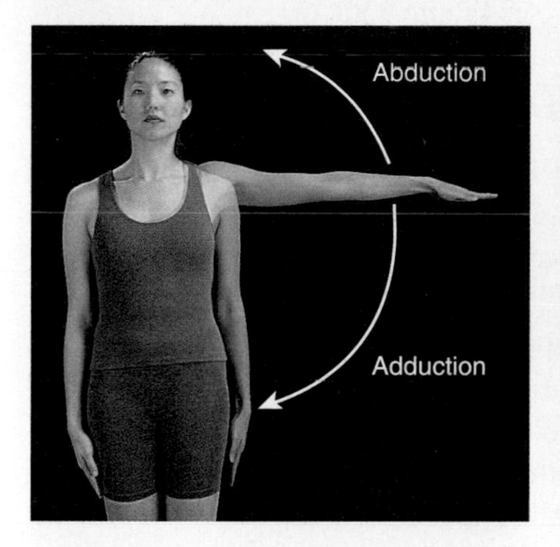

(a) Shoulder joint

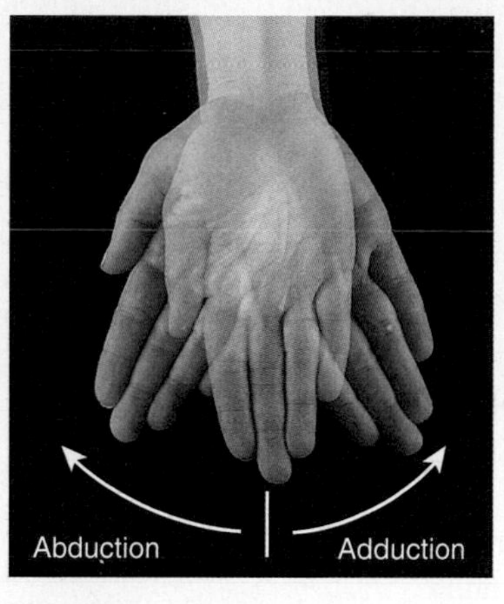

(b) Wrist joint

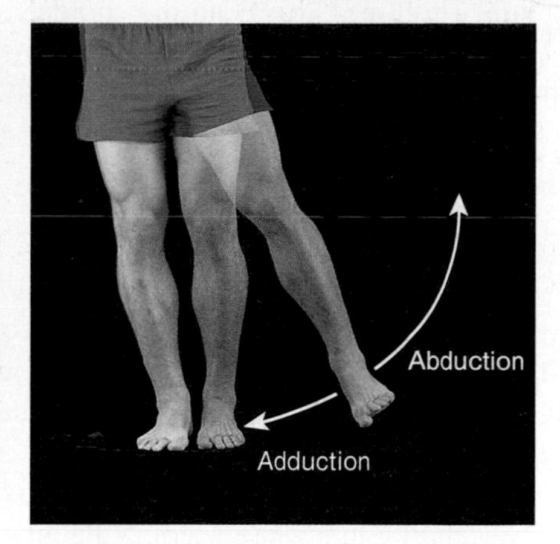

(c) Hip joint

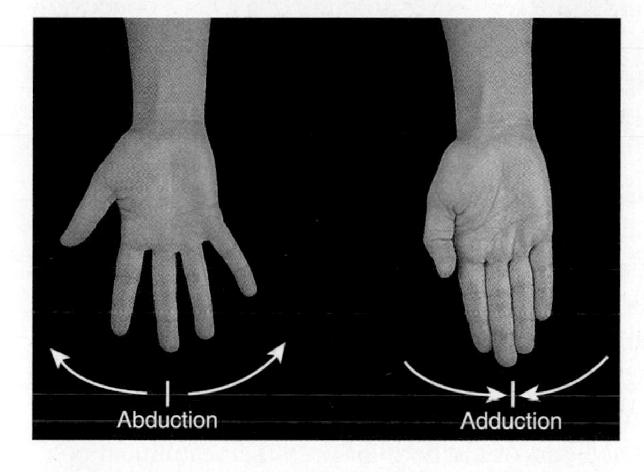

(d) Metacarpophalangeal joints of the fingers (not the thumb)

Is considering adduction as "adding your limb to your trunk" an effective learning device?

finger, and the fingers move away (spread out) from the middle finger (Figure 9.7d). In abduction of the thumb, the thumb moves away from the palm in the sagittal plane (see Figure 11.18c on page 358). Abduction of the toes is relative to an imaginary line drawn through the second toe. Adduction of the fingers and toes returns them to the anatomical position. Adduction of the thumb moves the thumb toward the palm in the sagittal plane (see Figure 11.18c).

Circumduction (ser-kum-DUK-shun; *circ-* = circle) is movement of the distal end of a body part in a circle (Figure 9.8). Circumduction occurs as a result of a continuous sequence of flexion, abduction, extension, and adduction. Examples of circumduction are moving the humerus in a circle at the shoulder joint (Figure 9.8a), moving the hand in a circle at the wrist joint,

Figure 9.8 Angular movements at synovial joints—circumduction.

Circumduction is the movement of the distal end of a body part in a circle.

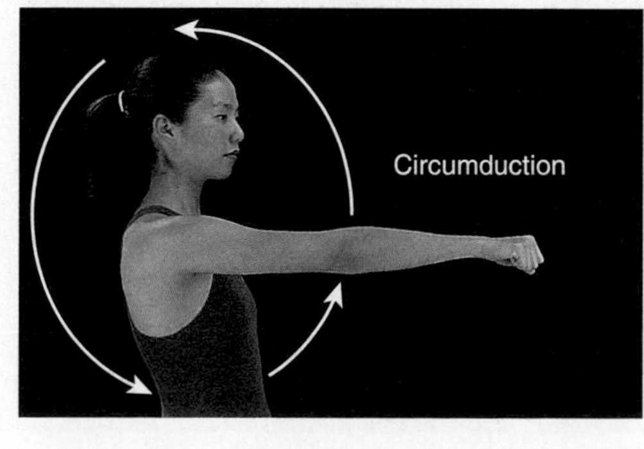

(a) Shoulder joint

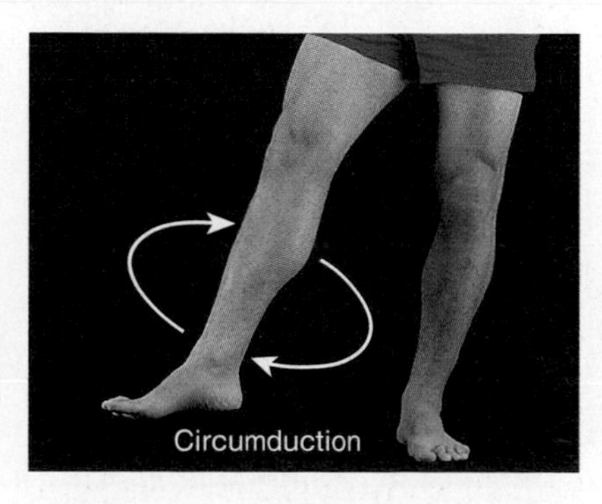

(b) Hip joint

Which movements in continuous sequence produce circumduction?

moving the thumb in a circle at the carpometacarpal joint, moving the fingers in a circle at the metacarpophalangeal joints (between the metacarpals and phalanges), and moving the femur in a circle at the hip joint (Figure 9.8b). Both the shoulder and hip joints permit circumduction. Flexion, abduction, extension, and adduction are more limited in the hip joints than in the shoulder joints due to the tension on certain ligaments and muscles (see Exhibits 9.1 and 9.3).

Rotation

In **rotation** (rō-TĀ-shun; *rota-* = revolve), a bone revolves around its own longitudinal axis. Pivot and ball-and-socket joints permit rotation. One example is turning the head from side-to-side at the atlanto-axial joint (between the atlas and axis), as in signifying "no" (Figure 9.9a). Another is turning the trunk from side-to-side at the intervertebral joints while keeping the hips and lower limbs in the anatomical position. In the limbs, rotation is defined relative to the midline, and specific qualifying terms are used. If the anterior surface of a bone of the limb is turned toward the midline, the movement is called *medial (internal) rotation.* You can medially rotate the humerus at the shoulder joint as follows: Starting in the anatomical position, flex your elbow and then draw your palm across the chest (Figure 9.9b). Medial rotation of the forearm at the radioulnar joints (between the radius and ulna) involves turning the palm medially from the anatomical position (see Figure 9.10h). You can medially rotate the femur at the hip joint as follows: Lie on your back, bend your knee, and then move your leg and foot laterally from the midline. Although you are moving your leg and foot laterally, the femur is rotating medially (Figure 9.9c). Medial rotation of the leg at the knee joint can be produced by sitting on a chair, bending your knee, raising your lower limb off the floor, and turning your toes medially. If the anterior surface of the bone of a limb is turned away from the midline, the movement is called *lateral (external) rotation* (see Figure 9.9b, c).

Special Movements

As noted previously, **special movements** occur only at certain joints. They include elevation, depression, protraction, retraction, inversion, eversion, dorsiflexion, plantar flexion, supination, pronation, and opposition (Figure 9.10):

- **Elevation** (el-e-VĀ-shun = to lift up) is an upward movement of a part of the body, such as closing the mouth at the temporomandibular joint (between the mandible and temporal bone) to elevate the mandible (Figure 9.10a) or shrugging the shoulders at the acromioclavicular joint to elevate the scapula.
- **Depression** (de-PRESH-un = to press down) is a downward movement of a part of the body, such as opening the mouth to depress the mandible (Figure 9.10b) or returning shrugged shoulders to the anatomical position to depress the scapula.
- **Protraction** (prō-TRAK-shun = to draw forth) is a movement of a part of the body anteriorly in the transverse plane. You can protract your mandible at the temporomandibular joint by thrusting it outward (Figure 9.10c) or protract your clavicles at the acromioclavicular and sternoclavicular joints by crossing your arms.
- **Retraction** (rē-TRAK-shun = to draw back) is a movement of a protracted part of the body back to the anatomical position (Figure 9.10d).

Figure 9.9 Rotation at synovial joints.

In rotation, a bone revolves around its own longitudinal axis.

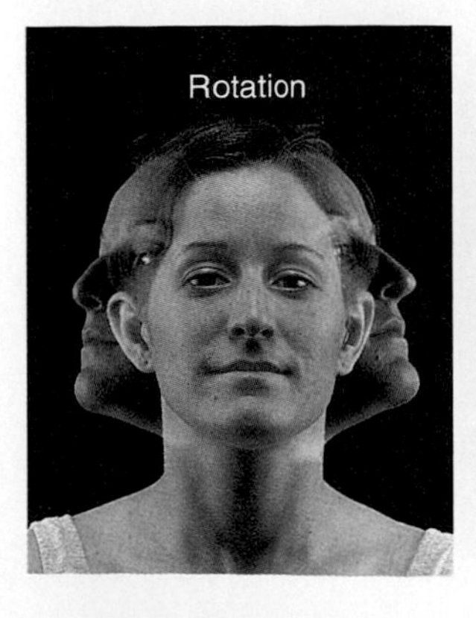

(a) Atlanto-axial joint

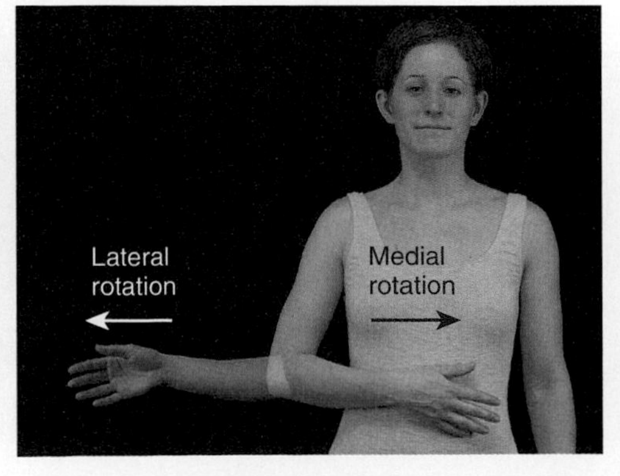

(b) Shoulder joint

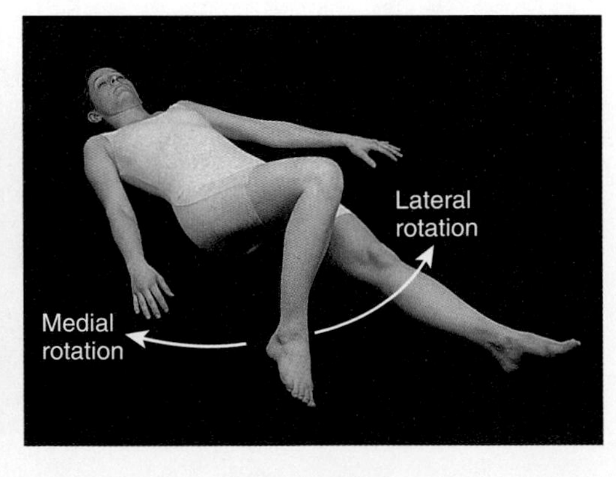

(c) Hip joint

How do medial and lateral rotation differ?

Figure 9.10 Special movements at synovial joints.

Special movements occur only at certain synovial joints.

(a) Temporomandibular joint (b)

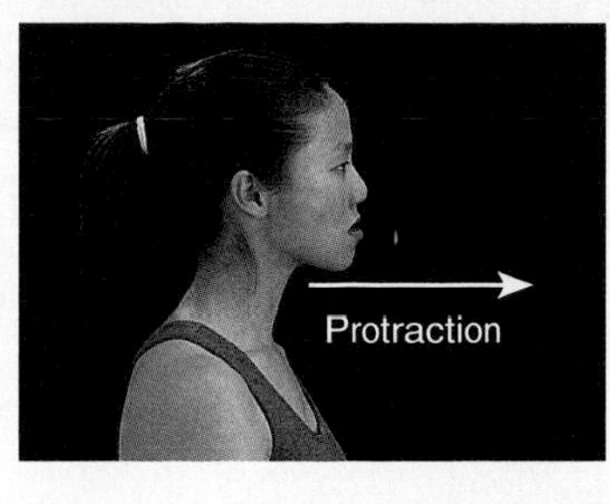

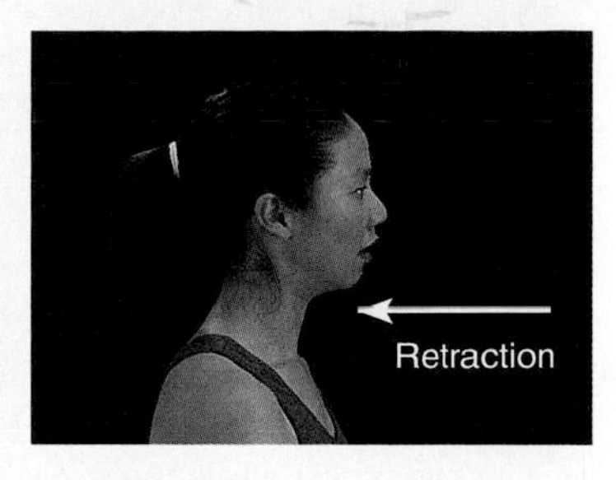

(c) Temporomandibular joint (d)

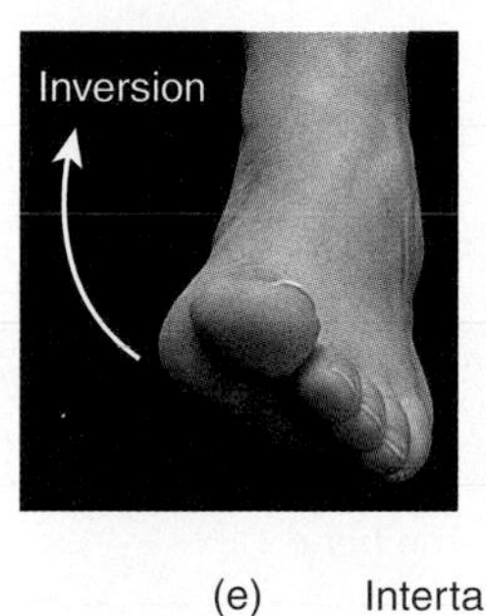

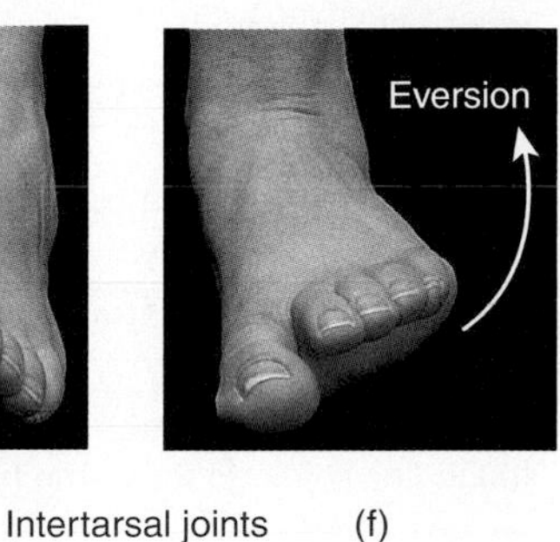

(e) Intertarsal joints (f)

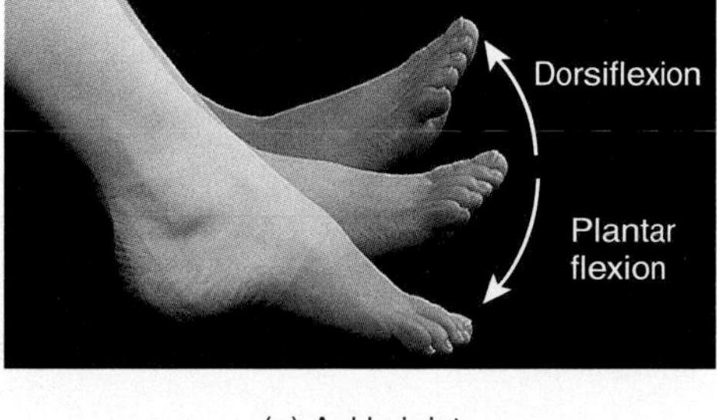

(g) Ankle joint

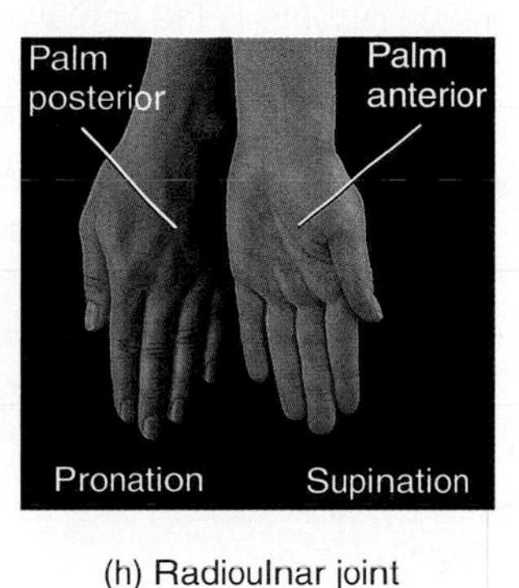

(h) Radioulnar joint

What movement of the shoulder girdle occurs while bringing the arms forward until the elbows touch?

- **Inversion** (in-VER-zhun = to turn inward) is movement of the soles medially at the intertarsal joints (between the tarsals) so that the soles face each other (Figure 9.10e).
- **Eversion** (ē-VER-zhun = to turn outward) is a movement of the soles laterally at the intertarsal joints so that the soles face away from each other (Figure 9.10f).
- **Dorsiflexion** (dor-si-FLEK-shun) refers to bending of the foot at the ankle or talocrural joint (between the tibia, fibula, and talus) in the direction of the dorsum (superior surface) (Figure 9.10g). Dorsiflexion occurs when you stand on your heels.
- **Plantar flexion** involves bending of the foot at the ankle joint in the direction of the plantar or inferior surface (see Figure 9.10g), as when standing on your toes.
- **Supination** (soo-pi-NĀ-shun) is a movement of the forearm at the proximal and distal radioulnar joints in which the palm is turned anteriorly or superiorly (Figure 9.10h). This position of the palms is one of the defining features of the anatomical position.
- **Pronation** (prō-NĀ-shun) is a movement of the forearm at the proximal and distal radioulnar joints in which the distal end of the radius crosses over the distal end of the ulna and the palm is turned posteriorly or inferiorly (Figure 9.10h).
- **Opposition** (op-ō-ZISH-un) is the movement of the thumb at the carpometacarpal joint (between the trapezium and metacarpal of the thumb) in which the thumb moves across the palm to touch the tips of the fingers on the same hand (see Figure 11.18c on page 358). This is the distinctive digital movement that gives humans and other primates the ability to grasp and manipulate objects very precisely.

A summary of the movements that occur at synovial joints is presented in Table 9.2.

Table 9.2 Summary of Movements at Synovial Joints

Movement	Description
Gliding	Movement of relatively flat bone surfaces back-and-forth and from side-to-side over one another; little change in the angle between bones.
Angular	Increase or decrease in the angle between bones.
Flexion	Decrease in the angle between articulating bones, usually in the sagittal plane.
Lateral flexion	Movement of the trunk in the frontal plane.
Extension	Increase in the angle between articulating bones, usually in the sagittal plane.
Hyperextension	Extension beyond the anatomical position.
Abduction	Movement of a bone away from the midline, usually in the frontal plane.
Adduction	Movement of a bone toward the midline, usually in the frontal plane.
Circumduction	Flexion, abduction, extension, and adduction in succession, in which the distal end of a body part moves in a circle.
Rotation	Movement of a bone around its longitudinal axis; in the limbs, it may be medial (toward midline) or lateral (away from midline).
Special	Occurs at specific joints.
Elevation	Superior movement of a body part.
Depression	Inferior movement of a body part.
Protraction	Anterior movement of a body part in the transverse plane.
Retraction	Posterior movement of a body part in the transverse plane.
Inversion	Medial movement of the soles so that they face each other.
Eversion	Lateral movement of the soles so that they face away from each other.
Dorsiflexion	Bending the foot in the direction of the dorsum (superior surface).
Plantar flexion	Bending the foot in the direction of the plantar surface (sole).
Supination	Movement of the forearm that turns the palm anteriorly or superiorly.
Pronation	Movement of the forearm that turns the palm posteriorly or inferiorly.
Opposition	Movement of the thumb across the palm to touch fingertips on the same hand.

Joint Dislocation

Dislocation (dis′-lō-KĀ-shun; *dis-* = apart) or **luxation** (luks-Ā-shun; *luxatio* = dislocation) is the displacement of a bone from a joint, with resultant tearing of ligaments, tendons, and articular capsules. It is usually caused by a blow or fall. A partial or incomplete dislocation is termed a **subluxation** (sub-luks-Ā-shun). The joint most commonly dislocated in adults is the shoulder joint because its socket is quite shallow. Usually, the head of the humerus becomes displaced inferiorly, where the articular capsule is least protected. Dislocations of the mandible, elbow, fingers, knee, or hip are less common. In children, the most common upper limb dislocation involves the elbow joint (see dislocation of the radial head on page 267). ■

▶ CHECKPOINT

9. On yourself or with a partner, demonstrate each movement listed in Table 9.2.

SELECTED JOINTS OF THE BODY

In previous chapters we discussed the major bones and their markings. In this chapter we have examined how joints are classified according to their structure and function, and we have introduced the movements that occur at joints. Table 9.3 (selected joints of the axial skeleton) and Table 9.4 (selected joints of the appendicular skeleton) will help you integrate the information you learned about bones with the information about joint classifications and movements. These tables list some of the major joints of the body according to their articular components (the bones that enter into their formation), their structural and functional classification, and the type(s) of movement that occurs at each joint.

Next we examine in detail four selected joints of the body in a series of exhibits. Each exhibit considers a specific synovial joint and contains (1) a definition—a description of the type of joint and the bones that form the joint; (2) the anatomical components—a description of the major connecting ligaments, articular disc, articular capsule, and other distinguishing features of the joint; and (3) the joint's possible movements. Each exhibit also refers you to a figure that illustrates the joint. The joints described are the shoulder (humeroscapular or glenohumeral) joint, elbow joint, hip (coxal) joint, and knee (tibiofemoral) joint. Because the shoulder, elbow, hip, and knee joints are described in Exhibits 9.1 through 9.4, they are not included in Tables 9.3 and 9.4.

Table 9.3 Selected Joints of the Axial Skeleton

Joint	Articular Components	Classification	Movements
Suture	Between skull bones.	*Structural:* fibrous. *Functional:* synarthrosis.	None.
Temporomandibular (TMJ)	Between condylar process of mandible and mandibular fossa and articular tubercle of temporal bone.	*Structural:* synovial (combined hinge and planar joint). *Functional:* diarthrosis.	Depression, elevation, protraction, retraction, lateral displacement, and slight rotation of mandible.
Atlanto-occipital	Between superior articular facets of atlas and occipital condyles of occipital bone.	*Structural:* synovial (condyloid). *Functional:* diarthrosis.	Flexion and extension of head and slight lateral flexion of head to either side.
Atlanto-axial	(1) Between dens of axis and anterior arch of atlas and (2) between lateral masses of atlas and axis.	*Structural:* synovial (pivot) between dens and anterior arch, and synovial (planar) between lateral masses. *Functional:* diarthrosis.	Rotation of head.
Intervertebral	(1) Between vertebral bodies and (2) between vertebral arches.	*Structural:* cartilaginous (symphysis) between vertebral bodies, and synovial (planar) between vertebral arches. *Functional:* amphiarthrosis between vertebral bodies, and diarthrosis between vertebral arches.	Flexion, extension, lateral flexion, and rotation of vertebral column.
Vertebrocostal	(1) Between facets of heads of ribs and facets of bodies of adjacent thoracic vertebrae and intervertebral discs between them and (2) between articular part of tubercles of ribs and facets of transverse processes of thoracic vertebrae.	*Structural:* synovial (planar). *Functional:* diarthrosis.	Slight gliding.
Sternocostal	Between sternum and first seven pairs of ribs.	*Structural:* cartilaginous (synchondrosis) between sternum and first pair of ribs, and synovial (planar) between sternum and second through seventh pairs of ribs. *Functional:* synarthrosis between sternum and first pair of ribs, and diarthrosis between sternum and second through seventh pairs of ribs.	None between sternum and first pair of ribs; slight gliding between sternum and second through seventh pairs of ribs.
Lumbosacral	(1) Between body of fifth lumbar vertebra and base of sacrum and (2) between inferior articular facets of fifth lumbar vertebra and superior articular facets of first vertebra of sacrum.	*Structural:* cartilaginous (symphysis) between body and base, and synovial (planar) between articular facets. *Functional:* amphiarthrosis between body and base, and diarthrosis between articular facets.	Flexion, extension, lateral flexion, and rotation of vertebral column.

Table 9.4 Selected Joints of the Appendicular Skeleton

Joint	Articular Components	Classification	Movements
Sternoclavicular	Between sternal end of clavicle, manubrium of sternum, and first costal cartilage.	*Structural:* synovial (planar and pivot). *Functional:* diarthrosis.	Gliding, with limited movements in nearly every direction.
Acromioclavicular	Between acromion of scapula and acromial end of clavicle.	*Structural:* synovial (planar). *Functional:* diarthrosis.	Gliding and rotation of scapula on clavicle.
Radioulnar	Proximal radioulnar joint between head of radius and radial notch of ulna; distal radioulnar joint between ulnar notch of radius and head of ulna.	*Structural:* synovial (pivot). *Functional:* diarthrosis.	Rotation of forearm.
Wrist (radiocarpal)	Between distal end of radius and scaphoid, lunate, and triquetrum of carpus.	*Structural:* synovial (condyloid). *Functional:* diarthrosis.	Flexion, extension, abduction, adduction, circumduction, and slight hyperextension of wrist.
Intercarpal	Between proximal row of carpal bones, distal row of carpal bones, and between both rows of carpal bones (midcarpal joints).	*Structural:* synovial (planar), except for hamate, scaphoid, and lunate (midcarpal) joint, which is synovial (saddle). *Functional:* diarthrosis.	Gliding plus flexion and abduction at midcarpal joints.
Carpometacarpal	Carpometacarpal joint of thumb between trapezium of carpus and first metacarpal; carpometacarpal joints of remaining digits formed between carpus and second through fifth metacarpals.	*Structural:* synovial (saddle) at thumb and synovial (planar) at remaining digits. *Functional:* diarthrosis.	Flexion, extension, abduction, adduction, and circumduction at thumb, and gliding at remaining digits.
Metacarpophalangeal and metatarsophalangeal	Between heads of metacarpals (or metatarsals) and bases of proximal phalanges.	*Structural:* synovial (condyloid). *Functional:* diarthrosis.	Flexion, extension, abduction, adduction, and circumduction of phalanges.
Interphalangeal	Between heads of phalanges and bases of more distal phalanges.	*Structural:* synovial (hinge). *Functional:* diarthrosis.	Flexion and extension of phalanges.
Sacroiliac	Between auricular surfaces of sacrum and ilia of hip bones.	*Structural:* synovial (planar). *Functional:* diarthrosis.	Slight gliding (even more so during pregnancy).
Pubic symphysis	Between anterior surfaces of hip bones.	*Structural:* cartilaginous (symphysis). *Functional:* amphiarthrosis.	Slight movements (even more so during pregnancy).
Tibiofibular	Proximal tibiofibular joint between lateral condyle of tibia and head of fibula; distal tibiofibular joint between distal end of fibula and fibular notch of tibia.	*Structural:* synovial (planar) at proximal joint, and fibrous (syndesmosis) at distal joint. *Functional:* diarthrosis at proximal joint, and amphiarthrosis at distal joint.	Slight gliding at proximal joint, and slight rotation of fibula during dorsiflexion of foot.
Ankle (talocrural)	(1) Between distal end of tibia and its medial malleolus and talus and (2) between lateral malleolus of fibula and talus.	*Structural:* synovial (hinge). *Functional:* diarthrosis.	Dorsiflexion and plantar flexion of foot.
Intertarsal	Subtalar joint between talus and calcaneus of tarsus; talocalcaneonavicular joint between talus and calcaneus and navicular of tarsus; calcaneocuboid joint between calcaneus and cuboid of tarsus.	*Structural:* synovial (planar) at subtalar and calcaneocuboid joints, and synovial at talocalcaneonavicular joint. *Functional:* diarthrosis.	Inversion and eversion of foot.
Tarsometatarsal	Between three cuneiforms of tarsus and bases of five metatarsal bones.	*Structural:* synovial (planar). *Functional:* diarthrosis.	Slight gliding.

Exhibit 9.1 Shoulder Joint (Figure 9.11)

▶ **OBJECTIVES**

- Describe the anatomical components of the shoulder joint.
- Explain the movements that can occur at this joint.

Definition

The **shoulder joint** is a ball-and-socket joint formed by the head of the humerus and the glenoid cavity of the scapula. It also is referred to as the *humeroscapular* or *glenohumeral joint.*

Anatomical Components

1. ***Articular capsule.*** Thin, loose sac that completely envelops the joint and extends from the glenoid cavity to the anatomical neck of the humerus. The inferior part of the capsule is its weakest area.
2. ***Coracohumeral ligament.*** Strong, broad ligament that strengthens the superior part of the articular capsule and extends from the coracoid process of the scapula to the greater tubercle of the humerus.
3. ***Glenohumeral ligaments.*** Three thickenings of the articular capsule over the anterior surface of the joint. They extend from the glenoid cavity to the lesser tubercle and anatomical neck of the humerus. These ligaments are often indistinct or absent and provide only minimal strength.
4. ***Transverse humeral ligament.*** Narrow sheet extending from the greater tubercle to the lesser tubercle of the humerus.
5. ***Glenoid labrum.*** Narrow rim of fibrocartilage around the edge of the glenoid cavity. It slightly deepens and enlarges the glenoid cavity.
6. ***Bursae.*** Four *bursae* are associated with the shoulder joint. They are the *subscapular bursa, subdeltoid bursa, subacromial bursa,* and *subcoracoid bursa.*

Movements

The shoulder joint allows flexion, extension, abduction, adduction, medial rotation, lateral rotation, and circumduction of the arm (see Figures 9.6–9.9). It has more freedom of movement than any other joint of the body. This freedom results from the looseness of the articular capsule and shallowness of the glenoid cavity in relation to the large size of the head of the humerus.

Although the ligaments of the shoulder joint strengthen it to some extent, most of the strength results from the muscles that surround the joint, especially the *rotator cuff muscles.* These muscles (supraspinatus, infraspinatus, teres minor, and subscapularis) join the scapula to the humerus (see also Figure 11.15 on pages 346–347). The tendons of the muscles, together called the **rotator cuff,** encircle the joint (except for the inferior portion) and fuse with the articular capsule. The rotator cuff muscles work as a group to hold the head of the humerus in the glenoid cavity.

▶ **CHECKPOINT**

Which tendons at the shoulder joint of a baseball pitcher are most likely to be torn due to excessive circumduction?

Figure 9.11 Right shoulder (humeroscapular or glenohumeral) joint. (See Tortora, *A Photographic Atlas of the Human Body,* Figures 4.1 and 4.2.)

Most of the stability of the shoulder joint results from the arrangement of the rotator cuff muscles.

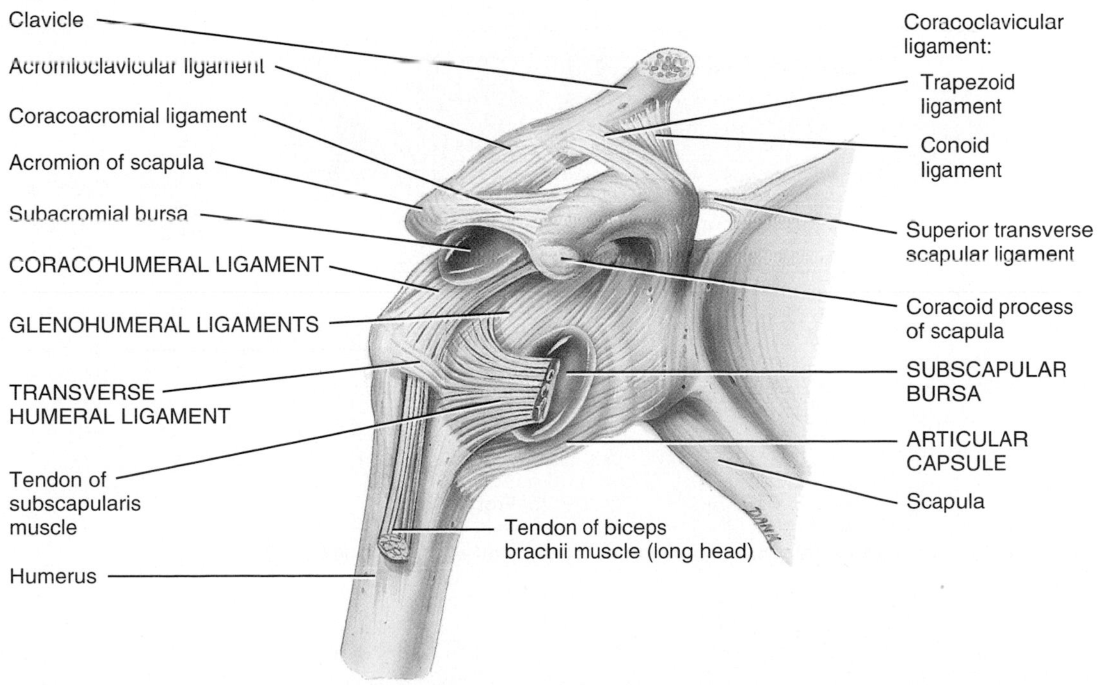

(a) Anterior view

(continues)

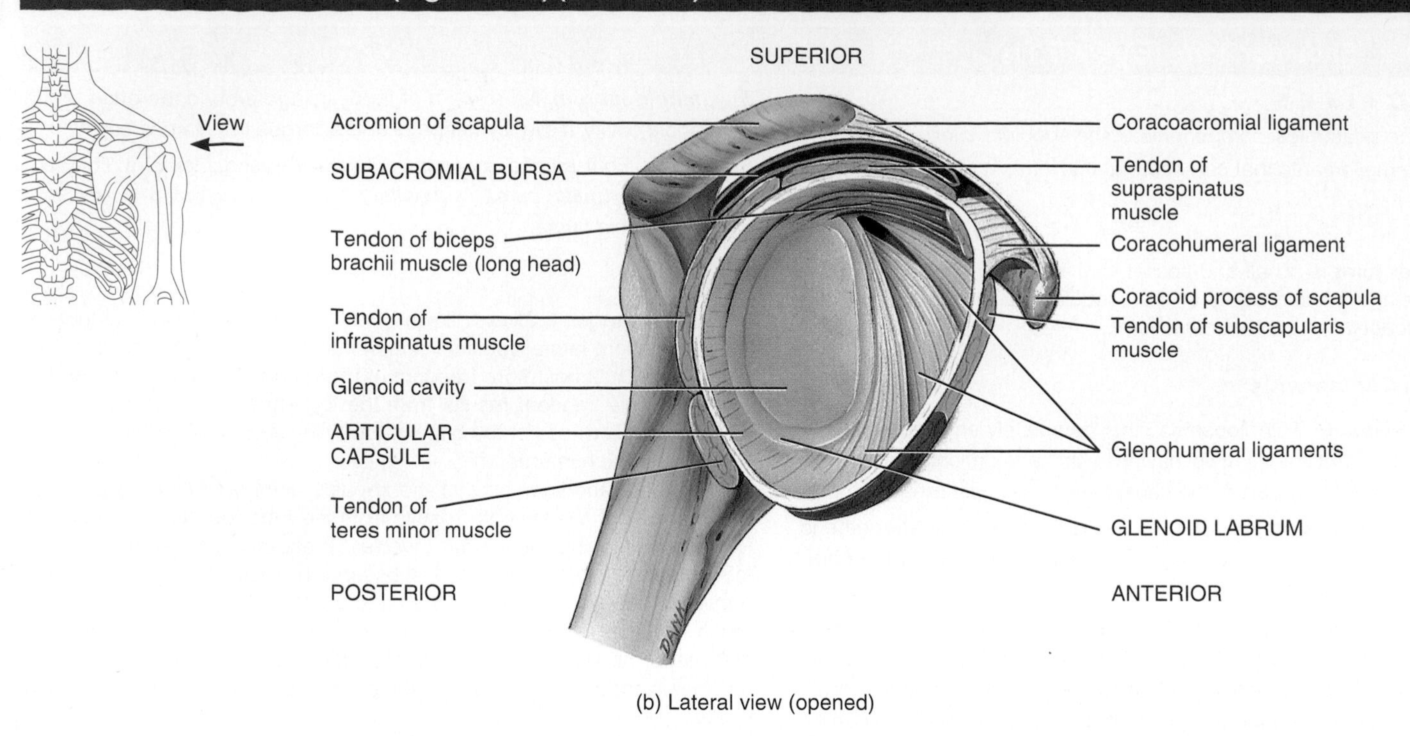

(b) Lateral view (opened)

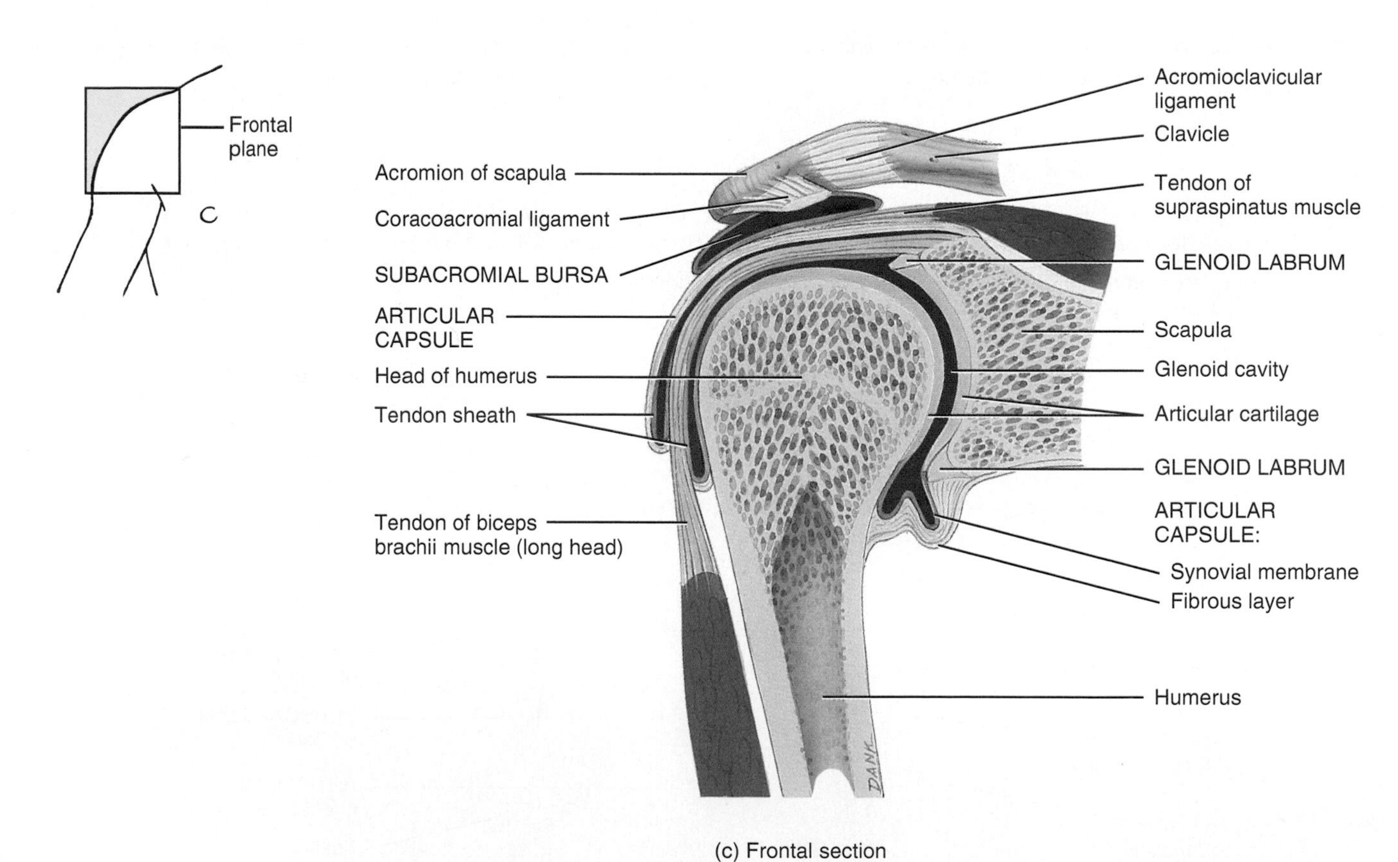

(c) Frontal section

Why does the shoulder joint have more freedom of movement than any other joint of the body?

Exhibit 9.2 Elbow Joint (Figure 9.12)

▶ OBJECTIVES

- Describe the anatomical components of the elbow joint.
- Explain the movements that can occur at this joint.

Definition

The **elbow joint** is a hinge joint formed by the trochlea of the humerus, the trochlear notch of the ulna, and the head of the radius.

Anatomical Components

1. ***Articular capsule.*** The anterior part covers the anterior part of the joint, from the radial and coronoid fossae of the humerus to the coronoid process of the ulna and the annular ligament of the radius. The posterior part extends from the capitulum, olecranon fossa, and lateral epicondyle of the humerus to the annular ligament of the radius, the olecranon of the ulna, and the ulna posterior to the radial notch.
2. ***Ulnar collateral ligament.*** Thick, triangular ligament that extends from the medial epicondyle of the humerus to the coronoid process and olecranon of the ulna.
3. ***Radial collateral ligament.*** Strong, triangular ligament that extends from the lateral epicondyle of the humerus to the annular ligament of the radius and the radial notch of the ulna.

Movements

The elbow joint allows flexion and extension of the forearm (see Figure 9.6c).

▶ CHECKPOINT

At the elbow joint, which ligaments connect (a) the humerus and the ulna, and (b) the humerus and the radius?

Figure 9.12 Right elbow joint. (See Tortora, *A Photographic Atlas of the Human Body,* Figures 4.3 and 4.4.)

The elbow joint is formed by parts of three bones: humerus, ulna, and radius.

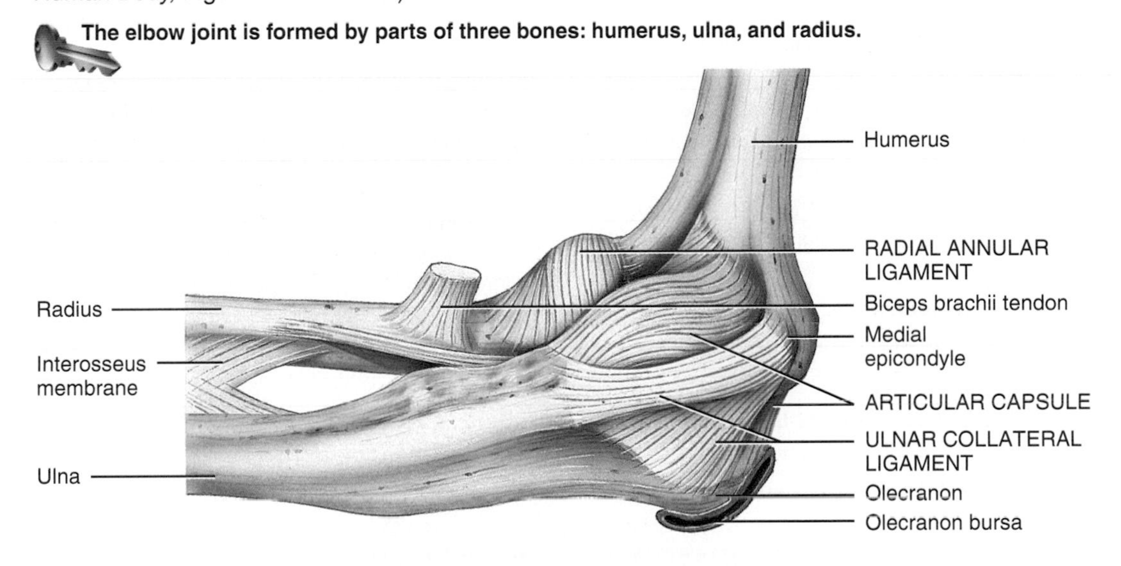

(a) Medial aspect

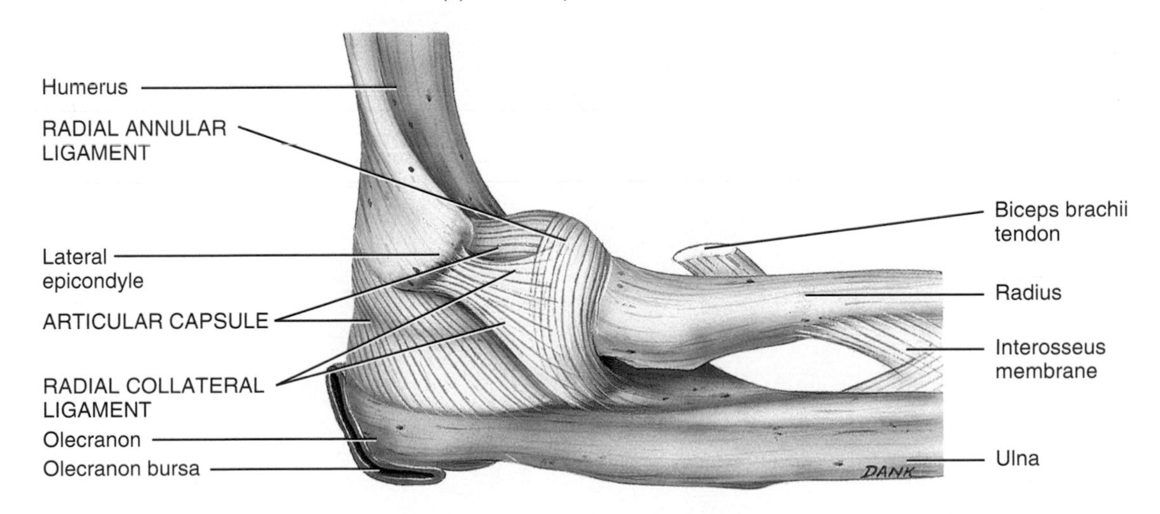

(b) Lateral aspect

Which movements are possible at a hinge joint?

Exhibit 9.3 Hip Joint (Figure 9.13)

▶ OBJECTIVES

- Describe the anatomical components of the hip joint.
- Explain the movements that can occur at this joint.

Definition

The **hip joint** *(coxal joint)* is a ball-and-socket joint formed by the head of the femur and the acetabulum of the hip bone.

Anatomical Components

1. ***Articular capsule.*** Very dense and strong capsule that extends from the rim of the acetabulum to the neck of the femur. One of the strongest structures of the body, the capsule consists of circular and longitudinal fibers. The circular fibers, called the *zona orbicularis,* form a collar around the neck of the femur. Accessory ligaments known as the iliofemoral ligament, pubofemoral ligament, and ischiofemoral ligament reinforce the longitudinal fibers.
2. ***Iliofemoral ligament.*** Thickened portion of the articular capsule that extends from the anterior inferior iliac spine of the hip bone to the intertrochanteric line of the femur.
3. ***Pubofemoral ligament.*** Thickened portion of the articular capsule that extends from the pubic part of the rim of the acetabulum to the neck of the femur.
4. ***Ischiofemoral ligament.*** Thickened portion of the articular capsule that extends from the ischial wall of the acetabulum to the neck of the femur.
5. ***Ligament of the head of the femur.*** Flat, triangular band that extends from the fossa of the acetabulum to the fovea capitis of the head of the femur.
6. ***Acetabular labrum.*** Fibrocartilage rim attached to the margin of the acetabulum that enhances the depth of the acetabulum. Because the diameter of the acetabular rim is smaller than that of the head of the femur, dislocation of the femur is rare.
7. ***Transverse ligament of the acetabulum.*** Strong ligament that crosses over the acetabular notch. It supports part of the acetabular labrum and is connected with the ligament of the head of the femur and the articular capsule.

Movements

The hip joint allows flexion, extension, abduction, adduction, circumduction, medial rotation, and lateral rotation of thigh (see Figures 9.6–9.9). The extreme stability of the hip joint is related to the very strong articular capsule and its accessory ligaments, the manner in which the femur fits into the acetabulum, and the muscles surrounding the joint. Although the shoulder and hip joints are both ball-and-socket joints, the movements at the hip joints do not have as wide a range of motion. Flexion is limited by the anterior surface of the thigh coming into contact with the anterior abdominal wall when the knee is flexed and by tension of the hamstring muscles when the knee is extended. Extension is limited by tension of the iliofemoral, pubofemoral, and ischiofemoral ligaments. Abduction is limited by the tension of the pubofemoral ligament, and adduction is limited by contact with the opposite limb and tension in the ligament of the head of the femur. Medial rotation is limited by the tension in the ischiofemoral ligament, and lateral rotation is limited by tension in the iliofemoral and pubofemoral ligaments.

▶ CHECKPOINT

What factors limit the degree of flexion and abduction at the hip joint?

Figure 9.13 Right hip (coxal) joint. (See Tortora, *A Photographic Atlas of the Human Body,* Figure 4.5.)

The articular capsule of the hip joint is one of the strongest structures in the body.

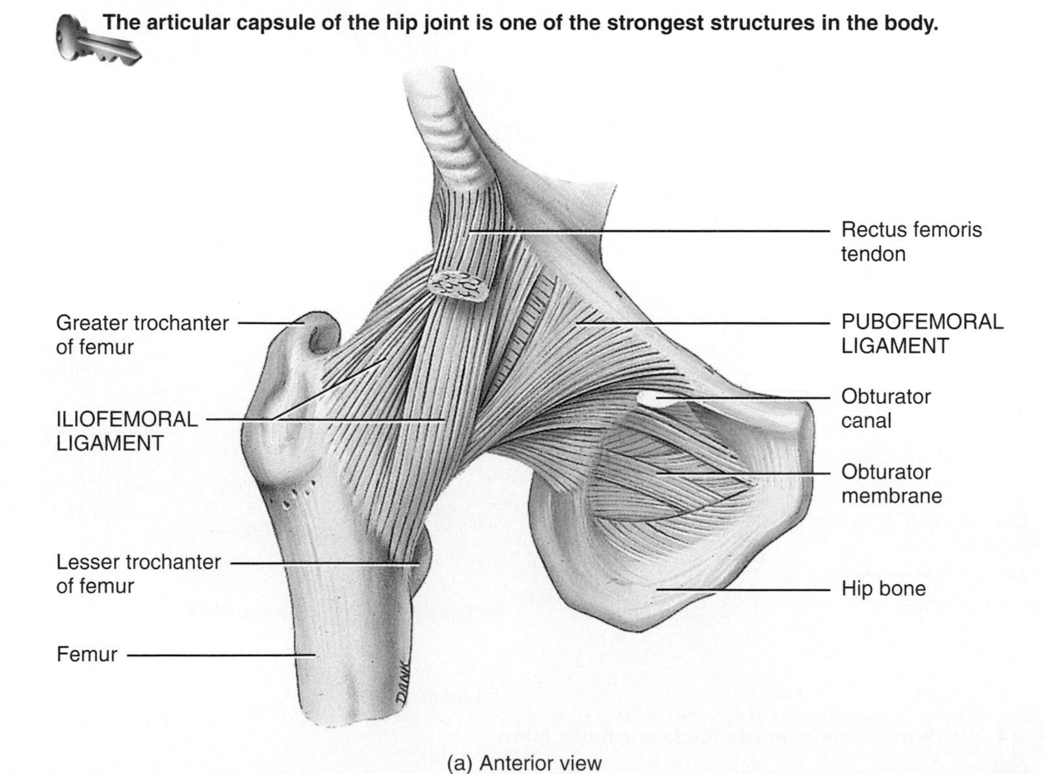

(a) Anterior view

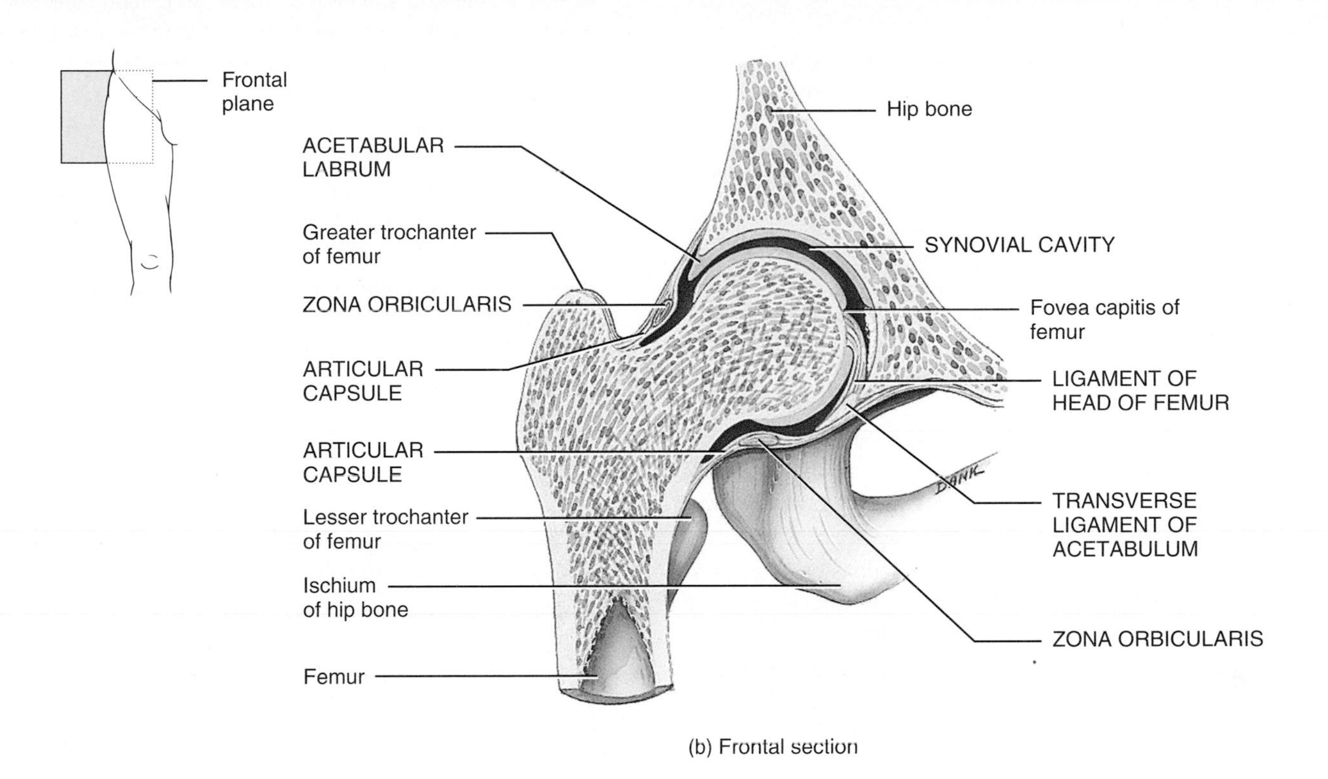

(b) Frontal section

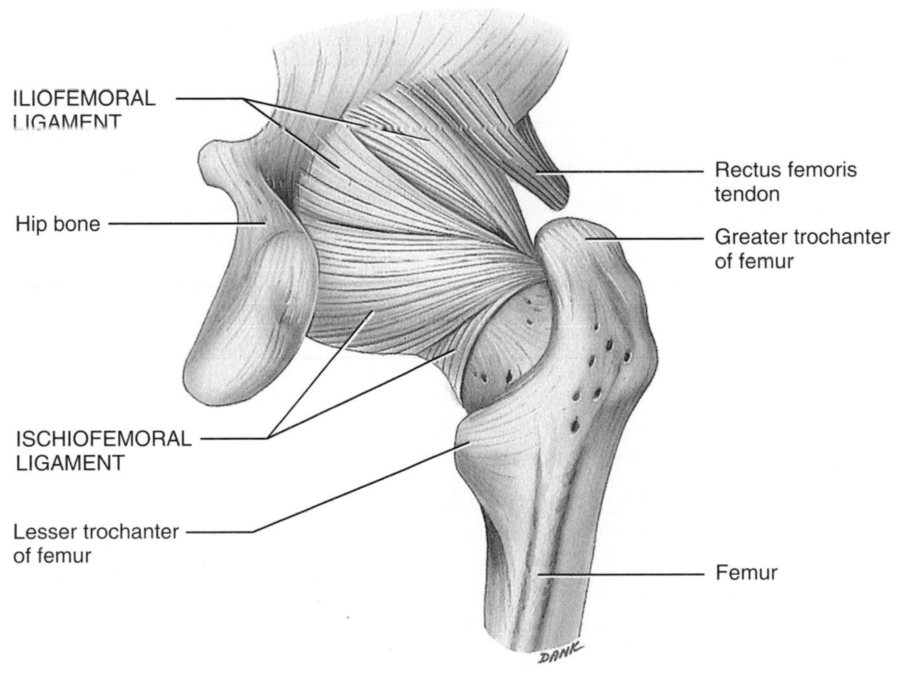

(c) Posterior view

Which ligaments limit the degree of extension that is possible at the hip joint?

Exhibit 9.4 Knee Joint (Figure 9.14)

▶ OBJECTIVES

- Describe the anatomical components of the knee joint.
- Explain the movements that can occur at this joint.

Definition

The **knee joint** *(tibiofemoral joint)* is the largest and most complex joint of the body, actually consisting of three joints within a single synovial cavity:

- Laterally is a tibiofemoral joint, between the lateral condyle of the femur, lateral meniscus, and lateral condyle of the tibia. It is a modified hinge joint.
- Medially is a second tibiofemoral joint, between the medial condyle of the femur, medial meniscus, and medial condyle of the tibia. It also a modified hinge joint.
- An intermediate patellofemoral joint, between the patella and the patellar surface of the femur, is a planar joint.

Anatomical Components

1. ***Articular capsule.*** No complete, independent capsule unites the bones. The ligamentous sheath surrounding the joint consists mostly of muscle tendons or expansions of them. There are, however, some capsular fibers connecting the articulating bones.
2. ***Medial and lateral patellar retinacula.*** Fused tendons of insertion of the quadriceps femoris muscle and the fascia lata (deep fascia of thigh) that strengthen the anterior surface of the joint.
3. ***Patellar ligament.*** Continuation of the common tendon of insertion of the quadriceps femoris muscle that extends from the patella to the tibial tuberosity. This ligament also strengthens the anterior surface of the joint. The posterior surface of the ligament is separated from the synovial membrane of the joint by an infrapatellar fat pad.
4. ***Oblique popliteal ligament.*** Broad, flat ligament that extends from the intercondylar fossa of the femur to the head of the tibia. The tendon of the semimembranosus muscle is superficial to the ligament and passes from the medial condyle of the tibia to the lateral condyle of the femur. The ligament and tendon strengthen the posterior surface of the joint.
5. ***Arcuate popliteal ligament.*** Extends from the lateral condyle of the femur to the styloid process of the head of the fibula. It strengthens the lower lateral part of the posterior surface of the joint.
6. ***Tibial collateral ligament.*** Broad, flat ligament on the medial surface of the joint that extends from the medial condyle of the femur to the medial condyle of the tibia. Tendons of the sartorius, gracilis, and semitendinosus muscles, all of which strengthen the medial aspect of the joint, cross the ligament. Because the tibial collateral ligament is firmly attached to the medial meniscus, tearing of the ligament frequently results in tearing of the meniscus and damage to the anterior cruciate ligament, described under 8a.
7. ***Fibular collateral ligament.*** Strong, rounded ligament on the lateral surface of the joint that extends from the lateral condyle of the femur to the lateral side of the head of the fibula. It strengthens the lateral aspect of the joint. The ligament is covered by the tendon of the biceps femoris muscle. The tendon of the popliteal muscle is deep to the ligament.
8. ***Intracapsular ligaments.*** Ligaments within the capsule that connect the tibia and femur. The anterior and posterior cruciate ligaments (KROO-shē-āt = shaped like a cross) are named based on their origins relative to the intercondylar area of the tibia. Following their originations, they cross on their way to their destinations on the femur.
   a. ***Anterior cruciate ligament (ACL).*** Extends posteriorly and laterally from a point *anterior* to the intercondylar area of the tibia to the posterior part of the medial surface of the lateral condyle of the femur. This ligament is stretched or torn in about 70% of all serious knee injuries. The ACL limits hyperextension of the knee and prevents the anterior sliding of the tibia on the femur.
   b. ***Posterior cruciate ligament (PCL).*** Extends anteriorly and medially from a depression on the *posterior* intercondylar area of the tibia and lateral meniscus to the anterior part of the lateral surface of the medial condyle of the femur. The PCL prevents the posterior sliding of the tibia (and anterior sliding of the femur) when the knee is flexed. This is very important when walking down stairs or a steep incline.
9. ***Articular discs (menisci).*** Two fibrocartilage discs between the tibial and femoral condyles that help compensate for the irregular shapes of the bones and circulate synovial fluid.
   a. ***Medial meniscus.*** Semicircular piece of fibrocartilage (C-shaped). Its anterior end is attached to the anterior intercondylar fossa of the tibia, anterior to the anterior cruciate ligament. Its posterior end is attached to the posterior intercondylar fossa of the tibia between the attachments of the posterior cruciate ligament and lateral meniscus.
   b. ***Lateral meniscus.*** Nearly circular piece of fibrocartilage (approaches an incomplete O in shape). Its anterior end is attached anterior to the intercondylar eminence of the tibia, and lateral and posterior to the anterior cruciate ligament. Its posterior end is attached posterior to the intercondylar eminence of the tibia, and anterior to the posterior end of the medial meniscus. The medial and lateral menisci are connected to each other by the *transverse ligament* and to the margins of the head of the tibia by the *coronary ligaments* (not illustrated).
10. The more important *bursae* of the knee include the following:
    a. *Prepatellar bursa* between the patella and skin.
    b. *Intrapatellar bursa* between superior part of tibia and patellar ligament.
    c. *Suprapatellar bursa* between inferior part of femur and deep surface of quadriceps femoris muscle.

Movements

The knee joint allows flexion, extension, slight medial rotation, and lateral rotation of leg in flexed position (see Figures 9.6f and 9.9c).

▶ CHECKPOINT

Why is the knee joint said to be "three joints in one"?

Figure 9.14 Right knee (tibiofemoral) joint. (See Tortora, *A Photographic Atlas of the Human Body,* Figures 4.6 through 4.8.)

The knee joint is the largest and most complex joint in the body.

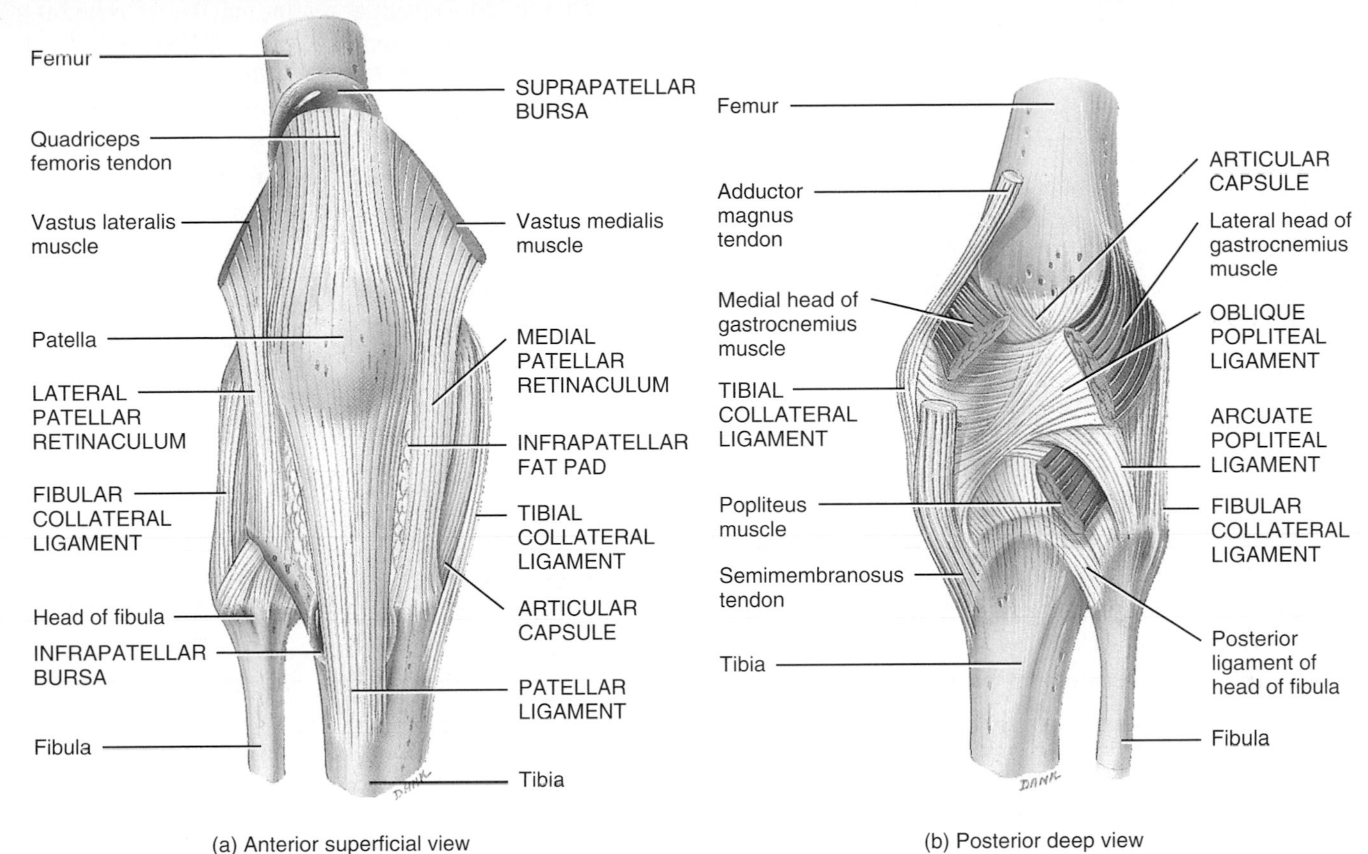

(a) Anterior superficial view

(b) Posterior deep view

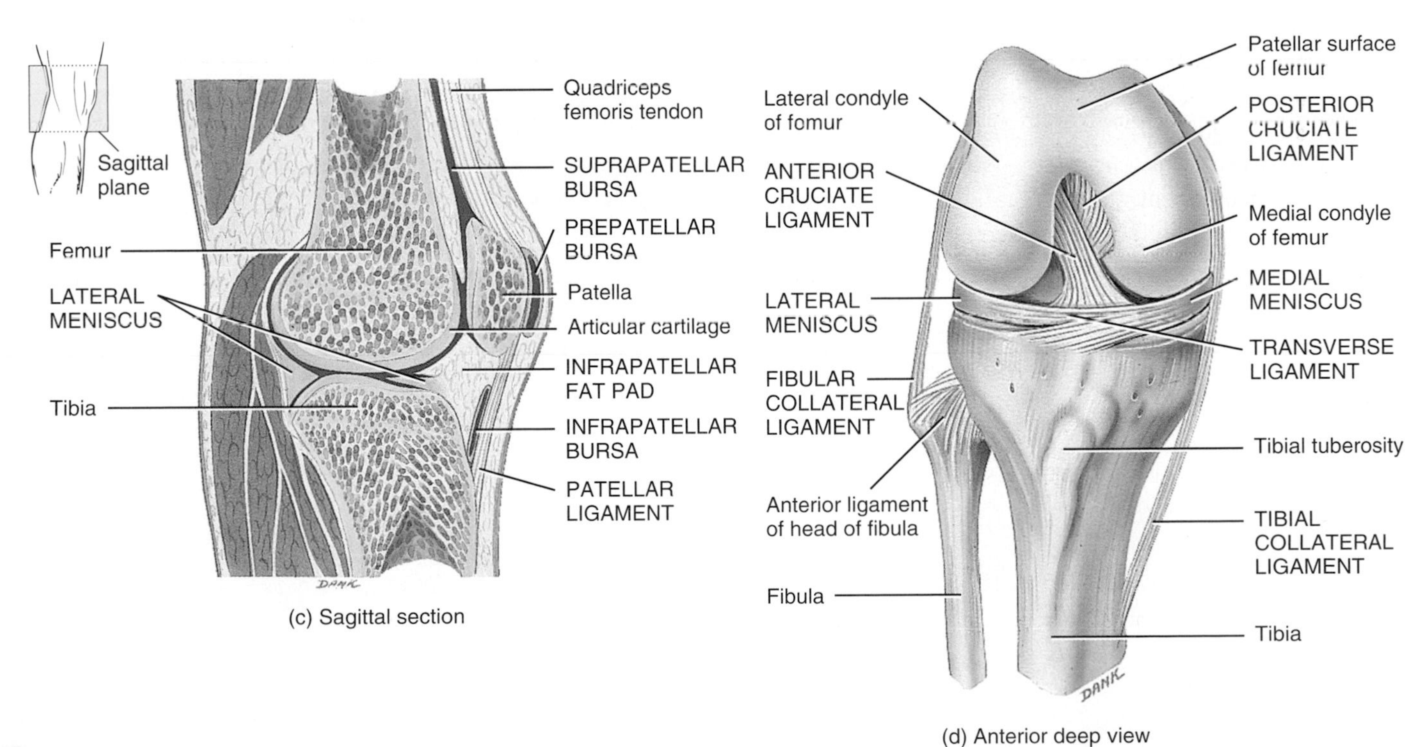

(c) Sagittal section

(d) Anterior deep view

What movement occurs at the knee joint when the quadriceps femoris (anterior thigh) muscles contract?

FACTORS AFFECTING CONTACT AND RANGE OF MOTION AT SYNOVIAL JOINTS

OBJECTIVE

- **Describe five factors that influence the type of movement and range of motion possible at a synovial joint.**

The articular surfaces of synovial joints contact one another and determine the type and range of motion that is possible. **Range of motion (ROM)** is the range, measured in degrees of a circle, through which the bones of a joint can be moved. The following factors contribute to keeping the articular surfaces in contact and affect range of motion:

1. ***Structure or shape of the articulating bones.*** The structure or shape of the articulating bones determines how closely they can fit together. The articular surfaces of some bones interlock with one another. This spatial relationship is very obvious at the hip joint, where the head of the femur articulates with the acetabulum of the hip bone. An interlocking fit allows rotational movement.
2. ***Strength and tension of the joint ligaments.*** The different components of a fibrous capsule are tense or taut only when the joint is in certain positions. Tense ligaments restrict the range of motion and guide the movement of the articulating bones. In the knee joint, for example, the anterior cruciate ligament is taut and the posterior cruciate ligament is loose when the knee is straightened, and the reverse occurs when the knee is bent.
3. ***Arrangement and tension of the muscles.*** Muscle tension reinforces the restraint placed on a joint by its ligaments, and thus restricts movement. A good example of the effect of muscle tension on a joint is seen at the hip joint. When the thigh is raised with the knee straight, the movement is restricted by the tension of the hamstring muscles on the posterior surface of the thigh. But if the knee is bent, the tension on the hamstring muscles is lessened, and the thigh can be raised farther.
4. ***Apposition of soft parts.*** The point at which one body surface contacts another may limit mobility. For example, if you bend your arm at the elbow, it can move no farther after the anterior surface of the forearm meets with and presses against the biceps brachii muscle of the arm.
5. ***Disuse.*** Movement at a joint may be restricted if a joint has not been used for an extended period. For example, if an elbow joint is immobilized by a cast, only a limited range of motion at the joint may be possible for a time after the cast is removed. Restricted movement due to disuse may result from decreased amounts of synovial fluid, from diminished flexibility of ligaments and tendons, and from muscular atrophy, a reduction in size or wasting of a muscle.

CHECKPOINT

10. Besides the hip joint, which joints have bones that interlock and allow rotational movement?

AGING AND JOINTS

OBJECTIVE

- **Explain the effects of aging on joints.**

Aging usually results in decreased production of synovial fluid in joints. In addition, the articular cartilage becomes thinner with age, and ligaments shorten and lose some of their flexibility. The effects of aging on joints, which vary considerably from one person to another, are affected by genetic factors and by wear and tear. Although degenerative changes in joints may begin in those as young as 20 years of age, most changes do not occur until much later. By age 80, almost everyone develops some type of degeneration in the knees, elbows, hips, and shoulders. Males also commonly develop degenerative changes in the vertebral column, resulting in a hunched-over posture and pressure on nerve roots. One type of arthritis, called osteoarthritis, is at least partially age-related. Nearly everyone over age 70 has evidence of some osteoarthritic changes.

CHECKPOINT

11. Which joints show evidence of degeneration in nearly all individuals as aging progresses?

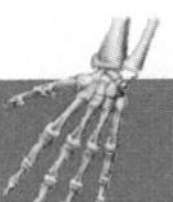

DISORDERS: HOMEOSTATIC IMBALANCES

Common Joint Injuries

Rotator cuff injury is a common injury among baseball pitchers and volleyball players due to shoulder movements that involve vigorous circumduction. Most often, there is tearing of the supraspinatus muscle tendon of the rotator cuff. This tendon is especially predisposed to wear-and-tear changes because of its location between the head of the humerus and acromion of the scapula, which compresses the tendon during shoulder movements. A **separated shoulder** is an injury of the acromioclavicular joint, the joint formed by the acromion of the scapula and the acromial end of the clavicle. It most often happens with forceful trauma, as may happen when the shoulder strikes the ground in a fall.

Tennis elbow most commonly refers to pain at or near the lateral epicondyle of the humerus, usually caused by an improperly executed backhand. The extensor muscles strain or sprain, resulting in pain. **Little-league elbow** typically develops because of a heavy pitching schedule or throwing many curve balls, especially in youngsters. In this injury, the elbow may enlarge, fragment, or separate.

A **dislocation of the radial head** is the most common upper limb dislocation in children. In this injury, the head of the radius slides past or ruptures the radial annular ligament, a ligament that forms a collar around the head of the radius at the proximal radioulnar joint. Dislocation is most apt to occur when a strong pull is applied to the forearm while it is extended and supinated, for instance while swinging a child around with outstretched arms.

The knee joint is the joint most vulnerable to damage because of the stresses to which it is subjected and because there is no interlocking of the articulating bones. A **swollen knee** may occur immediately or hours after an injury. Immediate swelling is due to escape of blood from damaged blood vessels adjacent to areas involving rupture of the anterior cruciate ligament, damage to synovial membranes, torn menisci, fractures, or collateral ligament sprains. Delayed swelling is due to excessive production of synovial fluid, a condition commonly referred to as "water on the knee." A common type of knee injury in football is **rupture of the tibial collateral ligaments,** often associated with tearing of the anterior cruciate ligament and medial meniscus (torn cartilage). Usually, a hard blow to the lateral side of the knee while the foot is fixed on the ground causes the damage. A **dislocated knee** refers to the displacement of the tibia relative to the femur. The most common type is dislocation anteriorly, resulting from hyperextension of the knee. A frequent consequence of a dislocated knee is damage to the popliteal artery.

Rheumatism and Arthritis

Rheumatism (ROO-ma-tizm) is any painful disorder of the supporting structures of the body—bones, ligaments, tendons, or muscles—that is not caused by infection or injury. **Arthritis** is a form of rheumatism in which the joints are swollen, stiff, and painful. It afflicts about 40 million people in the United States.

Rheumatoid Arthritis

Rheumatoid arthritis (RA) is an autoimmune disease in which the immune system of the body attacks its own tissues—in this case, its own cartilage and joint linings. RA is characterized by inflammation of the joint, which causes swelling, pain, and loss of function. Usually, this form of arthritis occurs bilaterally: If one wrist is affected, the other is also likely to be affected, although often not to the same degree.

The primary symptom of rheumatoid arthritis is inflammation of the synovial membrane. If untreated, the membrane thickens, and synovial fluid accumulates. The resulting pressure causes pain and tenderness. The membrane then produces an abnormal granulation tissue, called *pannus,* that adheres to the surface of the articular cartilage and sometimes erodes the cartilage completely. When the cartilage is destroyed, fibrous tissue joins the exposed bone ends. The fibrous tissue ossifies and fuses the joint so that it becomes immovable—the ultimate crippling effect of rheumatoid arthritis. The growth of the granulation tissue causes the distortion of the fingers that characterizes hands of RA sufferers.

Osteoarthritis

Osteoarthritis (OA) (os′-tē-ō-ar-THRĪ-tis) is a degenerative joint disease in which joint cartilage is gradually lost. It results from a combination of aging, irritation of the joints, and wear and abrasion. Commonly known as "wear-and-tear" arthritis, osteoarthritis is the leading cause of disability in older persons.

Osteoarthritis is a progressive disorder of synovial joints, particularly weight-bearing joints. Articular cartilage deteriorates and new bone forms in the subchondral areas and at the margins of the joint. The cartilage slowly degenerates, and as the bone ends become exposed, spurs (small bumps) of new osseous tissue are deposited on them. These spurs decrease the space of the joint cavity and restrict joint movement. Unlike rheumatoid arthritis, osteoarthritis affects mainly the articular cartilage, although the synovial membrane often becomes inflamed late in the disease. A major distinction between osteoarthritis and rheumatoid arthritis is that osteoarthritis first afflicts the larger joints (knees, hips), whereas rheumatoid arthritis first strikes smaller joints.

Gouty Arthritis

Uric acid (a substance that gives urine its name) is a waste product produced during the metabolism of nucleic acid (DNA and RNA) subunits. A person who suffers from **gout** (GOWT) either produces excessive amounts of uric acid or is not able to excrete as much as normal. The result is a buildup of uric acid in the blood. This excess acid then reacts with sodium to form a salt called sodium urate. Crystals of this salt accumulate in soft tissues such as the kidneys and in the cartilage of the ears and joints.

In **gouty arthritis,** sodium urate crystals are deposited in the soft tissues of the joints. The crystals irritate and erode the cartilage, causing inflammation, swelling, and acute pain. Eventually, the crystals destroy all joint tissues. If the disorder is untreated, the ends of the articulating bones fuse, and the joint becomes immovable.

Lyme Disease

A spiral-shaped bacterium called *Borrelia burgdorferi* causes **Lyme disease,** named for the town of Lyme, Connecticut, where it was first reported in 1975. The bacteria are transmitted to humans mainly by deer ticks *(Ixodes dammini)*. These ticks are so small that their bites often go unnoticed. Within a few weeks of the tick bite, a rash may appear at the site. Although the rash often resembles a bull's eye target, there are many variations, and some people never develop a rash. Other symptoms include joint stiffness, fever and chills, headache, stiff neck, nausea, and low back pain. In advanced stages of the disease, arthritis is the main complication. It usually afflicts the larger joints such as the knee, ankle, hip, elbow, or wrist. Antibiotics are generally effective against Lyme disease, especially if they are given promptly. However, some symptoms may linger for years.

MEDICAL TERMINOLOGY

Arthralgia (ar-THRAL-jē-a; *arthr-* = joint; *-algia* = pain) Pain in a joint.

Bursectomy (bur-SEK-tō-mē; *-ectomy* = removal of) Removal of a bursa.

Chondritis (kon-DRĪ-tis; *chondr-* = cartilage) Inflammation of cartilage.

Synovitis (sin′-ō-VĪ-tis) Inflammation of a synovial membrane in a joint.

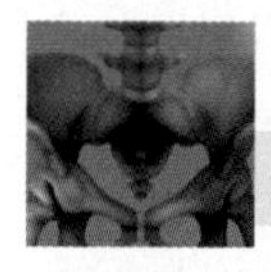

STUDY OUTLINE

INTRODUCTION (p. 244)

1. A joint (articulation or arthrosis) is a point of contact between two bones, between bone and cartilage, or between bone and teeth.
2. A joint's structure may permit no movement, slight movement, or free movement.

JOINT CLASSIFICATIONS (p. 244)

1. Structural classification is based on the presence or absence of a synovial cavity and the type of connecting tissue. Structurally, joints are classified as fibrous, cartilaginous, or synovial.
2. Functional classification of joints is based on the degree of movement permitted. Joints may be synarthroses (immovable), amphiarthroses (slightly movable), or diarthroses (freely movable).

FIBROUS JOINTS (p. 244)

1. Bones held together by fibrous connective tissue are fibrous joints.
2. These joints include immovable sutures (found between skull bones), slightly movable syndesmoses (such as the distal tibiofibular joint), and immovable gomphoses (roots of teeth in alveoli of mandible and maxilla).

CARTILAGINOUS JOINTS (p. 245)

1. Bones held together by cartilage are cartilaginous joints.
2. These joints include immovable synchondroses united by hyaline cartilage (epiphyseal plates between diaphyses and epiphyses) and slightly movable symphyses united by fibrocartilage (pubic symphysis).

SYNOVIAL JOINTS (p. 246)

1. Synovial joints contain a space between bones called the synovial cavity. All synovial joints are diarthroses.
2. Other characteristics of synovial joints are the presence of articular cartilage and an articular capsule, made up of a fibrous capsule and a synovial membrane.
3. The synovial membrane secretes synovial fluid, which forms a thin, viscous film over the surfaces within the articular capsule.

4. Many synovial joints also contain accessory ligaments (extracapsular and intracapsular) and articular discs (menisci).
5. Synovial joints contain an extensive nerve and blood supply. The nerves convey information about pain, joint movements, and the degree of stretch at a joint. Blood vessels penetrate the articular capsule and ligaments.
6. Bursae are saclike structures, similar in structure to joint capsules, that reduce friction in joints such as the shoulder and knee joints.
7. Tendon sheaths are tubelike bursae that wrap around tendons where there is considerable friction.
8. Subtypes of synovial joints are planar, hinge, pivot, condyloid, saddle, and ball-and-socket.
9. Planar joint: articulating surfaces are flat; bone glides back and forth and side to side (nonaxial); found between carpals and tarsals.
10. Hinge joint: convex surface of one bone fits into concave surface of another; motion is angular around one axis (monaxial); examples are the elbow, knee, and ankle joints.
11. Pivot joint: a round or pointed surface of one bone fits into a ring formed by another bone and a ligament; movement is rotation (monaxial); examples are the atlanto-axial and radioulnar joints.
12. Condyloid joint: an oval-shaped projection of one bone fits into an oval cavity of another; motion is angular around two axes (biaxial); examples are the wrist joint and metacarpophalangeal joints for the second through fifth digits.
13. Saddle joint: articular surface of one bone is shaped like a saddle and the other bone fits into the "saddle" like a sitting rider; motion is angular around two axes (biaxial); an example is the carpometacarpal joint between the trapezium and the metacarpal of the thumb.
14. Ball-and-socket joint: ball-shaped surface of one bone fits into cuplike depression of another; motion is angular and rotational around three axes and all directions in between (multiaxial); examples are the shoulder and hip joints.
15. Table 9.1 on page 250 summarizes the structural and functional categories of joints.

TYPES OF MOVEMENTS AT SYNOVIAL JOINTS (p. 251)

1. In a gliding movement, the nearly flat surfaces of bones move back-and-forth and from side-to-side. Gliding movements occur at planar joints.
2. In angular movements, a change in the angle between bones occurs. Examples are flexion–extension, lateral flexion, hyperextension, and abduction–adduction. Circumduction refers to flexion, abduction, extension, and adduction in succession. Angular movements occur at hinge, condyloid, saddle, and ball-and-socket joints.
3. In rotation, a bone moves around its own longitudinal axis. Rotation can occur at pivot and ball-and-socket joints.
4. Special movements occur at specific synovial joints. Examples are elevation–depression, protraction–retraction, inversion–eversion, dorsiflexion–plantar flexion, supination–pronation, and opposition.
5. Table 9.2 on page 256 summarizes the various types of movements at synovial joints.

SELECTED JOINTS OF THE BODY (p. 256)

1. Tables 9.3 and 9.4 on pages 257–258 provide a summary of selected joints of the body, including articular components, structural and functional classifications, and movements.
2. The shoulder (humeroscapular or glenohumeral) joint is between the head of the humerus and glenoid cavity of the scapula (Exhibit 9.1 on page 259).
3. The elbow joint is between the trochlea of the humerus, the trochlear notch of the ulna, and the head of the radius (Exhibit 9.2 on page 261).
4. The hip (coxal) joint is between the head of the femur and acetabulum of the hip bone (Exhibit 9.3 on page 262).
5. The knee (tibiofemoral) joint is between the patella and patellar surface of femur; lateral condyle of femur, lateral meniscus, and lateral condyle of tibia; and medial condyle of femur, medial meniscus, and medial condyle of tibia (Exhibit 9.4 on page 264).

FACTORS AFFECTING CONTACT AND RANGE OF MOTION AT SYNOVIAL JOINTS (p. 266)

1. The ways that articular surfaces of synovial joints contact one another determines the type of movement that is possible.
2. Factors that contribute to keeping the surfaces in contact and affect range of motion are structure or shape of the articulating bones, strength and tension of the joint ligaments, arrangement and tension of the muscles, apposition of soft parts, hormones, and disuse.

AGING AND JOINTS (p. 266)

1. With aging, a decrease in synovial fluid, thinning of articular cartilage, and decreased flexibility of ligaments occur.
2. Most individuals experience some degeneration in the knees, elbows, hips, and shoulders due to the aging process.

SELF-QUIZ QUESTIONS

Fill in the blanks in the following statements.

1. A point of contact between two bones, between bone and cartilage, or between bone and teeth is called a(n) ________.

2. ________ lubricates, reduces friction, and supplies nutrients to and removes wastes from joints.

3. A degenerative joint disease that is a leading cause of disability in the elderly is ________.

Choose the one best answer to the following questions.

4. A joint where bones are held together by collagen-rich fibrous connective tissue is a (a) diarthrotic joint, (b) synovial joint, (c) cartilaginous joint, (d) fibrous joint, (e) functional joint.

5. A slightly movable joint is a(n) (a) synarthrosis, (b) amphiarthrosis, (c) diarthrosis, (d) suture, (e) synchondrosis.

6. Which of the following is a cartilaginous joint? (a) symphysis, (b) gomphosis, (c) suture, (d) syndesmosis, (e) synovial joint.

7. Which of the following statements are *true*? (1) The bones at a synovial joint are covered by a mucous membrane. (2) The articular capsule surrounds a synovial joint, encloses the synovial cavity, and unites the articulating bones. (3) The fibrous capsule of the articular capsule permits considerable movement at a joint. (4) The tensile strength of the fibrous capsule helps prevent bones from disarticulating. (5) All joints contain a fibrous capsule. (a) 1, 2, 3, and 4, (b) 2, 3, 4, and 5, (c) 2, 3, and 4, (d) 1, 2, and 3, (e) 2, 4, and 5.

8. Which of the following keep the articular surfaces of synovial joints in contact and affect range of motion? (1) structure or shape of the articulating bones, (2) strength and tension of the joint ligaments, (3) arrangement and tension of muscles, (4) lack of use, (5) apposition of soft parts. (a) 1, 2, 3, and 5, (b) 2, 3, 4, and 5, (c) 1, 3, 4, and 5, (d) 1, 3, and 5, (e) 1, 2, 3, 4, and 5.

9. Which of the following does *not* reduce friction in joints? (a) bursae, (b) articular cartilage, (c) tendon sheaths, (d) synovial fluid, (e) accessory ligaments.

Indicate whether the following statements are true or false.

10. The fibers of some fibrous capsules are arranged in parallel bundles called ligaments.

11. Movement of the trunk sideways at the waist is an example of hyperextension.

12. Match the following:

____ (a) movement in which relatively flat bone surfaces move back-and-forth and side-to-side with respect to one another
____ (b) decrease in angle between bones
____ (c) increase in angle between bones
____ (d) movement of bone away from midline
____ (e) movement of bone toward midline
____ (f) movement of distal end of a part of the body in a circle
____ (g) a bone revolves around its own longitudinal axis

(1) abduction
(2) rotation
(3) flexion
(4) adduction
(5) gliding
(6) extension
(7) circumduction

13. Match the following:

____ (a) a fibrous joint that unites the bones of the skull; a synarthrosis
____ (b) a fibrous joint between the tibia and fibula; an amphiarthrosis
____ (c) the articulation between bone and teeth
____ (d) the epiphyseal plate
____ (e) joint between the two pubic bones
____ (f) a bony joint

(1) synostosis
(2) synchondrosis
(3) syndesmosis
(4) suture
(5) symphysis
(6) gomphosis

14. Match the following:

____ (a) rounded or pointed surface of one bone articulates with a ring formed by another bone and a ligament; allows rotation around its own axis
____ (b) articulating bone surfaces are flat or slightly curved; permit gliding movement
____ (c) convex, oval projection of one bone fits into oval depression of another bone; permits movement in two axes
____ (d) convex surface of one bone articulates with concave surface of another bone; permits flexion and extension
____ (e) ball-shaped surface of one bone articulates with cuplike depression of another bone; permits largest degree of movement
____ (f) modified condyloid joint where articulating bones resemble a rider sitting in a saddle

(1) hinge joint
(2) saddle joint
(3) ball-and-socket joint
(4) planar joint
(5) condyloid joint
(6) pivot joint

15. Match the following:

____ (a) upward movement of a body part
____ (b) downward movement of a body part
____ (c) movement of a body part anteriorly in the transverse plane
____ (d) movement of an anteriorly projected body part back to the anatomical position
____ (e) movement of the soles medially at the intertarsal joints
____ (f) movement of the soles laterally
____ (g) action that occurs when you stand on your heels
____ (h) action that occurs when you stand on your toes
____ (i) movement of the forearm to turn the palm anteriorly
____ (j) movement of the forearm to turn the palm posteriorly
____ (k) movement of thumb across the palm to touch the tips of the fingers of the same hand

(1) pronation
(2) plantar flexion
(3) eversion
(4) retraction
(5) opposition
(6) elevation
(7) depression
(8) inversion
(9) protraction
(10) dorsiflexion
(11) supination

CRITICAL THINKING QUESTIONS

1. Katie loves pretending that she's a human cannonball. As she jumps off the diving board, she assumes the proper position before she pounds into the water: head and thighs tucked against her chest; back rounded; arms pressed against her sides while her forearms, crossed in front of her shins, hold her legs tightly folded against her chest. Use the proper anatomical terms to describe the position of Katie's back, head, and limbs.
HINT ***She is about as far as she can get from the anatomical position.***

2. After the cast was removed from his arm, 85-year-old Mr. Singh was referred to physical therapy for a frozen elbow joint. When the therapist said she would be working to free up his "hinge," Mr. Singh complained, "You're supposed to be fixing my elbow, not my back door!" Explain the situation to Mr. Singh.
HINT ***Think about how a door moves.***

3. Kendal was sitting in the cafeteria about 30 minutes before his big anatomy and physiology exam. "I'm not worried," he told his lab partner while he drank his third cup of coffee. "I've got it all figured out. There are three types of joints: Cartilaginous joints are made out of hyaline cartilage, like the intervertebral discs and the suture; fibrous joints have an articular capsule; and the serous joints have cavities that do not open to the outside of the body." Kendal's lab partner was speechless. Can you explain the three types better?
HINT ***Synovial fluid lubricates diarthroses.***

ANSWERS TO FIGURE QUESTIONS

9.1 Functionally, sutures are classified as synarthroses because they are immovable; syndesmoses are classified as amphiarthroses because they are slightly movable.

9.2 The difference between a synchondrosis and a symphysis is the type of cartilage that holds the joint together: hyaline cartilage in a synchondrosis and fibrocartilage in a symphysis.

9.3 Functionally, synovial joints are diarthroses, freely movable joints.

9.4 Condyloid and saddle joints are biaxial joints.

9.5 Gliding movements occur at planar joints.

9.6 Two examples of flexion that do not occur in the sagittal plane are flexion of the thumb and lateral flexion of the trunk.

9.7 When you adduct your arm or leg, you bring it closer to the midline of the body, thus "adding" it to the trunk.

9.8 Circumduction involves flexion, abduction, extension, and adduction in continuous sequence.

9.9 The anterior surface of a bone or limb rotates toward the midline in medial rotation, and away from the midline in lateral rotation.

9.10 Bringing the arms forward until the elbows touch is an example of protraction.

9.11 The shoulder joint is so freely movable because of the looseness of its articular capsule and the shallowness of the glenoid cavity in relation to the size of the head of the humerus.

9.12 A hinge joint permits flexion and extension.

9.13 Tension in three ligaments—iliofemoral, pubofemoral, and ischiofemoral—limits the degree of extension at the hip joint.

9.14 Contraction of the quadriceps causes extension at the knee joint.

Muscle Tissue

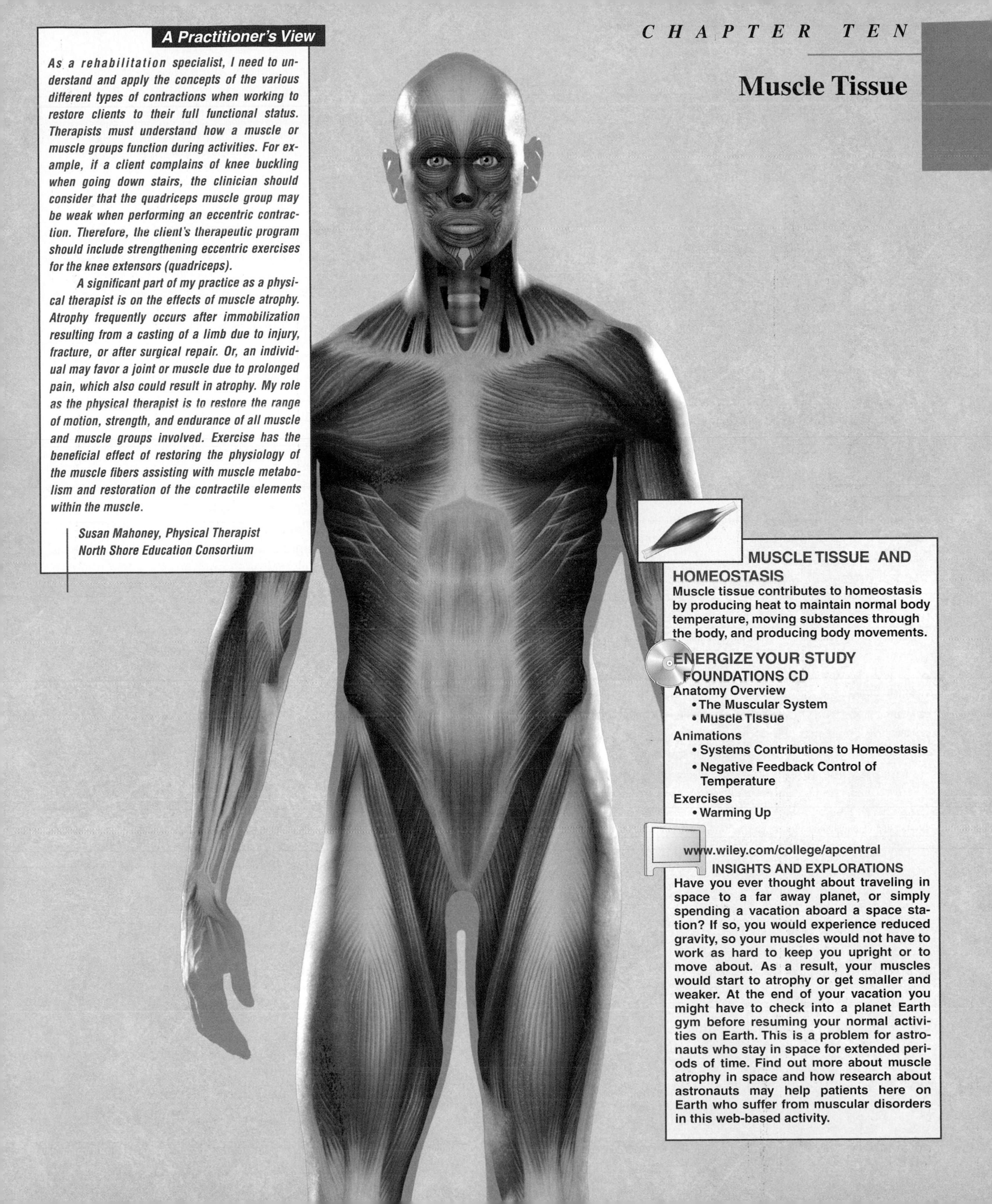

A Practitioner's View

As a rehabilitation specialist, I need to understand and apply the concepts of the various different types of contractions when working to restore clients to their full functional status. Therapists must understand how a muscle or muscle groups function during activities. For example, if a client complains of knee buckling when going down stairs, the clinician should consider that the quadriceps muscle group may be weak when performing an eccentric contraction. Therefore, the client's therapeutic program should include strengthening eccentric exercises for the knee extensors (quadriceps).

A significant part of my practice as a physical therapist is on the effects of muscle atrophy. Atrophy frequently occurs after immobilization resulting from a casting of a limb due to injury, fracture, or after surgical repair. Or, an individual may favor a joint or muscle due to prolonged pain, which also could result in atrophy. My role as the physical therapist is to restore the range of motion, strength, and endurance of all muscle and muscle groups involved. Exercise has the beneficial effect of restoring the physiology of the muscle fibers assisting with muscle metabolism and restoration of the contractile elements within the muscle.

Susan Mahoney, Physical Therapist
North Shore Education Consortium

MUSCLE TISSUE AND HOMEOSTASIS

Muscle tissue contributes to homeostasis by producing heat to maintain normal body temperature, moving substances through the body, and producing body movements.

ENERGIZE YOUR STUDY

FOUNDATIONS CD

Anatomy Overview
- The Muscular System
- Muscle TIssue

Animations
- Systems Contributions to Homeostasis
- Negative Feedback Control of Temperature

Exercises
- Warming Up

www.wiley.com/college/apcentral

INSIGHTS AND EXPLORATIONS

Have you ever thought about traveling in space to a far away planet, or simply spending a vacation aboard a space station? If so, you would experience reduced gravity, so your muscles would not have to work as hard to keep you upright or to move about. As a result, your muscles would start to atrophy or get smaller and weaker. At the end of your vacation you might have to check into a planet Earth gym before resuming your normal activities on Earth. This is a problem for astronauts who stay in space for extended periods of time. Find out more about muscle atrophy in space and how research about astronauts may help patients here on Earth who suffer from muscular disorders in this web-based activity.

Although bones provide leverage and form the framework of the body, they cannot move body parts by themselves. Motion results from the alternating contraction and relaxation of muscles, which constitute 40–50% of total body weight in adults. Your muscular strength reflects the prime function of muscle—changing chemical energy into mechanical energy to generate force, perform work, and produce movement. In addition, muscle tissues stabilize the body's position, regulate organ volume, generate heat, and propel fluids and food matter through various body systems. The scientific study of muscles is known as **myology** (mī-OL-ō-jē; *myo-* = muscle; *-logy* = study of).

OVERVIEW OF MUSCLE TISSUE

OBJECTIVE

- **Correlate the three types of muscle tissue with their functions and special properties.**

Types of Muscle Tissue

The three types of muscle tissue are skeletal, cardiac, and smooth (see the comparison in Table 4.4 on pages 130–131). Although the different types of muscle tissue share some properties, they differ from one another in their microscopic anatomy, location, and control by the nervous and endocrine systems.

Skeletal muscle tissue is so-named because most skeletal muscles function to move bones of the skeleton. (A few skeletal muscles attach to and move the skin or other skeletal muscles.) Skeletal muscle tissue is *striated:* Alternating light and dark bands *(striations)* are seen when the tissue is examined with a microscope (see Figure 10.5). Skeletal muscle tissue works mainly in a *voluntary* manner. Its activity can be consciously controlled by neurons (nerve cells) that are part of the somatic (voluntary) division of the nervous system. (Figure 12.1 on page 387 depicts the divisions of the nervous system.) Most skeletal muscles also are controlled subconsciously to some extent. For instance, the diaphragm continues to alternately contract and relax while we are asleep to power breathing. Also, we do not need to consciously think about contracting the skeletal muscles that maintain posture or stabilize body positions.

Only the heart contains **cardiac muscle tissue,** which forms most of the heart wall. Cardiac muscle is also *striated*, but its action is *involuntary.* The alternating contraction and relaxation of the heart is not consciously controlled. Rather, the heart beats because it has a pacemaker that initiates each contraction. This built-in or intrinsic rhythm is termed **autorhythmicity.** Several hormones and neurotransmitters can adjust heart rate by speeding or slowing the pacemaker.

Smooth muscle tissue is located in the walls of hollow internal structures, such as blood vessels, airways, and most organs in the abdominopelvic cavity. It is also found in the skin, attached to hair follicles. Under a microscope, this tissue lacks the striations of skeletal and cardiac muscle tissue. For this reason, it looks *nonstriated* or *smooth.* The action of smooth muscle is usually *involuntary,* and some smooth muscle tissue has autorhythmicity. Both cardiac muscle and smooth muscle are regulated by neurons that are part of the autonomic (involuntary) division of the nervous system and by hormones released by endocrine glands.

Functions of Muscle Tissue

Through sustained contraction or alternating contraction and relaxation, muscle tissue has four key functions. It produces body movements, stabilizes body positions, stores and moves substances within the body, and generates heat.

1. ***Producing body movements.*** Movements of the whole body such as walking and running, and localized movements such as grasping a pencil or nodding the head, rely on the integrated functioning of bones, joints, and skeletal muscles.
2. ***Stabilizing body positions.*** Skeletal muscle contractions stabilize joints and help maintain body positions, such as standing or sitting. Postural muscles contract continuously when a person is awake; for example, sustained contractions in neck muscles hold the head upright.
3. ***Storing and moving substances within the body.*** Sustained contractions of ringlike bands of smooth muscle called *sphincters* may prevent outflow of the contents of a hollow organ. Temporary storage of food in the stomach or urine in the urinary bladder is possible because smooth muscle sphincters close off the outlets of these organs. Cardiac muscle contractions pump blood through the body's blood vessels. Contraction and relaxation of smooth muscle in the walls of blood vessels help adjust their diameter and thus regulate the rate of blood flow. Smooth muscle contractions also move food and substances such as bile and enzymes through the gastrointestinal tract, push gametes (sperm and oocytes) through the passageways of the reproductive systems, and propel urine through the urinary system. Skeletal muscle contractions promote the flow of lymph and aid the return of blood to the heart.
4. ***Generating heat.*** As muscle tissue contracts, it produces heat, a process known as **thermogenesis.** Much of the heat generated by muscle is used to maintain normal body temperature. Involuntary contractions of skeletal muscle, known as shivering, can increase the rate of heat production severalfold.

Properties of Muscle Tissue

Muscle tissue has four special properties that enable it to function and contribute to homeostasis:

1. Electrical excitability, a property of both muscle and nerve cells, is the ability to respond to certain stimuli by producing electrical signals such as *action potentials*. (Chapter 12 describes how action potentials arise; see page 398.) Action potentials can propagate (travel) along a cell's plasma membrane due to the presence of specific voltage-gated channels. For muscle cells, two main types of stimuli trigger action potentials. One is autorhythmic electrical signals arising in the muscle tissue itself, such as occurs in the heart's pacemaker. The other is chemical stimuli, such as neurotransmitters released by neurons, hormones distributed by the blood, or even local changes in pH.

2. Contractility is the ability of muscle tissue to contract forcefully when stimulated by an action potential. When muscle contracts, it generates tension (force of contraction) while pulling on its attachment points. In an **isometric contraction** (*iso-* = equal; *-metric* = measure or length), the muscle develops tension but does not shorten. An example is holding a book in an outstretched hand. If the tension generated is great enough to overcome the resistance of the object to being moved, the muscle shortens and movement occurs. In an **isotonic contraction** (*-tonic* = tension), the tension developed by the muscle remains almost constant while the muscle shortens. An example is lifting a book off a table.

3. Extensibility is the ability of muscle to stretch without being damaged. Extensibility allows a muscle to contract forcefully even if it is already stretched. Normally, smooth muscle is subject to the greatest amount of stretching. For example, each time the stomach fills with food, the muscle in its wall is stretched. Cardiac muscle also is stretched each time the heart fills with blood. During normal activities, the stretch on skeletal muscle remains almost constant.

4. Elasticity is the ability of muscle tissue to return to its original length and shape after contraction or extension.

This chapter focuses mainly on the structure and function of skeletal muscle tissue. Cardiac muscle and smooth muscle are examined in detail later (in discussions of the autonomic nervous system in Chapter 17, of the heart in Chapter 20, and of the various organs that contain smooth muscle in several chapters).

▶ CHECKPOINT

1. What features distinguish the three types of muscle tissue?
2. What are the general functions of muscle tissue?

SKELETAL MUSCLE TISSUE

▶ OBJECTIVES

- **Explain the relation of connective tissue components, blood vessels, and nerves to skeletal muscles.**
- **Describe the microscopic anatomy of a skeletal muscle fiber.**

Each skeletal muscle is a separate organ composed of hundreds to thousands of cells called **muscle fibers** because of their elongated shapes. Thus, the terms *muscle cell* and *muscle fiber* are synonymous. Connective tissues surround muscle fibers and whole muscles, and blood vessels and nerves penetrate into muscle (Figure 10.1). To understand how contraction of skeletal muscle can generate tension, one first needs to understand its gross and microscopic anatomy.

Connective Tissue Components

Connective tissue surrounds and protects muscle tissue. A **fascia** (FASH-ē-a = bandage) is a sheet or broad band of fibrous connective tissue that supports and surrounds muscles and other organs of the body. The **superficial fascia** (or **subcutaneous layer**) separates muscle from skin (see Figure 11.21 on page 371). It is composed of areolar connective tissue and adipose tissue and provides a pathway for nerves, blood vessels, and lymphatic vessels to enter and exit muscles. The adipose tissue of superficial fascia stores most of the body's triglycerides, serves as an insulating layer that reduces heat loss, and protects muscles from physical trauma. **Deep fascia** is dense irregular connective tissue that lines the body wall and limbs and holds muscles with similar functions together. Deep fascia allows free movement of muscles, carries nerves, blood vessels, and lymphatic vessels, and fills spaces between muscles.

Three layers of connective tissue extend from the deep fascia to further protect and strengthen skeletal muscle (Figure 10.1). The outermost layer, encircling the whole muscle, is the **epimysium** (ep-i-MĪZ-ē-um; *epi-* = upon). **Perimysium** (per-i-MĪZ-ē-um; *peri-* = around) surrounds groups of 10 to 100 or more individual muscle fibers, separating them into bundles called **fascicles** (FAS-i-kuls = little bundles). Many fascicles are large enough to be seen with the naked eye. They give a cut of meat its characteristic "grain," and if you tear a piece of meat, it rips apart along the fascicles. Both epimysium and perimysium are dense irregular connective tissue. Penetrating the interior of each fascicle and separating individual muscle fibers (muscle cells) from one another is **endomysium** (en′-dō-MĪZ-ē-um; *endo-* = within), a thin sheath of areolar connective tissue.

The epimysium, perimysium, and endomysium all are continuous with the connective tissue that attaches skeletal muscle to other structures, such as bone or another muscle. All three connective tissue layers may extend beyond the muscle fibers to form a **tendon**—a cord of dense regular connective tissue that attaches a muscle to the periosteum of a bone. An example is the calcaneal (Achilles) tendon of the gastrocnemius (calf) muscle, which attaches the muscle to the calcaneus (shown in Figure 11.22 on page 374). When the connective tissue elements extend as a broad, flat layer, the tendon is called an **aponeurosis** (*apo-* = from; *neur-* = a sinew). An example is the epicranial aponeurosis on top of the skull between the frontal and occipital bellies of the occipitofrontalis muscle (shown in Figure 11.4c on page 321).

Figure 10.1 Organization of skeletal muscle and its connective tissue coverings.

A skeletal muscle consists of individual muscle fibers (cells) bundled into fascicles and surrounded by three connective tissue layers that are extensions of the deep fascia.

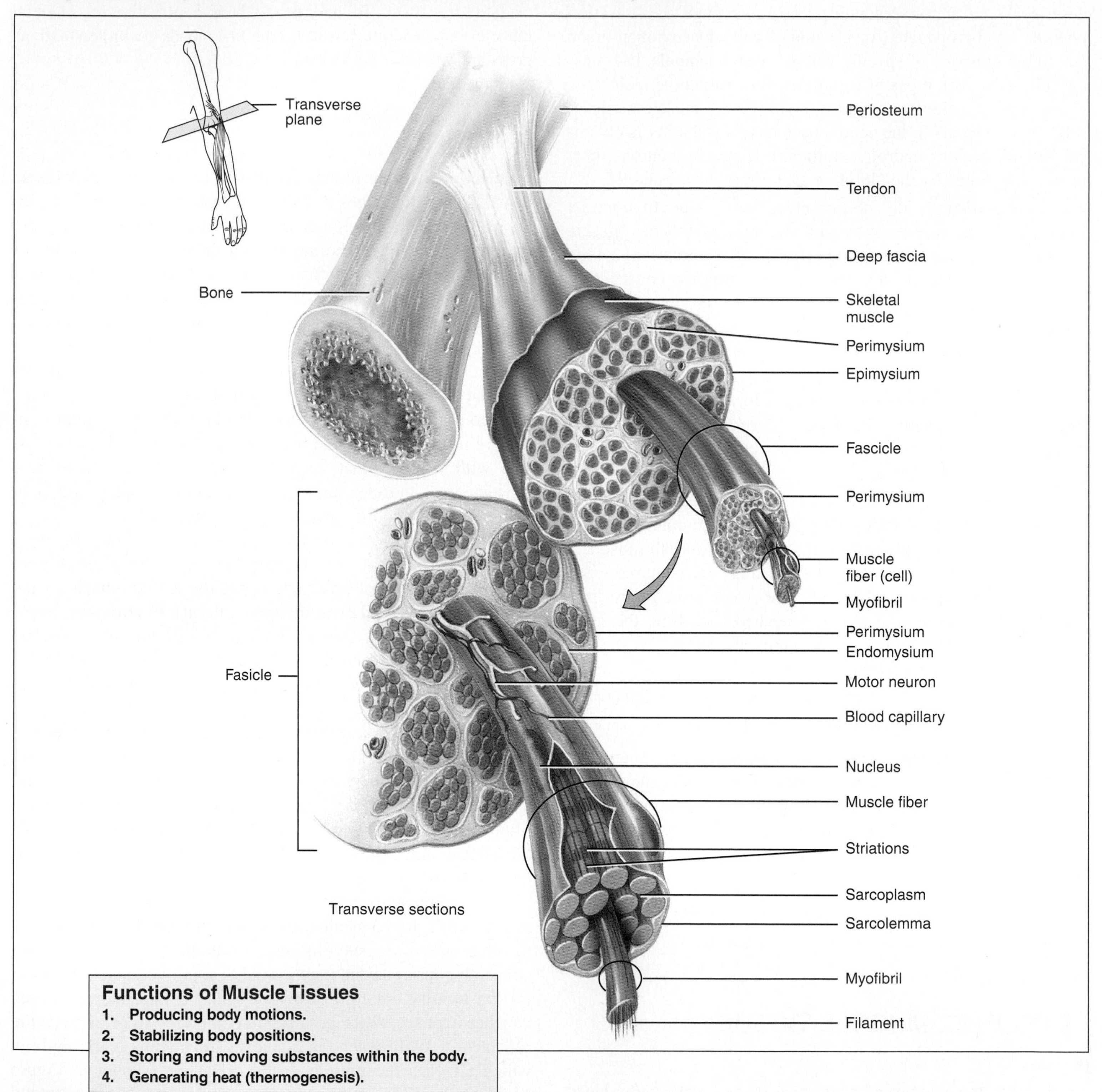

Functions of Muscle Tissues

1. **Producing body motions.**
2. **Stabilizing body positions.**
3. **Storing and moving substances within the body.**
4. **Generating heat (thermogenesis).**

Which connective tissue coat surrounds groups of muscle fibers, separating them into fascicles?

Nerve and Blood Supply

Skeletal muscles are well supplied with nerves and blood vessels. Generally, an artery and one or two veins accompany each nerve that penetrates a skeletal muscle. The neurons that stimulate skeletal muscle to contract are *somatic motor neurons*. Each somatic motor neuron has a threadlike axon that extends from the brain or spinal cord to a group of skeletal muscle fibers (Figure 10.2). The axon is wrapped with *myelin,* which is provided by nearby Schwann cells. The axon of a somatic motor neuron typically branches many times. Each axon collateral (branch) extends to a different skeletal muscle fiber. The structural point of contact and functional site of communication between the motor neuron and the muscle fiber is termed the **neuromuscular junction (NMJ).** At the NMJ, several axon terminals expand into clusters of *synaptic end bulbs.*

Microscopic blood vessels called capillaries are plentiful in muscle tissue; each muscle fiber is in close contact with one or more capillaries (Figure 10.2). The blood capillaries bring in oxygen and nutrients and remove heat and the waste products of muscle metabolism. Especially during contraction, a muscle fiber synthesizes and uses considerable ATP (adenosine triphosphate). These reactions require oxygen, glucose, fatty acids, and other substances that are supplied in the blood.

Microscopic Anatomy of a Skeletal Muscle Fiber

The most important components of a skeletal muscle are the muscle fibers themselves. During embryonic development, each skeletal muscle fiber arises from the fusion of a hundred or more small mesodermal cells called *myoblasts* (Figure 10.3a). Hence, each mature skeletal muscle fiber has a hundred or more nuclei. Once fusion has occurred, the muscle fiber loses its ability to undergo mitosis. Thus, the number of skeletal muscle fibers is set before birth, and most of these cells last a lifetime. The dramatic muscle growth that occurs after birth occurs mainly by enlargement of existing muscle fibers. During childhood, human growth hormone and other hormones stimulate an increase in the size of skeletal muscle fibers. The hormone testosterone (from the testes in males and in small amounts from other tissues in females) promotes further enlargement of muscle fibers. A few myoblasts do persist in mature skeletal muscle as *satellite cells* (Figure 10.3b). These cells retain the capacity to fuse with one another or with damaged muscle fibers to regenerate functional muscle fibers. Mature muscle fibers range from 10 to 100 μm* in diameter. In humans, although a typical length is about 10 cm (4 in.), some are up to 30 cm (12 in.) long.

Sarcolemma, T Tubules, and Sarcoplasm

The multiple nuclei of a skeletal muscle fiber are located just beneath the **sarcolemma** (*sarc-* = flesh; *-lemma* = sheath), the plasma membrane of a muscle cell (Figure 10.3b). Thousands of tiny invaginations of the sarcolemma, called **T (transverse) tubules,** tunnel in from the surface toward the center of each muscle fiber. T tubules are open to the outside of the fiber and

*One micrometer (μm) is 10^{-6} meter (1/25,000 in.).

Figure 10.2 Blood and nerve supply to skeletal muscle. The site of synaptic contact between a somatic motor neuron and a skeletal muscle fiber is the neuromuscular junction (NMJ). Two NMJs are present in this view.

Branches of one motor neuron typically contact many muscle fibers.

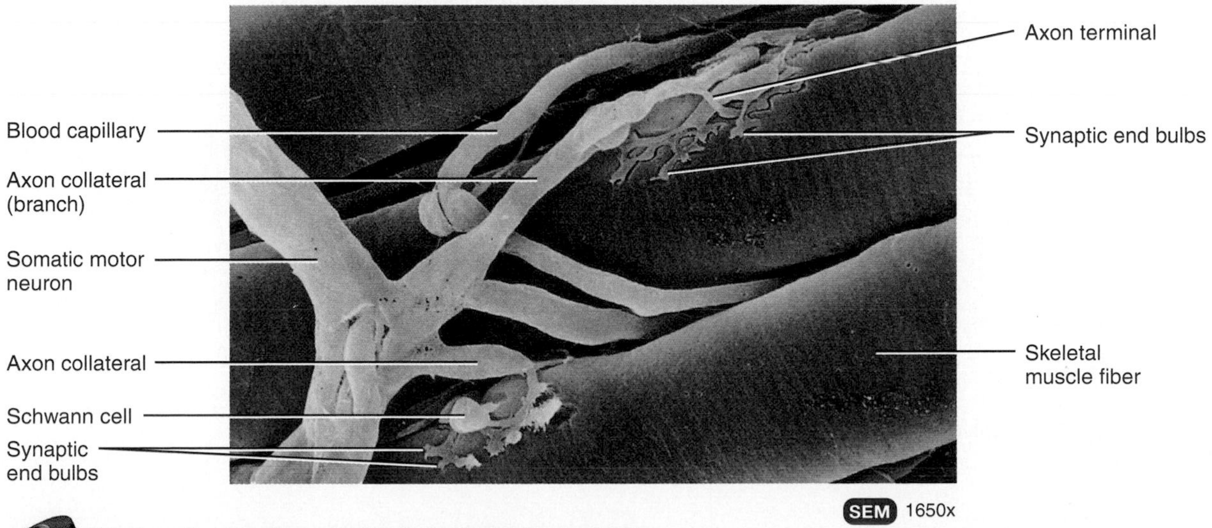

Axon terminal
Blood capillary
Synaptic end bulbs
Axon collateral (branch)
Somatic motor neuron
Axon collateral
Skeletal muscle fiber
Schwann cell
Synaptic end bulbs
SEM 1650x

What are the functions of blood capillaries in muscle?

Figure 10.3 Microscopic organization of skeletal muscle. (a) During embryonic development, many myoblasts fuse to form one skeletal muscle fiber. Once fusion has occurred, a skeletal muscle fiber loses the ability to undergo cell division, but satellite cells retain this ability. (b) The sarcolemma of the fiber encloses sarcoplasm and myofibrils, which are striated. Sarcoplasmic reticulum (SR) wraps around each myofibril. Thousands of T tubules, filled with interstitial fluid, invaginate from the sarcolemma toward the center of the muscle fiber. A triad is a T tubule and the two terminal cisterns of the SR on either side of it.

The contractile elements of muscle fibers are the myofibrils, which contain overlapping thick and thin filaments.

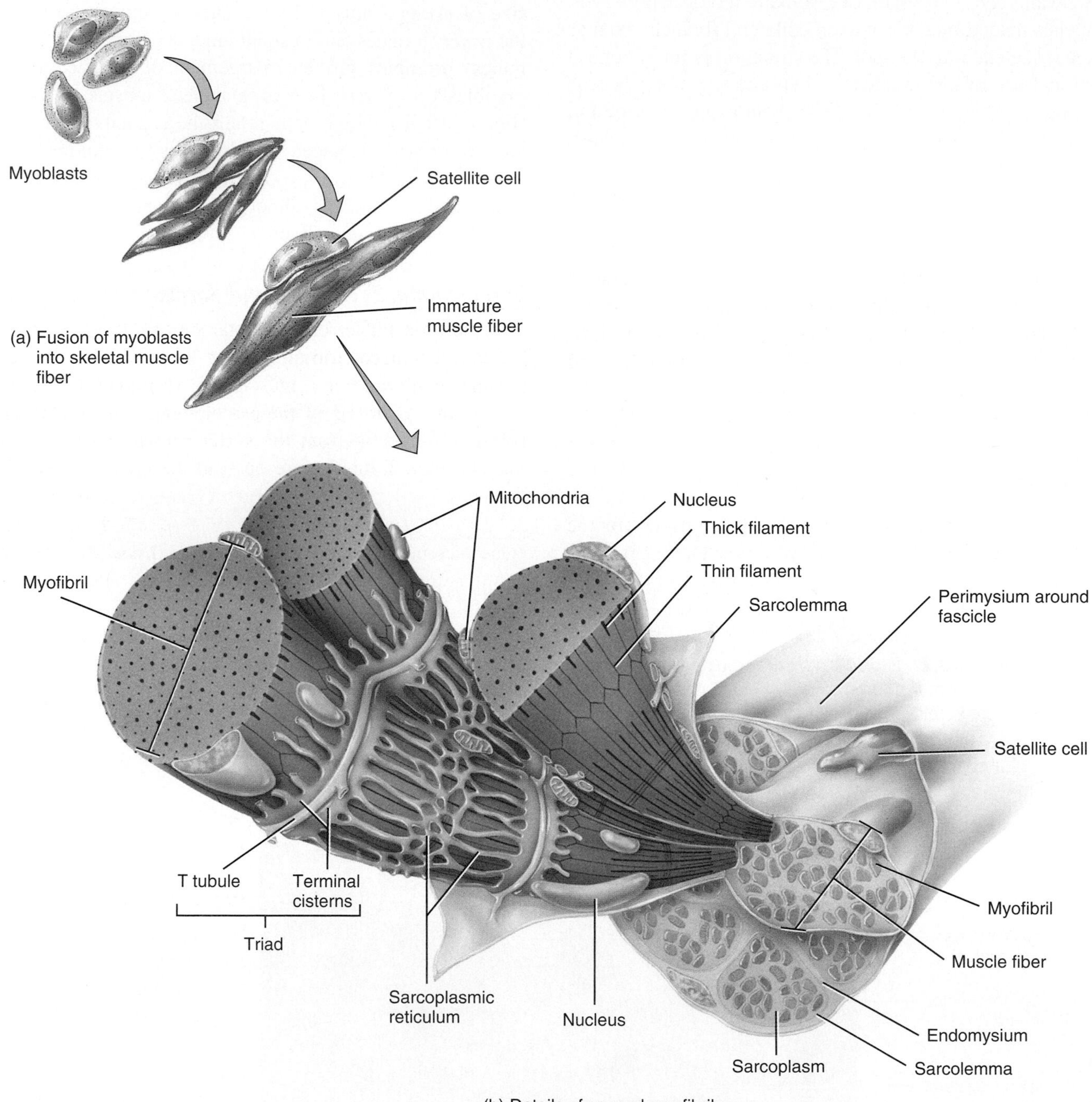

(b) Details of several myofibrils

Which structure shown here releases calcium ions to trigger muscle contraction?

thus are filled with interstitial fluid. Muscle action potentials propagate along the sarcolemma and through the T tubules, quickly spreading throughout the muscle fiber. This arrangement ensures that an action potential excites all parts of the muscle fiber at almost the same instant.

Within the sarcolemma is the **sarcoplasm,** the cytoplasm of a muscle fiber. Sarcoplasm includes a substantial amount of glycogen, which can be split into glucose that is used for synthesis of ATP. In addition, the sarcoplasm contains a red-colored protein called **myoglobin** (mī-ō-GLŌB-in). This protein, found only in muscle binds oxygen molecules that diffuse into muscle fibers from interstitial fluid. Myoglobin releases oxygen when mitochondria need it for ATP production. The mitochondria lie in rows throughout the muscle fiber, strategically close to the muscle proteins that use ATP during contraction.

Myofibrils and Sarcoplasmic Reticulum

At high magnification, the sarcoplasm appears stuffed with little threads. These small structures are the **myofibrils,** the contractile elements of skeletal muscle (Figure 10.3b). Myofibrils are about 2 μm in diameter and extend the entire length of a muscle fiber. Their prominent striations make an entire muscle fiber look striated.

A fluid-filled system of membranous sacs called the **sarcoplasmic reticulum (SR)** encircles each myofibril. This elaborate system is similar to smooth endoplasmic reticulum in nonmuscle cells. Dilated end sacs of the sarcoplasmic reticulum called *terminal cisterns* (= reservoirs) butt against the T tubule from both sides. A transverse tubule and the two terminal cisterns on either side of it form a *triad* (*tri-* = three). In a relaxed muscle fiber, the sarcoplasmic reticulum stores calcium ions (Ca^{2+}). Release of Ca^{2+} from the terminal cisterns of the sarcoplasmic reticulum triggers muscle contraction.

Muscular Atrophy and Hypertrophy

Muscular atrophy (A-trō-fē; *a-* = without, *-trophy* = nourishment) is a wasting away of muscles. Individual muscle fibers decrease in size because of progressive loss of myofibrils. The atrophy that occurs if muscles are not used is termed *disuse atrophy*. Bedridden individuals and people with casts experience disuse atrophy because the number of nerve impulses to inactive muscle is greatly reduced. If the nerve supply to a muscle is disrupted or cut, the muscle undergoes *denervation atrophy*. In about 6 months to 2 years, the muscle will be one-quarter of its original size, and the muscle fibers will be replaced by fibrous connective tissue. The transition to connective tissue, when complete, cannot be reversed.

Muscular hypertrophy (hī-PER-trō-fē; *hyper-* = above or excessive) is an increase in the diameter of muscle fibers owing to the production of more myofibrils, mitochondria, sarcoplasmic reticulum, and so forth. It results from very forceful, repetitive muscular activity, such as strength training. Because hypertrophied muscles contain more myofibrils, they are capable of contractions that are more forceful. ■

Filaments and the Sarcomere

Within myofibrils are two types of even smaller structures called **filaments,** which are only 1–2 μm long (Figure 10.3b). The diameter of the *thin filaments* is about 8 nm,* whereas that of the *thick filaments* is about 16 nm. Overall, there are two thin filaments for every thick filament. The filaments inside a myofibril do not extend the entire length of a muscle fiber. Instead, they are arranged in compartments called **sarcomeres** (*-mere* = part), which are the basic functional units of a myofibril (Figure 10.4a). Narrow, plate-shaped regions of dense material called **Z discs** separate one sarcomere from the next.

The thick and thin filaments overlap one another to a greater or lesser extent, depending on whether the muscle is contracted, relaxed, or stretched. The pattern of their overlap, consisting of a variety of zones and bands (Figure 10.4b), creates the striations that can be seen both in single myofibrils and in whole muscle fibers. The darker middle part of the sarcomere is the **A band,** which extends the entire length of the thick filaments (Figure 10.4b). Toward each end of the A band is a *zone of overlap,* where the thick and thin filaments lie side by side. The **I band** is a lighter, less dense area that contains the rest of the thin filaments but no thick filaments (Figure 10.4b). A Z disc passes through the center of each I band. A narrow **H zone** in the center of each A band contains thick but not thin filaments. Supporting proteins that hold the thick filaments together at the center of the H zone form the **M line,** so-named because it is at the *middle* of the sarcomere. Figure 10.5 shows the relations of the zones, bands, and lines as seen in a transmission electron micrograph.

Exercise-Induced Muscle Damage

Comparison of electron micrographs of muscle tissue taken from athletes before and after intense exercise reveal considerable exercise-induced muscle damage, including torn sarcolemmas in some muscle fibers, damaged myofibrils, and disrupted Z discs. Microscopic muscle damage after exercise also is indicated by increases in blood levels of proteins, such as myoglobin and the enzyme creatine kinase, that are normally confined within muscle fibers. From 12 to 48 hours after a period of strenuous exercise, skeletal muscles often become sore. Such **delayed onset muscle soreness (DOMS)** is accompanied by stiffness, tenderness, and swelling. Although the causes of DOMS are not completely understood, microscopic muscle damage appears to be a major factor. ■

*One nanometer (nm) is 10^{-9} meter (0.001 μm).

Figure 10.4 The arrangement of filaments within a sarcomere. A sarcomere extends from one Z disc to the next.

Myofibrils contain two types of filaments: thick filaments and thin filaments.

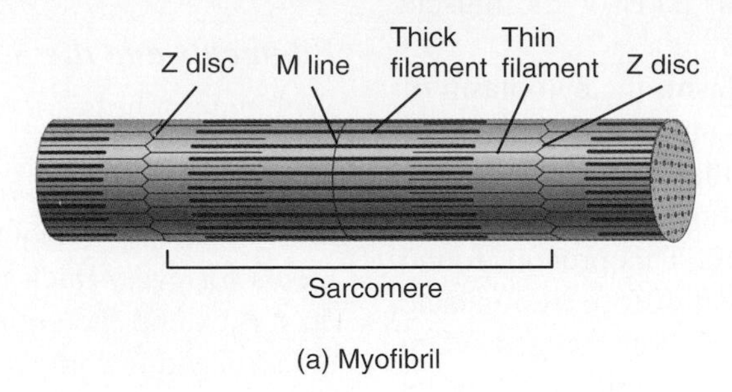

(a) Myofibril

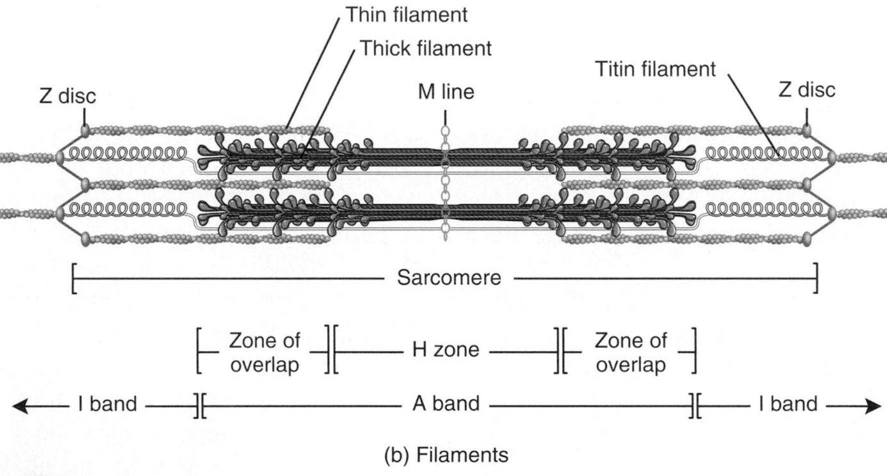

(b) Filaments

Among the following, which is smallest: muscle fiber, thick filament, or myofibril? Which is largest?

Figure 10.5 Characteristic zones and bands of a sarcomere.

The striations of skeletal muscle are alternating darker A bands and lighter I bands.

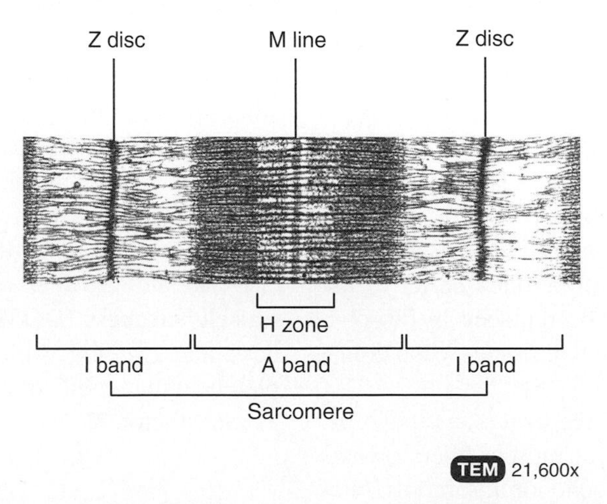

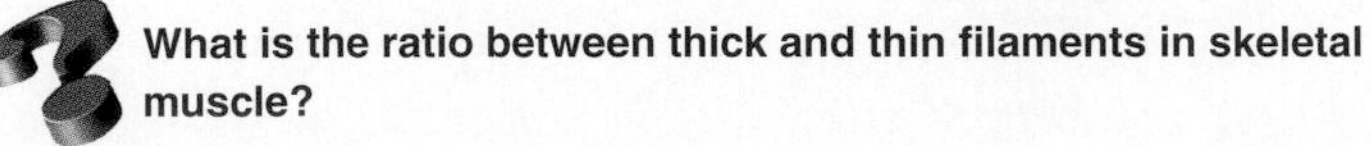

What is the ratio between thick and thin filaments in skeletal muscle?

Muscle Proteins

Myofibrils are built from three kinds of proteins: (1) contractile proteins, which generate force during contraction; (2) regulatory proteins, which help switch the contraction process on and off; and (3) structural proteins, which keep the thick and thin filaments in the proper alignment, give the myofibril elasticity and extensibility, and link the myofibrils to the sarcolemma and extracellular matrix.

The two *contractile proteins* in muscle are myosin and actin, which are the main components of thick and thin filaments, respectively. **Myosin** functions as a *motor protein* in all three types of muscle tissue. Motor proteins push or pull their cargo to achieve movement by converting the chemical energy in ATP to the mechanical energy of motion or force production. In skeletal muscle, about 300 molecules of myosin form a single thick filament. Each myosin molecule is shaped like two golf clubs twisted together (Figure 10.6a). The *myosin tail* (twisted golf club handles) points toward the M line in the center of the sarcomere. Tails of neighboring myosin molecules lie parallel to one another, forming the shaft of the thick filament. The two projections of each myosin molecule (golf club heads) are called *myosin heads* or *crossbridges.* The heads project outward from the shaft in a spiraling fashion, each extending toward one of the six thin filaments that surround each thick filament.

Thin filaments extend from anchoring points within the Z discs (see Figure 10.4b). Their main component is the protein **actin.** Individual actin molecules join to form an actin filament that is twisted into a helix (Figure 10.6b). On each actin molecule is a *myosin-binding site,* where a myosin head can attach. Smaller amounts of two *regulatory proteins*—**tropomyosin** and **troponin**—are also part of the thin filament. In relaxed muscle, myosin is blocked from binding to actin because strands of tropomyosin cover the *myosin-binding sites* on actin. The tropomyosin strands, in turn, are held in place by troponin molecules.

Besides contractile and regulatory proteins, muscle contains about a dozen *structural proteins,* which contribute to the alignment, stability, elasticity, and extensibility of myofibrils. Several key structural proteins are titin, myomesin, and dystrophin. **Titin** *(titan* = gigantic) is the third most plentiful protein in skeletal muscle (after actin and myosin). This molecule's name reflects its huge size. With a molecular weight of about 3 million daltons, titin is 50 times larger than an average-sized protein. Each titin molecule spans half a sarcomere, from a Z disc to an M line (see Figure 10.4b), a distance of 1 to 1.2 μm in relaxed muscle. Titin anchors a thick filament to both a Z disc and the M line, thereby helping stabilize the position of the thick filament. The part of the titin molecule that extends from the Z disc to the beginning of the thick filament is very elastic. Because it can stretch to at least four times its resting length and then spring back unharmed, titin accounts for much of the elasticity and extensibility of myofibrils. Titin probably helps the sarcomere return to its resting length after a muscle has contracted or been stretched.

Molecules of the protein *myomesin* form the M line. The M line proteins bind to titin and connect adjacent thick filaments to one another. *Dystrophin* is a cytoskeletal protein that links thin filaments of the sarcomere to integral membrane proteins of the sarcolemma. In turn, the membrane proteins attach to proteins in the connective tissue matrix that surrounds muscle fibers. Hence, dystrophin and its associated proteins are thought to reinforce the sarcolemma and help transmit the tension generated by the sarcomeres to the tendons.

▶ CHECKPOINT

3. What are the types of fascia that cover skeletal muscles?
4. Why is a rich blood supply important for muscle contraction?
5. How are the structures of thin and thick filaments different?

Figure 10.6 Structure of thick and thin filaments. (a) A thick filament contains about 300 myosin molecules, one of which is enlarged. The myosin tails form the shaft of the thick filament, whereas the myosin heads project outward toward the surrounding thin filaments. (b) Thin filaments contain actin, troponin, and tropomyosin.

Contractile proteins (myosin and actin) generate force during contraction, whereas regulatory proteins (troponin and tropomyosin) help switch contraction on and off.

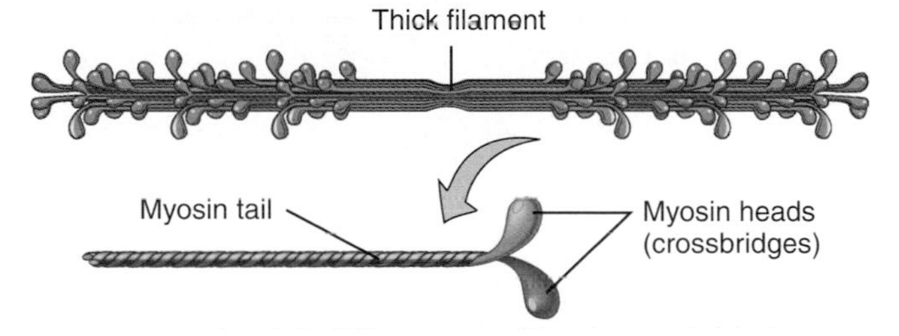

(a) One thick filament (above) and a myosin molecule (below)

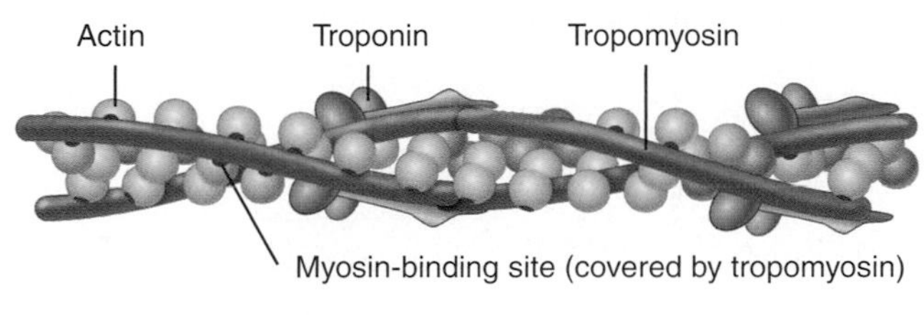

(b) Portion of a thin filament

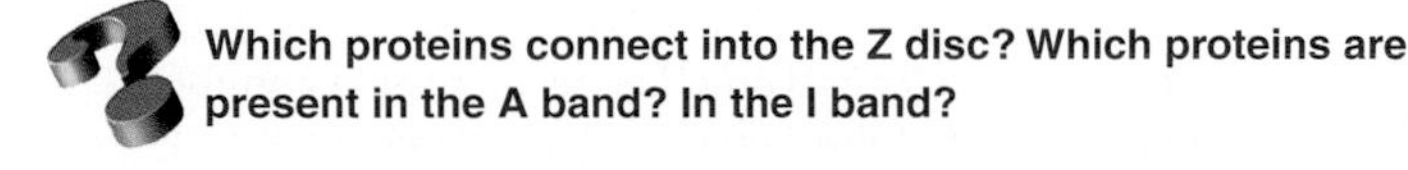

Which proteins connect into the Z disc? Which proteins are present in the A band? In the I band?

CONTRACTION AND RELAXATION OF SKELETAL MUSCLE FIBERS

OBJECTIVES

- **Outline the steps involved in the sliding filament mechanism of muscle contraction.**
- **Describe how muscle action potentials arise at the neuromuscular junction.**

When scientists examined the first electron micrographs of skeletal muscle in the mid-1950s, they were surprised to see that the lengths of the thick and thin filaments were the same in both relaxed and contracted muscle. It had been thought that muscle contraction must be a folding process, somewhat like closing an accordion. However, researchers discovered that skeletal muscle shortens during contraction because the thick and thin filaments slide past one another. The model describing the contraction of muscle is known as the **sliding filament mechanism.**

The Sliding Filament Mechanism

Muscle contraction occurs because myosin heads attach to and "walk" along the thin filaments at both ends of a sarcomere, progressively pulling the thin filaments toward the M line (Figure 10.7). As a result, the thin filaments slide inward and meet at the center of a sarcomere. They may even move so far inward that their ends overlap (Figure 10.7c). As the thin filaments slide inward, the Z discs come closer together, and the sarcomere shortens. However, the lengths of the individual thick and thin filaments do not change. Shortening of the sarcomeres causes shortening of the whole muscle fiber, which leads to shortening of the entire muscle.

The Contraction Cycle

At the onset of contraction, the sarcoplasmic reticulum releases calcium ions (Ca^{2+}) into the cytosol. There, they bind to troponin and cause the troponin–tropomyosin complexes to move away from the myosin-binding sites on actin. Once the binding sites are "free," the **contraction cycle**—the repeating sequence of events that causes the filaments to slide—begins. The contraction cycle consists of four steps (Figure 10.8):

1. ***ATP hydrolysis.*** The myosin head includes an ATP-binding pocket and an ATPase, an enzyme that hydrolyzes ATP into ADP (adenosine diphosphate) and a phosphate group. This hydrolysis reaction energizes the myosin head. Notice that the products of ATP hydrolysis—ADP and a phosphate group—are still attached to the myosin head.
2. ***Attachment of myosin to actin to form crossbridges.*** The energized myosin head attaches to the myosin-binding site on actin and then releases the previously hydrolyzed phosphate group.
3. ***Power stroke.*** The release of the phosphate group triggers the power stroke of contraction. During the power stroke, the pocket on the myosin head where ADP is still bound opens. As a result, the myosin head rotates and releases the ADP. The myosin head generates force as it rotates toward the center of the sarcomere, sliding the thin filament past the thick filament toward the M line.
4. ***Detachment of myosin from actin.*** At the end of the power stroke, the myosin head remains firmly attached to actin until it binds another molecule of ATP. As ATP binds to the ATP-binding pocket on the myosin head, the myosin head detaches from actin.

The contraction cycle repeats as the myosin ATPase again hydrolyzes ATP. The reaction reorients the myosin head and transfers energy from ATP to the myosin head. Then, the myosin head is again energized and ready to combine with the next myosin-binding site on the thin filament. The contraction cycle repeats for as long as ATP is available and the Ca^{2+} level near the thin filament is sufficiently high. The myosin heads keep

Figure 10.7 Sliding filament mechanism of muscle contraction, as it occurs in two adjacent sarcomeres.

During muscle contractions, thin filaments move toward the M line of each sarcomere.

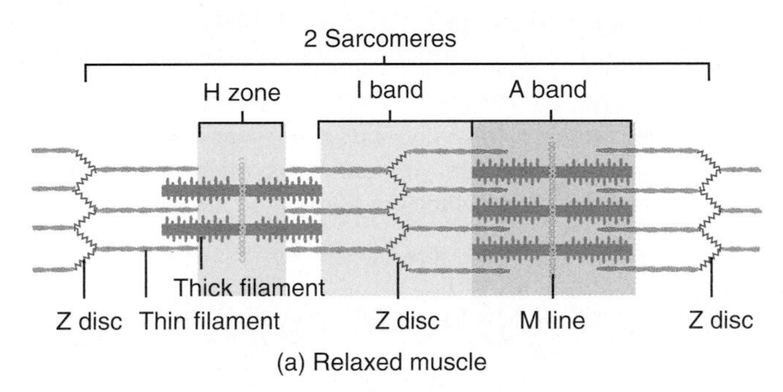

(a) Relaxed muscle

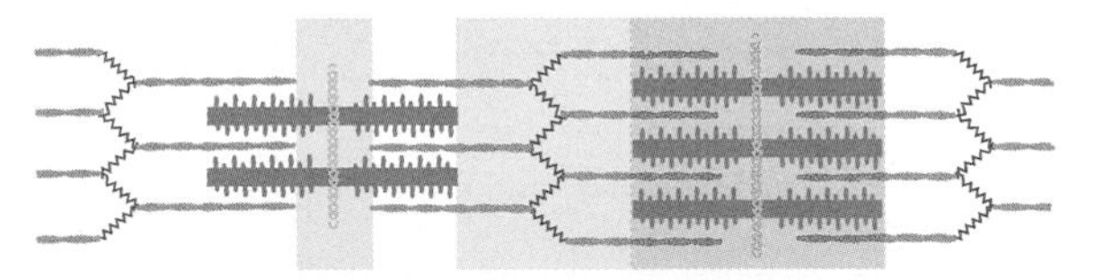

(b) Partially contracted muscle

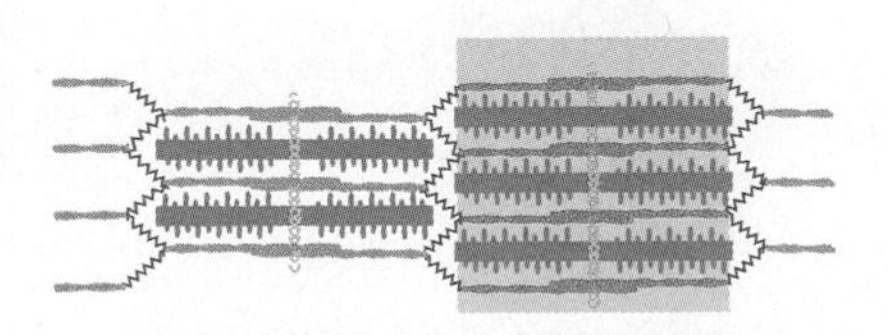

(c) Maximally contracted muscle

What happens to the I band and H zone as muscle contracts? Do the lengths of the thick and thin filaments change?

Figure 10.8 The contraction cycle. Sarcomeres exert force and shorten through repeated cycles during which the myosin heads (crossbridges) attach to actin, rotate, and detach.

During the power stroke of contraction, myosin heads rotate and move the thin filaments past the thick filaments toward the center of the sarcomere.

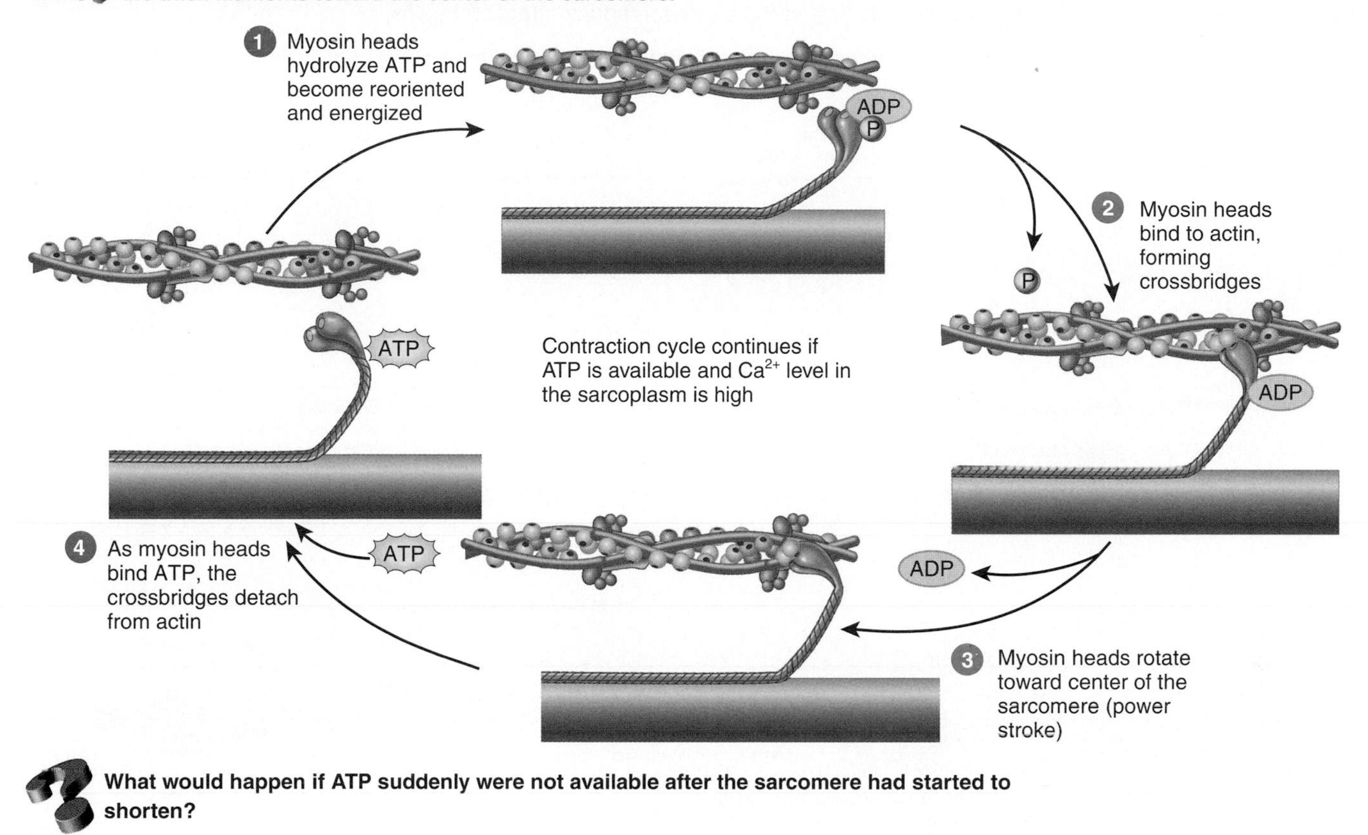

What would happen if ATP suddenly were not available after the sarcomere had started to shorten?

rotating back and forth with each power stroke, pulling the thin filaments toward the M line. Each of the 600 myosin heads in one thick filament attaches and detaches about five times per second. At any one instant, some of the myosin heads are attached to actin and generating force, whereas others are detached and getting ready to bind again.

Contraction is analogous to running on a foot-powered treadmill. One foot (myosin head) strikes the belt (thin filament) and pushes it backward (toward the M line). Then the other foot comes down and imparts a second push. The belt (thin filament) moves smoothly while the runner (thick filament) remains stationary. Each myosin head progressively "walks" along a thin filament, coming closer to the Z disc with each "step," while the thin filament moves toward the M line. And like the legs of a runner, the myosin head needs a constant supply of energy to keep going—one molecule of ATP for each contraction cycle!

As the contraction cycle continues, movement of myosin heads applies the force that draws the Z discs toward each other, and the sarcomere shortens. During a maximal muscle contraction, the distance between two Z discs can decrease to half the resting length. The Z discs, in turn, pull on neighboring sarcomeres, and the whole muscle fiber shortens. Some of the components of muscle are elastic: They stretch slightly before they transfer the tension generated by the sliding filaments. The elastic components include titin molecules, connective tissue around the muscle fibers (endomysium, perimysium, and epimysium), and tendons that attach muscle to bone. As the cells of a skeletal muscle start to shorten, they first pull on their connective tissue coverings and tendons. The coverings and tendons stretch and then become taut, and the tension passed through the tendons pulls on the bones to which they are attached. The result is movement of a part of the body. However, the contraction cycle does not always result in shortening of the muscle fibers and the whole muscle. In isometric contractions, the myosin heads rotate and generate tension, but the thin filaments cannot slide inward because the tension they generate is not large enough to move the load on the muscle (see Figure 10.17c).

Excitation–Contraction Coupling

An increase in Ca^{2+} concentration in the cytosol starts muscle contraction, whereas a decrease stops it. When a muscle fiber is relaxed, the concentration of Ca^{2+} in its cytosol is very low, only about 0.1 micromole per liter (0.1 μm/L). A huge amount

Figure 10.9 The role of Ca^{2+} in the regulation of contraction by troponin and tropomyosin. (a) During relaxation, the level of Ca^{2+} in the sarcoplasm is low, only 0.1 μM (0.001 mM), because calcium ions are pumped into the sarcoplasmic reticulum (SR) by Ca^{2+} active transport pumps. (b) A muscle action potential propagating along a transverse tubule opens Ca^{2+} release channels in the SR, calcium ions flow into the cytosol, and contraction begins.

An increase in the Ca^{2+} level in the sarcoplasm starts the sliding of thin filaments. When the level of Ca^{2+} in the sarcoplasm declines, sliding stops.

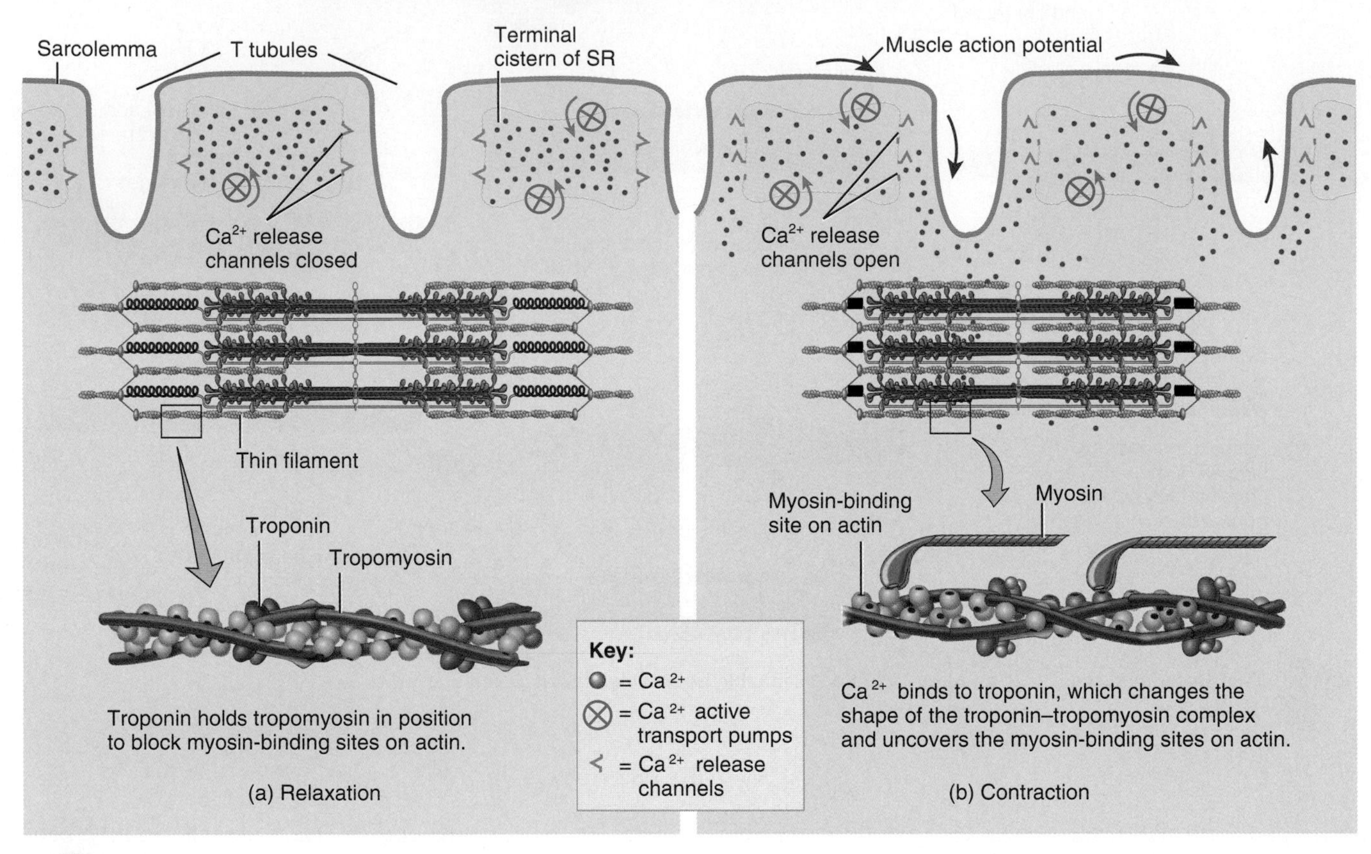

What are three functions of ATP in muscle contraction?

of Ca^{2+}, however, is stored inside the sarcoplasmic reticulum (Figure 10.9a). As a muscle action potential propagates along the sarcolemma and into the T tubules, it causes **Ca^{2+} release channels** in the SR membrane to open (Figure 10.9b). When these channels are open, Ca^{2+} flows out of the SR into the cytosol around the thick and thin filaments. As a result, the Ca^{2+} concentration in the cytosol rises tenfold or more. The released calcium ions combine with troponin, causing it to change shape. This conformational change moves the troponin–tropomyosin complex away from the myosin-binding sites on actin. Once these binding sites are free, myosin heads bind to them, and the contraction cycle begins. The events just described constitute **excitation–contraction coupling,** the steps that connect excitation (a muscle action potential propagating along the sarcolemma and into the T tubules) to contraction (sliding of the filaments).

The sarcoplasmic reticulum membrane also contains **Ca^{2+} active transport pumps** that use ATP to constantly move Ca^{2+} from the cytosol into the SR (Figure 10.9). While muscle action potentials continue to propagate through the T tubules, the Ca^{2+} release channels are open. Calcium ions flow into the cytosol more rapidly than they are transported back by the pumps. After the last action potential has propagated throughout the T tubules, the Ca^{2+} release channels close. As the pumps move Ca^{2+} back into the SR, the concentration of calcium ions in the cytosol quickly decreases. Inside the SR, molecules of a calcium-binding protein, appropriately called **calsequestrin,** bind to the Ca^{2+}, enabling even more Ca^{2+} to be sequestered within the SR. As a result, the concentration of Ca^{2+} inside the SR is 10,000 times higher than in the cytosol in a relaxed muscle fiber. As the Ca^{2+} level drops in the cytosol, the troponin–tropomyosin complexes slide back over and cover the myosin-binding sites, and the muscle fiber relaxes.

Rigor Mortis

After death, cellular membranes start to become leaky. Calcium ions leak out of the sarcoplasmic reticulum into the cytosol and allow myosin heads to bind to actin. ATP synthesis has ceased, however, so the crossbridges cannot detach from actin. The resulting condition, in which muscles are in a state of rigidity (cannot contract or stretch), is called **rigor mortis** (rigidity of death). Rigor mortis begins 3–4 hours after death and lasts about 24 hours; then it disappears as proteolytic enzymes from lysosomes digest the crossbridges. ■

Length–Tension Relationship

Figure 10.10 plots the **length–tension relationship** for skeletal muscle, which shows how the forcefulness of muscle contraction depends on the length of the sarcomeres within a muscle *before contraction begins.* At a sarcomere length of about 2.0–2.4 μm, the zone of overlap in each sarcomere is optimal, and the muscle fiber can develop maximum tension. Notice in Figure 10.10 that maximum tension (100%) occurs when the zone of overlap between a thick and thin filament extends from the edge of the H zone to one end of a thick filament.

As the sarcomeres of a muscle fiber are stretched to a longer length, the zone of overlap shortens, and fewer myosin heads can make contact with thin filaments. So, the tension the fiber can produce decreases. When a skeletal muscle fiber is stretched to 170% of its optimal length, there is no overlap between the thick and thin filaments. Because none of the myosin heads can bind to thin filaments, the muscle fiber cannot contract, and tension is zero. As sarcomere lengths become increasingly shorter than the optimum, the tension that can develop again decreases. This is because thick filaments crumple as they are compressed by the Z discs, resulting in fewer myosin heads making contact with thin filaments. Normally, resting muscle fiber length is held very close to the optimum by firm attachments of skeletal muscle to bones (via their tendons) and to other inelastic tissues, so that overstretching does not occur.

The Neuromuscular Junction

A muscle fiber contracts in response to one or more action potentials propagating along its sarcolemma and through its T tubule system. Muscle action potentials arise at the **neuromuscular junction (NMJ),** the region of synaptic contact between a somatic motor neuron and a skeletal muscle fiber (Figure 10.11a). A **synapse** is a region where communication occurs between two neurons, or between a neuron and a target cell—in this case, between a motor neuron and a skeletal muscle fiber. At most synapses a small gap, called the *synaptic cleft,* separates the two cells. Because the cells do not physically touch, the action potential from one cell cannot "jump the gap" to directly excite the next cell. Instead, the first cell communicates with the second indirectly, by releasing a chemical called a **neurotransmitter.**

Figure 10.10 Length–tension relationship in a skeletal muscle fiber. Maximum tension during contraction occurs when the resting sarcomere length is 2.0–2.4 μm.

A muscle fiber develops its greatest tension when there is an optimal zone of overlap between thick and thin filaments.

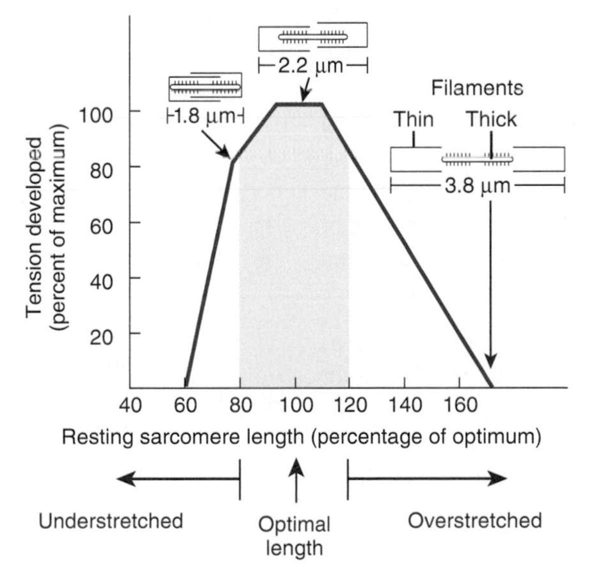

Why is tension maximal at a sarcomere length of 2.2 μm?

As you saw in Figure 10.2, each motor neuron axon terminal divides into a cluster of synaptic end bulbs (Figure 10.11a). Suspended in the cytosol within each end bulb are hundreds of membrane-enclosed sacs called **synaptic vesicles.** Inside each synaptic vesicle are thousands of molecules of **acetylcholine** (as′-ē-til-KŌ-lēn), abbreviated **ACh,** the neurotransmitter released at the NMJ.

The muscle fiber part of the NMJ is the sarcolemma opposite the synaptic end bulbs. This region of the sarcolemma is called the **motor end plate.** Within each motor end plate are 30 to 40 million **acetylcholine receptors,** which are integral transmembrane proteins that bind specifically to ACh. As you will see, the ACh receptors also are ligand-gated ion channels. A neuromuscular junction thus includes all the synaptic end bulbs on one side of the synaptic cleft, plus the motor end plate of the muscle fiber on the other side.

A nerve impulse elicits a muscle action potential in the following way (Figure 10.11c).

1. ***Release of acetylcholine.*** Arrival of the nerve impulse at the synaptic end bulbs causes many synaptic vesicles to undergo exocytosis. During exocytosis, the synaptic vesicles fuse with the motor neuron's plasma membrane, which liberates ACh into the synaptic cleft. The ACh then diffuses across the synaptic cleft between the motor neuron and the motor end plate.
2. ***Activation of ACh receptors.*** Binding of two molecules of ACh to the receptor opens the ion channel part of the ACh receptor. Once the channel is open, small cations, most importantly Na^+, can flow across the membrane.
3. ***Production of muscle action potential.*** The inflow of Na^+ (down its electrochemical gradient) makes the inside of the muscle fiber more positively charged. This change in the membrane potential is the signal that triggers a muscle action potential. Each nerve impulse normally elicits one muscle action potential. The muscle action potential then propagates along the sarcolemma and into the T tubule system.
4. ***Termination of ACh activity.*** The effect of ACh binding lasts only briefly because ACh is rapidly broken down by an enzyme called **acetylcholinesterase (AChE).** This enzyme is attached to collagen fibers in the extracellular matrix of the synaptic cleft. AChE breaks down ACh into acetyl and choline, products that cannot activate the ACh receptor.

If another nerve impulse releases more acetylcholine, steps 2 and 3 repeat. When action potentials cease in the motor neuron, ACh release stops, and AChE rapidly breaks down the ACh already present in the synaptic cleft. This ends the production of muscle action potentials, and the Ca^{2+} release channels in the sarcoplasmic reticulum membrane close.

The NMJ usually is near the midpoint of a skeletal muscle fiber. Muscle action potentials arise at the NMJ and then propagate toward both ends of the fiber. This arrangement permits nearly simultaneous activation (and thus contraction) of all parts of the muscle fiber.

Figure 10.12 summarizes the events that occur during contraction and relaxation of a skeletal muscle fiber.

Pharmacology of the NMJ

Several plant products and drugs selectively block certain events at the NMJ. *Botulinum toxin,* produced by the bacterium *Clostridium botulinum,* blocks exocytosis of synaptic vesicles at the NMJ. As a result, ACh is not released, and muscle contraction does not occur. The bacteria proliferate in improperly canned foods and their toxin is one of the most lethal chemicals known. A tiny amount can cause death by paralyzing skeletal muscles. Breathing stops due to paralysis of respiratory muscles, including the diaphragm. Yet, it is also the first bacterial toxin to be used as a medicine (Botox). Injections of Botox into the affected muscles can help patients who have strabismus (crossed eyes) or blepharospasm (uncontrollable blinking). It is also used as a cosmetic treatment to relax muscles that cause facial wrinkles and to alleviate chronic back pain due to muscle spasms in the lumbar region.

The plant derivative *curare,* a poison used by South American Indians on arrows and blowgun darts, causes muscle paralysis by binding to and blocking ACh receptors. In the presence of curare, the ion channels do not open. Curare-like drugs are often used during surgery to relax skeletal muscles. A family of chemicals called *anticholinesterase agents* have the property of slowing the enzymatic activity of acetylcholinesterase, thus slowing removal of ACh from the synaptic cleft. At low doses, these agents can strengthen weak muscle contractions. One example is neostigmine, which is used to treat patients with myasthenia gravis (see page 302). Neostigmine is also used as an antidote for curare poisoning and to terminate the effects of curare after surgery. ■

▶ CHECKPOINT

6. What roles do contractile, regulatory, and structural proteins play in muscle contraction and relaxation?
7. How do calcium ions and ATP contribute to muscle contraction and relaxation?
8. How does sarcomere length influence the maximum tension that is possible during muscle contraction?
9. How is the motor end plate different from other parts of the sarcolemma?

Figure 10.11 Structure of the neuromuscular junction (NMJ), the synapse between a somatic motor neuron and a skeletal muscle fiber.

Synaptic end bulbs at the tips of axon terminals contain synaptic vesicles filled with acetylcholine (ACh).

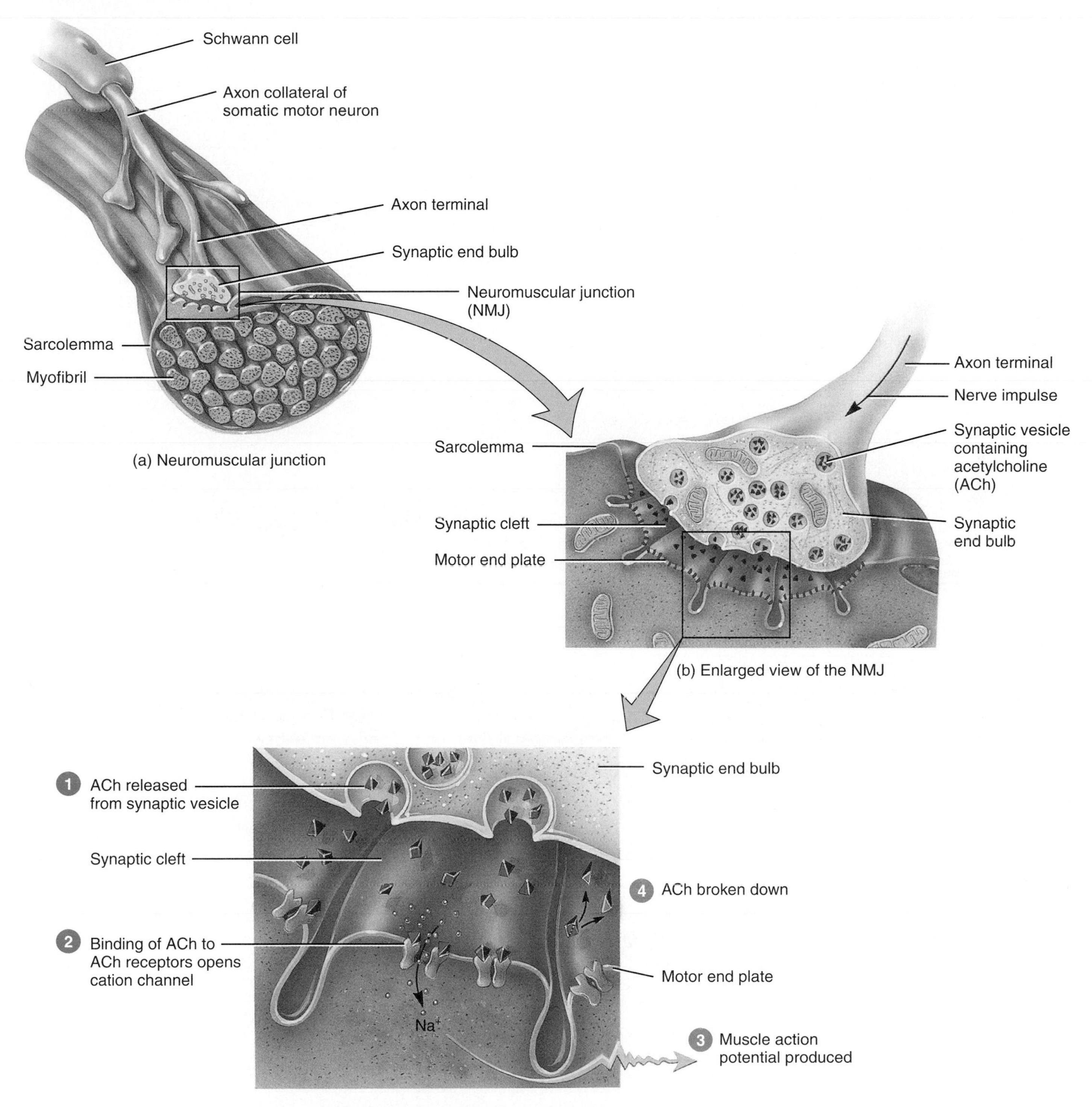

(a) Neuromuscular junction

(b) Enlarged view of the NMJ

(c) Binding of acetylcholine to ACh receptors in the motor end plate

What part of the sarcolemma contains acetylcholine receptors?

Figure 10.12 Summary of the events of contraction and relaxation in a skeletal muscle fiber.

Acetylcholine released at the neuromuscular junction triggers a muscle action potential, which leads to muscle contraction.

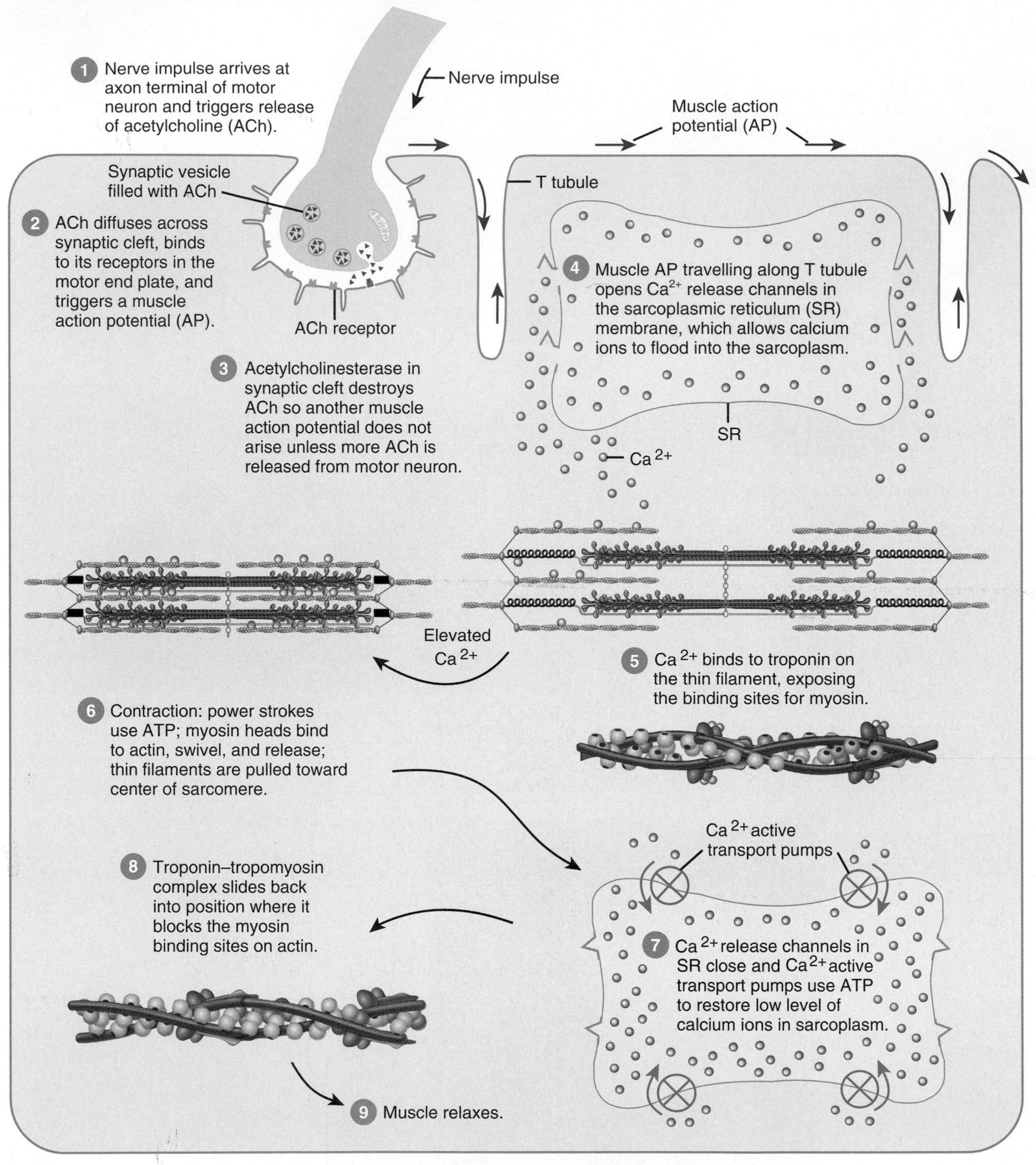

Which numbered steps in this figure are part of excitation–contraction coupling?

MUSCLE METABOLISM

OBJECTIVE

- **Describe the reactions by which muscle fibers produce ATP.**

Production of ATP in Muscle Fibers

Unlike most cells of the body, skeletal muscle fibers often switch between virtual inactivity, when they are relaxed and using only a modest amount of ATP, and great activity, when they are contracting and using ATP at a rapid pace. Contraction of muscle requires a huge amount of ATP for powering the contraction cycle, for pumping Ca^{2+} into the sarcoplasmic reticulum to achieve muscle relaxation, and for other metabolic reactions. However, the ATP present inside muscle fibers is enough to power contraction for only a few seconds. If strenuous exercise continues for more than a few seconds, more ATP is made. Muscle fibers have three ways to produce ATP: (1) from creatine phosphate, (2) by anaerobic cellular respiration, and (3) by aerobic cellular respiration (Figure 10.13). Whereas using creatine phosphate for ATP production is unique to muscle fibers, all body cells make ATP by the reactions of anaerobic and aerobic cellular respiration. We consider the events of cellular respiration briefly here and then in detail in Chapter 25.

Creatine Phosphate

While they are relaxed, muscle fibers produce more ATP than they need for resting metabolism. The excess ATP is used to synthesize **creatine phosphate,** an energy-rich molecule that is

Figure 10.13 Production of ATP for muscle contraction. (a) Creatine phosphate, formed from ATP while the muscle is relaxed, transfers a high-energy phosphate group to ADP, forming ATP, during muscle contraction. (b) Breakdown of muscle glycogen into glucose and production of pyruvic acid from glucose via glycolysis produce both ATP and lactic acid. Because no oxygen is needed, this is an anaerobic pathway. (c) Within mitochondria, pyruvic acid, fatty acids, and amino acids are used to produce ATP via aerobic cellular respiration, an oxygen-requiring set of reactions.

During a long-term event such as a marathon race, most ATP is produced aerobically.

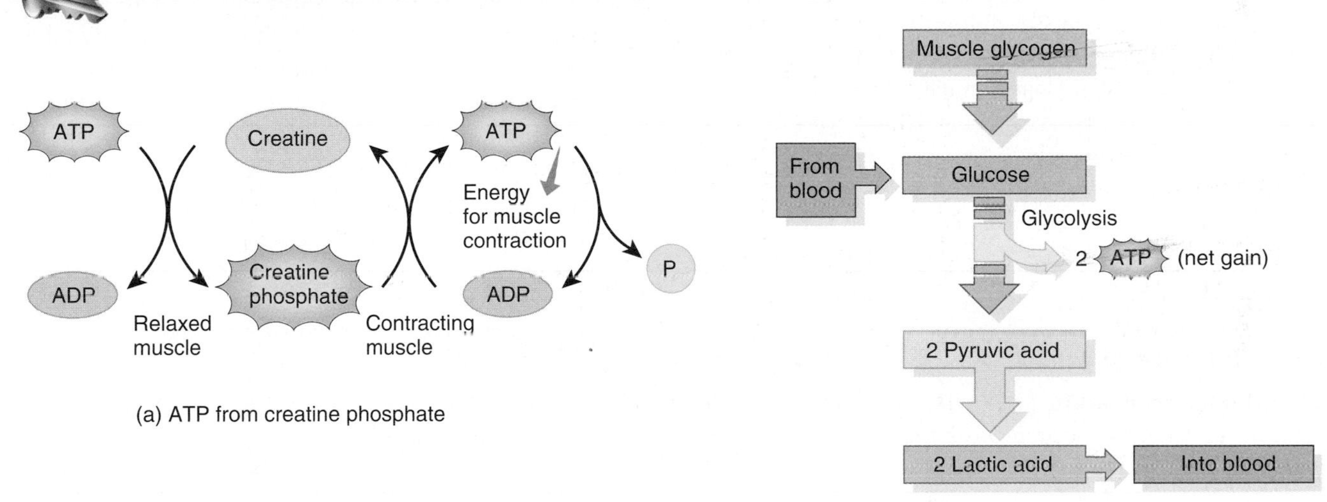

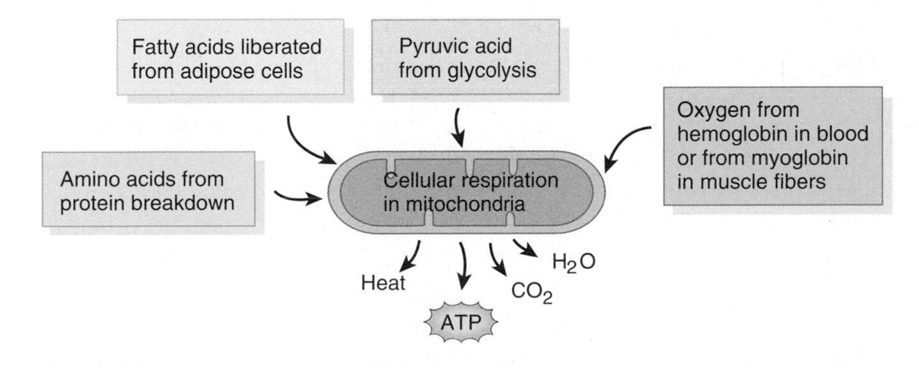

Where inside a skeletal muscle fiber are the events shown here occurring?

found only in muscle fibers (Figure 10.13a). The enzyme *creatine kinase (CK)* catalyzes the transfer of one of the high-energy phosphate groups from ATP to creatine, forming creatine phosphate and ADP. Creatine phosphate is three to six times more plentiful than ATP in the sarcoplasm of a relaxed muscle fiber. When contraction begins and the ADP level starts to rise, CK catalyzes the transfer of a high-energy phosphate group from creatine phosphate back to ADP. This direct phosphorylation reaction quickly forms new ATP molecules. Together, creatine phosphate and ATP provide enough energy for muscles to contract maximally for about 15 seconds. This amount of energy is sufficient for maximal short bursts of activity—for example, to run a 100-meter dash.

Creatine Supplementation

Creatine is a small, amino acid-like molecule that is both synthesized in the body and derived from foods. Adults need to synthesize and ingest a total of about 2 grams of creatine daily to make up for the urinary loss of creatinine, the breakdown product of creatine. Some studies have demonstrated improved performance during intense exercise in subjects who had ingested creatine supplements. For example, college football players who received supplements of 15 grams per day for 28 days gained more muscle mass and had larger gains in lifting power and sprinting performance than the control subjects. Other studies, however, have failed to find a performance-enhancing effect of creatine supplementation. Moreover, ingesting extra creatine decreases the body's own synthesis of creatine, and it is not known whether natural synthesis recovers after long-term creatine supplementation. Further research is needed to determine both the long-term safety and the value of creatine supplementation. ■

Anaerobic Cellular Respiration

Anaerobic cellular respiration is a series of ATP-producing reactions that do not require oxygen. When muscle activity continues and the supply of creatine phosphate is depleted, glucose is catabolized to generate ATP. Glucose easily passes from the blood into contracting muscle fibers via facilitated diffusion, and it is also produced by the breakdown of glycogen within muscle fibers (Figure 10.13b). Then, a series of ten reactions known as *glycolysis* quickly breaks down each glucose molecule into two molecules of pyruvic acid. (Figure 25.4 on page 912 shows the reactions of glycolysis.) These reactions use two ATP but form four ATP for a net gain of two.

Ordinarily, the pyruvic acid formed by glycolysis in the cytosol enters mitochondria. There, it enters a series of oxygen-requiring reactions that produce a large amount of ATP. During some activities, however, not enough oxygen is available. Then, anaerobic reactions convert most of the pyruvic acid to lactic acid in the cytosol. About 80% of the lactic acid produced in this way diffuses out of the skeletal muscle fibers into the blood. Liver cells can convert some of the lactic acid back to glucose. This conversion has two benefits: providing new glucose molecules and reducing acidity. In these ways, anaerobic cellular respiration can provide enough energy for about 30 to 40 seconds of maximal muscle activity. Together, creatine phosphate and glycolysis can provide enough ATP to run a 400-meter race.

Aerobic Cellular Respiration

Muscle activity that lasts longer than half a minute depends increasingly on **aerobic cellular respiration,** a series of oxygen-requiring reactions that produce ATP in mitochondria. If sufficient oxygen is present, pyruvic acid enters the mitochondria, where it is completely oxidized in reactions that generate ATP, carbon dioxide, water, and heat (Figure 10.13c). Although aerobic cellular respiration is slower than glycolysis, it yields much more ATP. Each glucose molecule yields about 36 molecules of ATP and a typical fatty acid molecule yields more than 100 molecules of ATP via aerobic cellular respiration.

Muscle tissue has two sources of oxygen: (1) oxygen that diffuses into muscle fibers from the blood and (2) oxygen released by myoglobin within muscle fibers. Both myoglobin (found only in muscle cells) and hemoglobin (found only in red blood cells) are oxygen-binding proteins. They bind oxygen when it is plentiful and release oxygen when it is scarce.

Aerobic cellular respiration provides enough ATP for prolonged activity so long as sufficient oxygen and nutrients are available. Besides pyruvic acid obtained from glycolysis of glucose, these nutrients include fatty acids (from the breakdown of triglycerides in adipose cells) and amino acids (from the breakdown of proteins). In activities that last more than 10 minutes, the aerobic system provides more than 90% of the needed ATP. At the end of an endurance event, such as a marathon race, nearly 100% of the ATP is being produced by aerobic cellular respiration.

Muscle Fatigue

The inability of a muscle to contract forcefully after prolonged activity is called **muscle fatigue.** Fatigue results mainly from changes within muscle fibers. Even before actual muscle fatigue occurs, a person may have feelings of tiredness and the desire to cease activity. This response, termed *central fatigue,* may be a protective mechanism in that it might cause a person to stop exercising before muscle becomes too damaged. As you will see, certain types of skeletal muscle fibers fatigue more quickly than others.

Although the precise mechanisms that cause muscle fatigue are still not clear, several factors are thought to contribute. One important factor is inadequate release of calcium ions from the SR, resulting in a decline of Ca^{2+} concentration in the sarcoplasm. Depletion of creatine phosphate also is associated with fatigue. Surprisingly, however, the ATP levels in fatigued muscle often are not much lower than those in resting muscle. Other factors that contribute to muscle fatigue include insufficient oxygen, depletion of glycogen and other nutrients, buildup of lactic acid and ADP, and failure of action potentials in the motor neuron to release enough acetylcholine.

Oxygen Consumption After Exercise

During prolonged periods of muscle contraction, increases in breathing effort and blood flow enhance oxygen delivery to muscle tissue. After muscle contraction has stopped, heavy breathing continues for a while, and oxygen consumption remains above the resting level. Depending on the intensity of the exercise, the recovery period may be just a few minutes or several hours. In 1922, A. V. Hill coined the term **oxygen debt** for the added oxygen, over and above the resting oxygen consumption, that is taken into the body after exercise. He proposed that this extra oxygen was used to "pay back" or restore metabolic conditions to the resting level in three ways: (1) to convert lactic acid back into glycogen stores in the liver, (2) to resynthesize creatine phosphate and ATP in muscle fibers, and (3) to replace the oxygen removed from myoglobin.

The metabolic changes that occur *during exercise,* however, can account for only some of the extra oxygen used *after exercise.* Only a small amount of glycogen resynthesis occurs from lactic acid. Instead, most glycogen is made much later from dietary carbohydrates. Much of the lactic acid that remains after exercise is converted back to pyruvic acid and used for ATP production via aerobic cellular respiration in the heart, liver, kidneys, and skeletal muscle. Postexercise oxygen use also is boosted by ongoing changes. First, the elevated body temperature after strenuous exercise increases the rate of chemical reactions throughout the body. Faster reactions use ATP more rapidly, and more oxygen is needed to produce the ATP. Second, the heart and muscles used in breathing are still working harder than they were at rest, and thus they consume more ATP. Third, tissue repair processes are occurring at an increased pace. For these reasons, **recovery oxygen uptake** is a better term than oxygen debt for the elevated use of oxygen after exercise.

► CHECKPOINT

10. Which ATP-producing reactions are aerobic and which are anaerobic?
11. Which sources provide ATP during a 1000-meter run?
12. What factors contribute to muscle fatigue?
13. Why is the term *recovery oxygen uptake* more accurate than *oxygen debt*?

CONTROL OF MUSCLE TENSION

OBJECTIVES

- **Describe the structure and function of a motor unit.**
- **Explain the phases of a twitch contraction.**
- **Describe how frequency of stimulation affects muscle tension.**

A single nerve impulse in a motor neuron elicits a single muscle action potential in all the muscle fibers with which it forms synapses. In contrast to action potentials, which always have the same size in a given neuron or muscle fiber, the contraction that results from a single muscle action potential has significantly smaller force than the maximum force the fiber is capable of producing. The total force or tension that a single fiber can produce depends mainly on the rate at which nerve impulses arrive at the neuromuscular junction. The number of impulses per second is the *frequency of stimulation.* In addition, as we saw in Figure 10.10, the amount of stretch before contraction influences the maximum tension that is possible during contraction. Finally, conditions such as nutrient and oxygen availability can influence the tension a single fiber can generate. When considering the contraction of a whole muscle, the total tension it can produce depends on the number of fibers that are contracting in unison.

Motor Units

Even though each skeletal muscle fiber has only a single neuromuscular junction, the axon of a motor neuron branches out and forms neuromuscular junctions with many different muscle fibers. A **motor unit** consists of a somatic motor neuron plus all the skeletal muscle fibers it stimulates (Figure 10.14). A single motor neuron makes contact with an average of 150 muscle fibers and all muscle fibers in one motor unit contract in unison. Typically, the muscle fibers of a motor unit are dispersed throughout a muscle rather than clustered together.

Muscles that control precise movements consist of many small motor units. For instance, muscles of the larynx (voice box) that control voice production have as few as two or three muscle fibers per motor unit, and muscles controlling eye movements may have 10 to 20 muscle fibers per motor unit. In contrast, skeletal muscles responsible for large-scale and powerful movements, such as the biceps brachii muscle in the arm and the gastrocnemius muscle in the calf of the leg, have as many as 2000 to 3000 muscle fibers in some motor units. Remember, all the muscle fibers of a motor unit contract and relax together. Accordingly, the total strength of a contraction depends, in part, on how large the motor units are and how many motor units are activated at the same time.

Figure 10.14 Motor units. Shown are two somatic motor neurons, one in purple and one in green, each supplying the muscle fibers of its motor unit.

A motor unit consists of a somatic motor neuron plus all the muscle fibers it stimulates.

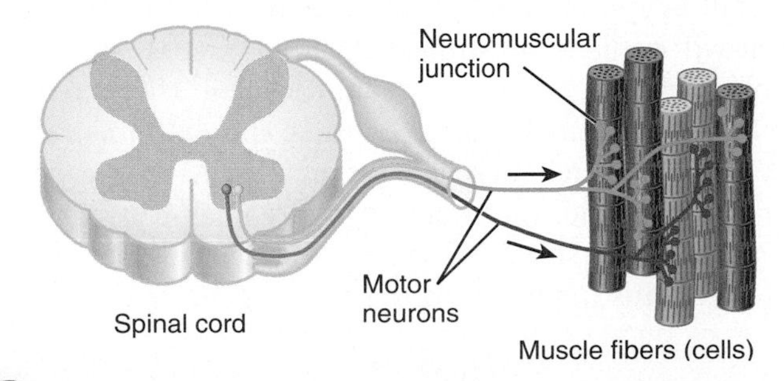

What is the effect of the size of a motor unit on its strength of contraction? (Assume that each muscle fiber can generate about the same amount of tension.)

Twitch Contraction

A **twitch contraction** is the brief contraction of all the muscle fibers in a motor unit in response to a single action potential in its motor neuron. In the laboratory, a twitch also can be produced by direct electrical stimulation of a motor neuron or its muscle fibers. The record of a muscle contraction, called a **myogram,** is shown in Figure 10.15. Twitches of skeletal muscle fibers last anywhere from 20 to 200 msec. This duration is very long compared to the brief 1–2 msec* duration of an action potential.

Note that a brief delay occurs between application of the stimulus (time zero on the graph) and the beginning of contraction. The delay, which lasts about two milliseconds, is termed the **latent period.** During the latent period, calcium ions are being released from the sarcoplasmic reticulum, the filaments start to exert tension, the elastic components stretch, and finally shortening begins. The second phase, the **contraction period,** lasts 10–100 msec. During the third phase, the **relaxation period,** also lasting 10–100 msec, the active transport of Ca^{2+} back into the sarcoplasmic reticulum causes relaxation. The actual duration of these periods depends on the type of muscle fiber. Some fibers, such as those that move the eyes, are fast-twitch fibers (described shortly); they have a contraction period as brief as 10 msec and an equally brief relaxation period. Others, such as those that move the legs, are slow-twitch fibers, with contraction and relaxation periods of about 100 msec each.

If two stimuli are applied, one immediately after the other, the muscle will respond to the first stimulus but not to the second. When a muscle fiber receives enough stimulation to contract, it temporarily loses its excitability and cannot respond for a time. The period of lost excitability, called the **refractory period,** is a characteristic of all muscle and nerve cells. The duration of the refractory period varies with the muscle involved. Skeletal muscle has a short refractory period of about five milliseconds, whereas cardiac muscle has a long refractory period of about 300 milliseconds.

Frequency of Stimulation

When the second of two stimuli occurs after the refractory period is over, the skeletal muscle will respond to both stimuli (Figure 10.16). If the second stimulus occurs after the refractory period but before the muscle fiber has relaxed, the second contraction will actually be stronger than the first (Figure 10.16b). This phenomenon, in which stimuli arriving at different times cause larger contractions, is called **wave summation.**

When a skeletal muscle is stimulated at a rate of 20 to 30 times per second, it can only partially relax between stimuli. The result is a sustained but wavering contraction called **unfused tetanus** (*tetan-* = rigid, tense; Figure 10.16c). Stimulation at a higher rate of 80 to 100 stimuli per second causes **fused tetanus,** a sustained contraction in which individual twitches cannot be discerned (Figure 10.16d). Wave summation and both kinds of tetanus result from the addition of Ca^{2+} released from the sarcoplasmic reticulum by the second, and subsequent, stimuli to the Ca^{2+} still in the sarcoplasm from the first stimulus. Because the Ca^{2+} level builds up, the peak tension generated during fused tetanus is 5 to 10 times larger than the peak tension produced during a single twitch. Even so, smooth, sustained voluntary muscle contractions are achieved mainly by out-of-synchrony unfused tetanus in different motor units.

*One millisecond (msec) is 10^{-3} seconds (0.001 sec).

Figure 10.15 Myogram of a twitch contraction. The arrow indicates the time at which the stimulus occurred.

A myogram is a record of a muscle contraction.

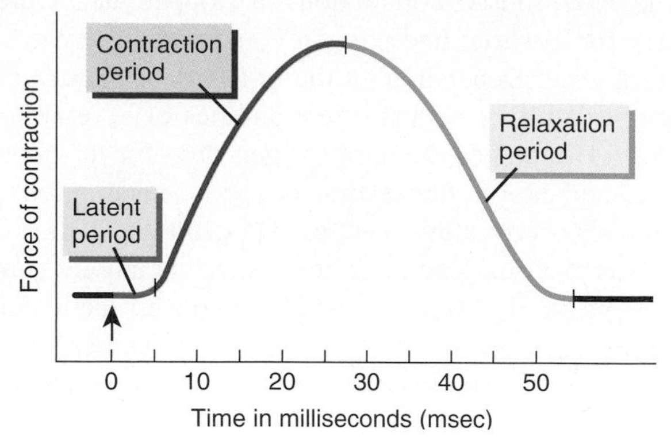

What events occur during the latent period?

The stretch of elastic components is also related to wave summation. During wave summation, elastic components are not given much time to spring back between contractions, and thus they remain taut. While in this state, the elastic components do not require very much stretching before the beginning of the next muscular contraction. The combination of the tautness of the elastic components and the partially contracted state of the filaments enables the force of another contraction to be added more quickly to the one before.

Motor Unit Recruitment

The process in which the number of active motor units increases is called **motor unit recruitment.** Typically, the different motor units in a whole muscle are not stimulated to contract in unison. While some motor units are contracting, others are relaxed. This pattern of motor unit activity delays muscle fatigue by allowing alternately contracting motor units to relieve one another. In this way, contraction of a whole muscle can be sustained for long periods. The weakest motor units are recruited first, with progressively stronger motor units being added if the task requires more force.

Recruitment is one factor responsible for producing smooth movements rather than a series of jerks. As mentioned, the num-

Figure 10.16 Myograms showing the effects of different frequencies of stimulation. (a) Single twitch. (b) When a second stimulus occurs before the muscle has relaxed, the second contraction is stronger than the first, a phenomenon called wave summation. (The dashed line indicates the force of contraction expected in a single twitch.) (c) In unfused tetanus, the curve looks jagged due to partial relaxation of the muscle between stimuli. (d) In fused tetanus, which occurs when there are 80–100 stimuli per second, the contraction force is steady and sustained.

Due to wave summation, the tension produced during a sustained contraction is greater than during a single twitch.

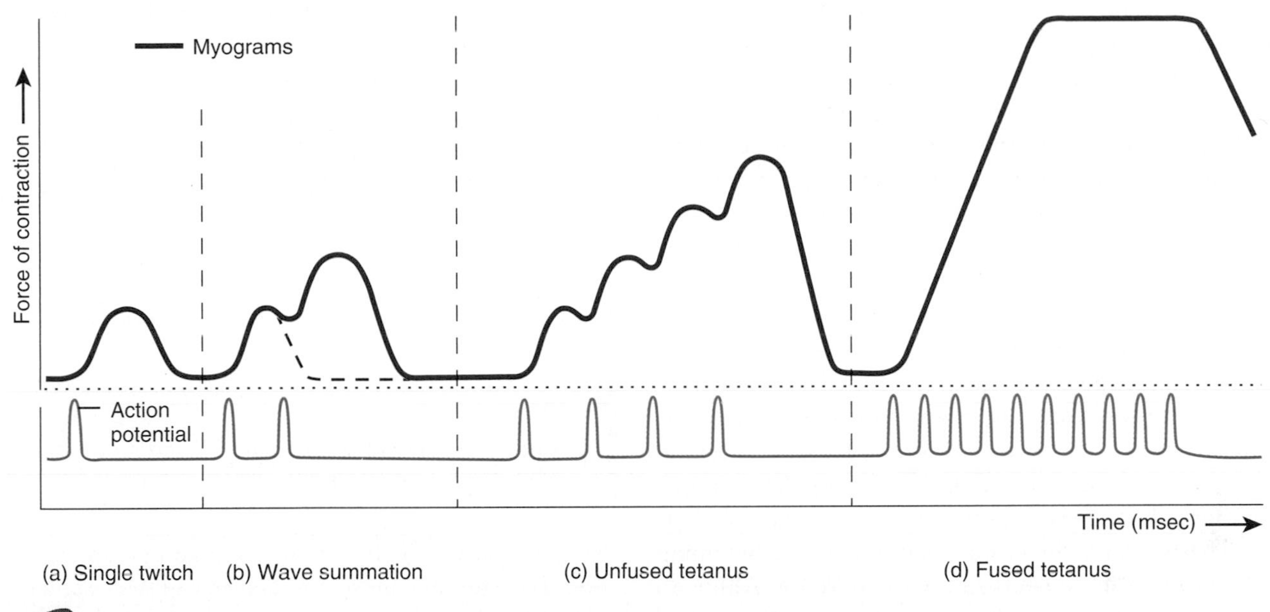

Would the peak force of the second contraction in (b) be larger or smaller if the second stimulus were applied a few milliseconds later?

ber of muscle fibers innervated by one motor neuron varies greatly. Precise movements are brought about by small changes in muscle contraction. Therefore, the muscles that produce precise movements are made up of small motor units. For this reason, when a motor unit is recruited or turned off, only slight changes occur in muscle tension. By contrast, large motor units are active where large tension is needed and precision is less important.

Aerobic Training versus Strength Training

Regular, repeated activities such as jogging or aerobic dancing increase the supply of oxygen-rich blood available to skeletal muscles for aerobic cellular respiration. By contrast, activities such as weight lifting rely more on anaerobic production of ATP through glycolysis. Such anaerobic activities stimulate synthesis of muscle proteins and result, over time, in increased muscle size (muscle hypertrophy). As a result, aerobic training builds endurance for prolonged activities, whereas anaerobic training builds muscle strength for short-term feats. **Interval training** is a workout regimen that incorporates both types of training—for example, alternating sprints with jogging. ■

Muscle Tone

Even at rest, a skeletal muscle exhibits **muscle tone** (*tonos* = tension), a small amount of tautness or tension in the muscle due to weak, involuntary contractions of its motor units. Recall that skeletal muscle contracts only after it is activated by acetylcholine released by nerve impulses in its motor neurons. Hence, muscle tone is established by neurons in the brain and spinal cord that excite the muscle's motor neurons. When the motor neurons serving a skeletal muscle are damaged or cut, the muscle becomes **flaccid** (FLAS-id = flabby), a state of limpness in which muscle tone is lost. To sustain muscle tone, small groups of motor units are alternately active and inactive in a constantly shifting pattern. Muscle tone keeps skeletal muscles firm, but it does not result in a force strong enough to produce movement. For example, when the muscles in the back of the neck are in normal tonic contraction, they keep the head upright and prevent it from slumping forward on the chest. Muscle tone also is important in smooth muscle tissues, such as those found in the gastrointestinal tract, where the walls of the digestive organs maintain a steady pressure on their contents. The tone of smooth muscle fibers in the walls of blood vessels plays a crucial role in regulating blood pressure.

Isotonic and Isometric Contractions

Isotonic contractions are used for body movements and for moving objects. The two types of isotonic contractions are concentric and eccentric. In a **concentric isotonic contraction,** a muscle shortens and pulls on another structure, such as a tendon, to produce movement and to reduce the angle at a joint. Picking a book up off a table involves concentric isotonic contractions of the biceps brachii muscle in the arm (Figure 10.17a). By contrast, as you lower the book to place it back on the table, the previously shortened biceps lengthens in a controlled manner while it continues to contract. When the length of a muscle increases during a contraction, the contraction is an **eccentric isotonic contraction** (Figure 10.17b). During an eccentric contraction, the tension exerted by the myosin crossbridges resists movement of a load (the book, in this case) and slows the lengthening process. For reasons that are not well understood, repeated eccentric isotonic contractions produce more muscle damage and more delayed-onset muscle soreness than do concentric isotonic contractions.

In **isometric contractions,** the myosin crossbridges generate considerable tension but the muscle doesn't shorten because the force of the load equals the muscle tension. An example would be holding a book steady using an outstretched arm (Figure 10.17c). These contractions are important for maintaining posture and for supporting objects in a fixed position. Although isometric contractions do not result in body movement, energy is still expended. The book pulls the arm downward, stretching the shoulder and arm muscles. The isometric contraction of the shoulder and arm muscles counteracts the stretch. Isometric contractions are important because they stabilize some joints as others are moved. Most activities include both isotonic and isometric contractions.

▶ CHECKPOINT

14. How are the sizes of motor units related to the degree of muscular control they allow?
15. What is motor unit recruitment?
16. Why is muscle tone important?
17. What do each of following terms mean: concentric isotonic contraction, eccentric isotonic contraction, and isometric contraction?

TYPES OF SKELETAL MUSCLE FIBERS

▶ OBJECTIVE

- **Compare the structure and function of the three types of skeletal muscle fibers.**

Skeletal muscle fibers are not all alike in composition and function. For example, muscle fibers vary in their content of myoglobin, the red-colored protein that binds oxygen in muscle fibers.

Figure 10.17 Comparison between isotonic (concentric and eccentric) and isometric contractions. Parts (a) and (b) show isotonic contraction of the biceps brachii muscle in the arm; part (c) shows isometric contraction of shoulder and arm muscles.

In an isotonic contraction, tension remains constant as muscle length decreases or increases; in an isometric contraction, tension increases greatly without a change in muscle length.

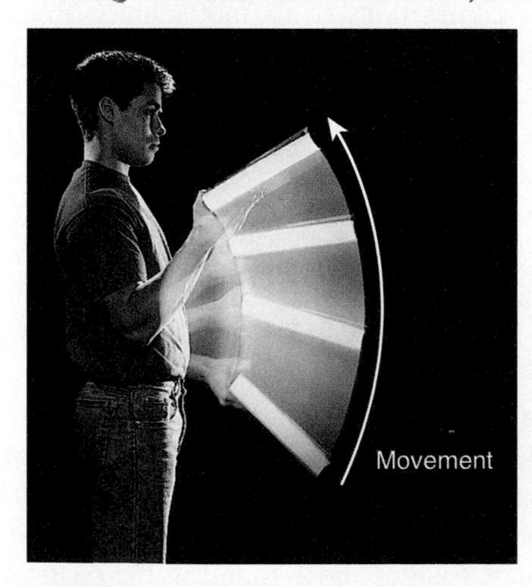

(a) Concentric contraction while picking up a book

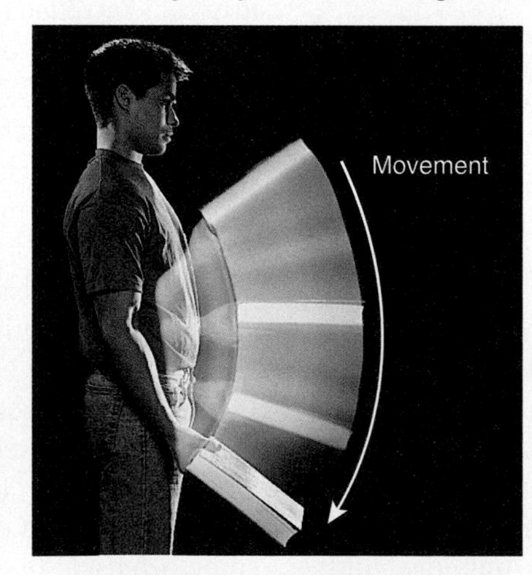

(b) Eccentric contraction while lowering a book

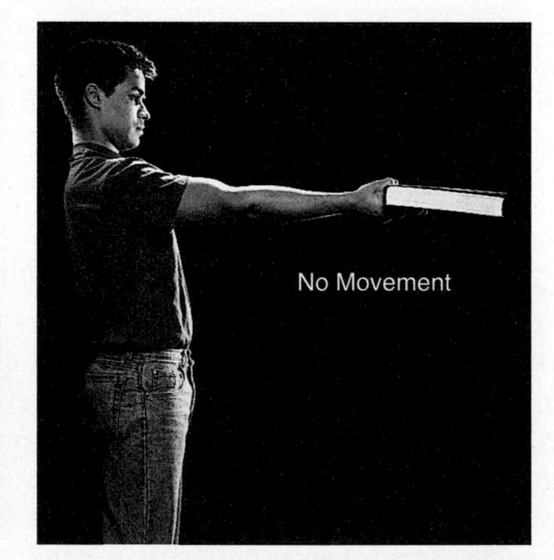

(c) Isometric contraction while holding a book steady

What type of contraction is occurring in your neck muscles while you are walking?

Skeletal muscle fibers that have a high myoglobin content are termed *red muscle fibers;* those that have a low content of myoglobin are called *white muscle fibers.* Red muscle fibers also contain more mitochondria and are supplied by more blood capillaries than are white muscle fibers.

Skeletal muscle fibers also contract and relax at different speeds. A fiber is categorized as either slow or fast depending on how rapidly the ATPase in its myosin heads hydrolyzes ATP. In addition, skeletal muscle fibers vary in which metabolic reactions they use to generate ATP and in how quickly they fatigue. Based on these structural and functional characteristics, skeletal muscle fibers are classified into three main types: (1) slow oxidative fibers, (2) fast oxidative-glycolytic fibers, and (3) fast glycolytic fibers.

Slow Oxidative Fibers

Slow oxidative (SO) fibers are smallest in diameter and thus are the least powerful type of muscle fibers. They appear dark red because they contain large amounts of myoglobin and many blood capillaries. Because they have many large mitochondria, SO fibers generate ATP mainly by aerobic cellular respiration, which is why they are called oxidative fibers. These fibers are said to be "slow" because the ATPase in the myosin heads hydrolyzes ATP relatively slowly and the contraction cycle proceeds at a slower pace than in "fast" fibers. As a result, SO fibers have a slow speed of contraction. Their twitch contractions last from 100 to 200 msec, and they take longer to reach peak tension. However, slow fibers are very resistant to fatigue and are capable of prolonged, sustained contractions for many hours. These slow-twitch, fatigue-resistant fibers are adapted for maintaining posture and for aerobic, endurance-type activities such as running a marathon.

Fast Oxidative-Glycolytic Fibers

Fast oxidative-glycolytic (FOG) fibers are intermediate in diameter between the other two types of fibers. Like slow oxidative fibers, they contain large amounts of myoglobin and many blood capillaries. Thus, they have a dark red appearance. FOG fibers can generate considerable ATP by aerobic cellular respiration, which gives them a moderately high resistance to fatigue. Because their intracellular glycogen level is high, they also generate ATP by anaerobic glycolysis. FOG fibers are "fast" because the ATPase in their myosin heads hydrolyzes ATP three to five times faster than the myosin ATPase in SO fibers, which makes their speed of contraction faster. Thus, twitches of FOG fibers reach peak tension more quickly than those of SO fibers but are briefer in duration—less than 100 msec. FOG fibers contribute to activities such as walking and sprinting.

Fast Glycolytic Fibers

Fast glycolytic (FG) fibers are largest in diameter and contain the most myofibrils. Hence, they can generate the most powerful contractions. FG fibers have low myoglobin content, relatively few blood capillaries, few mitochondria, and appear white in color. They contain large amounts of glycogen and generate ATP mainly by glycolysis. Due to their large size and their ability to hydrolyze ATP rapidly, FG fibers contract strongly and quickly. These fast-twitch fibers are adapted for intense anaerobic movements of short duration, such as weight lifting or throwing a ball, but they fatigue quickly. Strength training programs that engage a person in activities requiring great strength for short times produce increases in the size, strength, and glycogen content of fast glycolytic fibers. The FG fibers of a weight lifter may be 50% larger than those of a sedentary person or endurance athlete. The increase in size is due to increased synthesis of muscle proteins. The overall result is muscle enlargement due to hypertrophy of the FG fibers.

Distribution and Recruitment of Different Types of Fibers

Most skeletal muscles are a mixture of all three types of skeletal muscle fibers, about half of which are SO fibers. The proportions vary somewhat, depending on the action of the muscle, the person's training regimen, and genetic factors. For example, the continually active postural muscles of the neck, back, and legs have a high proportion of SO fibers. Muscles of the shoulders and arms, in contrast, are not constantly active but are used briefly now and then to produce large amounts of tension, such as in lifting and throwing. These muscles have a high proportion of FG fibers. Leg muscles, which not only support the body but are also used for walking and running, have large numbers of both SO and FOG fibers.

Even though most skeletal muscles are a mixture of all three types of skeletal muscle fibers, the skeletal muscle fibers of any given motor unit are all of the same type. However, the different motor units in a muscle are recruited in a specific order, depending on need. For example, if weak contractions suffice to perform a task, only SO motor units are activated. If more force is needed, the motor units of FOG fibers are also recruited. Finally, if maximal force is required, motor units of FG fibers are also called into action. Activation of various motor units is controlled by the brain and spinal cord.

Table 10.1 summarizes the characteristics of the three types of skeletal muscle fibers.

► CHECKPOINT

18. Why are some skeletal muscle fibers classified as "fast" whereas others are said to be "slow"?

Table 10.1 Characteristics of the Three Types of Skeletal Muscle Fibers

Slow oxidative fiber
Fast glycolytic fiber
Fast oxidative-glycolytic fiber

LM 440x

Transverse section of three types of skeletal muscle fibers

	Slow Oxidative (SO) Fibers	Fast Oxidative-Glycolytic (FOG) Fibers	Fast Glycolytic (FG) Fibers
Structural Characteristic			
Fiber diameter	Smallest.	Intermediate.	Largest.
Myoglobin content	Large amount.	Large amount.	Small amount.
Mitochondria	Many.	Many.	Few.
Capillaries	Many.	Many.	Few.
Color	Red.	Red-pink.	White (pale).
Functional Characteristic			
Capacity for generating ATP and method used	High capacity, by aerobic (oxygen-requiring) cellular respiration.	Intermediate capacity, by both aerobic (oxygen-requiring) cellular respirationand glycolysis (anaerobic).	Low capacity, by anaerobic cellular respiration (glycolysis).
Rate of ATP hydrolysis by myosin ATPase	Slow.	Fast.	Fast.
Contraction velocity	Slow.	Fast.	Fast.
Fatigue resistance	High.	Intermediate.	Low.
Creatine kinase	Lowest amount.	Intermediate amount.	Highest amount.
Glycogen stores	Low.	Intermediate.	High.
Order of recruitment	First.	Second.	Third.
Location where fibers are abundant	Postural muscles such as those of the neck.	Leg muscles.	Arm muscles.
Primary functions of fibers	Maintaining posture and aerobic endurance activities.	Walking, sprinting.	Rapid, intense movements of short duration.

EXERCISE AND SKELETAL MUSCLE TISSUE

▶ OBJECTIVE

- **Describe the effects of exercise on different types of skeletal muscle fibers.**

The relative ratio of fast-twitch and slow-twitch fibers in each muscle is genetically determined and helps account for individual differences in physical performance. For example, people with a higher proportion of fast glycolytic (FG) fibers often excel in activities that require periods of intense activity, such as weight lifting or sprinting. People with higher percentages of slow oxidative (SO) fibers are better at activities that require endurance, such as long-distance running.

Although the total number of skeletal muscle fibers usually does not increase, the characteristics of those present can change to some extent. Various type of exercises can induce changes in the fibers in a skeletal muscle. Endurance-type (aerobic) exercises, such as running or swimming, cause a gradual transformation of some fast glycolytic (FG) fibers into fast oxidative-glycolytic (FOG) fibers. The transformed muscle fibers show slight increases in diameter, number of mitochondria, blood supply, and strength. Endurance exercises also result in cardiovascular and respiratory changes that cause skeletal muscles to receive better supplies of oxygen and nutrients but do not increase muscle mass. By contrast, exercises that require great strength for short periods produce an increase in the size and strength of fast glycolytic (FG) fibers. The increase in size is due to increased synthesis of thick and thin filaments. The overall result is muscle enlargement (hypertrophy), as evidenced by the bulging muscles of body builders.

Anabolic Steroids

The illegal use of **anabolic steroids** by athletes has received widespread attention. These steroid hormones, similar to testosterone, are taken to increase muscle size and thus strength during athletic contests. The large doses needed to produce an effect, however, have damaging, sometimes even devastating side effects, including liver cancer, kidney damage, increased risk of heart disease, stunted growth, wide mood swings, and increased irritability and aggression. Additionally, females who take anabolic steroids may experience atrophy of the breasts and uterus, menstrual irregularities, sterility, facial hair growth, and deepening of the voice. Males may experience diminished testosterone secretion, atrophy of the testes, and baldness. ■

▶ CHECKPOINT

19. On a cellular level, what causes muscle hypertrophy?

CARDIAC MUSCLE TISSUE

▶ OBJECTIVE

- **Describe the main structural and functional characteristics of cardiac muscle tissue.**

The principal tissue in the heart wall is **cardiac muscle tissue** (described in more detail in Chapter 20 and illustrated in Figure 20.9 on page 673). Between the layers of **cardiac muscle fibers,** the contractile cells of the heart, are sheets of connective tissue that contain blood vessels, nerves, and the conduction system of the heart. Cardiac muscle fibers have the same arrangement of actin and myosin and the same bands, zones, and Z discs as skeletal muscle fibers. A unique feature of cardiac muscle fibers are their *intercalated discs* (in-TER-ka-lāt-ed; *intercal-* = to insert between). These microscopic structures are irregular transverse thickenings of the sarcolemma that connect the ends of cardiac muscle fibers to each other. The discs contain *desmosomes,* which hold the fibers together, and *gap junctions,* which allow muscle action potentials to spread from one cardiac muscle fiber to another (see Figure 4.1e on page 105).

In response to a single action potential, cardiac muscle tissue remains contracted 10 to 15 times longer than skeletal muscle tissue (see Figure 20.11 on page 676). The long contraction is due to prolonged delivery of Ca^{2+} into the sarcoplasm. In cardiac muscle fibers, calcium ions enter the sarcoplasm both from the sarcoplasmic reticulum (as in skeletal muscle fibers) and from the interstitial fluid that bathes the fibers. Because the channels that allow inflow of Ca^{2+} from interstitial fluid stay open for a relatively long time, a cardiac muscle contraction lasts much longer than a skeletal muscle twitch.

We have seen that skeletal muscle tissue contracts only when stimulated by acetylcholine released by a nerve impulse in a motor neuron. In contrast, cardiac muscle tissue contracts when stimulated by its own autorhythmic muscle fibers. Under normal resting conditions, cardiac muscle tissue contracts and relaxes about 75 times a minute. This continuous, rhythmic activity is a major physiological difference between cardiac and skeletal muscle tissue. Accordingly, cardiac muscle tissue requires a constant supply of oxygen, and the mitochondria in cardiac muscle fibers are larger and more numerous than in skeletal muscle fibers. This structural feature correctly suggests that cardiac muscle depends greatly on aerobic cellular respiration to generate ATP. Moreover, cardiac muscle fibers can use lactic acid produced by skeletal muscle fibers to make ATP, a benefit during exercise.

▶ CHECKPOINT

20. What are the similarities and differences between skeletal and cardiac muscle?

SMOOTH MUSCLE TISSUE

▶ OBJECTIVE

- **Describe the main structural and functional characteristics of smooth muscle tissue.**

Like cardiac muscle tissue, **smooth muscle tissue** is usually activated involuntarily. Of the two types of smooth muscle tissue, the more common type is **visceral (single-unit) smooth muscle tissue** (Figure 10.18a). It is found in wraparound sheets that form

Figure 10.18 Two types of smooth muscle tissue. In (a), one autonomic motor neuron synapses with several visceral smooth muscle fibers, and action potentials spread to neighboring fibers through gap junctions. In (b), three autonomic motor neurons synapse with individual multiunit smooth muscle fibers. Stimulation of one multiunit fiber causes contraction of that fiber only.

Visceral smooth muscle fibers connect to one another by gap junctions and contract as a single unit. Multiunit smooth muscle fibers lack gap junctions and contract independently.

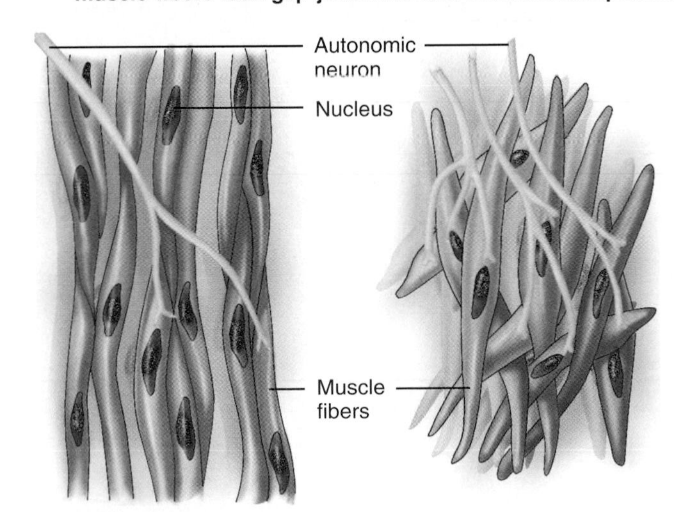

(a) Visceral (single-unit) smooth muscle tissue

(b) Multiunit smooth muscle tissue

Which type of smooth muscle is more like cardiac muscle than skeletal muscle, with respect to both its structure and function?

part of the walls of small arteries and veins and of hollow organs such as the stomach, intestines, uterus, and urinary bladder. Like cardiac muscle, visceral smooth muscle is autorhythmic. Because the fibers connect to one another by gap junctions, muscle action potentials spread throughout the network. When a neurotransmitter, hormone, or autorhythmic signal stimulates one fiber, the muscle action potential spreads to neighboring fibers, which then contract in unison, as a single unit.

The second type of smooth muscle tissue is **multiunit smooth muscle tissue** (Figure 10.18b). It consists of individual fibers, each with its own motor neuron terminals and with few gap junctions between neighboring fibers. Whereas stimulation of one visceral muscle fiber causes contraction of many adjacent fibers, stimulation of one multiunit fiber causes contraction of that fiber only. Multiunit smooth muscle tissue is found in the walls of large arteries, in airways to the lungs, in the arrector pili muscles that attach to hair follicles, in the muscles of the iris that adjust pupil diameter, and in the ciliary body that adjusts focus of the lens in the eye.

Microscopic Anatomy of Smooth Muscle

A single relaxed smooth muscle fiber is 30–200 μm long. It is thickest in the middle (3–8 μm) and tapers at each end (Figure 10.19). Within each fiber is a single, oval, centrally located nucleus.

Figure 10.19 Microscopic anatomy of a smooth muscle fiber. A photomicrograph of smooth muscle is shown in Table 4.4 on page 131.

Smooth muscle fibers have thick and thin filaments but no transverse tubules and scanty sarcoplasmic reticulum.

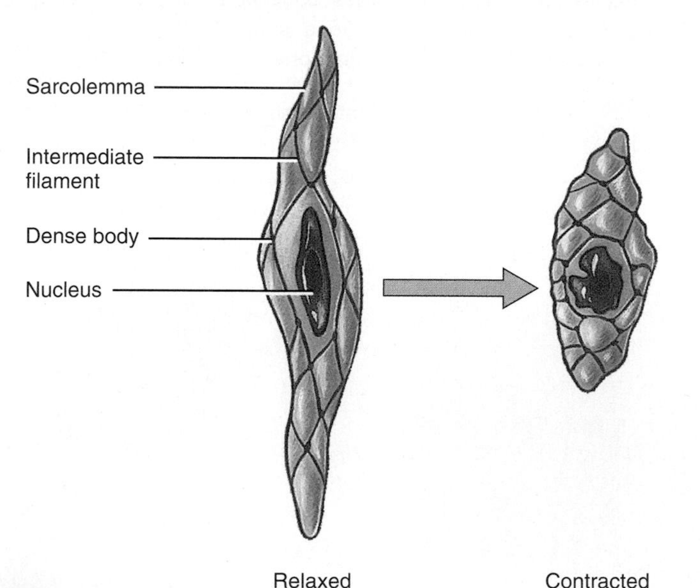

How does the speed of onset and duration of contraction in a smooth muscle fiber compare with that in a skeletal muscle fiber?

The sarcoplasm of smooth muscle fibers contains both *thick filaments* and *thin filaments,* in ratios between 1:10 and 1:15, but they are not arranged in orderly sarcomeres as in striated muscle. Smooth muscle fibers also contain *intermediate filaments.* Because the various filaments have no regular pattern of overlap, smooth muscle fibers do not exhibit striations (see Table 4.4 on page 131). This is the reason for the name *smooth.* Smooth muscle fibers also lack transverse tubules and have only scanty sarcoplasmic reticulum for storage of Ca^{2+}.

In smooth muscle fibers, intermediate filaments attach to structures called **dense bodies,** which are functionally similar to Z discs in striated muscle fibers. Some dense bodies are dispersed throughout the sarcoplasm; others are attached to the sarcolemma. Bundles of intermediate filaments stretch from one dense body to another (Figure 10.19). During contraction, the sliding filament mechanism involving thick and thin filaments generates tension that is transmitted to intermediate filaments. These, in turn, pull on the dense bodies attached to the sarcolemma, causing a lengthwise shortening of the muscle fiber. As a smooth muscle fiber contracts, it rotates as a corkscrew turns. The fiber twists in a helix as it contracts and rotates in the opposite direction as it relaxes.

Physiology of Smooth Muscle

Although the principles of contraction are similar in all three types of muscle tissue, smooth muscle tissue exhibits some important physiological differences. Compared with contraction in a skeletal muscle fiber, contraction in smooth muscle fiber starts more slowly and lasts much longer. Moreover, smooth muscle can both shorten and stretch to a greater extent than other muscle types.

An increase in the concentration of Ca^{2+} in smooth muscle cytosol initiates contraction, just as in striated muscle. Sarcoplasmic reticulum (the reservoir for Ca^{2+} in striated muscle) is scanty in smooth muscle. Calcium ions flow into smooth muscle cytosol from both the interstitial fluid and sarcoplasmic reticulum. Because there are no transverse tubules in smooth muscle fibers, however, it takes longer for Ca^{2+} to reach the filaments in the center of the fiber and trigger the contractile process. This accounts, in part, for the slow onset and prolonged contraction of smooth muscle.

Several mechanisms regulate contraction and relaxation of smooth muscle cells. In one, a regulatory protein called **calmodulin** (cal-MOD-ū-lin) binds to Ca^{2+} in the cytosol. (Recall that troponin has this role in striated muscle fibers.) After binding to Ca^{2+}, calmodulin activates an enzyme called *myosin light chain kinase.* This enzyme uses ATP to phosphorylate (add a phosphate group to) a portion of the myosin head. Once the phosphate group is attached, the myosin head can bind to actin, and contraction can occur. Myosin light chain kinase works rather slowly, also contributing to the slowness of smooth muscle contraction.

Not only do calcium ions enter smooth muscle fibers slowly, they also move slowly out of the muscle fiber when excitation declines, which delays relaxation. The prolonged presence of Ca^{2+} in the cytosol provides for **smooth muscle tone,** a state of continued partial contraction. Smooth muscle tissue can thus sustain long-term tone, which is important in the gastrointestinal tract, where the walls maintain a steady pressure on the contents of the tract, and in the walls of blood vessels called arterioles, which maintain a steady pressure on blood.

Most smooth muscle fibers contract or relax in response to action potentials from the autonomic nervous system. In addition, many smooth muscle fibers contract or relax in response to stretching, hormones, or local factors such as changes in pH, oxygen and carbon dioxide levels, temperature, and ion concentrations. For example, the hormone epinephrine, released by the adrenal medulla, causes relaxation of smooth muscle in the airways and in some blood vessel walls (those that have so-called β_2 receptors; see Table 17.4 on page 580).

Unlike striated muscle fibers, smooth muscle fibers can stretch considerably and still maintain their contractile function. When smooth muscle fibers are stretched, they initially contract, developing increased tension. Within a minute or so, the tension decreases. This phenomenon, which is called the **stress-relaxation response,** allows smooth muscle to undergo great changes in length while retaining the ability to contract effectively. Thus, even though smooth muscle in the walls of blood vessels and hollow organs such as the stomach, intestines, and urinary bladder can stretch, the pressure on the contents within them changes very little. After the organ empties, however, the smooth muscle in the wall rebounds, and the wall retains its firmness.

▶ CHECKPOINT

21. How do visceral and multiunit smooth muscle differ?
22. In what ways are the properties of skeletal and smooth muscle similar and different?

REGENERATION OF MUSCLE TISSUE

▶ OBJECTIVE

- **Explain how muscle fibers regenerate.**

Because mature skeletal muscle fibers have lost the ability to undergo cell division, growth of skeletal muscle after birth is due mainly to hypertrophy, the enlargement of existing cells, rather than to **hyperplasia,** an increase in the number of fibers. Satellite cells divide slowly and fuse with existing fibers to assist both in muscle growth and in repair of damaged fibers. The number of new skeletal muscle fibers formed, however, is not enough to compensate for significant skeletal muscle damage or degeneration. In such cases, skeletal muscle tissue undergoes **fibrosis,** the replacement of muscle fibers by fibrous scar tissue. For this reason, skeletal muscle tissue can regenerate only to a limited extent.

Until recently it was believed that damaged cardiac muscle fibers could not be replaced and that healing took place exclusively by fibrosis. New research described in Chapter 20 indicates that, under certain circumstances, cardiac muscle tissue can regenerate. In addition, cardiac muscle fibers can undergo hypertrophy in response to increased workload. Hence, many athletes have enlarged hearts.

Smooth muscle tissue, like skeletal and cardiac muscle tissue, can undergo hypertrophy. In addition, certain smooth muscle fibers, such as those in the uterus, retain their capacity for division and thus can grow by hyperplasia. Also, new smooth muscle fibers can arise from cells called *pericytes,* stem cells found in association with blood capillaries and small veins. Smooth muscle fibers can also proliferate in certain pathological conditions, such as occur in the development of atherosclerosis (see page 689). Compared with the other two types of muscle tissue, smooth muscle tissue has considerably greater powers of regeneration. Such powers are still limited when compared with other tissues, such as epithelium.

Table 10.2 summarizes the major characteristics of the three types of muscle tissue.

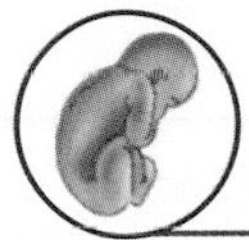

DEVELOPMENT OF MUSCLE

▶ OBJECTIVE

- **Describe the development of muscles.**

Except for the muscles of the iris of the eyes and the arrector pili muscles attached to hairs, all muscles of the body are derived from **mesoderm.** As the mesoderm develops, part of it becomes arranged in dense columns on either side of the developing nervous system. These columns of mesoderm undergo segmentation into a series of blocks of cells called **somites** (Figure 10.20a on page 301). The first pair of somites appears on the 20th day of embryonic development. Eventually, 42 to 44 pairs of somites are formed by the end of the fifth week.

With the exception of the skeletal muscles of the head and limbs, *skeletal muscles* develop from the **mesoderm of somites.** Because there are very few somites in the head region of the embryo, most of the skeletal muscles there develop from the **general mesoderm** in the head region. The skeletal muscles of the limbs develop from masses of general mesoderm around developing bones in embryonic limb buds (origins of future limbs; see Figure 6.13a on page 181).

The cells of a somite differentiate into three regions: (1) a **myotome,** which forms the skeletal muscles of the head, neck, and limbs; (2) a **dermatome,** which forms the connective

Table 10.2 Summary of the Major Features of the Three Types of Muscle Tissue

Characteristic	Skeletal Muscle	Cardiac Muscle	Smooth Muscle
Microscopic appearance and features	Long cylindrical fiber with many peripherally located nuclei; striated.	Branched cylindrical fiber with one centrally located nucleus; intercalated discs join neighboring fibers; striated.	Fiber is thickest in the middle, tapered at each end, and has one centrally positioned nucleus; not striated.
Location	Attached primarily to bones by tendons.	Heart.	Walls of hollow viscera, airways, blood vessels, iris and ciliary body of eye, arrector pili of hair follicles.
Fiber diameter	Very large (10–100 μm).	Large (14 μm).	Small (3–8 μm).
Connective tissue components	Endomysium, perimysium, and epimysium.	Endomysium.	Endomysium.
Fiber length	100 μm–30 cm.	50–100 μm.	30–200 μm.
Contractile proteins organized into sarcomeres	Yes.	Yes.	No.
Sarcoplasmic reticulum	Abundant.	Some.	Scanty.
Transverse tubules present	Yes, aligned with each A–I band junction.	Yes, aligned with each Z disc.	No.
Junctions between fibers	None.	Intercalated discs contain gap junctions and desmosomes.	Gap junctions in visceral smooth muscle; none in multiunit smooth muscle.
Autorhythmicity	No.	Yes.	Yes, in visceral smooth muscle.
Source of Ca^{2+} for contraction	Sarcoplasmic reticulum.	Sarcoplasmic reticulum and interstitial fluid.	Sarcoplasmic reticulum and interstitial fluid.
Regulator proteins for contraction	Troponin and tropomyosin.	Troponin and tropomyosin.	Calmodulin and myosin light chain kinase.
Speed of contraction	Fast.	Moderate.	Slow.
Nervous control	Voluntary (somatic nervous system).	Involuntary (autonomic nervous system).	Involuntary (autonomic nervous system).
Contraction regulated by	Acetylcholine released by somatic motor neurons.	Acetylcholine and norepinephrine released by autonomic motor neurons; several hormones.	Acetylcholine and norepinephrine released by autonomic motor neurons; several hormones, local chemical changes; stretching.
Capacity for regeneration	Limited, via satellite cells.	Limited, under certain conditions.	Considerable, via pericytes (compared with other muscle tissues, but limited compared with epithelium).

tissues, including the dermis; and (3) a **sclerotome,** which gives rise to the vertebrae (Figure 10.20b).

Cardiac muscle develops from **mesodermal cells** that migrate to and envelop the developing heart while it is still in the form of primitive heart tubes (see in Figure 20.18 on page 688).

Smooth muscle develops from **mesodermal cells** that migrate to and envelop the developing gastrointestinal tract and viscera.

Figure 10.20 Location and structure of somites, key structures in the development of the muscular system.

Most muscles are derived from mesoderm.

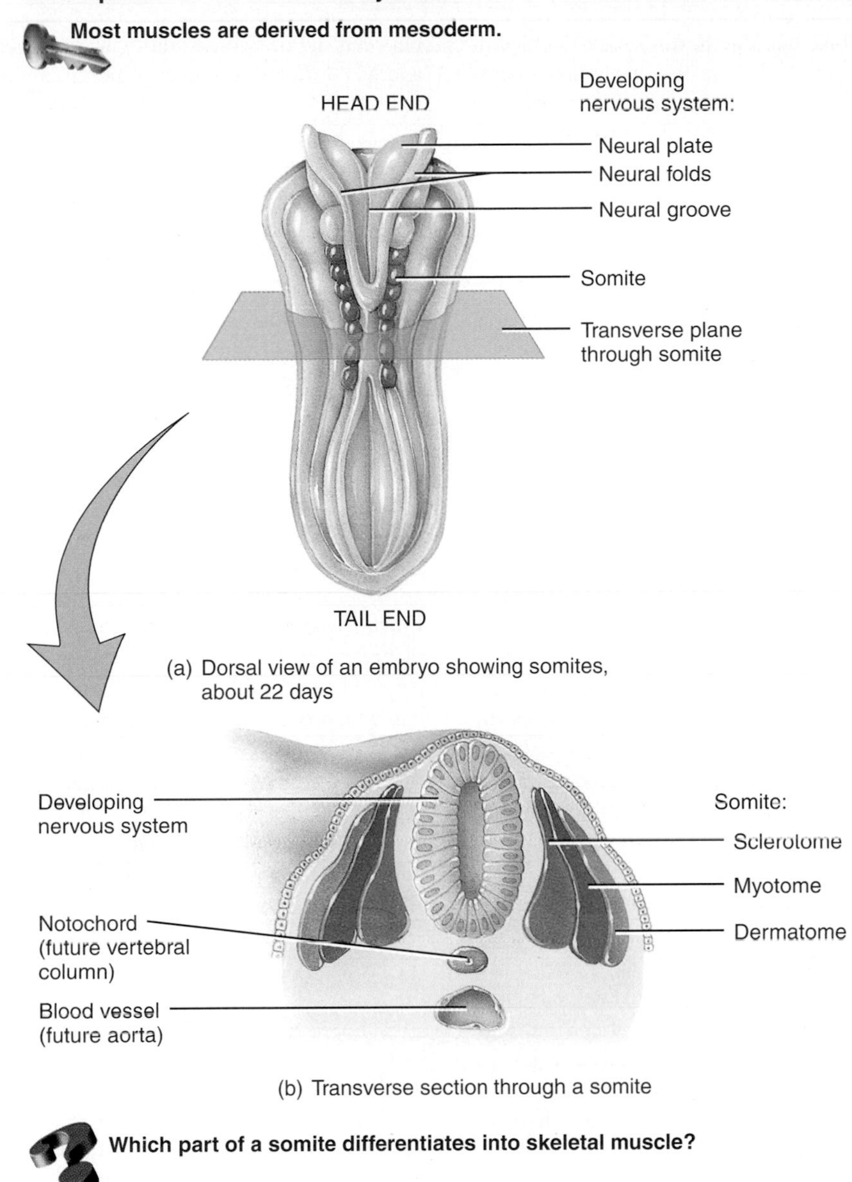

Which part of a somite differentiates into skeletal muscle?

AGING AND MUSCLE TISSUE

OBJECTIVE

- **Explain how aging affects skeletal muscle.**

Beginning at about 30 years of age, humans undergo a slow, progressive loss of skeletal muscle mass that is replaced largely by fibrous connective tissue and adipose tissue. In part, this decline is due to increasing inactivity. Accompanying the loss of muscle mass is a decrease in maximal strength, a slowing of muscle reflexes, and a loss of flexibility. In some muscles, a selective loss of muscle fibers of a given type may occur. With aging, the relative number of slow oxidative fibers appears to increase. This could be due either to atrophy of the other fiber types or their conversion into slow oxidative fibers. Whether this is an effect of aging itself or mainly reflects the more limited physical activity of older people is still an unresolved question. Nevertheless, aerobic activities and strength training programs are effective in older people and can slow or even reverse the age-associated decline in muscular performance.

CHECKPOINT

23. Which type of muscle fibers can regenerate most effectively after damage?
24. Why does muscle strength decrease with aging?

DISORDERS: HOMEOSTATIC IMBALANCES

Skeletal muscle function may be abnormal due to disease or damage of any of the components of a motor unit: somatic motor neuron, neuromuscular junctions, or muscle fibers. The term **neuromuscular disease** encompasses problems at all three sites, whereas the term **myopathy** (mī-OP-a-thē; *-pathy* = disease) signifies a disease or disorder of the skeletal muscle tissue itself.

Myasthenia Gravis

Myasthenia gravis (mī-as-THĒ-nē-a GRAV-is) is an autoimmune disease that causes chronic, progressive damage of the neuromuscular junction. In people with myasthenia gravis, the immune system inappropriately produces antibodies that bind to and block some ACh receptors, thereby decreasing the number of functional ACh receptors at the motor end plates of skeletal muscles (see Figure 10.12). Because 75% of patients with myasthenia gravis have hyperplasia or tumors of the thymus, it is possible that thymic abnormalities cause the disorder. As the disease progresses, more ACh receptors are lost. Thus, muscles become increasingly weaker, fatigue more easily, and may eventually cease to function.

Myasthenia gravis occurs in about 1 in 10,000 people and is more common in women, who typically are ages 20 to 40 at onset, than in men, who usually are ages 50 to 60 at onset. The muscles of the face and neck are most often affected. Initial symptoms include weakness of the eye muscles, which may produce double vision, and difficulty in swallowing. Later, the person has difficulty chewing and talking. Eventually the muscles of the limbs may become involved. Death may result from paralysis of the respiratory muscles, but often the disorder does not progress to this stage.

Anticholinesterase drugs such as pyridostigmine (Mestinon) or neostigmine are the first line of treatment for myasthenia gravis. They act as inhibitors of acetylcholinesterase, the enzyme that breaks down ACh. Thus, the inhibitors raise the level of ACh that is available to bind with still-functional receptors. More recently, steroid drugs, such as prednisone, have been used with success to reduce antibody levels. Another treatment is plasmapheresis, a procedure that removes the antibodies from the blood. Often, surgical removal of the thymus (thymectomy) is helpful.

Muscular Dystrophy

The term **muscular dystrophy** refers to a group of inherited muscle-destroying diseases that cause progressive degeneration of skeletal muscle fibers. The most common form of muscular dystrophy is *DMD—Duchenne muscular dystrophy* (doo-SHĀN). Because the mutated gene is on the X chromosome, which males have only one of, DMD strikes boys almost exclusively. (Sex-linked inheritance is described in Chapter 29.) Worldwide, about 1 in every 3500 male babies—21,000 in all—are born with DMD each year. The disorder usually becomes apparent between the ages of 2 and 5, when parents notice the child falls often and has difficulty running, jumping, and hopping. By age 12 most boys with DMD are unable to walk. Respiratory or cardiac failure usually causes death between the ages of 20 and 30.

In DMD, the gene that codes for the protein dystrophin is mutated, and little or no dystrophin is present. Without the reinforcing effect of dystrophin, the sarcolemma easily tears during muscle contraction. Because their plasma membrane is damaged, muscle fibers slowly rupture and die. The dystrophin gene was discovered in 1987, and by 1990 the first attempts were made to treat DMD patients with gene therapy. The muscles of three boys with DMD were injected with myoblasts bearing functional dystrophin genes, but only a few muscle fibers gained the ability to produce dystrophin. Similar clinical trials with additional patients have also failed. An alternate approach to the problem is to find a way to induce muscle fibers to produce the protein utrophin, which is similar to dystrophin. Experiments with dystrophin-deficient mice suggest this approach may work.

Fibromyalgia

Fibromyalgia (*algia* = painful condition) is a painful, nonarticular rheumatic disorder that usually appears between the ages of 25 and 50. An estimated 3 million people in the United States suffer from fibromyalgia, which is 15 times more common in women than in men. The disorder affects the fibrous connective tissue components of muscles, tendons, and ligaments. A striking sign is pain that results from gentle pressure at specific "tender points." Even without pressure, there is pain, tenderness, and stiffness of muscles, tendons, and surrounding soft tissues. Besides muscle pain, those with fibromyalgia report severe fatigue, poor sleep, headaches, depression, and inability to carry out their daily activities. Often, a gentle aerobic fitness program is beneficial.

Abnormal Contractions of Skeletal Muscle

One kind of abnormal muscular contraction is a **spasm,** a sudden involuntary contraction of a single muscle in a large group of muscles. A painful spasmodic contraction is known as a **cramp.** A **tic** is a spasmodic twitching made involuntarily by muscles that are ordinarily under voluntary control. Twitching of the eyelid and facial muscles are examples of tics. A **tremor** is a rhythmic, involuntary, purposeless contraction that produces a quivering or shaking movement. A **fasciculation** is an involuntary, brief twitch of an entire motor unit that is visible under the skin; it occurs irregularly and is not associated with movement of the affected muscle. Fasciculations may be seen in multiple sclerosis (see page 413) or in amyotrophic lateral sclerosis (Lou Gehrig's disease, see page 514). A **fibrillation** is a spontaneous contraction of a single muscle fiber that is not visible under the skin but can be recorded by electromyography. Fibrillations may signal destruction of motor neurons.

MEDICAL TERMINOLOGY

Hypertonia Increased muscle tone, either rigidity (muscle stiffness) or spasticity (muscle stiffness associated with an increase in tendon reflexes).

Hypotonia (*hypo-* = below or deficient) Decreased or lost muscle tone, usually due to damage to somatic motor neurons.

Muscle strain Tearing of a muscle because of forceful impact, accompanied by bleeding and severe pain. Also known as a Charley horse or pulled muscle. It often occurs in contact sports and typically affects the quadriceps femoris muscle on the anterior surface of the thigh. The condition is treated by RICE therapy: rest

(R), ice immediately after the injury (I), compression via a supportive wrap (C), and elevation of the limb (E).

Myalgia (mī-AL-jē-a; *-algia* = painful condition) Pain in or associated with muscles.

Myoma (mī-Ō-ma; *-oma* = tumor) A tumor consisting of muscle tissue.

Myomalacia (mī′-ō-ma-LĀ-shē-a; *-malacia* = soft) Pathological softening of muscle tissue.

Myositis (mī′-ō-SĪ-tis; *-itis* = inflammation of) Inflammation of muscle fibers (cells).

Myotonia (mī′-ō-TŌ-nē-a; *-tonia* = tension) Increased muscular excitability and contractility, with decreased power of relaxation; tonic spasm of the muscle.

Volkmann's contracture (FŌLK-manz kon-TRAK-tur; *contra-* = against) Permanent shortening (contracture) of a muscle due to replacement of destroyed muscle fibers by fibrous connective tissue, which lacks extensibility. Destruction of muscle fibers may occur from interference with circulation caused by a tight bandage, a piece of elastic, or a cast.

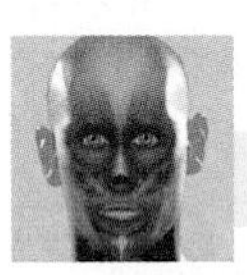

STUDY OUTLINE

INTRODUCTION (p. 274)

1. Motion results from alternating contraction and relaxation of muscles, which constitute 40–50% of total body weight.
2. The prime function of muscle is changing chemical energy into mechanical energy to perform work.

OVERVIEW OF MUSCLE TISSUE (p. 274)

1. The three types of muscle tissue are skeletal, cardiac, and smooth. Skeletal muscle tissue is primarily attached to bones; it is striated and voluntary. Cardiac muscle tissue forms the wall of the heart; it is striated and involuntary. Smooth muscle tissue is located primarily in internal organs; it is nonstriated (smooth) and involuntary.
2. Through contraction and relaxation, muscle tissue performs four important functions: producing body movements; stabilizing body positions; moving substances within the body and regulating organ volume; and producing heat.
3. Four special properties of muscle tissues are electrical excitability, the property of responding to stimuli by producing action potentials; contractility, the ability to generate tension to do work; extensibility, the ability to be extended (stretched); and elasticity, the ability to return to original shape after contraction or extension.
4. An isotonic contraction occurs when a muscle shortens and moves a constant load; tension remains almost constant. An isometric contraction occurs without shortening of the muscle; tension increases greatly.

SKELETAL MUSCLE TISSUE (p. 275)

1. Connective tissues surrounding muscle are the epimysium, covering the entire muscle; perimysium, covering fascicles; and the endomysium, covering muscle fibers. Superficial fascia separates muscle from skin.
2. Tendons and aponeuroses are extensions of connective tissue beyond muscle fibers that attach the muscle to bone or to other muscle.
3. Skeletal muscles are well supplied with nerves and blood vessels. Generally, an artery and one or two veins accompany each nerve that penetrates a skeletal muscle.
4. Somatic motor neurons provide the nerve impulses that stimulate skeletal muscle to contract.
5. Blood capillaries bring in oxygen and nutrients and remove heat and waste products of muscle metabolism.
6. The major cells of skeletal muscle tissue are termed skeletal muscle fibers. Each muscle fiber has 100 or more nuclei because it arises from fusion of many myoblasts. Satellite cells are myoblasts that persist after birth. The sarcolemma is a muscle fiber's plasma membrane; it surrounds the sarcoplasm. T tubules are invaginations of the sarcolemma.
7. Each muscle fiber (cell) contains hundreds of myofibrils, the contractile elements of skeletal muscle. Sarcoplasmic reticulum surrounds each myofibril. Within a myofibril are thin and thick filaments, arranged in compartments called sarcomeres.
8. The overlapping of thick and thin filaments produces striations. Darker A bands alternate with lighter I bands.
9. Myofibrils are built from three types of proteins: contractile, regulatory, and structural. The contractile proteins are myosin (thick filament) and actin (thin filament). Regulatory proteins are tropomyosin and troponin (both are part of the thin filament). Structural proteins include titin (links Z disc to M line and stabilizes thick filament), myomesin (forms M line), and dystrophin (links thin filaments to sarcolemma).
10. Projecting myosin heads (crossbridges) contain actin-binding and ATP-binding sites and are the motor proteins that power muscle contraction.

CONTRACTION AND RELAXATION OF SKELETAL MUSCLE FIBERS (p. 282)

1. Muscle contraction occurs because myosin heads attach to and "walk" along the thin filaments at both ends of a sarcomere, progressively pulling the thin filaments toward the center of a sarcomere. As the thin filaments slide inward, the Z discs come closer together, and the sarcomere shortens.
2. The contraction cycle is the repeating sequence of events that causes sliding of the filaments: (1) myosin ATPase hydrolyzes ATP and becomes energized, (2) the myosin head attaches to actin, (3) the myosin head generates force as it rotates toward the center

of the sarcomere (power stroke), and (4) binding of ATP to myosin detaches myosin from actin. The myosin head again hydrolyzes the ATP, returns to its original position, and binds to a new site on actin as the cycle continues.

3. An increase in Ca^{2+} concentration in the cytosol starts filament sliding, whereas a decrease turns off the sliding process.
4. The muscle action potential propagating into the T tubule system causes opening of Ca^{2+} release channels in the SR membrane. Calcium ions diffuse from the SR into the cytosol and combine with troponin. This binding causes the troponin–tropomyosin complex to move away from the myosin-binding sites on actin.
5. Ca^{2+} active transport pumps continually remove Ca^{2+} from the sarcoplasm into the SR. When the concentration of calcium ions in the cytosol decreases, the troponin–tropomyosin complexes slide back over and block the myosin-binding sites, and the muscle fiber relaxes.
6. A muscle fiber develops its greatest tension when there is an optimal zone of overlap between thick and thin filaments. This dependence is the length–tension relationship.
7. The neuromuscular junction (NMJ) is the synapse between a somatic motor neuron and a skeletal muscle fiber. The NMJ includes the axon terminals and synaptic end bulbs of a motor neuron, plus the adjacent motor end plate of the muscle fiber sarcolemma.
8. When a nerve impulse reaches the synaptic end bulbs of a somatic motor neuron, it triggers exocytosis of the synaptic vesicles, which releases acetylcholine (ACh). ACh diffuses across the synaptic cleft and binds to ACh receptors, initiating a muscle action potential. Acetylcholinesterase then quickly destroys ACh.

MUSCLE METABOLISM (p. 289)

1. Muscle fibers have three sources for ATP production: creatine, anaerobic cellular respiration, and aerobic cellular respiration.
2. Creatine kinase catalyzes the transfer of a high-energy phosphate group from creatine phosphate to ADP to form new ATP molecules. Together, creatine phosphate and ATP provide enough energy for muscles to contract maximally for about 15 seconds.
3. Glucose is converted to pyruvic acid in the reactions of glycolysis, which yield two ATPs without using oxygen. Such anaerobic cellular respiration can provide enough energy for 30–40 seconds of maximal muscle activity.
4. Muscular activity that lasts longer than half a minute depends on aerobic cellular respiration, mitochondrial reactions that require oxygen to produce ATP.
5. The inability of a muscle to contract forcefully after prolonged activity is muscle fatigue.
6. Elevated oxygen use after exercise is called recovery oxygen uptake.

CONTROL OF MUSCLE TENSION (p. 291)

1. A motor neuron and the muscle fibers it stimulates form a motor unit. A single motor unit may contain as few as two or as many as 3000 muscle fibers.
2. Recruitment is the process of increasing the number of active motor units.
3. A twitch contraction is a brief contraction of all the muscle fibers in a motor unit in response to a single action potential.
4. A record of a contraction is called a myogram. It consists of a latent period, a contraction period, and a relaxation period. The refractory period is the time when a muscle fiber has temporarily lost excitability. Skeletal muscles have a short refractory period, whereas cardiac muscle has a long refractory period.
5. Wave summation is the increased strength of a contraction that occurs when a second stimulus arrives before the muscle has completely relaxed after a previous stimulus.
6. Repeated stimuli can produce unfused tetanus, a sustained muscle contraction with partial relaxation between stimuli. More rapidly repeating stimuli produce fused tetanus, a sustained contraction without partial relaxation between stimuli.
7. Continual involuntary activation of a small number of motor units produces muscle tone, which is essential for maintaining posture.
8. In a concentric isotonic contraction, the muscle shortens to produce movement and to reduce the angle at a joint. During an eccentric isotonic contraction, the muscle lengthens.
9. Isometric contractions, in which tension is generated without muscle shortening, are important because they stabilize some joints as others are moved.

TYPES OF SKELETAL MUSCLE FIBERS (p. 295)

1. On the basis of their structure and function, skeletal muscle fibers are classified as slow oxidative (SO), fast oxidative-glycolytic (FOG), and fast glycolytic (FG) fibers.
2. Most skeletal muscles contain a mixture of all three fiber types. Their proportions vary with the typical action of the muscle.
3. The motor units of a muscle are recruited in the following order: first SO fibers, then FOG fibers, and finally FG fibers.
4. Table 10.1 on page 296 summarizes the three types of skeletal muscle fibers.

EXERCISE AND SKELETAL MUSCLE TISSUE (p. 296)

1. Various type of exercises can induce changes in the fibers in a skeletal muscle. Endurance-type (aerobic) exercises cause a gradual transformation of some fast glycolytic (FG) fibers into fast oxidative-glycolytic (FOG) fibers.
2. Exercises that require great strength for short periods produce an increase in the size and strength of fast-glycolytic (FG) fibers. The increase in size is due to increased synthesis of thick and thin filaments.

CARDIAC MUSCLE TISSUE (p. 297)

1. Cardiac muscle is found only in the heart. Cardiac muscle fibers have the same arrangement of actin and myosin and the same bands, zones, and Z discs as skeletal muscle fibers. The fibers connect to one another through intercalated discs, which contain both desmosomes and gap junctions.
2. Cardiac muscle tissue remains contracted 10 to 15 times longer than skeletal muscle tissue due to prolonged delivery of Ca^{2+} into the sarcoplasm.
3. Cardiac muscle tissue contracts when stimulated by its own autorhythmic fibers. Due to its continuous, rhythmic activity, cardiac muscle depends greatly on aerobic cellular respiration to generate ATP.

SMOOTH MUSCLE TISSUE (p. 297)

1. Smooth muscle is nonstriated and involuntary.
2. Smooth muscle fibers contain intermediate filaments and dense

bodies (whose function is similar to that of the Z discs in striated muscle).

3. Visceral (single-unit) smooth muscle is found in the walls of hollow viscera and of small blood vessels. Many fibers form a network that contracts in unison.
4. Multiunit smooth muscle is found in large blood vessels, large airways to the lungs, arrector pili muscles, and the eye (for adjusting pupil diameter and lens focus). The fibers operate independently rather than in unison.
5. The duration of contraction and relaxation of smooth muscle is longer than in skeletal muscle.
6. Smooth muscle fibers contract in response to nerve impulses, hormones, and local factors.
7. Smooth muscle fibers can stretch considerably and still maintain their contractile function.

REGENERATION OF MUSCLE TISSUE (p. 299)

1. Skeletal muscle fibers cannot divide and have limited powers of regeneration; cardiac muscle fibers can neither divide nor regenerate; smooth muscle fibers have limited capacity for division and regeneration.
2. Table 10.2 on page 300 summarizes the major characteristics of the three types of muscle tissue.

DEVELOPMENT OF MUSCLE (p. 299)

1. With few exceptions, muscles develop from mesoderm.
2. Skeletal muscles of the head and limbs develop from general mesoderm. Other skeletal muscles develop from the mesoderm of somites.

AGING AND MUSCLE TISSUE (p. 301)

1. Beginning at about 30 years of age, there is a slow, progressive loss of skeletal muscle, which is replaced by fibrous connective tissue and fat.
2. Aging also results in a decrease in muscle strength, slower muscle reflexes, and loss of flexibility.

SELF-QUIZ QUESTIONS

Fill in the blanks in the following statements.

1. The properties of muscle that enable it to carry out its functions and contribute to homeostasis are ________, ________, ________, and ________.
2. The contractile proteins in muscles are ________ and ________.
3. A single somatic motor neuron and all of the muscle fibers it stimulates is known as a ________.

Indicate whether the following statements are true or false.

4. The key functions of muscle are motion of the body, movement of substances within the body, body position stability, organ volume regulation, heat production, and energy formation.
5. The sequence of events resulting in skeletal muscle contraction are (a) generation of a nerve impulse, (b) release of the neurotransmitter acetylcholine, (c) generation of a muscle action potential, (d) release of calcium ions from the sarcoplasmic reticulum, (e) calcium ion binding to the troponin–tropomyosin complex, (f) power stroke with actin and myosin binding and release.

Choose the one best answer to the following questions.

6. In muscle physiology, the latent period refers to (a) the period of lost excitability that occurs when two stimuli are applied immediately one after the other, (b) the brief contraction of a motor unit, (c) the period of elevated oxygen use after exercise, (d) an inability of a muscle to contract forcefully after prolonged activity, (e) a brief delay that occurs between application of a stimulus and the beginning of contraction.
7. In muscle contraction, what starts filament sliding? (a) an increase in Ca^{2+} concentration in the cytosol, (b) a decrease in Na^{+} concentration in the cytosol, (c) the bending of the myosin head, (d) the release of energy from the breakdown of ATP, (e) the power stroke.
8. What would happen if ATP were suddenly unavailable after the sarcomere had begun to shorten? (a) Nothing. The contraction would proceed normally. (b) The myosin heads would be unable to detach from actin. (c) Troponin would bind with the myosin heads. (d) Actin and myosin filaments would separate completely and be unable to recombine. (e) The myosin heads would detach completely from actin and bind to the troponin–tropomyosin complex.
9. The calcium-binding protein in the sarcoplasmic reticulum of skeletal muscle is (a) calmodulin, (b) acetylcholinesterase, (c) troponin, (d) titin, (e) calsequestrin.
10. Which of the following are sources of ATP for muscle contraction? (1) creatine phosphate, (2) glycolysis, (3) anaerobic cellular respiration, (4) aerobic cellular respiration, (5) acetylcholine. (a) 1, 2, and 3, (b) 2, 3, and 4, (c) 2, 3, and 5, (d) 1, 2, 3, and 4, (e) 2, 3, 4, and 5.
11. Which of the following statements are true? (1) Skeletal muscles with a high myoglobin content are termed white muscle fibers. (2) Slow oxidative (SO) fibers generate ATP primarily by aerobic cellular respiration, are resistant to fatigue, and are capable of prolonged, sustained contractions. (3) Fast glycolytic (FG) fibers generate the most powerful contractions, generate ATP mainly by glycolysis, and are adapted for intense anaerobic movements. (4) Fast oxidative-glycolytic (FOG) fibers generate considerable ATP by aerobic cellular respiration but can also generate ATP by anaerobic glycolysis. (5) Most skeletal muscles are a mixture of SO, FOG, and FG fibers in various proportions. (6) Skeletal muscle fibers of any one motor unit can vary in muscle fiber types. (a) 1, 2, 3, and 4, (b) 2, 3, 4, 5, and 6, (c) 2, 3, 4, and 5, (d) 3, 4, 5, and 6, (e) 2, 4, and 6.

12. Match the following:

____ (a) a sheath of areolar connective tissue that wraps around individual skeletal muscle fibers
____ (b) dense irregular connective tissue that separates a muscle into groups of individual muscle fibers
____ (c) bundles of muscle fibers
____ (d) the outermost connective tissue layer that encircles an entire skeletal muscle
____ (e) dense irregular connective tissue that lines the body wall and limbs and holds muscles with similar function together
____ (f) a cord of dense regular connective tissue that attaches muscle to the periosteum of bone
____ (g) areolar and adipose connective tissue that separates muscle from skin
____ (h) connective tissue elements extended as a broad, flat layer

(1) aponeurosis
(2) deep fascia
(3) superficial fascia
(4) tendon
(5) endomysium
(6) perimysium
(7) epimysium
(8) fascicles

13. Match the following:

____ (a) point of contact between a motor neuron and a muscle fiber
____ (b) invaginations of the sarcolemma from the surface toward the center of the muscle fiber
____ (c) myoblasts that persist in mature skeletal muscle
____ (d) plasma membrane of a muscle fiber
____ (e) oxygen-binding protein found only in muscle fibers
____ (f) Ca^{2+}-storing tubular system similar to smooth endoplasmic reticulum
____ (g) the contracting unit of a skeletal muscle fiber
____ (h) middle area in the sarcomere where thick and thin filaments are found
____ (i) area in the sarcomere where thin filaments are present but thick filaments are not
____ (j) separates the sarcomeres from each other

(1) A band
(2) I band
(3) Z disc
(4) sarcomere
(5) neuromuscular junction
(6) myoglobin
(7) satellite cells
(8) T tubules
(9) sarcoplasmic reticulum
(10) sarcolemma

14. Match the following (some questions have more than one answer):

____ (a) has fibers joined by intercalated discs
____ (b) thick and thin filaments are not arranged as orderly sarcomeres
____ (c) uses satellite cells to repair damaged muscle fibers
____ (d) striated
____ (e) contraction begins slowly but lasts for long periods
____ (f) has a prolonged contraction due to prolonged calcium delivery from both the sarcoplasmic reticulum and the interstitial fluid
____ (g) does not exhibit autorhythmicity
____ (h) uses pericytes to repair damaged muscle fibers
____ (i) uses troponin as a regulatory protein
____ (j) can be classified as single-unit or multiunit

(1) skeletal muscle
(2) cardiac muscle
(3) smooth muscle

15. Match the following:

____ (a) the smooth muscle action that allows the fibers to maintain their contractile function even when stretched
____ (b) a brief contraction of all the muscle fibers in a motor unit of a muscle in response to a single action potential in its motor neuron
____ (c) sustained contraction of a muscle
____ (d) larger contractions resulting from stimuli arriving at different times
____ (e) process of increasing the number of activated motor units
____ (f) contraction in which the muscle shortens
____ (g) inability of a muscle to maintain its strength of contraction or tension during prolonged activity
____ (h) produced by the continual involuntary activation of a small number of skeletal muscle motor units; results in firmness in skeletal muscle
____ (i) contraction in which muscle tension is generated without shortening of the muscle
____ (j) contraction in which a muscle lengthens

(1) muscle fatigue
(2) twitch contraction
(3) wave summation
(4) tetanus
(5) concentric isotonic contraction
(6) motor unit recruitment
(7) muscle tone
(8) eccentric isotonic contraction
(9) isometric contraction
(10) stress-relaxation response

CRITICAL THINKING QUESTIONS

1. How can smooth muscle contract if it has no striations or sarcomeres?
   HINT ***Look for similarities in the structure of smooth and striated muscle.***

2. Peng is an amateur body builder with a huge collection of weight-lifting equipment in his basement. His little sister Ming tried to lift the 100-pound barbell—it didn't budge. Peng lifted the same barbell over his head, held it there, and then lowered it back to the ground. Ming was very impressed with her big brother. Describe the types of contractions performed by Ming and her brother.
   HINT ***Ming's contraction is the same type used by her brother to hold the barbell up.***

3. As the students from Mr. Klopfer's A&P class filed out of the room, they were massaging their writing hands. "Man, he was on a roll today! Two hours of notes with no break. My hand is killing me!" What's causing the muscle pain?
   HINT ***After two hours of A&P, the brain is not the only part of the anatomy that is fatigued.***

ANSWERS TO FIGURE QUESTIONS

10.1 Perimysium bundles groups of muscle fibers into fascicles.

10.2 Blood capillaries in muscle deliver oxygen and nutrients and remove heat and waste products of muscle metabolism.

10.3 The sarcoplasmic reticulum releases calcium ions to trigger muscle contraction.

10.4 Size, from smallest to largest: thick filament, myofibril, and muscle fiber.

10.5 There are two thin filaments for each thick filament in skeletal muscle.

10.6 Actin, titin, and nebulin connect into the Z disc. A bands contain myosin, actin, troponin, tropomyosin, titin, and nebulin; I bands contain actin, troponin, tropomyosin, titin, and nebulin.

10.7 The I bands and H zones disappear. The lengths of the filaments do not change.

10.8 If ATP were not available, the myosin heads would not be able to detach from actin.

10.9 Three functions of ATP in muscle contraction are (1) its hydrolysis by an ATPase activates the myosin head so it can bind to actin and rotate; (2) its binding to myosin causes detachment from actin after the power stroke; and (3) it powers the pumps that transport Ca^{2+} from the cytosol back into the sarcoplasmic reticulum.

10.10 A sarcomere length of 2.2 μm gives a generous zone of overlap between the parts of the thick filaments that have myosin heads and the thin filaments without the overlap being so extensive that sarcomere shortening is limited.

10.11 The portion of the sarcolemma that contains acetylcholine receptors is the motor end plate.

10.12 Steps 4 through 6 are part of excitation–contraction coupling (muscle action potential through binding of myosin heads to actin).

10.13 Glycolysis, exchange of phosphate between creatine phosphate and ADP, and glycogen breakdown occur in the cytosol. Oxidation of pyruvic acid, amino acids, and fatty acids (aerobic cellular respiration) occurs in mitochondria.

10.14 Motor units having many muscle fibers are capable of more forceful contractions than those having only a few fibers.

10.15 Events occurring during the latent period are the events of excitation–contraction coupling: Release of Ca^{2+} from SR and binding to troponin result in attachment of myosin heads to actin and the onset of rotation.

10.16 If the second stimulus were applied a little later, the second contraction would be smaller than the one illustrated in (b).

10.17 Holding your head upright without movement involves mainly isometric contractions.

10.18 Visceral smooth muscle and cardiac muscle are similar in that both contain gap junctions, which allow action potentials to spread from one cell to its neighbors.

10.19 Contraction in a smooth muscle fiber starts more slowly and lasts much longer than contraction in a skeletal muscle fiber.

10.20 The myotome of a somite differentiates into skeletal muscle.

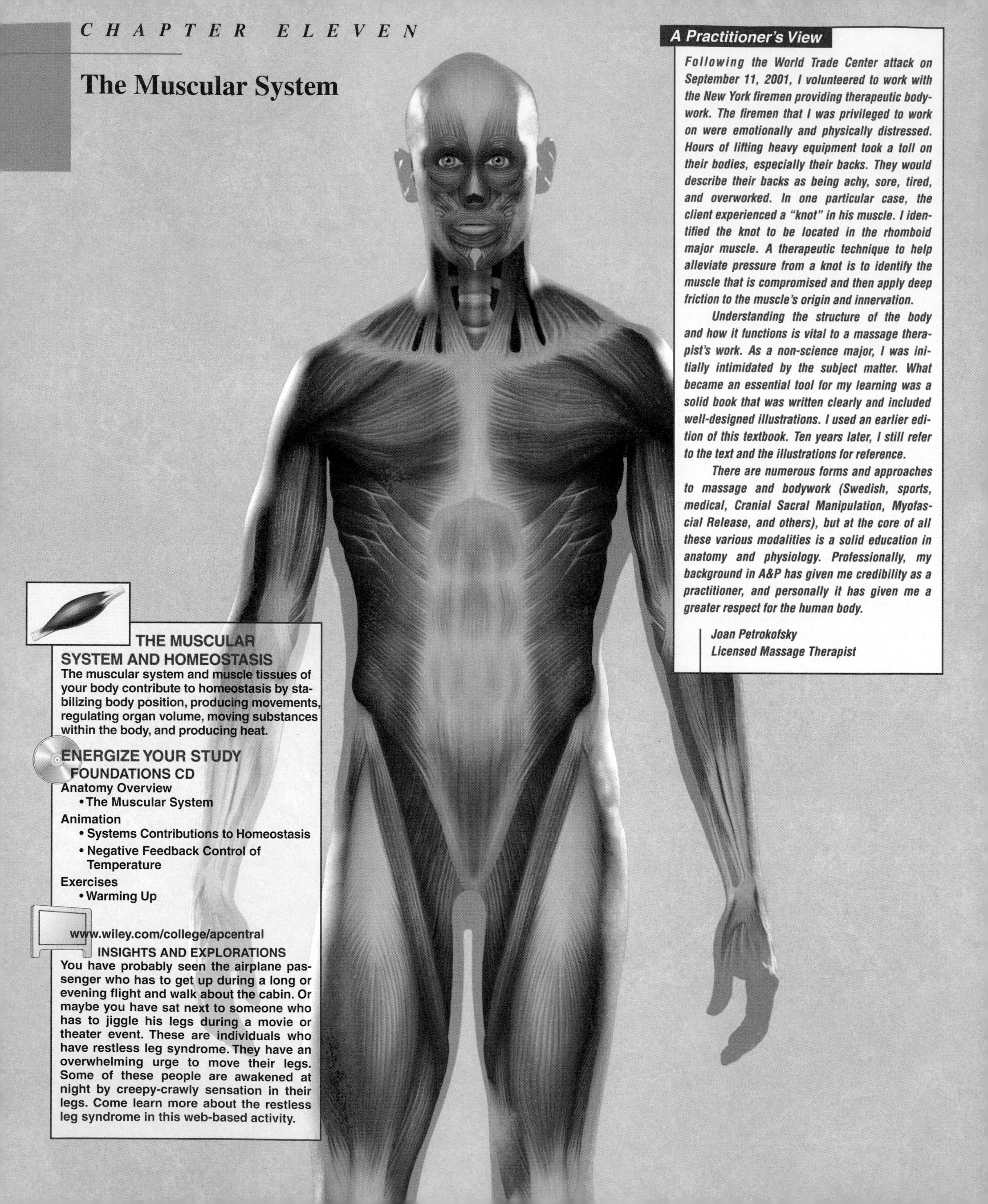

C H A P T E R   E L E V E N

The Muscular System

A Practitioner's View

Following the World Trade Center attack on September 11, 2001, I volunteered to work with the New York firemen providing therapeutic bodywork. The firemen that I was privileged to work on were emotionally and physically distressed. Hours of lifting heavy equipment took a toll on their bodies, especially their backs. They would describe their backs as being achy, sore, tired, and overworked. In one particular case, the client experienced a "knot" in his muscle. I identified the knot to be located in the rhomboid major muscle. A therapeutic technique to help alleviate pressure from a knot is to identify the muscle that is compromised and then apply deep friction to the muscle's origin and innervation.

Understanding the structure of the body and how it functions is vital to a massage therapist's work. As a non-science major, I was initially intimidated by the subject matter. What became an essential tool for my learning was a solid book that was written clearly and included well-designed illustrations. I used an earlier edition of this textbook. Ten years later, I still refer to the text and the illustrations for reference.

There are numerous forms and approaches to massage and bodywork (Swedish, sports, medical, Cranial Sacral Manipulation, Myofascial Release, and others), but at the core of all these various modalities is a solid education in anatomy and physiology. Professionally, my background in A&P has given me credibility as a practitioner, and personally it has given me a greater respect for the human body.

Joan Petrokofsky
Licensed Massage Therapist

THE MUSCULAR SYSTEM AND HOMEOSTASIS

The muscular system and muscle tissues of your body contribute to homeostasis by stabilizing body position, producing movements, regulating organ volume, moving substances within the body, and producing heat.

ENERGIZE YOUR STUDY FOUNDATIONS CD

Anatomy Overview
- The Muscular System

Animation
- Systems Contributions to Homeostasis
- Negative Feedback Control of Temperature

Exercises
- Warming Up

www.wiley.com/college/apcentral

INSIGHTS AND EXPLORATIONS

You have probably seen the airplane passenger who has to get up during a long or evening flight and walk about the cabin. Or maybe you have sat next to someone who has to jiggle his legs during a movie or theater event. These are individuals who have restless leg syndrome. They have an overwhelming urge to move their legs. Some of these people are awakened at night by creepy-crawly sensation in their legs. Come learn more about the restless leg syndrome in this web-based activity.

Together, the voluntarily controlled muscles of your body comprise the **muscular system.** The nearly 700 individual muscles that make up the muscular system, for instance, the deltoid muscle, include both skeletal muscle tissue and connective tissue. Most muscles have important functions in producing movements of body parts. A few muscles function mainly to stabilize bones so that other skeletal muscles can execute a movement more effectively. This chapter presents many of the major muscles in the body. Most of them are found on both the right and left sides. For each muscle, we'll identify its attachment sites and its innervation—the nerve or nerves that stimulate it to contract. Developing a working knowledge of these key aspects of skeletal muscle anatomy will enable you to understand how normal movements occur. This knowledge is especially crucial for allied health and physical rehabilitation professionals who work with people whose normal patterns of movement and physical mobility have been disrupted by physical trauma, surgery, or muscular paralysis.

HOW SKELETAL MUSCLES PRODUCE MOVEMENT

OBJECTIVES

- **Describe the relationship between bones and skeletal muscles in producing body movements.**
- **Define lever and fulcrum, and compare the three types of levers based on location of the fulcrum, effort, and load.**
- **Identify the types of fascicle arrangements in a skeletal muscle, and relate the arrangements to strength of contraction and range of motion.**
- **Explain how the prime mover, antagonist, synergist, and fixator in a muscle group work together to produce movement.**

Muscle Attachment Sites: Origin and Insertion

Skeletal muscles that produce movement do so by exerting force on tendons, which in turn pull on bones or other structures, for instance, skin. Most muscles cross at least one joint and typically attach to the articulating bones that form the joint (Figure 11.1a).

When a muscle contracts, it pulls one of the articulating bones toward the other. Typically, the two bones do not move equally in response to contraction. Rather, one bone remains stationary or near its original position. Either other muscles stabilize that bone by contracting and pulling it in the opposite direction or the bone's structure makes it less movable. The site of a muscle's attachment to the more stationary bone is its **origin.** The site of a muscle's attachment to the more movable bone is its **insertion.** Usually, the insertion moves toward the origin as a muscle contracts. A good analogy is a spring on a screen door. The spring has its origin on the doorframe, the stationary attachment, and its insertion on the door, the moveable attachment. Often, the origin is proximal to the insertion, especially for muscles in the limbs. The fleshy part of the muscle between the tendons of the origin and insertion is the **belly.**

Muscles that move a body part often do not lie over the moving part. For example, one of the functions of the biceps brachii muscle is to move the forearm (Figure 11.1b). Note, however, that the belly of the muscle is over the humerus, not the forearm. You will also see that muscles crossing two joints, such as the rectus femoris and sartorius, have more complex actions than muscles that cross only one joint.

Tenosynovitis (ten′-ō-sin-ō-VĪ-tis), commonly known as **tendinitis,** is a painful inflammation of the tendons, tendon sheaths, and synovial membranes of joints. The tendons most often affected are at the wrists, shoulders, elbows (resulting in *tennis elbow*), finger joints (resulting in *trigger finger*), ankles, and feet. The affected sheaths sometimes become visibly swollen due to fluid accumulation. The joint is tender and movement of the body part often causes pain. Trauma, strain, or excessive exercise may cause tenosynovitis. For instance, tying shoelaces too tightly may cause tenosynovitis of the dorsum of the foot. Also, gymnasts are prone to developing the condition because of chronic, repetitive, and maximum hyperextension at the wrists. ■

Lever Systems and Leverage

In producing movement, bones act as levers, and joints function as the fulcrums of these levers. A **lever** is a rigid structure that can move around a fixed point called a **fulcrum,** symbolized by △. A lever is acted on at two different points by two different forces: the **effort** [E], which causes movement, and the **load** [L] or **resistance,** which opposes movement. The effort is the force exerted by muscular contraction, whereas the load is typically the weight of the body part that is moved. Motion occurs when the effort applied to the bone at the insertion exceeds the load. Consider the biceps brachii flexing the forearm at the elbow as an object is lifted (Figure 11.1b). When the forearm is raised, the elbow is the fulcrum. The weight of the forearm plus the weight of the object in the hand is the load. The force of contraction of the biceps brachii pulling the forearm up is the effort.

Levers produce trade-offs between effort and the speed and range of motion. In one situation, a lever operates at a *mechanical advantage*—has **leverage**—when a smaller effort can move a heavier load. Here the trade-off is that the effort must move a greater distance (must have a longer range of motion) and faster

Figure 11.1 Relationship of skeletal muscles to bones. (a) Muscles are attached to bones by tendons known as the origin and insertion. (b) Skeletal muscles produce movements by pulling on bones. Bones serve as levers, and joints act as fulcrums for the levers. Here the movement of the forearm illustrates the lever-fulcrum principle. Note where the load (resistance) and effort are applied in this example.

In the limbs, the origin of a muscle is usually proximal and the insertion is usually distal.

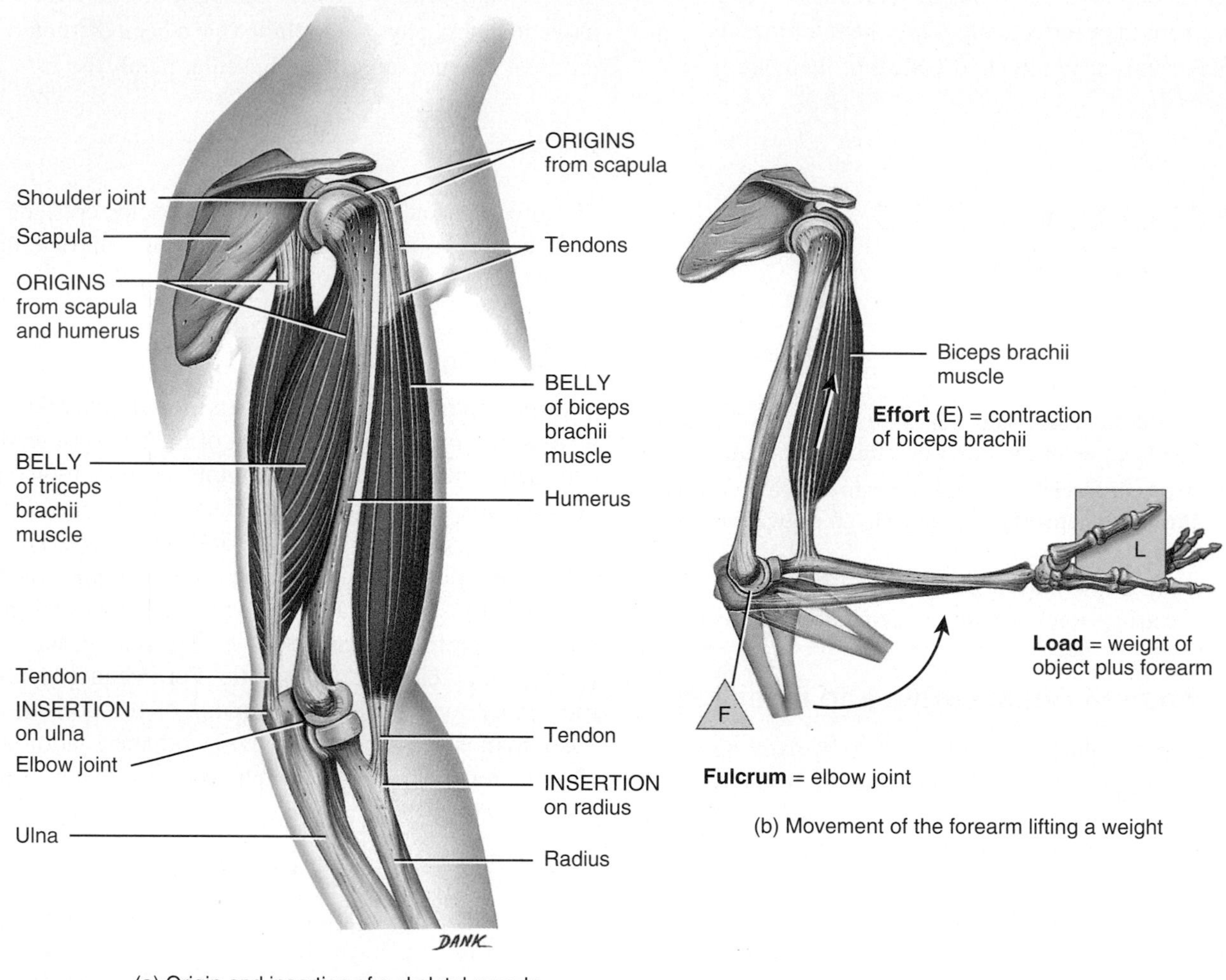

(a) Origin and insertion of a skeletal muscle

(b) Movement of the forearm lifting a weight

Where is the belly of the muscle that extends the forearm located?

than the load. Recall that range of motion refers to the range, measured in degrees of a circle, through which the bones of a joint can be moved. The lever formed by the mandible at the temporomandibular joints (fulcrums) and the effort provided by contraction of the jaw muscles produce a powerful mechanical advantage that crushes food. In another situation, a lever operates at a *mechanical disadvantage* when a larger effort moves a lighter load. In this case the trade-off is that the effort must move a shorter distance and slower than the load. The lever formed by the humerus at the shoulder joint (fulcrum) and the effort provided by the back and shoulder muscles produces a mechanical "disadvantage" that enables a major-league pitcher to hurl a baseball at nearly 100 miles per hour!

The positions of the effort, load, and fulcrum on the lever determine whether the lever operates at a mechanical advantage or disadvantage. When the load is close to the fulcrum and the effort is applied farther away, the lever operates at a mechanical advantage. When you chew food, the load (the food) is positioned close to the fulcrums (your temporomandibular joints) while your jaw muscles exert effort farther out from the joints. By contrast, when the effort is applied close to the fulcrum and the load is farther away, the lever operates at a mechanical disadvantage. When a pitcher throws a baseball, the back and shoulder muscles apply intense effort very close to the fulcrum (the shoulder joint) while the lighter load (the ball) is propelled at the far end of the lever (the arm bone).

Levers are categorized into three types according to the positions of the fulcrum, the effort, and the load:

1. The fulcrum is between the effort and the load in **first-class levers** (Figure 11.2a). (Think EFL.) Scissors and seesaws are examples of first-class levers. A first-class lever can produce either a mechanical advantage or disadvantage depending on whether the effort or the load is closer to the fulcrum. As we've seen in the preceding examples, if the effort is farther from the fulcrum than the load, a heavy load can be moved, but not very far or fast. If the effort is closer to the fulcrum than the load, only a lighter load can be moved, but it moves far and fast.

There are few first-class levers in the body. One example is the lever formed by the head resting on the vertebral column (Figure 11s.2sa). When the head is raised, the weight of the anterior portion of the skull is the load. The joint between the atlas and the occipital bone (atlanto-occipital joint) forms the fulcrum. The contraction of the posterior neck muscles provides the effort.

2. The load is between the fulcrum and the effort in **second-class levers** (Figure 11.2b). (Think FLE.) They operate like a wheelbarrow. Second-class levers always produce a mechanical advantage because the load is always closer to the fulcrum than the effort. This arrangement sacrifices speed and range of motion for force. Most authorities believe that there are no second-class levers in the body.

3. The effort is between the fulcrum and the load in **third-class levers** (Figure 11.2c). (Think FEL.) These levers operate like a pair of forceps and are the most common levers in the body. Third-class levers always produce a mechanical disadvantage because the effort is always closer to the fulcrum than the load. In the body, this arrangement favors speed and range of motion over force. The elbow joint, the bones of the arm and

Figure 11.2 Types of levers.

Levers are divided into three types based on the locations of the fulcrum, effort, and load (resistance).

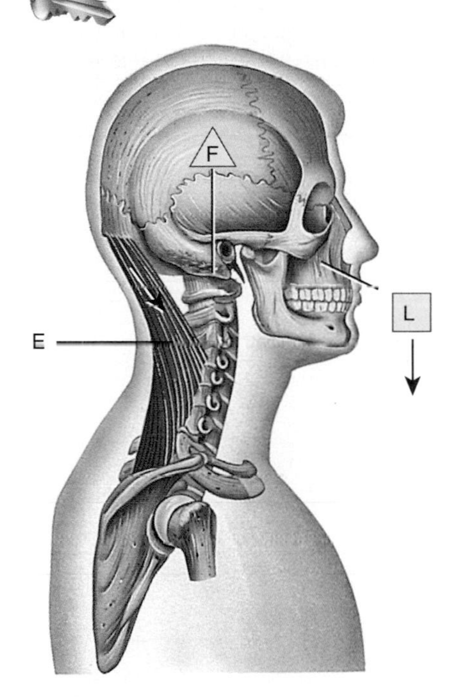

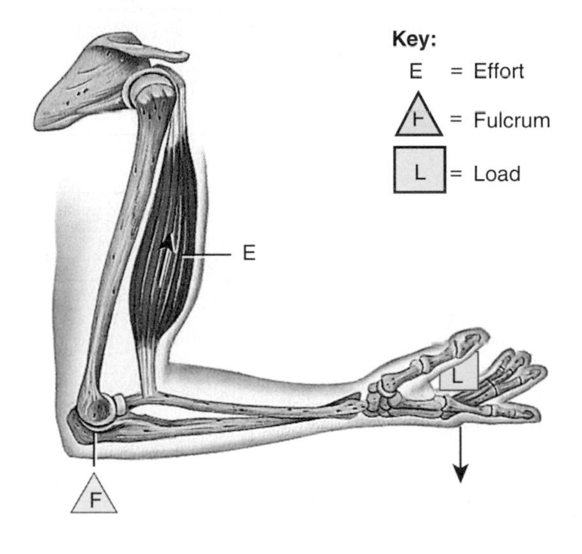

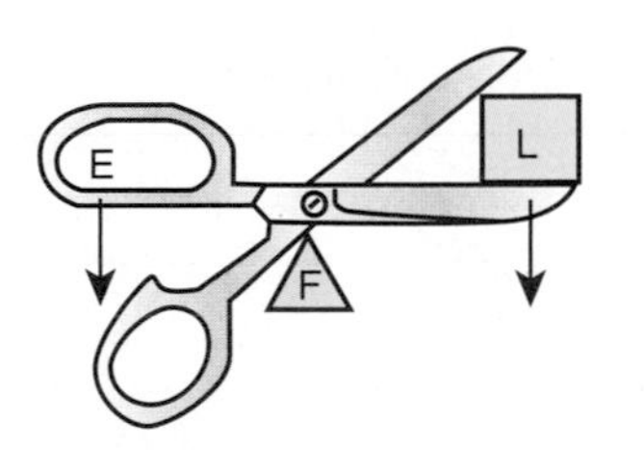

(a) First-class lever

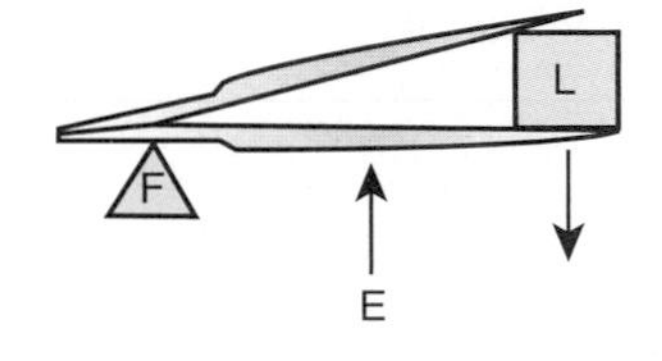

(b) Second-class lever

L
F
E

(c) Third-class lever

In which class of lever is the effort applied between the fulcrum and the load?

forearm, and the biceps brachii muscle are one example of a third-class lever (Figure 11.2c). As we have seen, in flexing the forearm at the elbow, the weight of the hand and forearm is the load, the elbow joint is the fulcrum, and the contraction of the biceps brachii muscle provides the effort. Another example of the action of a third-class lever is adduction of the thigh, in which the thigh is the load, the hip joint is the fulcrum, and the contraction of the adductor muscles is the effort.

Effects of Fascicle Arrangement

Recall from Chapter 10 that the skeletal muscle fibers (cells) within a muscle are arranged in bundles known as **fascicles.** Within a fascicle, all muscle fibers are parallel to one another. The fascicles, however, may form one of five patterns with respect to the tendons: parallel, fusiform (shaped like a cigar), circular, triangular, or pennate (shaped like a feather) (Table 11.1).

Fascicular arrangement affects a muscle's power and range of motion. As a muscle fiber contracts, it shortens to about 70% of its resting length. Thus, the longer the fibers in a muscle, the greater the range of motion it can produce. By contrast, the power of a muscle depends on its total cross-sectional area, because a short fiber can contract as forcefully as a long one. Fascicular arrangement often represents a compromise between power and range of motion. Pennate muscles, for instance, have a large number of fascicles distributed over their tendons, giving them greater power but a smaller range of motion. Parallel muscles, in contrast, have comparatively few fascicles that extend the length of the muscle. Thus, they have a greater range of motion but less power.

Coordination Within Muscle Groups

Movements often are the result of several skeletal muscles acting as a group rather than acting alone. Most skeletal muscles are arranged in opposing (antagonistic) pairs at joints—that is, flexors–extensors, abductors–adductors, and so on. Within opposing pairs, one muscle, called the **prime mover** or **agonist**

Table 11.1 Arrangement of Fascicles

Parallel

Fascicles parallel to longitudinal axis of muscle; terminate at either end end in flat tendons.

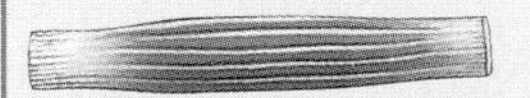

Example: Stylohyoid muscle (see Figure 11.8)

Fusiform

Fascicles nearly parallel to longitudinal axis of muscle; terminate in flat tendons; muscle tapers toward tendons, where diameter is less than at belly.

Example: Digastric muscle (see Figure 11.8)

Circular

Fascicles in concentric circular arrangements form sphincter muscles that enclose an orifice (opening).

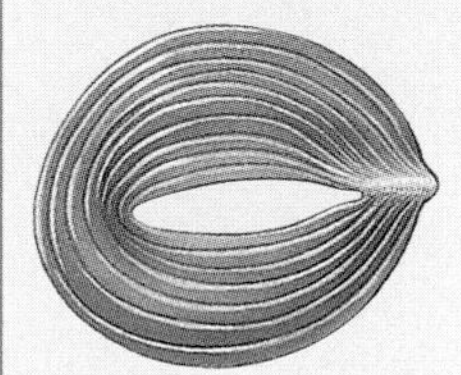

Example: Orbicularis oculi muscle (see Figure 11.4)

Triangular

Fascicles spread over broad area converge at thick central tendon; gives muscle a triangular appearance.

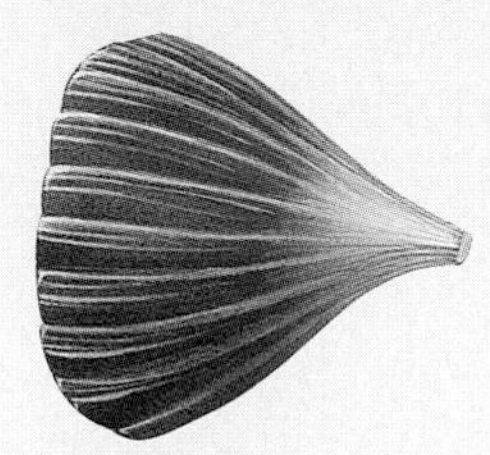

Example: Pectoralis major muscle (see Figure 11.3a)

Pennate

Short fascicles in relation to total muscle length; tendon extends nearly entire length of muscle.

Unipennate

Fascicles are arranged on only one side of tendon.

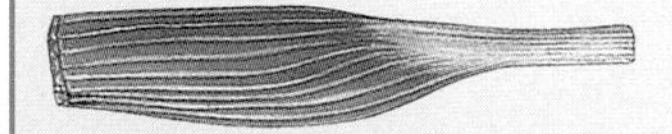

Example: Extensor digitorum longus muscle (see Figure 11.22b)

Bipennate

Fascicles are arranged on both sides of centrally positioned tendons.

Example: Rectus femoris muscle (see Figure 11.20a)

Multipennate

Fascicles attach obliquely from many directions to several tendons.

Example: Deltoid muscle (see Figure 11.10a

(= leader), contracts to cause an action while the other muscle, the **antagonist** (*anti-* = against), stretches and yields to the effects of the prime mover. In the process of flexing the forearm at the elbow, for instance, the biceps brachii is the prime mover, and the triceps brachii is the antagonist (see Figure 11.1). The antagonist and prime mover are usually located on opposite sides of the bone or joint, as is the case in this example.

With an opposing pair of muscles, the roles of the prime mover and antagonist can switch for different movements. For example, while extending the forearm at the elbow, the triceps brachii is the prime mover, and the biceps brachii is the antagonist. The roles of the two muscles reverse during flexion of the elbow. If a prime mover and its antagonist contract at the same time with equal force, there will be no movement.

Sometimes a prime mover crosses other joints before it reaches the joint at which its primary action occurs. The biceps brachii, for example, spans both the shoulder and elbow joints, and its primary action is on the forearm. To prevent unwanted movements at intermediate joints or to otherwise aid the movement of the prime mover, muscles called **synergists** (SIN-er-gists; *syn* = together; *ergon* = work) contract and stabilize the intermediate joints. As an example, muscles that flex the fingers (prime movers) cross the intercarpal and radiocarpal joints (intermediate joints). If movement at these intermediate joints was unrestrained, you would not be able to flex your fingers without flexing the wrist at the same time. Synergistic contraction of the wrist extensors stabilizes the wrist joint and prevents it from moving (unwanted movement) while the flexor muscles of the fingers contract to bring about efficient flexion of the fingers (primary action). Synergists are usually located close to the prime mover.

Some muscles in a group also act as **fixators,** which stabilize the origin of the prime mover so that the prime mover can act more efficiently. Fixators steady the proximal end of a limb while movements occur at the distal end. For example, the scapula in the pectoral (shoulder) girdle is a freely movable bone that serves as the origin for several muscles that move the arm. However, when the arm muscles contract, the scapula must be held steady. Fixator muscles accomplish this by holding the scapula firmly against the back of the chest. In abduction of the arm, the deltoid muscle serves as the prime mover, whereas fixators (pectoralis minor, trapezius, subclavius, serratus anterior muscles, and others) hold the scapula firmly (see Figure 11.14). These fixators stabilize the scapula, which serves as the attachment site for the origin of the deltoid muscle, whereas the insertion of the muscle pulls on the humerus to abduct the arm. Under different conditions—that is, for different movements—many muscles may act, at various times, as prime movers, antagonists, synergists, or fixators.

In the limbs, a **compartment** is a group of skeletal muscles, and their associated blood vessels and nerves, that have a common function. In the upper limbs, for example, flexor compartment muscles are on the anterior surface whereas extensor compartment muscles are on the posterior surface.

Benefits of Stretching

The overall goal of **stretching** is to achieve normal range of motion of joints and mobility of soft tissues surrounding the joints. For most individuals, the best stretching routine involves *static stretching,* that is, slow sustained stretching that holds a muscle in a lengthened position. The muscles should be stretched to the point of slight discomfort (not pain) and held for about 15–30 seconds. Stretching should be done after warming up as this will increase range of motion. Among the benefits of stretching are the following:

1. *Improved physical performance.* A flexible joint has the ability to move through a greater range of motion, which improves physical performance.
2. *Decreased risk of injury.* Stretching decreases resistance in various soft tissues and thus there is less likelihood of exceeding maximum tissue extensibility during an activity. This decreases the risk of injury.
3. *Reduced muscle soreness.* Stretching can reduce some of the muscle soreness that results from delayed onset mucle soreness (DOMS) after exercise (see page 279).
4. *Improved posture.* Poor posture results from improper posture and gravity over a number of years. Stretching can help re-align soft tissues to improve and maintain good posture. ■

▶ CHECKPOINT

1. With respect to muscles, what do the terms origin, insertion, and belly mean?
2. When you flex your knee, where are the fulcrum and load? Where is the muscle that provides the effort located?
3. Why can a parallel muscle have a greater range of motion than a pennate muscle?

HOW SKELETAL MUSCLES ARE NAMED

▶ OBJECTIVE

- **Explain seven features used in naming skeletal muscles.**

Several features of skeletal muscles provide descriptive ways to name muscles. The names of most of the nearly 700 skeletal muscles contain combinations of word roots for their distinctive features. Learning the terms that refer to these features will help you remember the names of muscles. Such muscle features include the pattern of the muscle's fascicles; the size, shape, action, number of origins, and location of the muscle; and the sites of origin and insertion of the muscle. Study Table 11.2 to become familiar with the terms used in muscle names.

▶ CHECKPOINT

4. Select ten muscles in Figure 11.3 on pages 316–317 and identify the features on which their names are based. (Hint: *Use the prefix, suffix, and root of each muscle's name as a guide.*)

Table 11.2 Characteristics Used to Name Muscles

Name	Meaning	Example	Figure
DIRECTION: Orientation of muscle fascicles relative to the body's midline.			
Rectus	Parallel to midline	Rectus abdominis	11.10b
Transverse	Perpendicular to midline	Transverse abdominis	11.10b
Oblique	Diagonal to midline	External oblique	11.10a
SIZE: Relative size of the muscle.			
Maximus	Largest	Gluteus maximus	11.20c
Minimus	Smallest	Gluteus minimus	11.20c
Longus	Longest	Adductor longus	11.20a
Brevis	Shortest	Adductor brevis	11.20b
Latissimus	Widest	Latissimus dorsi	11.15b
Longissimus	Longest	Longissimus capitis	11.19a
Magnus	Large	Adductor magnus	11.20a
Major	Larger	Pectoralis major	11.15a
Minor	Smaller	Pectoralis minor	11.14a
Vastus	Huge	Vastus lateralis	11.20a
SHAPE: Relative shape of the muscle.			
Deltoid	Triangular	Deltoid	11.10a
Trapezius	Trapezoid	Trapezius	11.3b
Serratus	Saw-toothed	Serratus anterior	11.14b
Rhomboid	Diamond-shaped	Rhomboid major	11.15c
Orbicularis	Circular	Orbicularis oculi	11.4a
Pectinate	Comblike	Pectineus	11.20a
Piriformis	Pear-shaped	Piriformis	11.20c
Platys	Flat	Platysma	11.4a
Quadratus	Square, four-sided	Quadratus femoris	11.20c
Gracilis	Slender	Gracilis	11.20a
ACTION: Principal action of the muscle.			
Flexor	Decreases a joint angle	Flexor carpi radialis	11.17a
Extensor	Increases a joint angle	Extensor carpi ulnaris	11.17c
Abductor	Moves a bone away from the midline	Abductor pollicis longus	11.17c
Adductor	Moves a bone closer to the midline	Adductor longus	11.20a
Levator	Raises or elevates a body part	Levator scapulae	11.14a
Depressor	Lowers or depresses a body part	Depressor labii inferioris	11.4b
Supinator	Turns palm superiorly or anteriorly	Supinator	11.17b
Pronator	Turns palm inferiorly or posteriorly	Pronator teres	11.17a
Sphincter	Decreases the size of an opening	External anal sphincter	11.12
Tensor	Makes a body part rigid	Tensor fasciae latae	11.20a
Rotator	Rotates a bone around its longitudinal axis	Obturator externus	11.20b
NUMBER OF ORIGINS: Number of tendons of origin.			
Biceps	Two origins	Biceps brachii	11.16a
Triceps	Three origins	Triceps brachii	11.16b
Quadriceps	Four origins	Quadriceps femoris	11.20a
LOCATION: Structure near which a muscle is found. *Example:* Temporalis, a muscle near the temporal bone.			11.4c
ORIGIN AND INSERTION: Sites where muscle originates and inserts. *Example:* Sternocleido-mastoid, originating on the sternum and clavicle and inserting on mastoid process of temporal bone.			11.3a

PRINCIPAL SKELETAL MUSCLES

Exhibits 11.1 through 11.20 will assist you in learning the names of the principal skeletal muscles in various regions of the body. The muscles in the exhibits are divided into groups according to the part of the body on which they act. As you study groups of muscles in the exhibits, refer to Figure 11.3 to see how each group is related to the others.

The exhibits contain the following elements:

- ***Objective.*** This statement describes the main outcomes to expect after studying the exhibit.
- ***Overview.*** These paragraphs provide a general orientation to the muscles under consideration and emphasize how the muscles are organized within various regions. The discussion also highlights any distinguishing or interesting features about the muscles.
- ***Muscle names.*** The word roots indicate how the muscles are named. Once you have mastered the naming of the muscles, their actions will have more meaning.
- ***Origins, insertions, and actions.*** For each muscle, you are also given its origin, insertion, and actions, the main movements that occur when the muscle contracts.
- ***Relating muscles to movements.*** These exercises will help you organize the muscles in a body region under consideration according to the actions they produce.
- ***Innervation.*** This section lists the nerve or nerves that cause contraction of each muscle. In general, cranial nerves, which arise from the lower parts of the brain, serve muscles in the head region. By contrast, spinal nerves, which arise from the spinal cord within the vertebral column, innervate muscles in the rest of the body. Cranial nerves are designated by both a name and a Roman numeral—for example, the facial nerve (cranial nerve VII). Spinal nerves are numbered in groups according to the part of the spinal cord from which they arise: C = cervical (neck region), T = thoracic (chest region), L = lumbar (lower back region), and S = sacral (buttocks region). An example is T1, the first thoracic spinal nerve.
- ***Questions.*** These review, thought-provoking, and application of knowledge questions relate specifically to information in each exhibit.
- ***Figures.*** The figures in the exhibits present superficial and deep, anterior and posterior, or medial and lateral views to show each muscle's position as clearly as possible. The muscle names in all capital letters are specifically referred to in the tabular part of the exhibit.

Here is a list of the exhibits and accompanying figures that describe the principal skeletal muscles:

• • •

To appreciate the many ways the muscular system contributes to homeostasis of other body systems, examine *Focus on Homeostasis: The Muscular System* on page 379. Next, in Chapter 12, you will see how the nervous system is organized, how neurons generate nerve impulses that activate muscle tissues as well as other neurons, and how synapses function.

Figure 11.3 Principal superficial skeletal muscles.

Most movements require several skeletal muscles acting in groups.

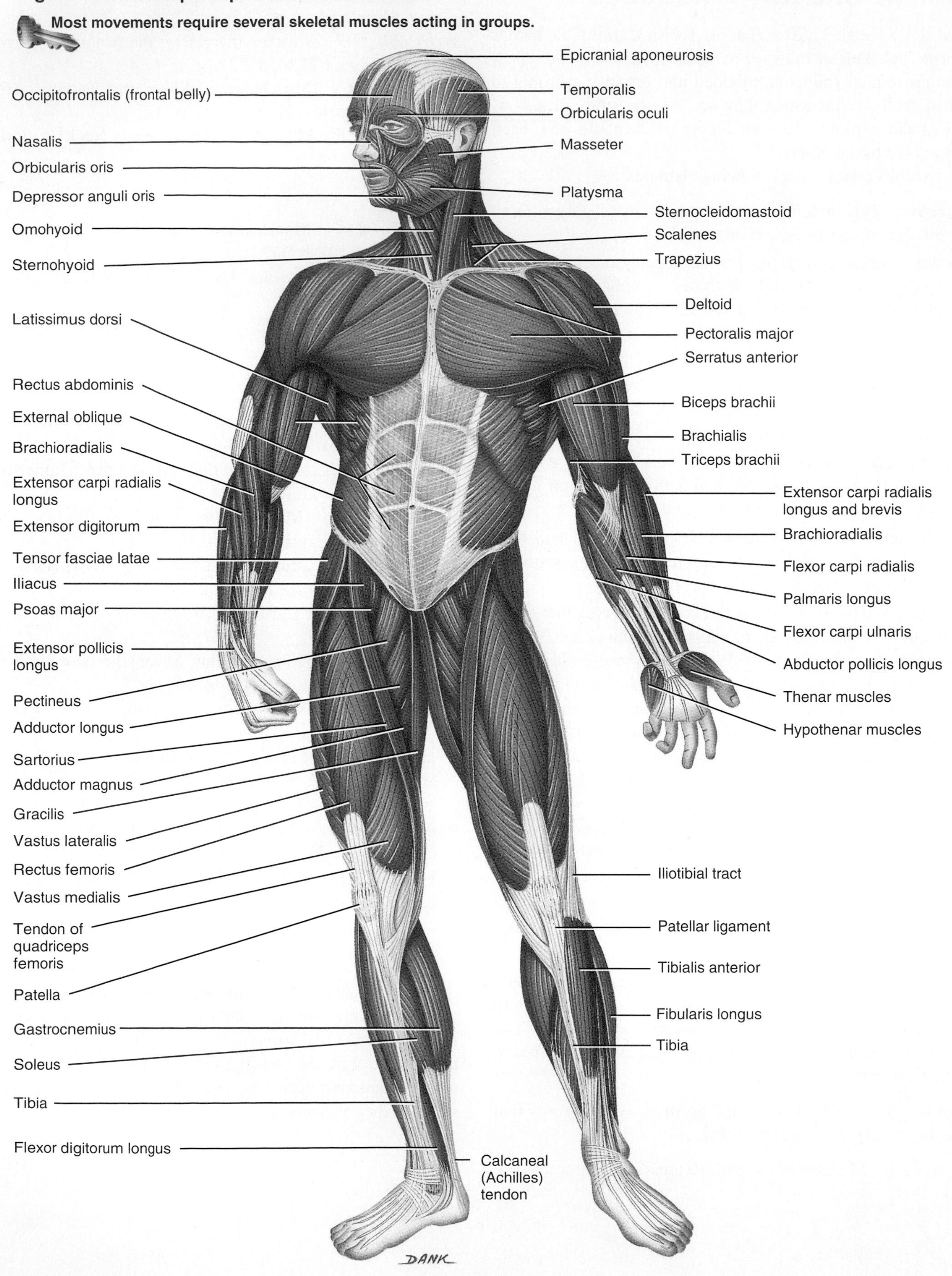

(a) Anterior view

Figure 11.3

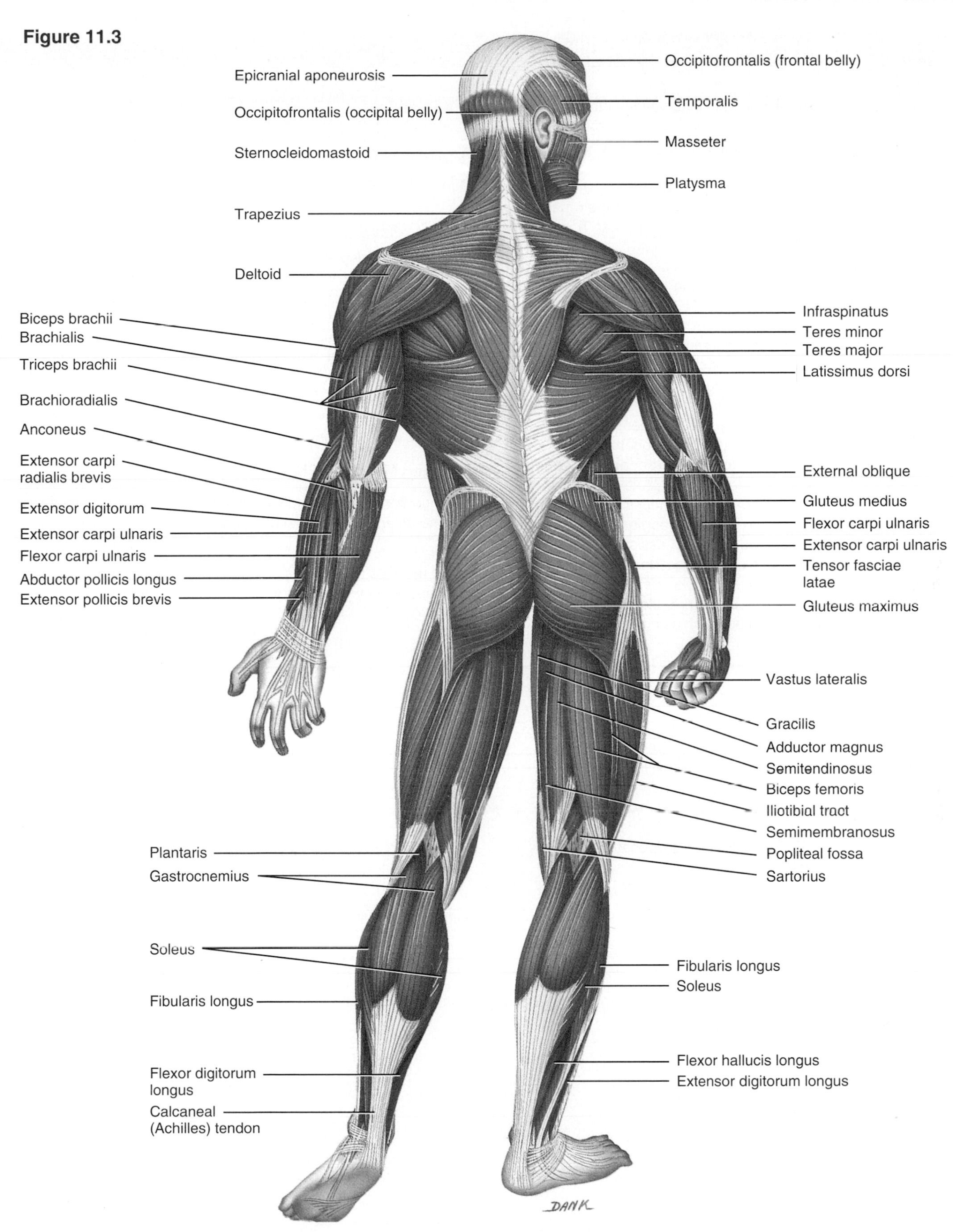

(b) Posterior view

Give an example of a muscle named for each of the following characteristics: direction of fascicles, muscle shape, muscle action, muscle size, origin, insertion, location, and number of tendons of origin.

Exhibit 11.1 Muscles of Facial Expression (Figure 11.4)

▶ OBJECTIVE

- Describe the origin, insertion, action, and innervation of the muscles of facial expression.

The muscles of facial expression provide humans with the ability to express a wide variety of emotions. The muscles themselves lie within the layers of superficial fascia. They usually originate in the fascia or bones of the skull and insert into the skin. Because of their insertions, the muscles of facial expression move the skin rather than a joint when they contract.

Among the noteworthy muscles in this group are those surrounding the orifices (openings) of the head such as the eyes, nose, and mouth. These muscles function as *sphincters* (SFINGK-ters), which close the orifices, and *dilators,* which open the orifices. For example, the **orbicularis oculi** muscle closes the eye, whereas the **levator palpebrae superioris** muscle opens the eye. The **occipitofrontalis** is an unusual muscle in this group because it is made up of two parts: an anterior part called the **frontal belly,** which is superficial to the frontal bone, and a posterior part called the **occipital belly,** which is superficial to the occipital bone. The two muscular portions are held together by a strong aponeurosis (sheetlike tendon), the **epicranial aponeurosis** (ep-i-KRĀ-nē-al ap-ō-noo-RŌ-sis) or **galea aponeurotica** (GĀ-lē-a ap-ō-noo′-RŌ-ti-ka), which covers the superior and lateral surfaces of the skull. The **buccinator** muscle forms the major muscular portion of the cheek. It functions in whistling, blowing, and sucking and assists in chewing. The duct of the parotid gland (a salivary gland) pierces the buccinator muscle to reach the oral cavity. The buccinator muscle is so-named because it compresses the cheeks (*bucc-* = cheek) during blowing—for example, when a musician plays a wind instrument such as a trumpet.

Innervation

- All muscles of facial expression except levator palpebrae superioris: facial nerve (cranial nerve VII)
- Levator palpebrae superioris: oculomotor nerve (cranial nerve III)

Bell's Palsy

Bell's palsy, also known as **facial paralysis,** is a unilateral paralysis of the muscles of facial expression. It is due to damage or disease of the facial nerve (cranial nerve VII). Although the cause is unknown, inflammation of the facial nerve or infection by the herpes simplex virus have been suggested. The paralysis causes the entire side of the face to droop in severe cases. The person cannot wrinkle the forehead, close the eye, or pucker the lips on the affected side. Difficulty in swallowing and drooling also occur. Eighty percent of patients recover completely within a few weeks to a few months. For others, paralysis is permanent. ■

Relating Muscles to Movements

Arrange the muscles in this exhibit into two groups: (1) those that act on the mouth and (2) those that act on the eyes.

▶ CHECKPOINT

What muscles would you use to do the following: show surprise, express sadness, bare your upper teeth, pucker your lips, squint, and blow up a balloon?

Muscle	Origin	Insertion	Action
Scalp Muscles			
Occipitofrontalis (ok-sip′-i-tō-frun-TĀ-lis)			
Frontal belly	Epicranial aponeurosis.	Skin superior to supraorbital margin.	Draws scalp anteriorly, raises eyebrows, and wrinkles skin of forehead horizontally.
Occipital belly (*occipit-* = back of the head)	Occipital bone and mastoid process of temporal bone.	Epicranial aponeurosis.	Draws scalp posteriorly.

Muscle	Origin	Insertion	Action
Mouth Muscles			
Orbicularis oris (or-bi′-kū-LAR-ioos OR-is; *orb-* = circular; *oris* = of the mouth)	Muscle fibers surrounding opening of mouth.	Skin at corner of mouth.	Closes and protrudes lips, compresses lips against teeth, and shapes lips during speech.
Zygomaticus major (zī-gō-MΛ-ti-kus; *zygomatic* = cheek bone; *major* = greater)	Zygomatic bone.	Skin at angle of mouth and orbicularis oris.	Draws angle of mouth superiorly and laterally, as in smiling or laughing.
Zygomaticus minor (*minor* = lesser)	Zygomatic bone.	Upper lip.	Raises (elevates) upper lip, exposing maxillary teeth.
Levator labii superioris (le-VĀ-tor LĀ-bē-ī soo-per′-ē-OR-is; *levator* = raises or elevates; *labii* = lip; *superioris* = upper)	Superior to infraorbital foramen of maxilla.	Skin at angle of mouth and orbicularis oris.	Raises upper lip.
Depressor labii inferioris (de-PRE-sor LĀ-bē-ī in-fer′-ē-OR-is; *depressor* = depresses or lowers; *inferioris* = lower)	Mandible.	Skin of lower lip.	Depresses (lowers) lower lip.
Depressor anguli oris (*angul* = angle or corner)	Mandible.	Angle of mouth.	Draws angle of mouth laterally and inferiorly, as in opening mouth.
Buccinator (BUK-si-nā′-tor; *bucc-* = cheek)	Alveolar processes of maxilla and mandible and pterygomandibular raphe (fibrous band extending from the pterygoid process to the mandible).	Orbicularis oris.	Presses cheeks against teeth and lips, as in whistling, blowing, and sucking; draws corner of mouth laterally; and assists in mastication (chewing) by keeping food between the teeth (and not between teeth and cheeks).
Risorius (ri-ZOR-ē-us; *risor* = laughter)	Fascia over parotid (salivary) gland.	Skin at angle of mouth.	Draws angle of mouth laterally, as in tenseness.
Mentalis (men-TĀ-lis; *ment-* = the chin)	Mandible.	Skin of chin.	Elevates and protrudes lower lip and pulls skin of chin up, as in pouting.
Neck Muscle			
Platysma (pla-TIZ-ma; *platy* = flat, broad)	Fascia over deltoid and pectoralis major muscles.	Mandible, muscles around angle of mouth, and skin of lower face.	Draws outer part of lower lip inferiorly and posteriorly as in pouting; depresses mandible.
Orbit and Eyebrow Muscles			
Orbicularis oculi (or-bi′-kū-LAR-is OK-ū-lī; *oculi* = of the eye)	Medial wall of orbit.	Circular path around orbit.	Closes eye.
Corrugator supercilii (KOR-a- gā′-tor soo-per-SI-lē-ī; *corrugat* = wrinkle; *supercilii* = of the eyebrow)	Medial end of superciliary arch of frontal bone.	Skin of eyebrow.	Draws eyebrow inferiorly and wrinkles skin of forehead vertically as in frowning.
Levator palpebrae superioris (le-VĀ-tor PAL-pe-brē soo-per′-ē-OR-is; *palpebrae* = eyelids) (see also Figure 11.5a)	Roof of orbit (lesser wing of sphenoid bone).	Skin of upper eyelid.	Elevates upper eyelid (opens eye).

(continues)

Exhibit 11.1 Muscles of Facial Expression (Figure 11.4) *(continued)*

Figure 11.4 Muscles of facial expression. (See Tortora, *A Photographic Atlas of the Human Body,* Figures 5.2–5.4.)

When they contract, muscles of facial expression move the skin rather than a joint.

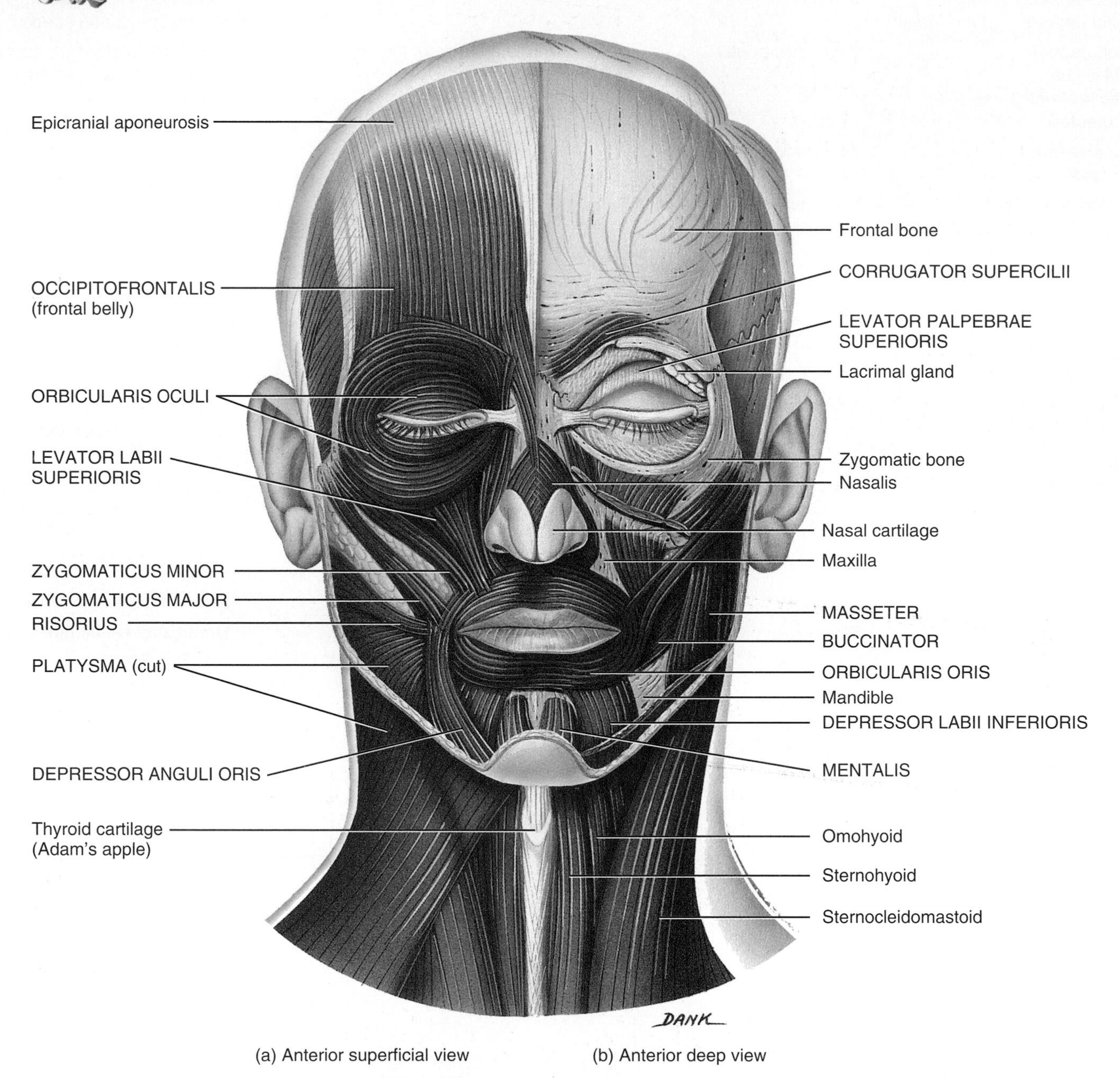

(a) Anterior superficial view (b) Anterior deep view

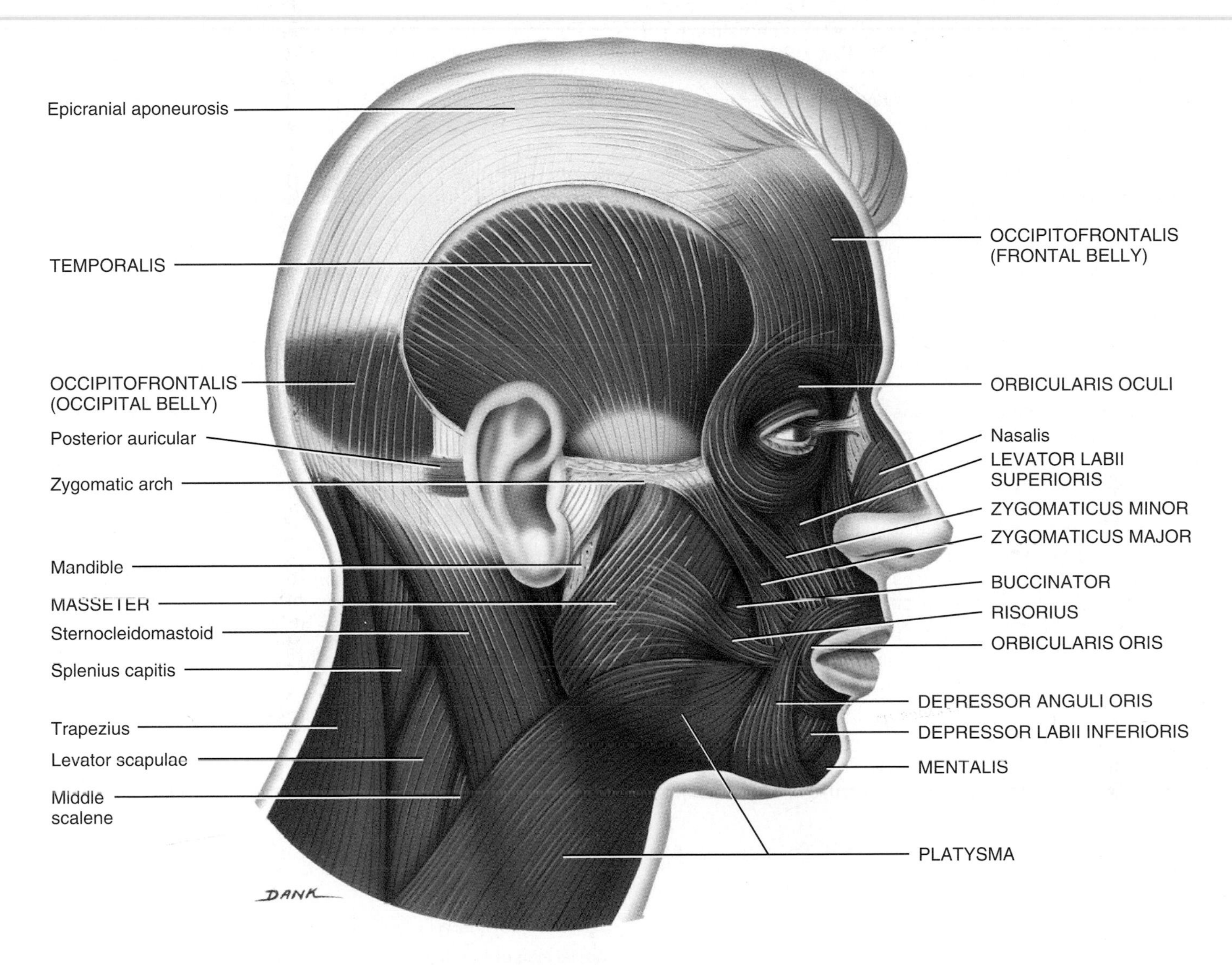

(c) Right lateral superficial view

Which muscles of facial expression cause frowning, smiling, pouting, and squinting?

Exhibit 11.2 Muscles that Move the Eyeballs—Extrinsic Eye Muscles (Figure 11.5)

▶ OBJECTIVE

- Describe the origin, insertion, action, and innervation of the extrinsic eye muscles.

Muscles that move the eyeballs are called **extrinsic eye muscles** because they originate outside the eyeballs (in the orbit) and insert on the outer surface of the sclera ("white of the eye"). The extrinsic eye muscles are some of the fastest contracting and most precisely controlled skeletal muscles in the body.

Three pairs of extrinsic eye muscles control movements of the eyeballs: (1) superior and inferior recti, (2) lateral and medial recti, and (3) superior and inferior oblique. The four recti muscles (superior, inferior, lateral, and medial) arise from a tendinous ring in the orbit and insert into the sclera of the eye. Whereas the **superior** and **inferior recti** lie in the same vertical plane, the **medial** and **lateral recti** lie in the same horizontal plane. The actions of the recti muscles can be deduced from their insertions on the sclera. The superior and inferior recti move the eyeballs superiorly and inferiorly, respectively; the lateral and medial recti move the eyeballs laterally and medially, respectively. It should be noted that neither the superior nor the inferior rectus muscle pulls directly parallel to the long axis of the eyeballs, and as a result, both muscles move the eyeballs medially.

It is not as easy to deduce the actions of the oblique muscles (superior and inferior) because of their paths through the orbits. For example, the **superior oblique** muscle originates posteriorly near the tendinous ring, then passes anteriorly, and ends in a round tendon. The tendon extends through a pulleylike loop called the *trochlea* (= pulley) in the anterior and medial part of the roof of the orbit. Finally, the tendon turns and inserts on the posterolateral aspect of the eyeballs. Accordingly, the superior oblique muscle moves the eyeballs inferiorly and laterally. The **inferior oblique** muscle originates on the maxilla at the anteromedial aspect of the floor of the orbit. It then passes posteriorly and laterally and inserts on the posterolateral aspect of the eyeballs. Because of this arrangement, the inferior oblique muscle moves the eyeballs superiorly and laterally.

Innervation

- Superior rectus, inferior rectus, medial rectus, and inferior oblique: oculomotor nerve (cranial nerve III)
- Lateral rectus: abducens nerve (cranial nerve VI)
- Superior oblique: trochlear nerve (cranial nerve IV)

Strabismus

Strabismus is a condition in which the two eyes are not properly aligned. A lesion of the oculomotor nerve (cranial nerve III), which controls the superior, inferior, and medial recti and the inferior oblique muscles, causes the eyeball to move laterally when at rest. The person cannot move the eyeball medially and inferiorly. A lesion in the abducens nerve (cranial nerve VI), which innervates the lateral rectus muscle, causes the eyeball to move medially when at rest with inability to move the eyeball laterally. ■

Relating Muscles to Movements

Arrange the muscles in this exhibit according to their actions on the eyeballs: (1) elevation, (2) depression, (3) abduction, (4) adduction, (5) medial rotation, and (6) lateral rotation. The same muscle may be mentioned more than once.

▶ CHECKPOINT

Which muscles contract and relax in each eye as you gaze to your left without moving your head?

Muscle	Origin	Insertion	Action
Superior rectus (*rectus* = fascicles parallel to midline)	Common tendinous ring (attached to orbit around optic foramen).	Superior and central part of eyeball.	Moves eyeball superiorly (elevation) and medially (adduction), and rotates it medially.
Inferior rectus	Same as above.	Inferior and central part of eyeball.	Moves eyeball inferiorly (depression) and medially (adduction), and rotates it laterally.
Lateral rectus	Same as above.	Lateral side of eyeball.	Moves eyeball laterally (abduction).
Medial rectus	Same as above.	Medial side of eyeball.	Moves eyeball medially (adduction).
Superior oblique (*oblique* = fascicles diagonal to midline)	Sphenoid bone, superior and medial to the tendinous ring in the orbit.	Eyeball between superior and lateral recti. The muscle inserts into the superior and lateral surfaces of the eyeball via a tendon that passes through the trochlea.	Moves eyeball inferiorly (depression) and laterally (abduction), and rotates it medially.
Inferior oblique	Maxilla in floor of orbit.	Eyeball between inferior and lateral recti.	Moves eyeball superiorly (elevation) and laterally (abduction), and rotates it laterally.

Figure 11.5 Extrinsic muscles of the eyeball. (See Tortora, *A Photographic Atlas of the Human Body,* Figures 5.4 and 5.5.)

The extrinsic muscles of the eyeball are some of the fastest contracting and most precisely controlled skeletal muscles in the body.

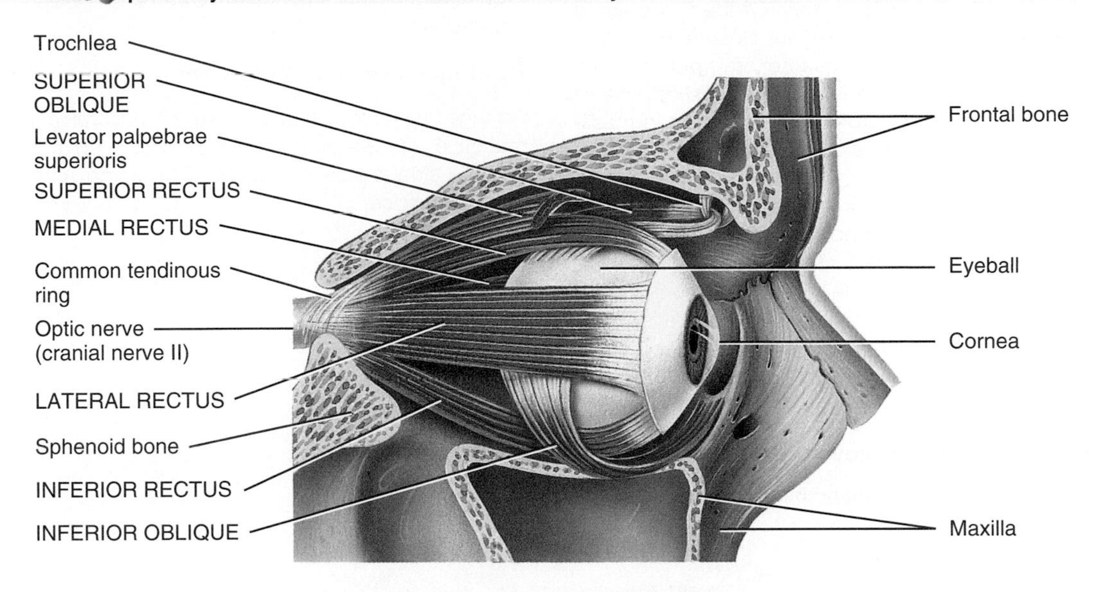

(a) Lateral view of right eyeball

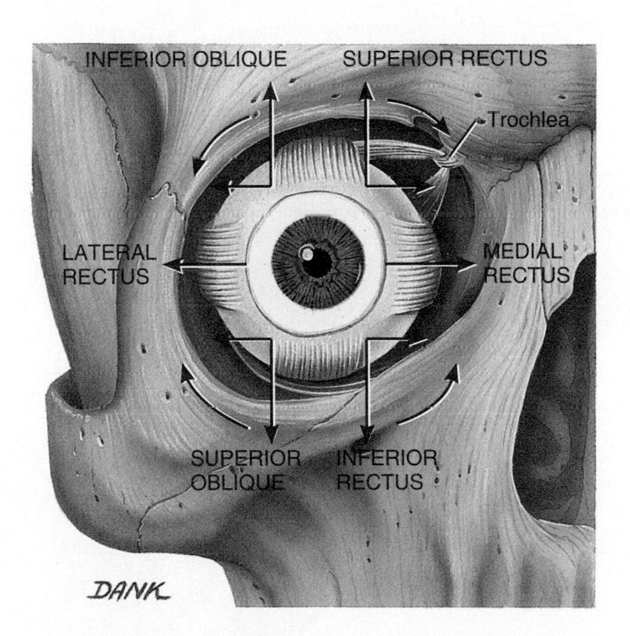

(b) Movements of right eyeball in response to contraction of extrinsic muscles

Why does the inferior oblique muscle move the eyeball superiorly and laterally?

Exhibit 11.3 Muscles that Move the Mandible (Lower Jaw Bone) (Figure 11.6)

▶ OBJECTIVE

- Describe the origin, insertion, action, and innervation of the muscles that move the mandible.

The muscles that move the mandible (lower jaw) at the temporomandibular joint (TMJ) are known as the muscles of mastication because they are involved in chewing (mastication). Of the four pairs of muscles involved in mastication, three are powerful closers of the jaw and account for the strength of the bite: **masseter, temporalis,** and **medial pterygoid.** Of these, the masseter is the strongest muscle of mastication. The medial and **lateral pterygoid** muscles assist in mastication by moving the mandible from side to side to help grind food. Additionally, these muscles protrude the mandible.

In 1996, researchers at the University of Maryland identified a new muscle in the skull and named it the **sphenomandibularis muscle.** It extends from the lateral surface of the sphenoid bone to the medial surface of the condylar process and ramus of the mandible (Figure 11.6b). The muscle is believed to be either a fifth muscle of mastication or a previously unidentified component of an already identified muscle (temporalis or medial pterygoid).

Innervation

- Masseter, temporalis, medial pterygoid, lateral pterygoid: mandibular division of the trigeminal nerve (cranial nerve V)
- Sphenomandibularis: maxillary branch of the trigeminal nerve (cranial nerve V)

Relating Muscles to Movements

Arrange the muscles in this exhibit according to their actions on the mandible: (1) elevation, (2) depression, (3) retraction, (4) protraction, and (5) side-to-side movement. The same muscle may be mentioned more than once.

▶ CHECKPOINT

What would happen if you lost tone in the masseter and temporalis muscles?

Figure 11.6 Muscles that move the mandible (lower jaw bone).

The muscles that move the mandible are also known as muscles of mastication.

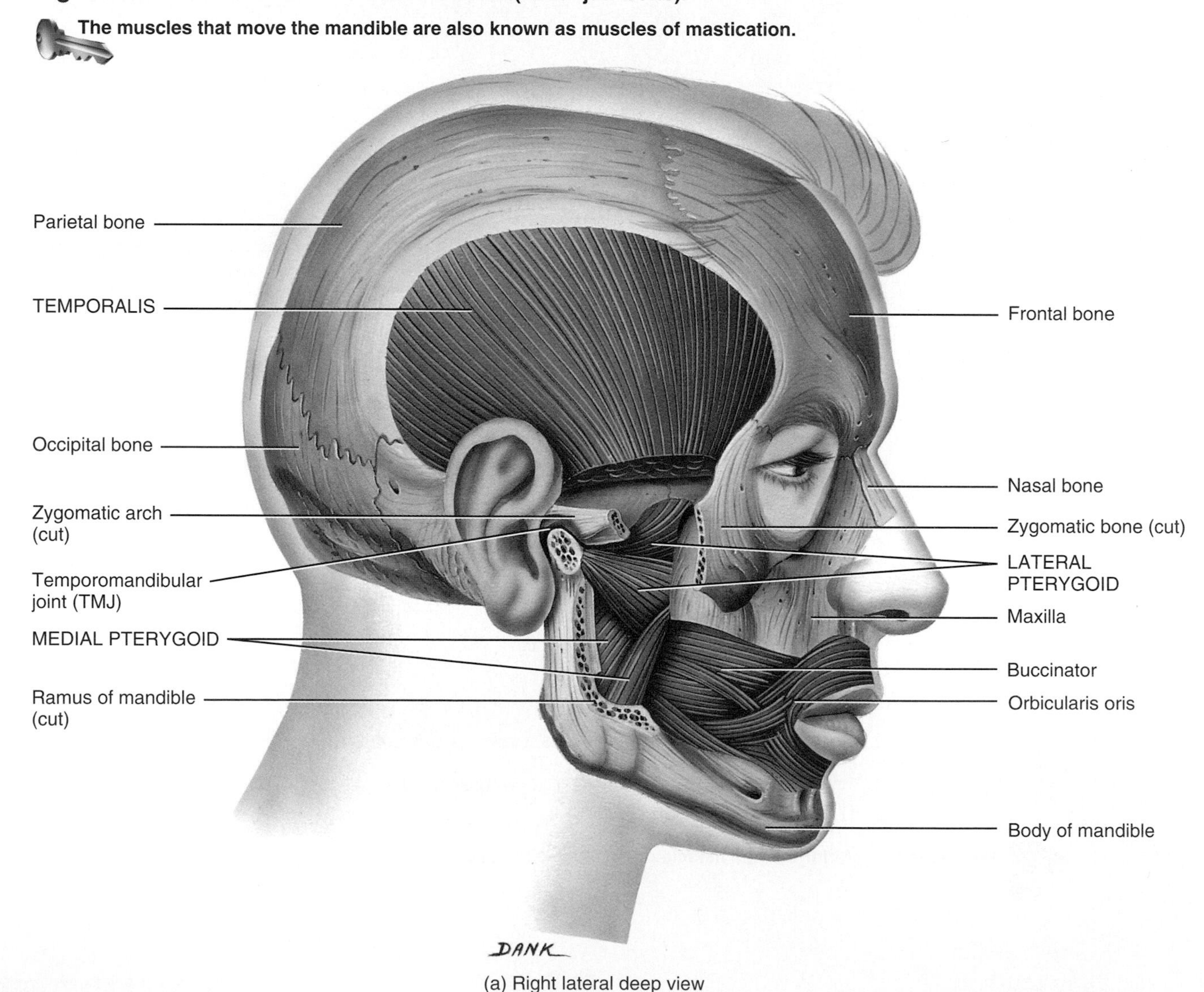

(a) Right lateral deep view

Muscle	Origin	Insertion	Action
Masseter (MA-se-ter = a chewer) (see Figure 11.4c)	Maxilla and zygomatic arch.	Angle and ramus of mandible.	Elevates mandible, as in closing mouth, and retracts (draws back) mandible.
Temporalis (tem′-pō-RĀ-lis; *tempor-* = temples)	Temporal bone.	Coronoid process and ramus of mandible.	Elevates and retracts mandible.
Medial pterygoyd (TER-i-goyd; *medial* = closer to midline; *pterygoid* = like a wing)	Medial surface of lateral portion of pterygoid process of sphenoid bone; maxilla.	Angle and ramus of mandible.	Elevates and protracts (protrudes) mandible and moves mandible from side to side.
Lateral pterygoid (TER-i-goyd; *lateral* = farther from midline)	Greater wing and lateral surface of lateral portion of pterygoid process of sphenoid bone.	Condyle of mandible; temporo-mandibular joint (TMJ).	Protracts mandible, depresses mandible as in opening mouth, and moves mandible from side to side.

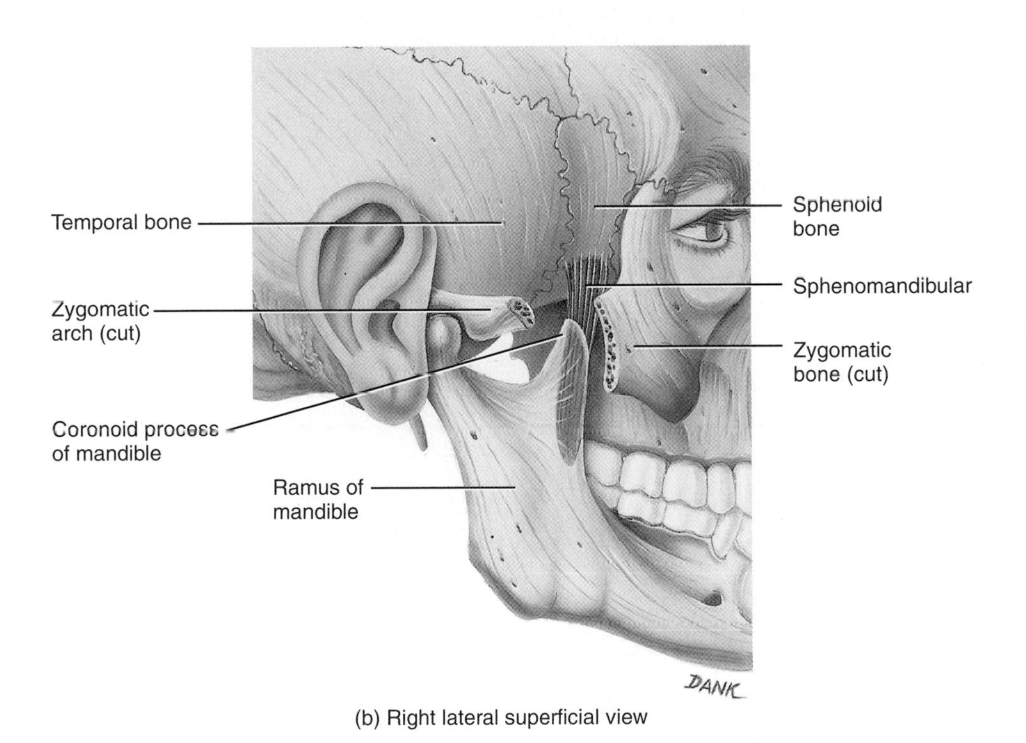

(b) Right lateral superficial view

Which is the strongest muscle of mastication?

Exhibit 11.4 Muscles that Move the Tongue—Extrinsic Tongue Muscles (Figure 11.7)

► **OBJECTIVE**

- Describe the origin, insertion, action, and innervation of the extrinsic muscles of the tongue.

The tongue is a highly mobile structure that is vital to digestive functions such as mastication, perception of taste, and deglutition (swallowing). It is also important in speech. The tongue's mobility is greatly aided by its suspension from the mandible, styloid process of the temporal bone, and hyoid bone.

The tongue is divided into lateral halves by a median fibrous septum. The septum extends throughout the length of the tongue. Inferiorly, the septum attaches to the hyoid bone. Muscles of the tongue are of two principal types: extrinsic and intrinsic. **Extrinsic tongue muscles** originate outside the tongue and insert into it. They move the entire tongue in various directions, such as anteriorly, posteriorly, and laterally. **Intrinsic tongue muscles** originate and insert within the tongue. These muscles alter the shape of the tongue rather than moving the entire tongue. The extrinsic and intrinsic muscles of the tongue insert into both lateral halves of the tongue.

When you study the extrinsic tongue muscles, you will notice that all of their names end in *glossus,* meaning tongue. You will also notice that the actions of the muscles are obvious, considering the positions of the mandible, styloid process, hyoid bone, and soft palate, which serve as origins for these muscles. For example, the **genioglossus** (originates on the mandible) pulls the tongue downward and forward, the **styloglossus** (originates on the styloid process) pulls the tongue upward and backward, the **hyoglossus** (originates on the hyoid bone) pulls the tongue downward and flattens it, and the **palatoglossus** (originates on the soft palate) raises the back portion of the tongue.

Innervation

- Genioglossus, styloglossus, and hyoglossus: hypoglossal nerve (cranial nerve XII)
- Palatoglossus: the pharyngeal plexus, which contains axons from both the vagus nerve (cranial nerve X) and the accessory nerve (cranial nerve XI)

Intubation During Anesthesia

When general anesthesia is administered during surgery, a total relaxation of the genioglossus muscle results. This causes the tongue to fall posteriorly, which may obstruct the airway to the lungs. To avoid this, the mandible is either manually thrust forward and held in place, or a tube is inserted from the lips through the laryngopharynx (inferior portion of the throat) into the trachea (endotracheal intubation). ■

Relating Muscles to Movements

Arrange the muscles in this exhibit according to the following actions on the tongue: (1) depression, (2) elevation, (3) protraction, and (4) retraction. The same muscle may be mentioned more than once.

► **CHECKPOINT**

When your physician says, "Open your mouth, stick out your tongue, and say *ahh,*" so she can examine the inside of your mouth for possible signs of infection, which muscles do you contract?

Muscle	Origin	Insertion	Action
Genioglossus (jē′-nē-ō-GLOS-us; *genio-* = the chin; *glossus* = tongue)	Mandible.	Undersurface of tongue and hyoid bone.	Depresses tongue and thrusts it anteriorly (protraction).
Styloglossus (stī′-lō-GLOS-us; *stylo* = stake or pole; styloid process of temporal bone)	Styloid process of temporal bone.	Side and undersurface of tongue.	Elevates tongue and draws it posteriorly (retraction).
Palatoglossus (pal′-a-tō-GLOS- us; *palato-* = the roof of the mouth or palate)	Anterior surface of soft palate.	Side of tongue.	Elevates posterior portion of tongue and draws soft palate down on tongue.
Hyoglossus (hī′-ō-GLOS-us)	Greater horn and body of hyoid bone.	Side of tongue.	Depresses tongue and draws down its sides.

Figure 11.7 Muscles that move the tongue.

The extrinsic and intrinsic muscles of the tongue are arranged in both lateral halves of the tongue.

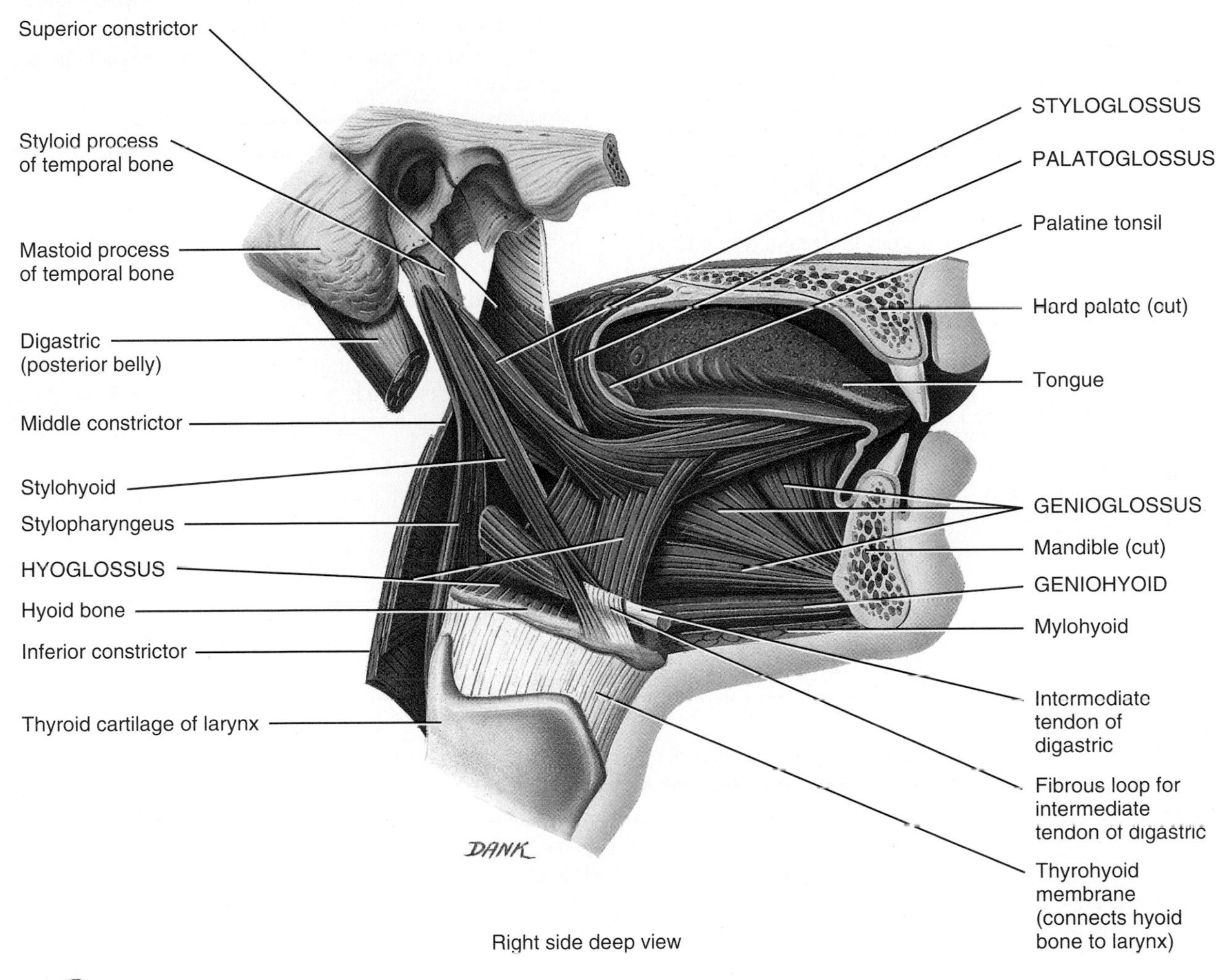

Right side deep view

What are the functions of the tongue?

Exhibit 11.5 Muscles of the Anterior Neck (Figure 11.8)

▶ **OBJECTIVE**

- Describe the origin, insertion, action, and innervation of the muscles of the anterior neck.

Two groups of muscles are associated with the anterior aspect of the neck: (1) the **suprahyoid muscles,** so-called because they are located superior to the hyoid bone, and (2) the **infrahyoid muscles,** so-named because they lie inferior to the hyoid bone. Acting with the infrahyoid muscles, the suprahyoid muscles stabilize the hyoid bone. Thus, the hyoid bone serves as a firm base upon which the tongue can move.

As a group, the suprahyoid muscles elevate the hyoid bone, floor of the oral cavity, and tongue during swallowing. As its name suggests, the **digastric** muscle has two bellies, anterior and posterior, united by an intermediate tendon that is held in position by a fibrous loop (see Figure 11.7). This muscle elevates the hyoid bone and larynx (voice box) during swallowing and speech and depresses the mandible. The **stylohyoid** muscle elevates and draws the hyoid bone posteriorly, thus elongating the floor of the oral cavity during swallowing. The **mylohyoid** muscle elevates the hyoid bone and helps press the tongue against the roof of the oral cavity during swallowing to move food from the oral cavity into the throat. The **geniohyoid** muscle (see Figure 11.7) elevates and draws the hyoid bone anteriorly to shorten the floor of the oral cavity and to widen the throat to receive food that is being swallowed. It also depresses the mandible.

The infrahyoid muscles are sometimes called "strap" muscles because of their ribbonlike appearance. Most of the infrahyoid muscles depress the hyoid bone and some move the larynx during swallowing and speech. The **omohyoid** muscle, like the digastric muscle, is composed of two bellies connected by an intermediate tendon. In this case, however, the two bellies are referred to as *superior* and *inferior,* rather than anterior and posterior. Together, the omohyoid, **sternohyoid,** and **thyrohyoid** muscles depress the hyoid bone. In addition, the **sternothyroid** muscle depresses the thyroid cartilage (Adam's apple) of the larynx, whereas the thyrohyoid muscle elevates the thyroid cartilage. The actions are necessary during the production of low and high tones, respectively, during phonation.

Innervation

- Anterior belly of the digastric and the mylohyoid muscles: mandibular division of the trigeminal nerve (cranial nerve V)
- Posterior belly of the digastric and the stylohyoid: facial nerve (cranial nerve VII)
- Geniohyoid muscle: first cervical spinal nerve
- Omohyoid, sternohyoid, and sternothyroid: branches of ansa cervicalis (spinal nerves C1–C3)
- Thyrohyoid: branches of ansa cervicalis (spinal nerves C1–C2) and descending hypoglossal nerve (cranial nerve XII)

Relating Muscles to Movements

Arrange the muscles in this exhibit according to the following actions on the hyoid bone: (1) elevating it, (2) drawing it anteriorly, (3) drawing it posteriorly, and (4) depressing it; and on the thyroid cartilage: (1) elevating it and (2) depressing it. The same muscle may be mentioned more than once.

▶ **CHECKPOINT**

Which tongue, facial, and mandibular muscles do you use for chewing?

Muscle	Origin	Insertion	Action
Suprahyoid muscles			
Digastric (dī′-GAS-trik; *di-* = two; *gastr-* = belly)	Anterior belly from inner side of inferior border of mandible; posterior belly from mastoid process of temporal bone.	Body of hyoid bone via an intermediate tendon.	Elevates hyoid bone and depresses mandible, as in opening the mouth.
Stylohyoid (stī′-lō-HĪ-oid; *stylo* = stake or pole, styloid process of temporal bone; *hyo-* = U-shaped, pertaining to hyoid bone)	Styloid process of temporal bone.	Body of hyoid bone.	Elevates hyoid bone and draws it posteriorly.
Mylohyoid (mī′-lō-HĪ-oid) (*mylo-* = mill)	Inner surface of mandible.	Body of hyoid bone.	Elevates hyoid bone and floor of mouth and depresses mandible.
Geniohyoid (jē′-nē-ō-HĪ-oid; *genio* = chin) (see also Figure 11.7)	Inner surface of mandible.	Body of hyoid bone.	Elevates hyoid bone, draws hyoid bone and tongue anteriorly, and depresses mandible.
Infrahyoid muscles			
Omohyoid (ō-mō-HĪ-oid; *omo-* = relationship to the shoulder)	Superior border of scapula and superior transverse ligament.	Body of hyoid bone.	Depresses hyoid bone.
Sternohyoid (ster′-no-HĪ-oid; *sterno-* = sternum)	Medial end of clavicle and manubrium of sternum.	Body of hyoid bone.	Depresses hyoid bone.
Sternothyroid (ster′-nō-THĪ-roid; *thyro-* = thyroid gland)	Manubrium of sternum.	Thyroid cartilage of larynx.	Depresses thyroid cartilage of larynx.
Thyrohyoid (thī′-rō-HĪ-oid)	Thyroid cartilage of larynx.	Greater horn of hyoid bone.	Elevates thyroid cartilage and depresses hyoid bone.

Figure 11.8 Muscles of the anterior neck.

The suprahyoid muscles elevate the hyoid bone, the floor of the oral cavity, and the tongue during swallowing.

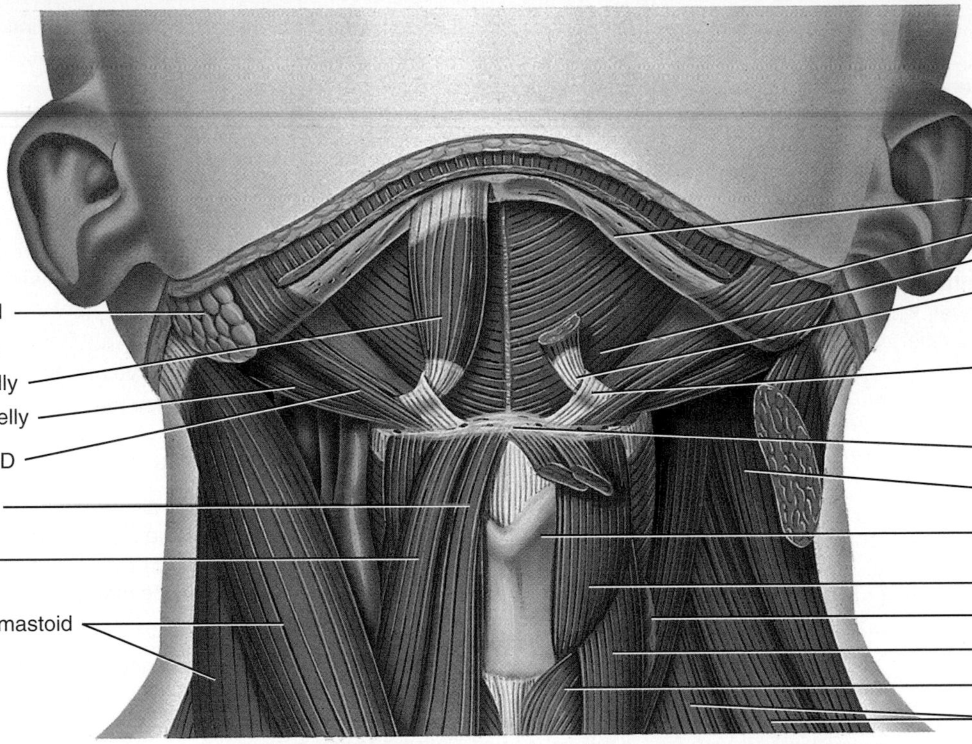

(a) Anterior superficial view (b) Anterior deep view

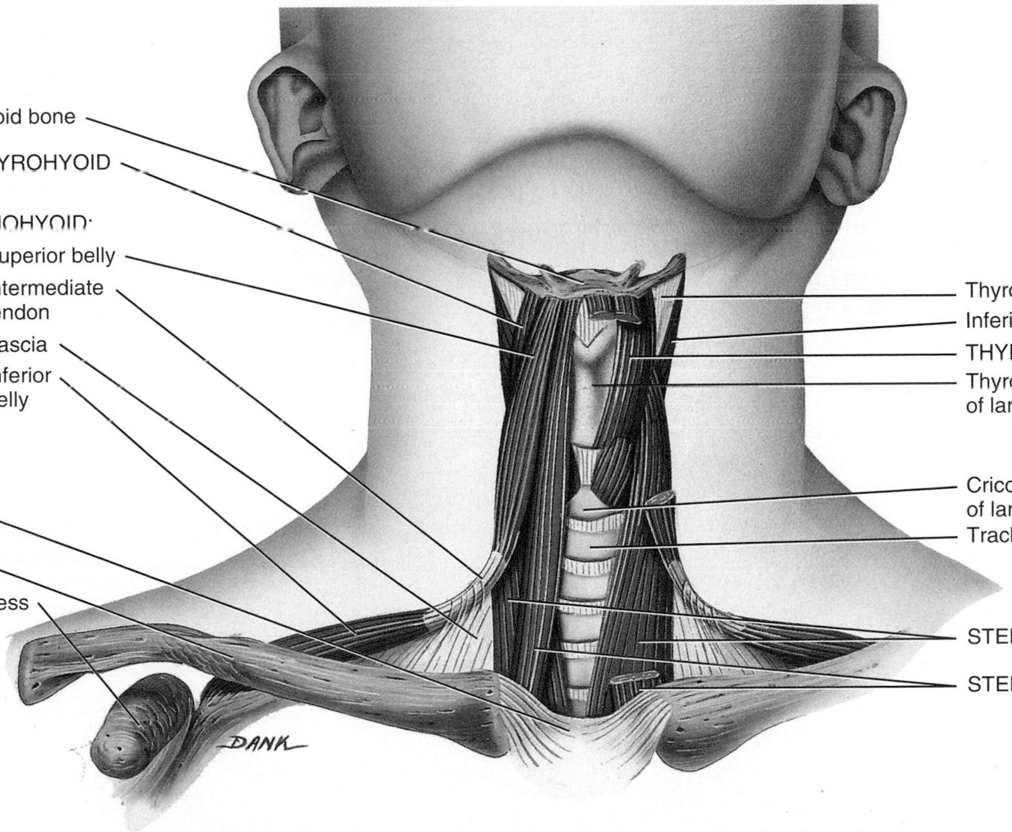

Anterior superficial view (c) Anterior deep view

What is the combined action of the suprahoid and infrahyoid muscles?

Exhibit 11.6 Muscles that Move the Head (Figure 11.9)

▶ OBJECTIVE

- Describe the origin, insertion, action, and innervation of the muscles that move the head.

The head is attached to the vertebral column at the atlanto-occipital joint formed by the atlas and occipital bone. Balance and movement of the head on the vertebral column involves the action of several neck muscles. For example, contraction of the two **sternocleidomastoid** muscles together (bilaterally) flexes the cervical portion of the vertebral column and extends the head. Acting singly (unilaterally), the sternocleidomastoid muscle laterally flexes and rotates the head. Bilateral contraction of the **semispinalis capitis, splenius capitis,** and **longissimus capitis** muscles extends the head. However, when these same muscles contract unilaterally, their actions are quite different, involving primarily rotation of the head.

The sternocleidomastoid muscle is an important landmark that divides the neck into two major triangles: anterior and posterior. The triangles are important because of the structures that lie within their boundaries.

The **anterior triangle** is bordered superiorly by the mandible, inferiorly by the sternum, medially by the cervical midline, and laterally by the anterior border of the sternocleidomastoid muscle. The anterior triangle is subdivided into an unpaired submental triangle and three paired triangles: submandibular, carotid, and muscular. The anterior triangle contains submental, submandibular, and deep cervical lymph nodes; the submandibular salivary gland and a portion of the parotid salivary gland; the facial artery and vein; carotid arteries and internal jugular vein; and cranial nerves IX (glossopharyngeal), X (vagus), XI (accessory), and XII (hypoglossal).

The **posterior triangle** is bordered inferiorly by the clavicle, anteriorly by the posterior border of the sternocleidomastoid muscle, and posteriorly by the anterior border of the trapezius muscle. The posterior triangle is subdivided into two triangles, occipital and supraclavicular, by the inferior belly of the omohyoid muscle. The posterior triangle contains part of the subclavian artery, external jugular vein, cervical lymph nodes, brachial plexus, and the accessory nerve (cranial nerve XI).

Innervation

- Sternocleidomastoid: accessory nerve (cranial nerve XI)
- All capitis muscles: cervical spinal nerves

Relating Muscles to Movements

Arrange the muscles in this exhibit according to the following actions on the head: (1) flexion, (2) lateral flexion, (3) extension, (4) rotation to side opposite contracting muscle, and (5) rotation to same side as contracting muscle. The same muscle may be mentioned more than once.

▶ CHECKPOINT

What muscles do you contract to signify "yes" and "no"?

Muscle	Origin	Insertion	Action
Sternocleidomastoid (ster′-nō-klī′-dō-MAS-toid; *sterno-* = breastbone; *cleido-* = clavicle; *mastoid* = mastoid process of temporal bone)	Sternum and clavicle.	Mastoid process of temporal bone.	Acting together (bilaterally), flex cervical portion of vertebral column, extend head, and elevate sternum during forced inhalation; acting singly (unilaterally), laterally flex and rotate head to side opposite contracting muscle.
Semispinalis capitis (se′-mē-spi-NĀ-lis KAP-i-tis; *semi-* = half; *spine* = spinous process; *capit-* = head) (see Figure 11.19a)	Transverse processes of first six or seven thoracic vertebrae and seventh cervical vertebra, and articular processes of fourth, fifth, and sixth cervical vertebrae.	Occipital bone between superior and inferior nuchal lines.	Acting together, extend head; acting singly, rotate head to side opposite contracting muscle.
Splenius capitis (SPLĒ-nē-us KAP-i-tis; *splen-* = bandage) (see Figure 11.19a)	Ligamentum nuchae and spinous processes of seventh cervical vertebra and first three or four thoracic vertebrae.	Occipital bone and mastoid process of temporal bone.	Acting together, extend head; acting singly, laterally flex and rotate head to same side as contracting muscle.
Longissimus capitis (lon-JIS-i-mus KAP-i-tis; *longissimus* = longest) (see Figure 11.19a)	Transverse processes of upper four thoracic vertebrae and articular processes of last four cervical vertebrae.	Mastoid process of temporal bone.	Acting together, extend head; acting singly, laterally flex and rotate head to same side as contracting muscle.

Figure 11.9 Triangles of the neck.

The sternocleidomastoid muscle divides the neck into two principal triangles: anterior and posterior.

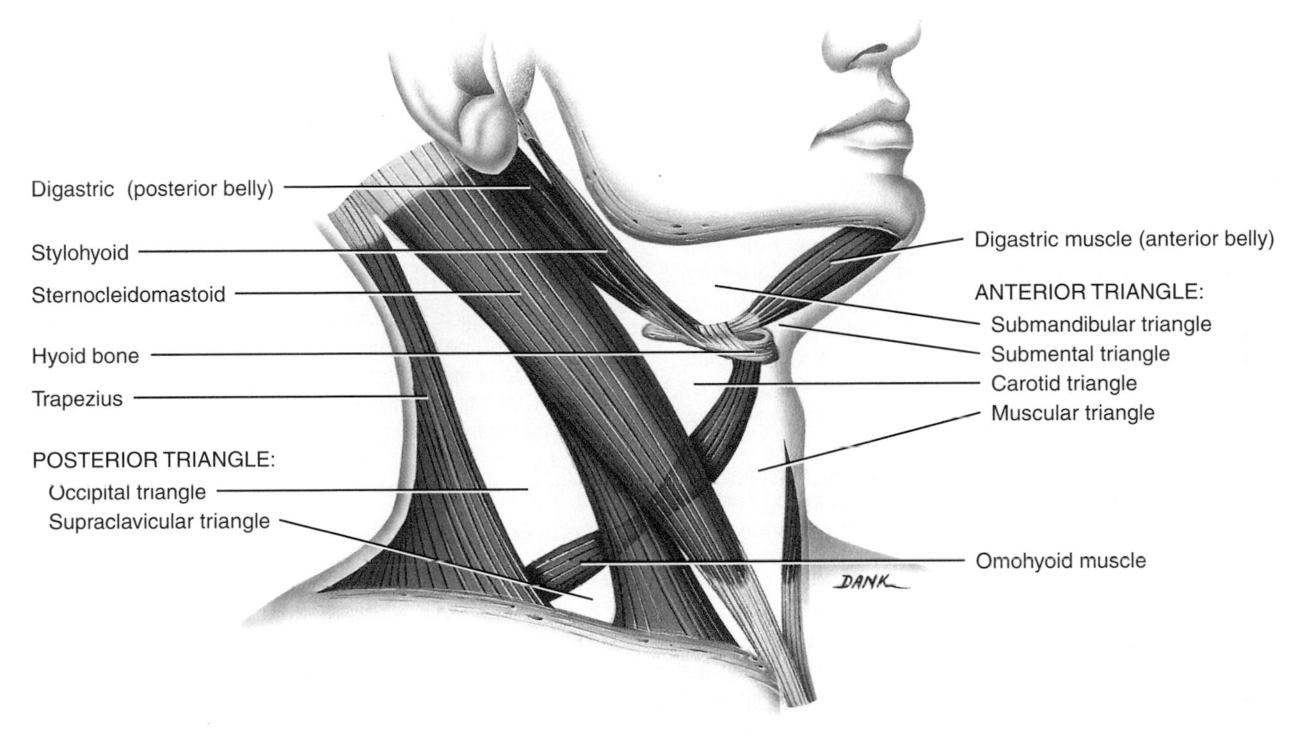

Right lateral view

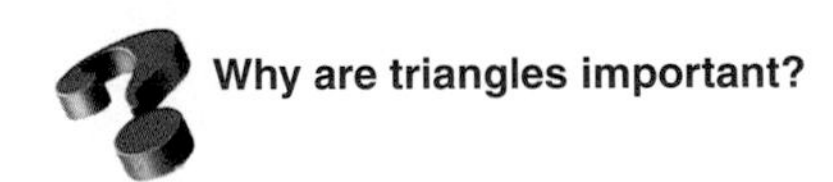

Why are triangles important?

Exhibit 11.7 Muscles that Act on the Abdominal Wall (Figure 11.10)

▶ OBJECTIVE

- Describe the origin, insertion, action, and innervation of the muscles that act on the abdominal wall.

The anterolateral abdominal wall is composed of skin, fascia, and four pairs of muscles: the external oblique, internal oblique, transversus abdominis, and rectus abdominis. The first three muscles are arranged from superficial to deep. The **external oblique** is the superficial muscle. Its fascicles extend inferiorly and medially. The **internal oblique** is the intermediate flat muscle. Its fascicles extend at right angles to those of the external oblique. The **transversus abdominis** is the deep muscle, with most of its fascicles directed transversely around the abdominal wall. Together, the external oblique, internal oblique, and transversus abdominis form three layers of muscle around the abdomen. In each layer, the muscle fascicles extend in a different direction. This is a structural arrangement that affords considerable protection to the abdominal viscera, especially when the muscles have good tone.

The **rectus abdominis** muscle is a long muscle that extends the entire length of the anterior abdominal wall, from the pubic crest and pubic symphysis to the cartilages of ribs 5–7 and the xiphoid process of the sternum. The anterior surface of the muscle is interrupted by three transverse fibrous bands of tissue called **tendinous intersections,** believed to be remnants of septa that separated myotomes during embryological development (see Figure 10.20b on page 301).

As a group, the muscles of the anterolateral abdominal wall help contain and protect the abdominal viscera; flex, laterally flex, and rotate the vertebral column at the intervertebral joints; compress the abdomen during forced exhalation; and produce the force required for defecation, urination, and childbirth.

The aponeuroses of the external oblique, internal oblique, and transversus abdominis muscles form the **rectus sheaths,** which enclose the rectus abdominis muscles. The sheaths meet at the midline to form the **linea alba** (= white line), a tough, fibrous band that extends from the xiphoid process of the sternum to the pubic symphysis. In the latter stages of pregnancy, the linea alba stretches to increase the distance between the rectus abdominis muscles. The inferior free border of the external oblique aponeurosis, plus some collagen fibers, forms the **inguinal ligament,** which runs from the anterior superior iliac spine to the pubic tubercle (see Figure 11.20a). Just superior to the medial end of the inguinal ligament is a triangular slit in the aponeurosis referred to as the **superficial inguinal ring,** the outer opening of the **inguinal canal** (see Figure 28.2). The canal contains the spermatic cord and ilioinguinal nerve in males, and the round ligament of the uterus and ilioinguinal nerve in females.

The posterior abdominal wall is formed by the lumbar vertebrae, parts of the ilia of the hip bones, psoas major and iliacus muscles (described in Exhibit 11.17), and quadratus lumborum muscle. Whereas the anterolateral abdominal wall can contract and distend, the posterior abdominal wall is bulky and stable by comparison.

Innervation

- Rectus abdominis: thoracic spinal nerves T7–T12
- External oblique: thoracic spinal nerves T7–T12 and the iliohypogastric nerve
- Internal oblique and transversus abdominis: thoracic spinal nerves T8–T12, the iliohypogastric nerve, and the ilioinguinal nerve
- Quadratus lumborum: thoracic spinal nerve T12 and lumbar spinal nerves L1–L3 or L1–L4

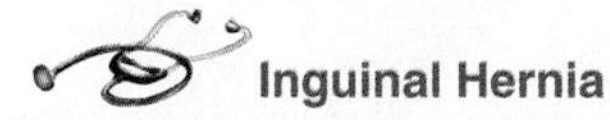

Inguinal Hernia

The inguinal region is a weak area in the abdominal wall. It is often the site of an **inguinal hernia,** a rupture or separation of a portion of the inguinal area of the abdominal wall resulting in the protrusion of a part of the small intestine. Hernia is much more common in males than in females because the inguinal canals in males are larger to accommodate the spermatic cord and ilioinguinal nerve. They are especially weak points in the abdominal wall. ■

Relating Muscles to Movements

Arrange the muscles in this exhibit according to the following actions on the vertebral column: (1) flexion, (2) lateral flexion, (3) extension, and (4) rotation. The same muscle may be mentioned more than once.

▶ CHECKPOINT

Which muscles do you contract when you "suck in your tummy," thereby compressing the anterior abdominal wall?

Muscle	Origin	Insertion	Action
Rectus abdominis (REK-tus ab-DOM-in-is; *rectus* = fascicles parallel to midline; *abdomin* = abdomen)	Pubic crest and pubic symphysis.	Cartilage of fifth to seventh ribs and xiphoid process.	Flexes vertebral column, especially lumbar portion, and compresses abdomen to aid in defecation, urination, forced exhalation, and childbirth.
External oblique (ō-BLĒK; *external* = closer to surface; *oblique* = fascicles diagonal to midline)	Inferior eight ribs.	Iliac crest and linea alba.	Acting together (bilaterally), compress abdomen and flex vertebral column; acting singly (unilaterally), laterally flex vertebral column, especially lumbar portion, and rotate vertebral column.
Internal oblique (ō-BLĒK; *internal* = farther from surface)	Iliac crest, inguinal ligament, and thoracolumbar fascia.	Cartilage of last three or four ribs and linea alba.	Acting together, compress abdomen and flex vertebral column; acting singly, laterally flex vertebral column, especially lumbar portion, and rotate vertebral column.
Transversus abdominis (tranz-VER-sus ab-DOM-in-is; *transverse* = fascicles perpendicular to midline)	Iliac crest, inguinal ligament, lumbar fascia, and cartilages of inferior six ribs.	Xiphoid process, linea alba, and pubis.	Compresses abdomen.
Quadratus lumborum (kwod-RĀ-tus lum-BOR-um; *quad* = four; *lumbo-* = lumbar region) (see Figure 11.11)	Iliac crest and iliolumbar ligament.	Inferior border of twelfth rib and transverse processes of first four lumbar vertebrae.	Acting together, pull twelfth ribs inferiorly during forced exhalation, fix twelfth ribs to prevent their elevation during deep inhalation, and help extend lumbar portion of vertebral column; acting singly, laterally flex vertebral column, especially lumbar portion.

(continues)

Exhibit 11.7 Muscles that Act on the Abdominal Wall (Figure 11.10) *(continued)*

Figure 11.10 Muscles of the male anterolateral abdominal wall. (See Tortora, *A Photographic Atlas of the Human Body,* Figure 5.7.)

The anterolateral abdominal muscles protect the abdominal viscera, move the vertebral column, and assist in forced exhalation, defecation, urination, and childbirth.

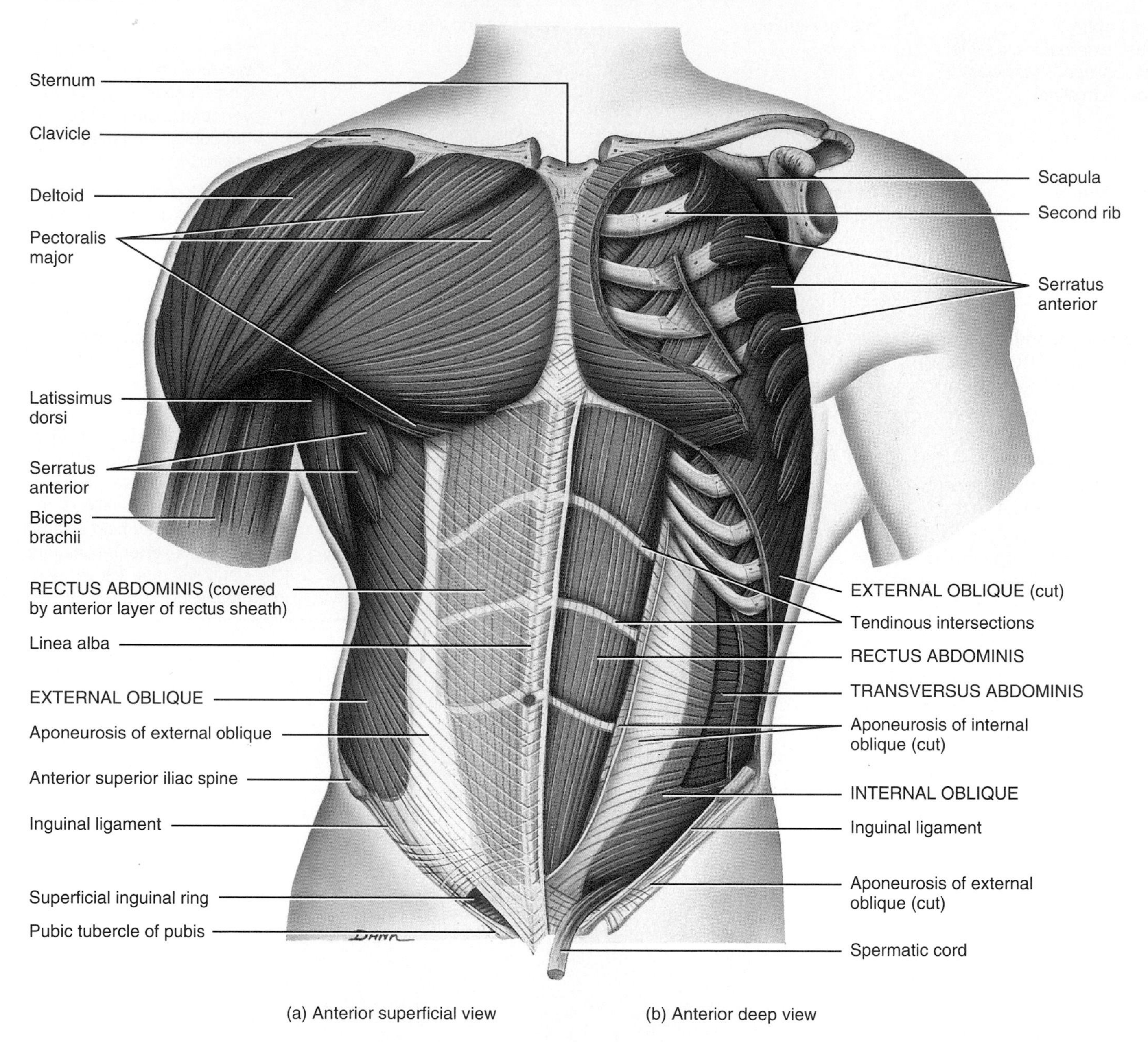

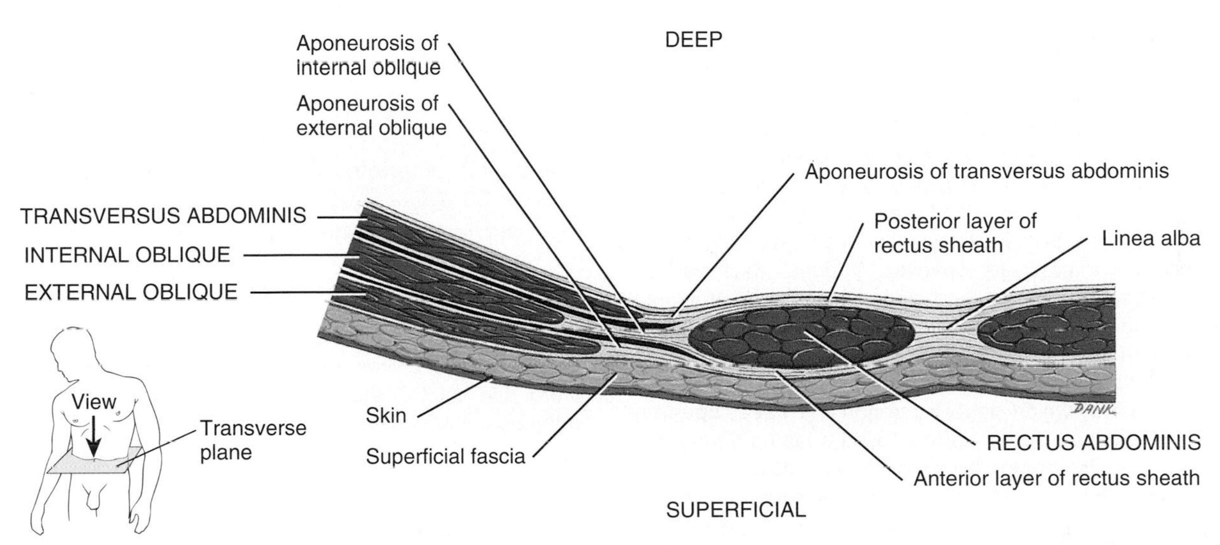

(c) Transverse section of anterior abdominal wall superior to umbilicus (navel)

Which abdominal muscle aids in urination?

Exhibit 11.8 Muscles Used in Breathing (Figure 11.11)

▶ OBJECTIVE

- Describe the origin, insertion, action, and innervation of the muscles used in breathing.

The muscles described here alter the size of the thoracic cavity so that breathing can occur. Inhalation (breathing in) occurs when the thoracic cavity increases in size, and exhalation (breathing out) occurs when the thoracic cavity decreases in size.

The dome-shaped **diaphragm** is the most important muscle that powers breathing. It also separates the thoracic and abdominal cavities. The diaphragm is composed of a peripheral muscular portion and a central portion called the central tendon. The **central tendon** is a strong aponeurosis that serves as the tendon of insertion for all peripheral muscular fibers of the diaphragm. It fuses with the inferior surface of the fibrous pericardium (external covering of the heart) and the parietal pleurae (external coverings of the lungs).

Movements of the diaphragm also help return venous blood passing through the abdomen to the heart. Together with the anterolateral abdominal muscles, the diaphragm helps to increase intra-abdominal pressure to evacuate the pelvic contents during defecation, urination, and childbirth. This mechanism is further assisted when you take a deep breath and close the rima glottidis (the space between vocal folds). The trapped air in the respiratory system prevents the diaphragm from elevating. The increase in intra-abdominal pressure as just described will also help support the vertebral column and prevent flexion during weight lifting. This greatly assists the back muscles in lifting a heavy weight.

The diaphragm has three major openings through which various structures pass between the thorax and abdomen. These structures include the aorta, along with the thoracic duct and azygos vein, which pass through the **aortic hiatus;** the esophagus with accompanying vagus (X) cranial nerves, which pass through the **esophageal hiatus;** and the inferior vena cava, which passes through the **caval opening (foramen for the vena cava).** In a condition called a hiatus hernia, the stomach protrudes superiorly through the esophageal hiatus.

Other muscles involved in breathing are called **intercostal** muscles. They span the intercostal spaces, the spaces between ribs. These muscles are arranged in three layers. The 11 **external intercostal muscles** occupy the superficial layer, and their fibers run obliquely inferiorly and anteriorly from the rib above to the rib below. They elevate the ribs during inhalation to help expand the thoracic cavity. The 11 **internal intercostal muscles** occupy the intermediate layer of the intercostal spaces. The fibers of these muscles run obliquely inferiorly and posteriorly from the inferior border of the rib above to the superior border of the rib below. They draw adjacent ribs together during forced exhalation to help decrease the size of the thoracic cavity. The deepest muscle layer is made up of the **transversus thoracis.** These poorly developed muscles (not illustrated) extend in the same direction as the internal intercostals and may have the same role.

As you will see in Chapter 23, the diaphragm and external intercostal muscles are used during quiet inhalation and exhalation. However, during deep, forceful inhalation, the sternocleidomastoid, scalene, and pectoralis minor muscles are also used and during deep, forceful exhalation, the external oblique, internal oblique, transversus abdominis, rectus abdominis, and internal intercostals are also used.

Innervation

- Diaphragm: phrenic nerve, which contains axons from cervical spinal nerves C3–C5
- External and internal intercostals: thoracic spinal nerves T2–T12

Relating Muscles to Movements

Arrange the muscles in this exhibit according to the following actions on the size of the thorax: (1) increase in vertical length, (2) increase in lateral and anteroposterior dimensions, and (3) decrease in lateral and anteroposterior dimensions.

▶ CHECKPOINT

What are the names of the three openings in the diaphragm, and which structures pass through each?

Muscle	Origin	Insertion	Action
Diaphragm (DĪ-a-fram; *dia* = across; *-phragm* = wall)	Xiphoid process of the sternum, costal cartilages of inferior six ribs, and lumbar vertebrae.	Central tendon.	Contraction of the diaphragm causes it to flatten and increases the vertical dimension of the thoracic cavity, resulting in inhalation; relaxation of the diaphragm causes it to move superiorly and decreases the vertical dimension of the thoracic cavity, resulting in exhalation.
External intercostals (in′-ter-KOS-tals; *external* = closer to surface; *inter-* = between; *costa* = rib)	Inferior border of rib above.	Superior border of rib below.	Contraction elevates the ribs and increases the anteroposterior and lateral dimensions of the thoracic cavity, resulting in inhalation; relaxation depresses the ribs and decreases the anteroposterior and lateral dimensions of the thoracic cavity, resulting in exhalation.
Internal intercostals (in′-ter-KOS-tals; *internal* = father from surface)	Superior border of rib below.	Inferior border of rib above.	Contraction draws adjacent ribs together to further decrease the anteroposterior and lateral dimensions of the thoracic cavity during forced exhalation.

Figure 11.11 Muscles used in breathing, as seen in a male.

Openings in the diaphragm permit passage of the aorta, esophagus, and inferior vena cava.

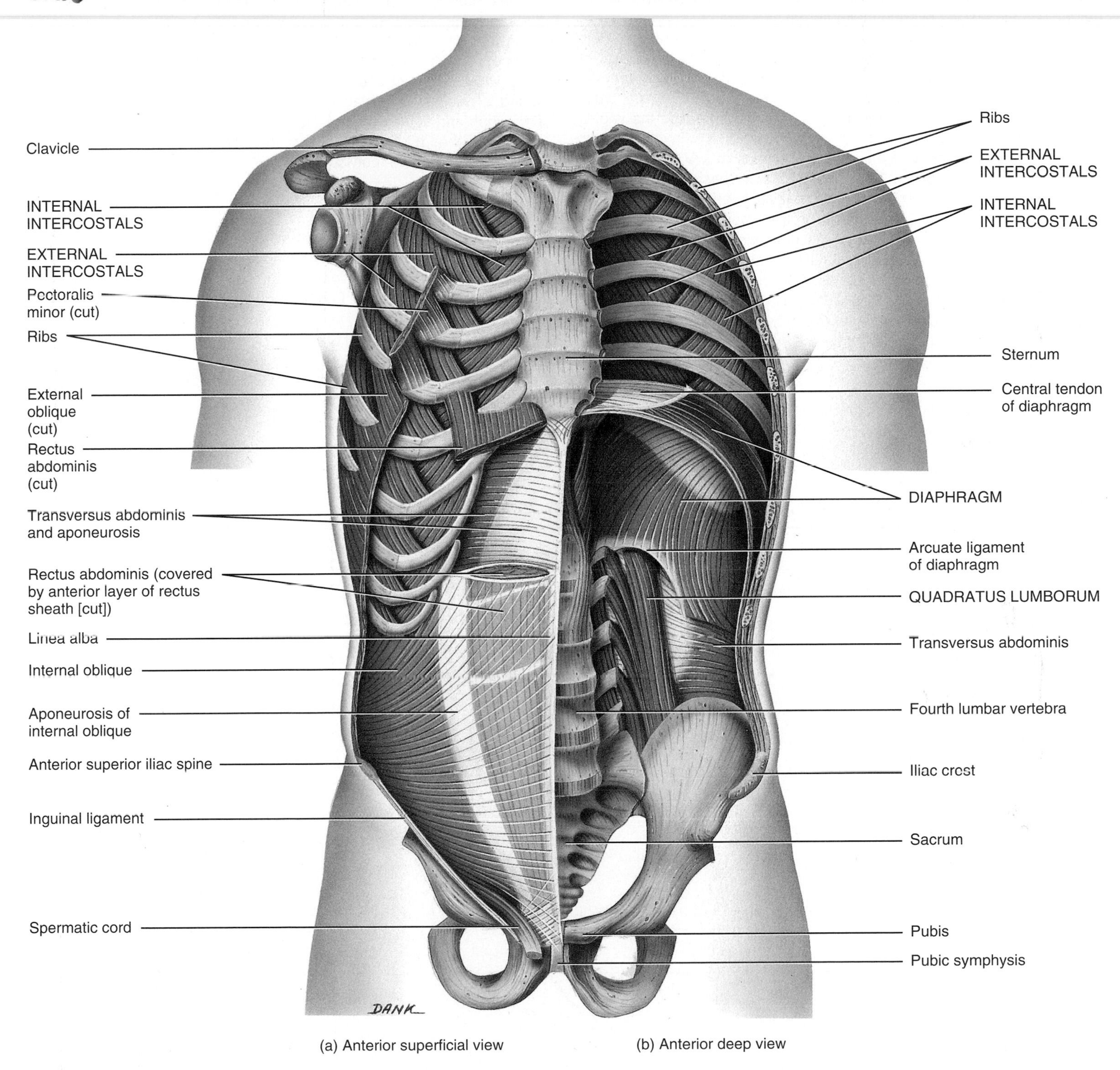

(a) Anterior superficial view (b) Anterior deep view

Which muscle associated with breathing is innervated by the phrenic nerve?

Exhibit 11.9 Muscles of the Pelvic Floor (Figure 11.12)

▶ OBJECTIVE

- Describe the origin, insertion, action, and innervation of the muscles of the pelvic floor.

The muscles of the pelvic floor are the levator ani and coccygeus. Together with the fascia covering their internal and external surfaces, these muscles are referred to as the **pelvic diaphragm,** which stretches from the pubis anteriorly to the coccyx posteriorly, and from one lateral wall of the pelvis to the other. This arrangement gives the pelvic diaphragm the appearance of a funnel suspended from its attachments. The anal canal and urethra pierce the pelvic diaphragm in both sexes, and the vagina goes through it in females.

The two components of the **levator ani** muscle are the **pubococcygeus** and **iliococcygeus.** Figure 11.12 shows these muscles in the female and Figure 11.13 in the male. The levator ani is the largest and most important muscle of the pelvic floor. It supports the pelvic viscera and resists the inferior thrust that accompanies increases in intra-abdominal pressure during functions such as forced exhalation, coughing, vomiting, urination, and defecation. The muscle also functions as a sphincter at the anorectal junction, urethra, and vagina. In addition to assisting the levator ani, the **coccygeus** pulls the coccyx anteriorly after it has been pushed posteriorly during defecation or childbirth.

Innervation

- Pubococcygeus and iliococcygeus: sacral spinal nerves S2–S4
- Coccygeus: sacral spinal nerves S4–S5

Injury of Levator Ani and Stress Incontinence

During childbirth, the levator ani muscle supports the head of the fetus, and the muscle may be injured during a difficult childbirth or traumatized during an episiotomy (a cut made with surgical scissors to prevent or direct tearing of the perineum during the birth of a baby). The consequence of such injury may be urinary stress incontinence, leakage of urine whenever intra-abdominal pressure is increased—for example, during coughing. One way to treat urinary stress incontinence is to strengthen and tighten the muscles that support the pelvic viscera. This is accomplished by *Kegel exercises,* the alternate contraction and relaxation of muscles of the pelvic floor. ■

Relating Muscles to Movements

Arrange the muscles in this exhibit according to the following actions on the pelvic viscera: supporting and maintaining their position, and resisting an increase in intra-abdominal pressure; and according to the following action on the anus, urethra, and vagina: constriction. The same muscle may be mentioned more than once.

▶ CHECKPOINT

Which muscles are strengthened by Kegel exercises?

Muscle	Origin	Insertion	Action
Levator ani (le-VĀ-tor Ā-nē; *levator* = raises; *ani* = anus)	This muscle is divisible into two parts, the pubococcygeus muscle and the iliococcygeus muscle.		
Pubococcygeus (pū′-bō-kok-SIJ-ē-us; *pubo-* = pubis; *coccygeus* = coccyx)	Pubis.	Coccyx, urethra, anal canal, central tendon of perineum, and anococcygeal raphe (narrow fibrous band that extends from anus to coccyx).	Supports and maintains position of pelvic viscera; resists increase in intra-abdominal pressure during forced exhalation, coughing, vomiting, urination, and defecation; constricts anus, urethra, and vagina; and supports fetal head during childbirth.
Iliococcygeus (il′-ē-ō-kok-SIJ-ē-us; *ilio-* = ilium)	Ischial spine.	Coccyx.	As above.
Coccygeus (kok-SIJ-ē-us)	Ischial spine.	Lower sacrum and upper coccyx.	Supports and maintains position of pelvic viscera; resists increase in intra-abdominal pressure during forced exhalation, coughing, vomiting, urination, and defecation; and pulls coccyx anteriorly following defecation or childbirth.

Figure 11.12 Muscles of the pelvic floor, as seen in the female perineum.

The pelvic diaphragm supports the pelvic viscera.

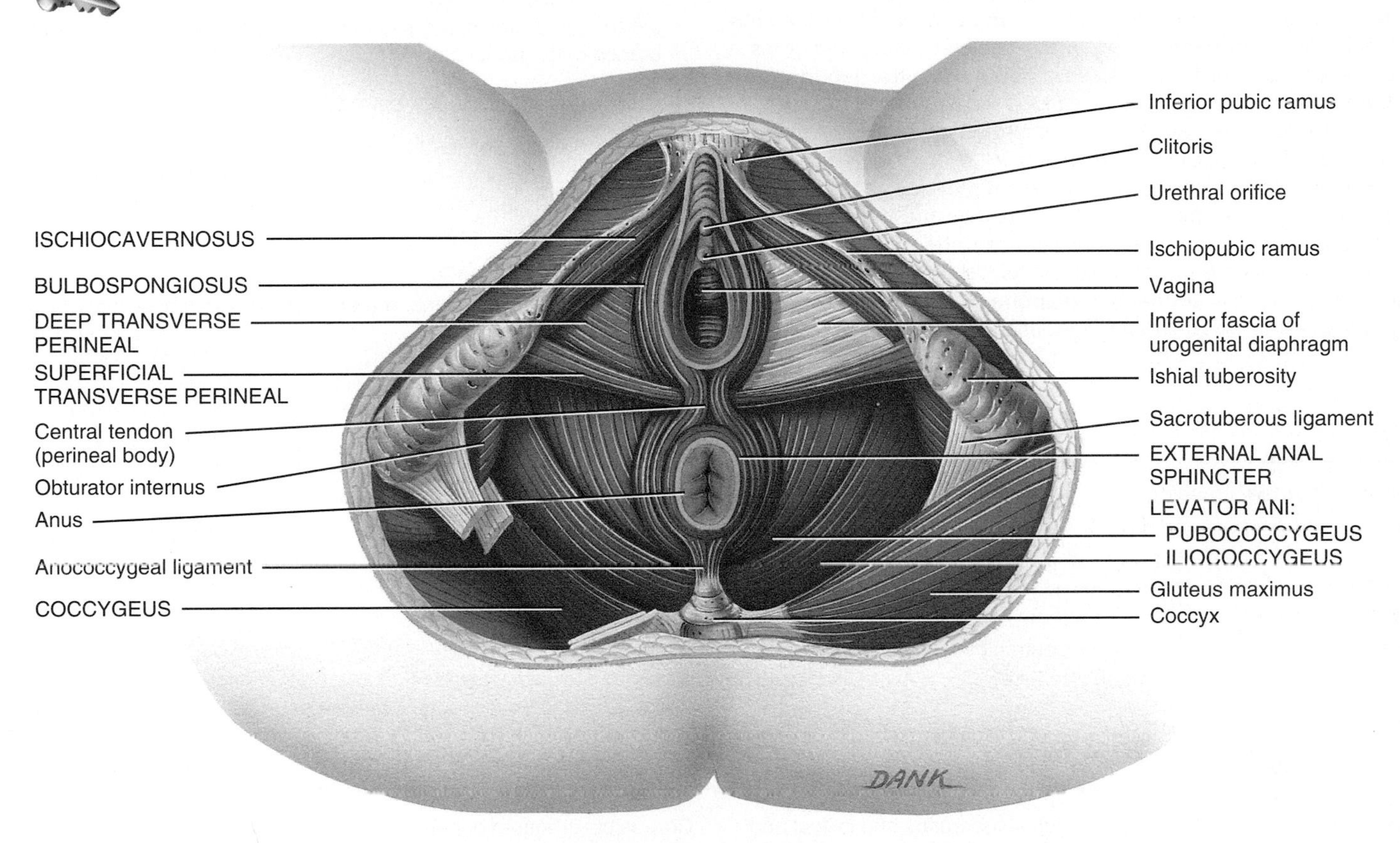

Inferior superficial view

What are the borders of the pelvic diaphragm?

Exhibit 11.10 Muscles of the Perineum (Figures 11.12 and 11.13)

▶ OBJECTIVE

- Describe the origin, insertion, action, and innervation of the muscles of the perineum.

The **perineum** is the region of the trunk inferior to the pelvic diaphragm. A diamond-shaped area extends from the pubic symphysis anteriorly, to the coccyx posteriorly, and to the ischial tuberosities laterally. The female and the male perineums may be compared in Figures 11.12 and 11.13, respectively. A transverse line drawn between the ischial tuberosities divides the perineum into an anterior **urogenital triangle** that contains the external genitals and a posterior **anal triangle** that contains the anus (see Figure 28.23 on page 1039). In the center of the perineum is a wedge-shaped mass of fibrous tissue called the **central tendon (perineal body).** Several perineal muscles insert into this strong tendon.

The muscles of the perineum are arranged in two layers: **superficial** and **deep.** The muscles of the superficial layer are the **superficial transverse perineal muscle,** the **bulbospongiosus,** and the **ischiocavernosus.** The deep muscles of the perineum are the **deep transverse perineal muscle** and the **external urethral sphincter**. The deep transverse perineal muscle, external urethral sphincter, and their fascia are known as the **urogenital diaphragm.** The muscles of this diaphragm assist in urination and ejaculation in males and urination in females. The **external anal sphincter** closely adheres to the skin around the margin of the anus and keeps the anal canal and anus closed except during defecation.

Innervation

- All muscles of the perineum except external anal sphincter: perineal branch of the pudendal nerve of the sacral plexus (shown in Exhibit 13.4 on page 444).
- External anal sphincter: sacral spinal nerve S4 and the inferior rectal branch of the pudendal nerve.

Relating Muscles to Movements

Arrange the muscles in this exhibit according to the following actions: (1) expulsion of urine and semen, (2) erection of the clitoris and penis, (3) closing the anal orifice, and (4) constricting the vaginal orifice. The same muscle may be mentioned more than once.

▶ CHECKPOINT

What are the borders and contents of the urogenital triangle and the anal triangle?

Muscle	Origin	Insertion	Action
Superficial Perineal Muscles			
Superficial transverse perineal (per-i-NĒ-al; *superficial* = closer to surface; *transverse* = across; *perineus* = perineum)	Ischial tuberosity.	Central tendon of perineum.	Helps stabilize central tendon of perineum.
Bulbospongiosus (bul′-bō-spon′-jē-Ō-sus; *bulb* = a bulb; *spongio-* = sponge)	Central tendon of perineum.	Inferior fascia of urogenital diaphragm, corpus spongiosum of penis, and deep fascia on dorsum of penis in male; pubic arch and root and dorsum of clitoris in female.	Helps expel urine during urination, helps propel semen along urethra, assists in erection of the penis in male; constricts vaginal orifice and assists in erection of clitoris in female.
Ischiocavernosus (is′-kē-ō-ka′-ver-NŌ-sus; *ischio-* = the hip)	Ischial tuberosity and ischial and pubic rami.	Corpus cavernosum of penis in male and clitoris in female.	Maintains erection of penis in male and clitoris in female.
Deep Perineal Muscles			
Deep transverse perineal (per-i-NĒ-al; *deep* = farther from surface)	Ischial rami.	Central tendon of perineum.	Helps expel last drops of urine and semen in male and urine in female.
External urethral sphincter (ū-RĒ-thral SFINGK-ter)	Ischial and pubic rami.	Median raphe in male and vaginal wall in female.	Helps expel last drops of urine and semen in male and urine in female.
External anal sphincter (Ā-nal)	Anococcygeal ligament.	Central tendon of perineum.	Keeps anal canal and anus closed.

Figure 11.13 Muscles of the male perineum.

The urogenital diaphragm assists in urination in females and males, ejaculation in males, and helps strengthen the pelvic floor.

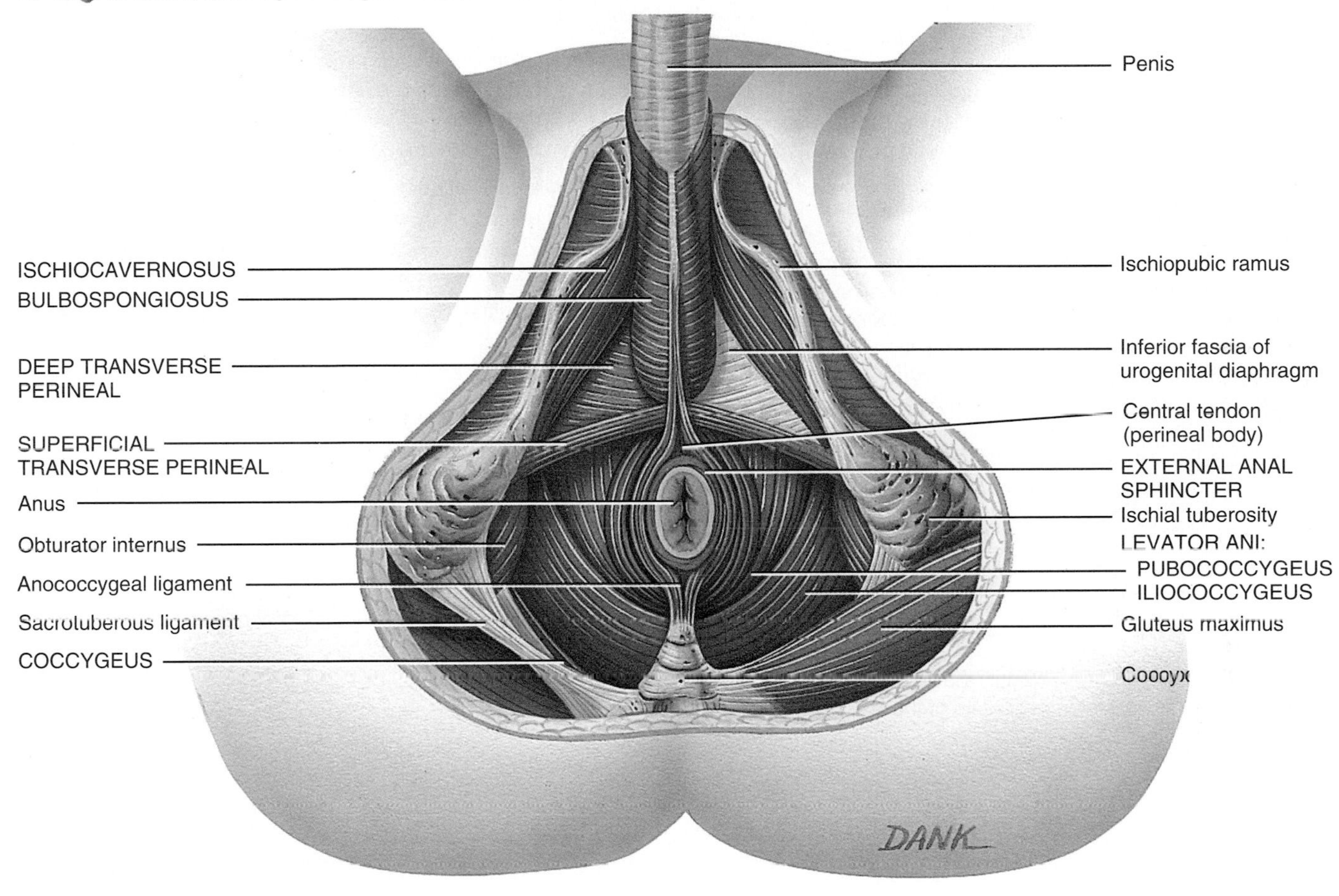

Inferior superficial view

What are the borders of the perineum?

Exhibit 11.11 Muscles that Move the Pectoral (Shoulder) Girdle (Figure 11.14)

▶ OBJECTIVE

- Describe the origin, insertion, action, and innervation of the muscles that move the pectoral girdle.

The main action of the muscles that move the pectoral girdle is to stabilize the scapula so it can function as a steady origin for most of the muscles that move the humerus. Because scapular movements usually accompany humeral movements in the same direction, the muscles also move the scapula to increase the range of motion of the humerus. For example, it would not be possible to abduct the humerus past the horizontal if the scapula did not move with the humerus. During abduction, the scapula follows the humerus by rotating upward.

Muscles that move the pectoral girdle can be classified into two groups based on their location in the thorax: **anterior** and **posterior thoracic muscles.** The anterior thoracic muscles are the subclavius, pectoralis minor, and serratus anterior. The **subclavius** is a small, cylindrical muscle under the clavicle that extends from the clavicle to the first rib. It steadies the clavicle during movements of the pectoral girdle. The **pectoralis minor** is a thin, flat, triangular muscle that is deep to the pectoralis major. Besides its role in movements of the scapula, the pectoralis minor muscle assists in forced inhalation. The **serratus anterior** is a large, flat, fan-shaped muscle between the ribs and scapula. It is named because of the saw-toothed appearance of its origins on the ribs.

The posterior thoracic muscles are the trapezius, levator scapulae, rhomboid major, and rhomboid minor. The **trapezius** is a large, flat, triangular sheet of muscle extending from the skull and vertebral column medially to the pectoral girdle laterally. It is the most superficial back muscle and covers the posterior neck region and superior portion of the trunk. The two trapezius muscles form a trapezium (diamond-shaped quadrangle)—hence its name. The **levator scapulae** is a narrow, elongated muscle in the posterior portion of the neck. It is deep to the sternocleidomastoid and trapezoid muscles. As its name suggests, one of its actions is to elevate the scapula. The **rhomboid major** and **rhomboid minor** lie deep to the trapezius and are not always distinct from each other. They appear as parallel bands that pass inferiolaterally from the vertebrae to the scapula. They are named based on their shape—that is, a rhomboid (an oblique parallelogram). The rhomboid major is about two times wider than the rhomboid minor. Both muscles are used when forcibly lowering the raised upper limbs, as in driving a stake with a sledgehammer.

To understand the actions of muscles that move the scapula, it is first helpful to describe the various movements of the scapula:

- **Elevation:** superior movement of the scapula, such as shrugging the shoulders or lifting a weight over the head.
- **Depression:** inferior movement of the scapula, as in doing a "pull-up."
- **Abduction (protraction):** movement of the scapula laterally and anteriorly, as in doing a "push-up" or punching.
- **Adduction (retraction):** movement of the scapula medially and posteriorly, as in pulling the oars in a rowboat.
- **Upward rotation:** movement of the inferior angle of the scapula laterally so that the glenoid cavity is moved upward. This movement is required to abduct the humerus past the horizontal.
- **Downward rotation:** movement of the inferior angle of the scapula medially so that the glenoid cavity is moved downward. This movement is seen when a gymnast on parallel bars supports the weight of the body on the hands.

Innervation

Muscles that move the shoulder receive their innervation mainly from nerves that emerge from the cervical and brachial plexuses (shown in Exhibits 13.1 and 13.2 on pages 436 and 438).

- Subclavius: subclavian nerve
- Pectoralis minor: medial pectoral nerve
- Serratus anterior: long thoracic nerve
- Trapezius: accessory nerve (cranial nerve XI) and cervical spinal nerves C3–C5
- Levator scapulae: dorsal scapular nerve and cervical spinal nerves C3–C5
- Rhomboid major and rhomboid minor: dorsal scapular nerve

Relating Muscles to Movements

Arrange the muscles in this exhibit according to the following actions on the scapula: (1) depression, (2) elevation, (3) abduction, (4) adduction, (5) upward rotation, and (6) downward rotation. The same muscle may be mentioned more than once.

▶ CHECKPOINT

What muscles in this exhibit are used to raise your shoulders, lower your shoulders, join your hands behind your back, and join your hands in front of your chest?

Muscle	Origin	Insertion	Action
Anterior Thoracic Muscles			
Subclavius (sub-KLĀ-vē-us; *sub-* = under; *clavius* = clavicle)	First rib.	Clavicle.	Depresses and moves clavicle anteriorly and helps stabilize pectoral girdle.
Pectoralis minor (pek′-tō-RĀ-lis; *pector-* = the breast, chest, thorax; *minor* = lesser)	Second through fifth, third through fifth, or second through fourth ribs.	Coracoid process of scapula.	Abducts scapula and rotates it downward; elevates third through fifth ribs during forced inhalation when scapula is fixed.
Serratus anterior (ser-Ā-tus; *serratus* = saw-toothed; *anterior* = front)	Superior eight or nine ribs.	Vertebral border and inferior angle of scapula.	Abducts scapula and rotates it upward; elevates ribs when scapula is stabilized; known as "boxer's muscle."

Posterior Thoracic Muscles			
Trapezius (tra-PĒ-zē-us; *trapezi-* = trapezoid-shaped)	Superior nuchal line of occipital bone, ligamentum nuchae, and spines of seventh cervical and all thoracic vertebrae.	Clavicle and acromion and spine of scapula.	Superior fibers elevate scapula and can help extend head; middle fibers adduct scapula; inferior fibers depress scapula; superior and inferior fibers together rotate scapula upward; stabilizes scapula.
Levator scapulae (le-VĀ-tor SKA-pū-lē; *levator* = raises; *scapulae* = of the scapula)	Superior four or five cervical vertebrae.	Superior vertebral border of scapula.	Elevates scapula and rotates it downward.
Rhomboid major (rom-BOYD; *rhomboid* = rhomboid or diamond-shaped) (see Figure 11.15c)	Spines of second to fifth thoracic vertebrae.	Vertebral border of scapula inferior to spine.	Elevates and adducts scapula and rotates it downward; stabilizes scapula.
Rhomboid minor (rom-BOYD) (see Figure 11.15c)	Spines of seventh cervical and first thoracic vertebrae.	Vertebral border of scapula superior to spine.	Elevates and adducts scapula and rotates it downward; stabilizes scapula.

Figure 11.14 Muscles that move the pectoral (shoulder) girdle. (See Tortora, *A Photographic Atlas of the Human Body,* Figure 5.8.)

Muscles that move the pectoral girdle originate on the axial skeleton and insert on the clavicle or scapula.

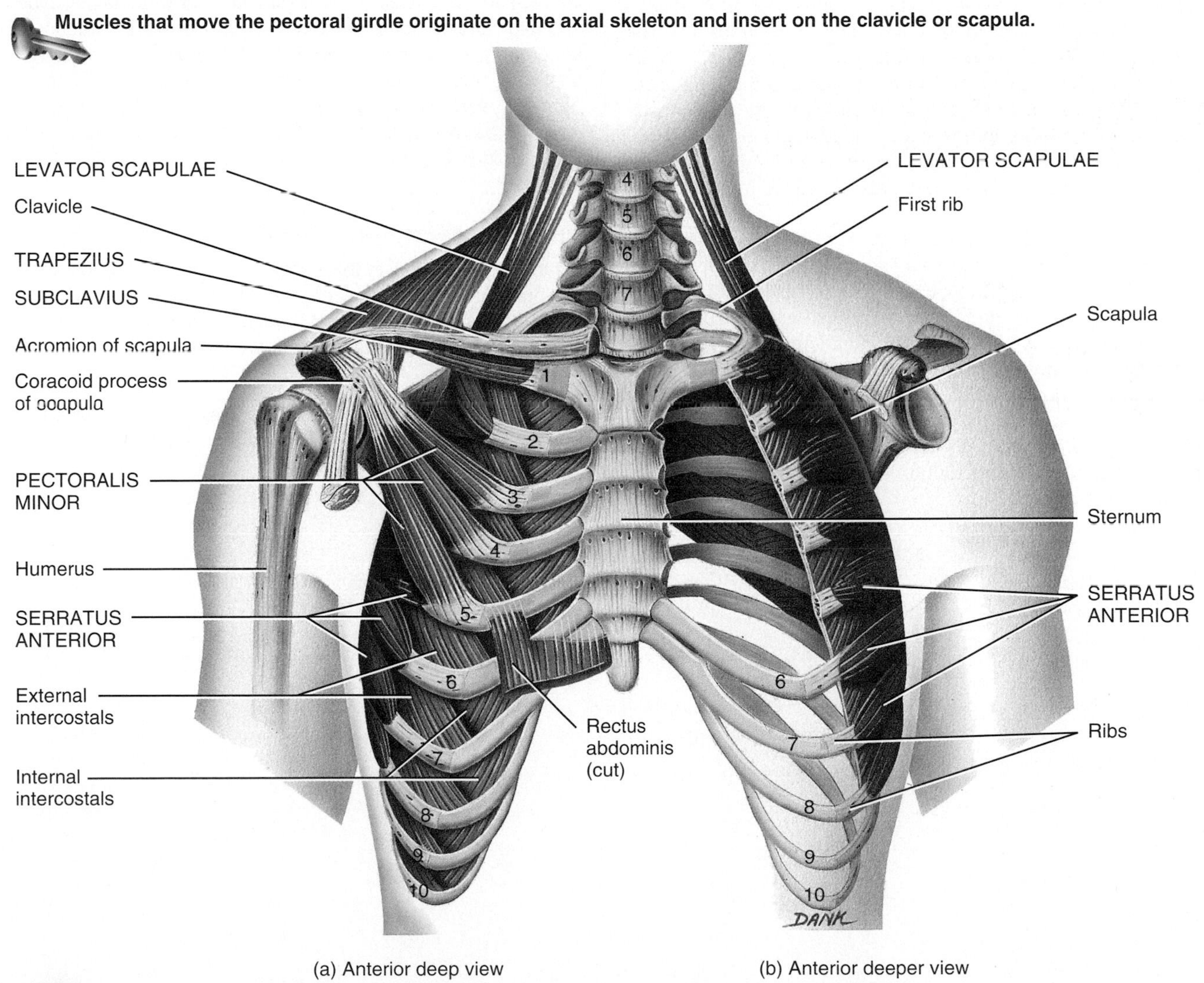

What is the main action of the muscles that move the pectoral girdle?

Exhibit 11.12 Muscles that Move the Humerus (Arm Bone) (Figure 11.15)

► OBJECTIVE

- Describe the origin, insertion, action, and innervation of the muscles that move the humerus.

Of the nine muscles that cross the shoulder joint, all except the pectoralis major and latissimus dorsi originate on the scapula. The pectoralis major and latissimus dorsi thus are called **axial muscles** because they originate on the axial skeleton. The remaining seven muscles, the **scapular muscles,** arise from the scapula.

Of the two axial muscles that move the humerus, the **pectoralis major** is a large, thick, fan-shaped muscle that covers the superior part of the thorax. It has two origins: a smaller clavicular head and a larger sternocostal head. The **latissimus dorsi** is a broad, triangular muscle located on the inferior part of the back. It is commonly called the "swimmer's muscle" because its many actions are used while swimming.

Among the scapular muscles, the **deltoid** is a thick, powerful shoulder muscle that covers the shoulder joint and forms the rounded contour of the shoulder. This muscle is a frequent site of intramuscular injections. As you study the deltoid, note that its fascicles originate from three different points and that each group of fascicles moves the humerus differently. The **subscapularis** is a large triangular muscle that fills the subscapular fossa of the scapula and forms part of the posterior wall of the axilla. The **supraspinatus** is a rounded muscle, named for its location in the supraspinous fossa of the scapula. It lies deep to the trapezius. The **infraspinatus** is a triangular muscle, also named for its location in the infraspinous fossa of the scapula. The **teres major** is a thick, flattened muscle inferior to the teres minor that also helps form part of the posterior wall of the axilla. The **teres minor** is a cylindrical, elongated muscle, often inseparable from the infraspinatus, which lies along its superior border. The **coracobrachialis** is an elongated, narrow muscle in the arm.

The strength and stability of the shoulder joint are not provided by the shape of the articulating bones or its ligaments. Instead, four deep muscles of the shoulder—subscapularis, supraspinatus, infraspinatus, and teres minor—strengthen and stabilize the shoulder joint. These muscles join the scapula to the humerus. Their flat tendons fuse together to form the **rotator (musculotendinous) cuff,** a nearly complete circle of tendons around the shoulder joint, like the cuff on a shirtsleeve. The supraspinatus muscle is especially predisposed to wear and tear because of its location between the head of the humerus and acromion of the scapula, which compress its tendon during shoulder movements, especially abduction of the arm.

Innervation

Nerves that emerge from the brachial plexus (shown in Exhibit 13.2 on page 438) innervate muscles that move the arm.

- Pectoralis major: medial and lateral pectoral nerves
- Latissimus dorsi: thoracodorsal nerve
- Deltoid and teres minor: axillary nerve
- Subscapularis: upper and lower subscapular nerves
- Supraspinatus and infraspinatus: suprascapular nerve
- Teres major: lower subscapular nerve
- Coracobrachialis: musculocutaneous nerve

Impingement Syndrome

One of the most common causes of shoulder pain and dysfunction in athletes is known as **impingement syndrome.** The repetitive movement of the arm over the head that is common in baseball, overhead racquet sports, lifting weights over the head, spiking a volleyball, and swimming puts these athletes at risk for developing this syndrome. It may also be caused by a direct blow or stretch injury. Continual pinching of the supraspinatus tendon as a result of overhead motions causes it to become inflamed and results in pain. If movement is continued despite the pain, the tendon may degenerate near the attachment to the humerus and ultimately may tear away from the bone (rotator cuff injury). Treatment consists of resting the injured tendons, strengthening the shoulder through exercise, and surgery if the injury is particularly severe. ■

Relating Muscles to Movements

Arrange the muscles in this exhibit according to the following actions on the humerus at the shoulder joint: (1) flexion, (2) extension, (3) abduction, (4) adduction, (5) medial rotation, and (6) lateral rotation. The same muscle may be mentioned more than once.

► CHECKPOINT

Why are two muscles that cross the shoulder joint called axial muscles, and the seven others called scapular muscles?

Muscle	Origin	Insertion	Action
Axial Muscles that Move the Humerus			
Pectoralis major (pek′-tō-RĀ-lis; *pector-* = chest; *major* = larger) (see also Figure 11.10a)	Clavicle (clavicular head), sternum, and costal cartilages of second to sixth ribs and sometimes first to seventh ribs (sternocostal head).	Greater tubercle and intertubercular sulcus of humerus.	As a whole, adducts and medially rotates arm at shoulder joint; clavicular head alone flexes arm, and sternocostal head alone extends arm at shoulder joint.
Latissimus dorsi (la-TIS-i-mus DOR-sī; *latissimus* = widest; *dorsi* = of the back)	Spines of inferior six thoracic vertebrae, lumbar vertebrae, crests of sacrum and ilium, inferior four ribs.	Intertubercular sulcus of humerus.	Extends, adducts, and medially rotates arm at shoulder joint; draws arm inferiorly and posteriorly.
Scapular Muscles that Move the Humerus			
Deltoid (DEL-toyd = triangularly shaped)	Acromial extremity of clavicle (anterior fibers), acromion of scapula (lateral fibers), and spine of scapula (posterior fibers).	Deltoid tuberosity of humerus.	Lateral fibers abduct arm at shoulder joint; anterior fibers flex and medially rotate arm at shoulder joint; posterior fibers extend and laterally rotate arm at shoulder joint.
Subscapularis (sub-scap′-ū-LA-ris; *sub-* = below; *scapularis* = scapula).	Subscapular fossa of scapula.	Lesser tubercle of humerus.	Medially rotates arm at shoulder joint.
Supraspinatus (soo-pra-spī-NĀ-tus; *supra-* = above; *spina-* = spine [of the scapula])	Supraspinous fossa of scapula.	Greater tubercle of humerus.	Assists deltoid muscle in abducting arm at shoulder joint.
Infraspinatus (in′-fra-spī-NĀ-tus; *infra-* = below)	Infraspinous fossa of scapula.	Greater tubercle of humerus.	Laterally rotates and adducts arm at shoulder joint.
Teres major (TE-rēz; *teres* = long and round)	Inferior angle of scapula.	Intertubercular sulcus of humerus.	Extends arm at shoulder joint and assists in adduction and medial rotation of arm at shoulder joint.
Teres minor (TE-rēz)	Inferior lateral border of scapula.	Greater tubercle of humerus.	Laterally rotates, extends, and adducts arm at shoulder joint.
Coracobrachialis (kor′-a-kō-brā-kē-Ā′-lis; *coraco-* = coracoid process [of the scapula]; *brachi-* = arm)	Coracoid process of scapula.	Middle of medial surface of shaft of humerus.	Flexes and adducts arm at shoulder joint.

(continues)

Exhibit 11.12 Muscles that Move the Humerus (Arm Bone) (Figure 11.15) *(continued)*

Figure 11.15 Muscles that move the humerus (arm bone). (See Tortora, *A Photographic Atlas of the Human Body,* Figures 5.9 and 5.10.)

The tendons that form the rotator cuff provide the strength and stability of the shoulder joint.

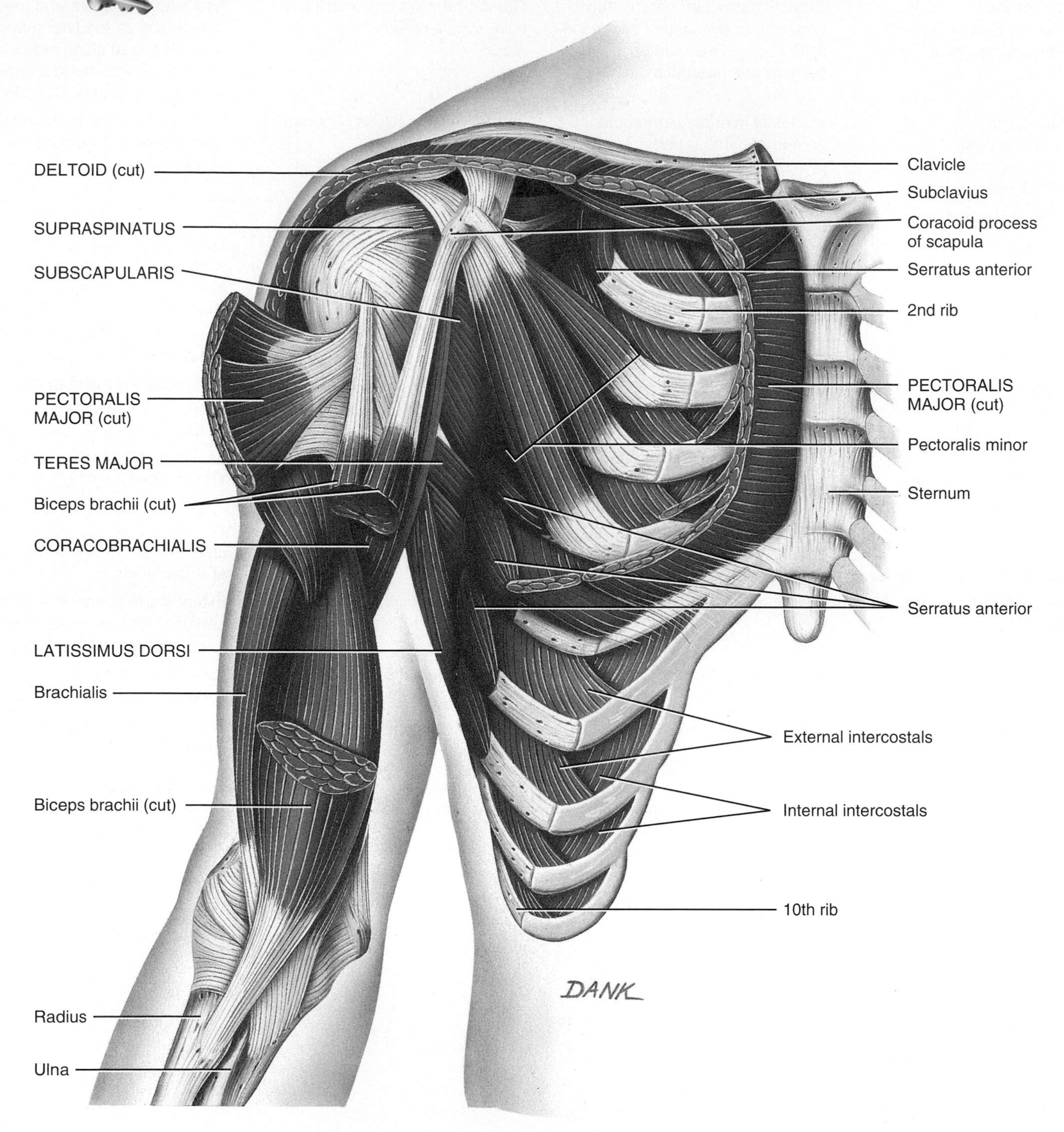

(a) Anterior deep view (the intact pectoralis major muscle is shown in Figure 11.10a)

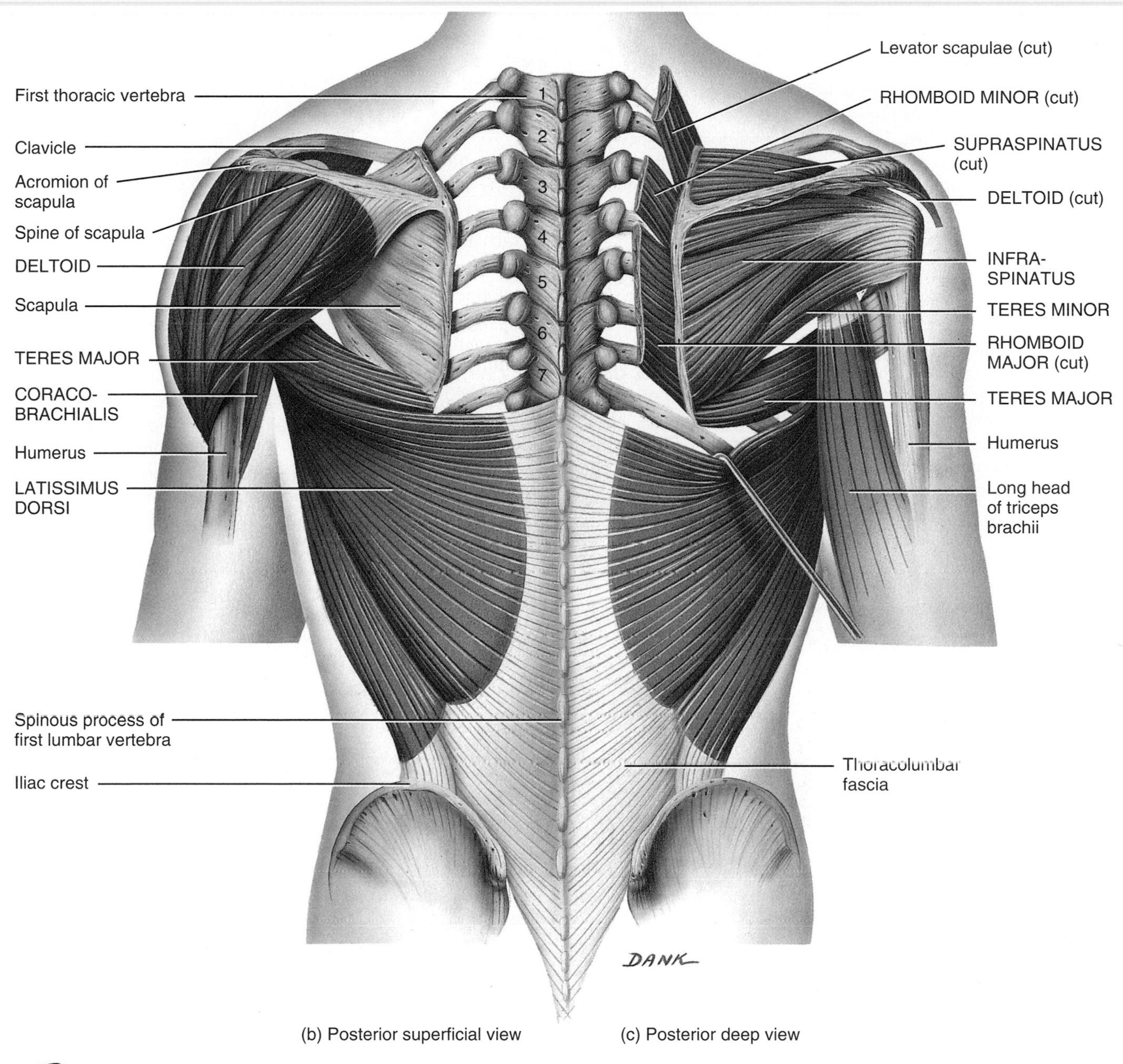

(b) Posterior superficial view (c) Posterior deep view

Which tendons make up the rotator cuff?

Exhibit 11.13 Muscles that Move the Radius and Ulna (Forearm Bones) (Figure 11.16)

▶ OBJECTIVE

- Describe the origin, insertion, action, and innervation of the muscles that move the radius and ulna.

Most of the muscles that move the radius and ulna (forearm bones) cause flexion and extension at the elbow, which is a hinge joint. The biceps brachii, brachialis, and brachioradialis muscles are the flexor muscles. The extensor muscles are the triceps brachii and the anconeus.

The **biceps brachii** is the large muscle located on the anterior surface of the arm. As indicated by its name, it has two heads of origin (long and short), both from the scapula. The muscle spans both the shoulder and elbow joints. In addition to its role in flexing the forearm at the elbow joint, it also supinates the forearm at the radioulnar joints and flexes the arm at the shoulder joint. The **brachialis** is deep to the biceps brachii muscle. It is the most powerful flexor of the forearm at the elbow joint. For this reason, it is called the "workhorse" of the elbow flexors. The **brachioradialis** flexes the forearm at the elbow joint, especially when a quick movement is required or when a weight is lifted slowly during flexion of the forearm.

The **triceps brachii** is the large muscle located on the posterior surface of the arm. It is the more powerful of the extensors of the forearm at the elbow joint. As its name implies, it has three heads of origin, one from the scapula (long head) and two from the humerus (lateral and medial heads). The long head crosses the shoulder joint; the other heads do not. The **anconeus** is a small muscle located on the lateral part of the posterior aspect of the elbow that assists the triceps brachii in extending the forearm at the elbow joint.

Some muscles that move the radius and ulna are involved in pronation and supination at the radioulnar joints. The pronators, as suggested by their names, are the **pronator teres** and **pronator quadratus** muscles. The supinator of the forearm is the aptly named the **supinator** muscle. You use the powerful action of the supinator when you twist a corkscrew or turn a screw with a screwdriver.

In the limbs, functionally related skeletal muscles and their associated blood vessels and nerves are grouped together by fascia into regions called **compartments.** In the arm, the biceps brachii, brachialis, and coracobrachialis muscles comprise the **anterior (flexor) compartment.** The triceps brachii muscle forms the **posterior (extensor) compartment.**

Innervation

Muscles that move the radius and ulna are innervated by nerves derived from the brachial plexus (shown in Exhibit 13.2 on page 438).

- Biceps brachii: musculocutaneous nerve
- Brachialis: musculocutaneous and radial nerves
- Brachioradialis, triceps brachii, and anconeus: radial nerve
- Pronator teres and pronator quadratus: median nerve
- Supinator: deep radial nerve

Relating Muscles to Movements

Arrange the muscles in this exhibit according to the following actions on the elbow joint: (1) flexion and (2) extension; the following actions on the forearm at the radioulnar joints: (1) supination and (2) pronation; and the following actions on the humerus at the shoulder joint: (1) flexion and (2) extension. The same muscle may be mentioned more than once.

▶ CHECKPOINT

Which muscles are in the anterior and posterior compartments of the arm?

Muscle	Origin	Insertion	Action
Forearm Flexors			
Biceps brachii (BĪ-ceps BRĀ-kē-ī ; *biceps* = two heads [of origin]; *brachii* = arm)	*Long head* originates from tubercle above glenoid cavity of scapula (supraglenoid tubercle); *short head* originates from coracoid process of scapula.	Radial tuberosity of radius and bicipital aponeurosis.*	Flexes forearm at elbow joint, supinates forearm at radioulnar joints, and flexes arm at shoulder joint.
Brachialis (brā-kē-Ā-lis)	Distal, anterior surface of humerus.	Ulnar tuberosity and coronoid process of ulna.	Flexes forearm at elbow joint.
Brachioradialis (bra′-kē-ō-rā-dē-Ā-lis; *radi-* = radius) (see Figure 11.17a)	Lateral border of distal end of humerus.	Superior to styloid process of radius.	Flexes forearm at elbow joint; supinates and pronates forearm at radioulnar joints to neutral position.

Forearm Extensors			
Triceps brachii (TRĪ-ceps BRĀ-kē-ī; *triceps* = three heads [of origin])	*Long head:* infraglenoid tubercle, a projection inferior to glenoid cavity of scapula. *Lateral head:* lateral and posterior surface of humerus superior to radial groove. *Medial head:* entire posterior surface of humerus inferior to a groove for the radial nerve.	Olecranon of ulna.	Extends forearm at elbow joint and extends arm at shoulder joint.
Anconeus (an-KŌ-nē-us; *ancon* = the elbow) (see Figure 11.17c)	Lateral epicondyle of humerus.	Olecranon and superior portion of shaft of ulna.	Extends forearm at elbow joint.
Forearm Pronators			
Pronator teres (PRŌ-nā-tor TE-rēz; *pronator* = turns palm downward or posteriorly) (see Figure 11.17a)	Medial epicondyle of humerus and coronoid process of ulna.	Midlateral surface of radius.	Pronates forearm at radioulnar joints and weakly flexes forearm at elbow joint.
Pronator quadratus (PRŌ-nā-tor kwod-RĀ-tus; *quadratus* = square, four-sided) (see Figure 11.17a)	Distal portion of shaft of ulna.	Distal portion of shaft of radius.	Pronates forearm at radioulnar joints.
Forearm Supinator			
Supinator (SOO-pi-nā-tor; *supinator* = turns palm upward or anteriorly) (see Figure 11.17b)	Lateral epicondyle of humerus and ridge near radial notch of ulna (supinator crest).	Lateral surface of proximal one-third of radius.	Supinates forearm at radioulnar joints.

*The *bicipital aponeurosis* is a broad aponeurosis from the tendon of insertion of the biceps brachii muscle that descends medially across the brachial artery and fuses with deep fascia over the forearm flexor muscles.

(continues)

Exhibit 11.13 Muscles that Move the Radius and Ulna (Forearm Bones) (Figure 11.16) *(continued)*

Figure 11.16 Muscles that move the radius and ulna (forearm bones). (See Tortora, *A Photographic Atlas of the Human Body,* Figure 5.11.)

Whereas the anterior arm muscles flex the forearm, the posterior arm muscles extend it.

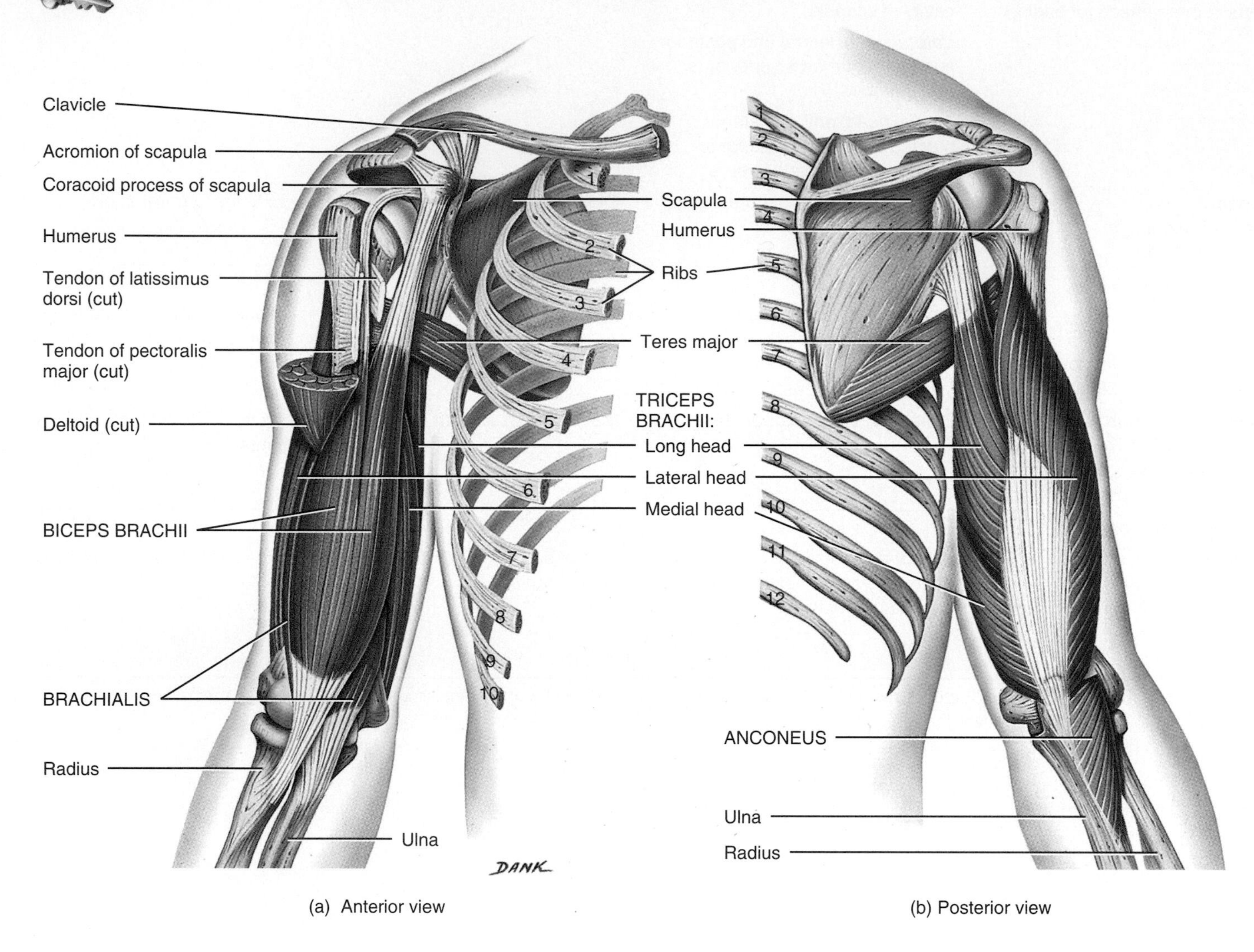

(a) Anterior view

(b) Posterior view

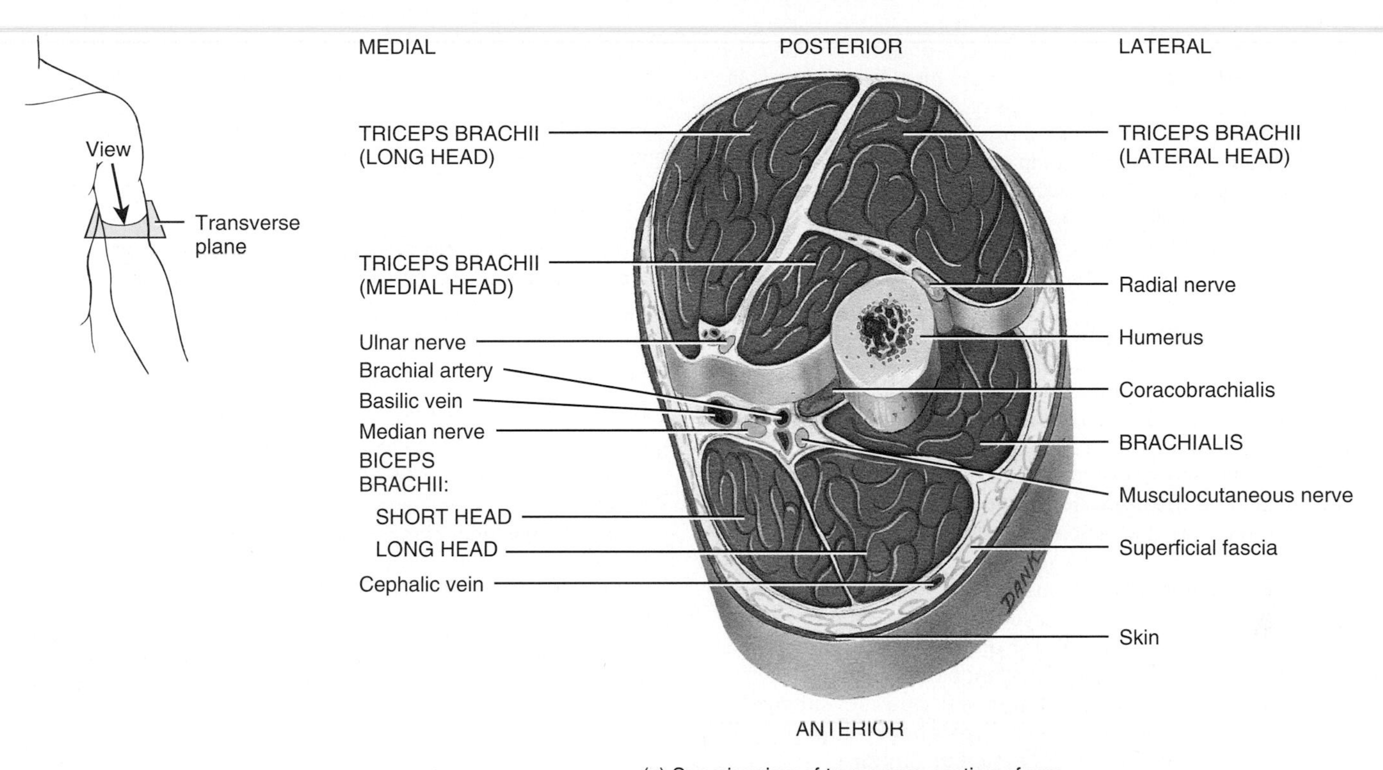

(c) Superior view of transverse section of arm

 **Which muscles are the most powerful flexor and the most powerful extensor of the forearm?**

Exhibit 11.14 Muscles that Move the Wrist, Hand, Thumb, and Fingers (Figure 11.17)

► **OBJECTIVE**

- Describe the origin, insertion, action, and innervation of the muscles that move the wrist, hand, thumb, and fingers.

Muscles of the forearm that move the wrist, hand, thumb, and fingers are many and varied. Those in this group that act on the digits are known as **extrinsic hand muscles** because they originate *outside* the hand and insert within it. As you will see, the names for the muscles that move the wrist, hand, and digits give some indication of their origin, insertion, or action. Based on location and function, the muscles of the forearm are divided into two groups: (1) anterior compartment muscles and (2) posterior compartment muscles. The **anterior (flexor) compartment** muscles of the forearm originate on the humerus, typically insert on the carpals, metacarpals, and phalanges, and function as flexors. The bellies of these muscles form the bulk of the forearm. The **posterior (extensor) compartment** muscles of the forearm originate on the humerus, insert on the metacarpals and phalanges, and function as extensors. Within each compartment, the muscles are grouped as superficial or deep.

The **superficial anterior compartment** muscles are arranged in the following order from lateral to medial: **flexor carpi radialis, palmaris longus** (absent in about 10% of the population), and **flexor carpi ulnaris** (the ulnar nerve and artery are just lateral to the tendon of this muscle at the wrist). The **flexor digitorum superficialis** muscle is deep to the other three muscles and is the largest superficial muscle in the forearm.

The **deep anterior compartment** muscles are arranged in the following order from lateral to medial: **flexor pollicis longus** (the only flexor of the distal phalanx of the thumb) and **flexor digitorum profundus** (ends in four tendons that insert into the distal phalanges of the fingers).

The **superficial posterior compartment** muscles are arranged in the following order from lateral to medial: **extensor carpi radialis longus, extensor carpi radialis brevis, extensor digitorum** (occupies most of the posterior surface of the forearm and divides into four tendons that insert into the middle and distal phalanges of the fingers), **extensor digiti minimi** (a slender muscle generally connected to the extensor digitorum), and the **extensor carpiulnaris.**

The **deep posterior compartment** muscles are arranged in the following order from lateral to medial: **abductor pollicis longus, extensor pollicis brevis, extensor pollicis longus,** and **extensor indicis.**

The tendons of the muscles of the forearm that attach to the wrist or continue into the hand, along with blood vessels and nerves, are held close to bones by strong fasciae. The tendons are also surrounded by tendon sheaths. At the wrist, the deep fascia is thickened into fibrous bands called **retinacula** (*retinacul* = a holdfast). The **flexor retinaculum** is located over the palmar surface of the carpal bones. The long flexor tendons of the digits and wrist and the median nerve pass through the flexor retinaculum. The **extensor retinaculum** is located over the dorsal surface of the carpal bones. The extensor tendons of the wrist and digits pass through it.

Innervation

Nerves derived from the brachial plexus (shown in Exhibit 13.2 on page 438) innervate muscles that move the wrist, hand, thumb, and fingers.

- Flexor carpi radialis, palmaris longus, flexor digitorum superficialis, and flexor pollicis longus: median nerve
- Flexor carpi ulnaris: ulnar nerve
- Flexor digitorum profundus: median and ulnar nerves
- Extensor carpi radialis longus, extensor carpi radialis brevis, and extensor digitorum: radial nerve
- Extensor digiti minimi, extensor carpi ulnaris, and all of the deep muscles of the posterior compartment: deep radial nerve

Relating Muscles to Movements

Arrange the muscles in this exhibit according to the following actions on the wrist joint: (1) flexion, (2) extension, (3) abduction, and (4) adduction; the following actions on the fingers at the metacarpophalangeal joints: (1) flexion and (2) extension; the following actions on the fingers at the interphalangeal joints: (1) flexion and (2) extension; the following actions on the thumb at the carpometacarpal, metacarpophalangeal, and interphalangeal joints: (1) extension and (2) abduction; and the following action on the thumb at the interphalangeal joint: flexion. The same muscle may be mentioned more than once.

► **CHECKPOINT**

Which muscles and actions of the wrist, hand, and digits are used when writing?

Muscle	Origin	Insertion	Action
Superficial Anterior (Flexor) Compartment of the Forearm			
Flexor carpi radialis (FLEK-sor KAR-pē rā′-dē -Ā-lis; *flexor* = decreases angle at joint; *carpi* = of the wrist; *radi-* = radius)	Medial epicondyle of humerus.	Second and third metacarpals.	Flexes and abducts hand (radial deviation) at wrist joint.
Palmaris longus (pal-MA-ris LON-gus; *palma* = palm; *longus* = long)	Medial epicondyle of humerus.	Flexor retinaculum and palmar aponeurosis (deep fascia in center of palm).	Weakly flexes hand at wrist joint.
Flexor carpi ulnaris FLEK-sor KAR-pē ul-NAR-is; *ulnar-* = ulna)	Medial epicondyle of humerus and superior posterior border of ulna.	Pisiform, hamate, and base of fifth metacarpal.	Flexes and adducts hand (ulnar deviation) at wrist joint.
Flexor digitorum superficialis FLEK-sor di-ji-TOR-um soo′- per-fish′-ē -Ā-lis; *digit* = finger or toe; *superficialis* = closer to surface)	Medial epicondyle of humerus, coronoid process of ulna, and a ridge along lateral margin of anterior surface (anterior oblique line) of radius.	Middle phalanx of each finger.*	Flexes middle phalanx of each finger at proximal interphalangeal joint, proximal phalanx of each finger at metacarpophalangeal joint, and hand at wrist joint.

Muscle	Origin	Insertion	Action
Deep Anterior (Flexor) Compartment of the Forearm			
Flexor pollicis longus (FLEK-sor POL-li-sis LON-gus; *pollic-* = thumb)	Anterior surface of radius and interosseous membrane (sheet of fibrous tissue that holds shafts of ulna and radius together).	Base of distal phalanx of thumb.	Flexes distal phalanx of thumb at interphalangeal joint.
Flexor digitorum profundus (FLEK-sor di′-ji-TOR-um prō- FUN-dus; *profundus* = deep)	Anterior medial surface of body of ulna.	Base of distal phalanx of each finger.	Flexes distal and middle phalanges of each finger at interphalangeal joints, proximal phalanx of each finger at metacarpophalangeal joint, and hand at wrist joint.
Superficial Posterior (Extensor) Compartment of the Forearm			
Extensor carpi radialis longus (eks-TEN-sor KAR-pē rā′-dē-Ā-lis LON-gus; *extensor* = increases angle at joint)	Lateral supracondylar ridge of humerus.	Second metacarpal.	Extends and abducts hand at wrist joint.
Extensor carpi radialis brevis (eks-TEN-sor KAR-pē rā′-dē -Ā-lis BREV-is; *brevis* = short)	Lateral epicondyle of humerus.	Third metacarpal.	Extends and abducts hand at wrist joint.
Extensor digitorum (eks-TEN-sor di′-ji-TOR-um)	Lateral epicondyle of humerus.	Distal and middle phalanges of each finger.	Extends distal and middle phalanges of each finger at interphalangeal joints, proximal phalanx of each finger at metacarpophalangeal joint, and hand at wrist joint.
Extensor digiti minimi (eks-TEN-sor DIJ-i-tē MIN-i-mē; *digit* = finger or toe; *minimi* = little finger)	Lateral epicondyle of humerus.	Tendon of extensor digitorum on fifth phalanx.	Extends proximal phalanx of little finger at metacarpophalangeal joint and hand at wrist joint.
Extensor carpi ulnaris (eks-TEN-sor KAR-pē ul-NAR-is)	Lateral epicondyle of humerus and posterior border of ulna.	Fifth metacarpal.	Extends and adducts hand at wrist joint.
Deep Posterior (Extensor) Compartment of the Forearm			
Abductor pollicis longus (ab-DUK-tor POL-li-sis LON-gus; *abductor* = moves part away from midline)	Posterior surface of middle of radius and ulna and interosseous membrane.	First metacarpal.	Abducts and extends thumb at carpometacarpal joint and abducts hand at wrist joint.
Extensor pollicis brevis (eks-TEN-sor POL-li-sis BREV- is)	Posterior surface of middle of radius and interosseous membrane.	Base of proximal phalanx of thumb.	Extends proximal phalanx of thumb at metacarpophalangeal joint, first metacarpal of thumb at carpometacarpal joint, and hand at wrist joint.
Extensor pollicis longus (eks-TEN-sor POL-li-sis LON-gus)	Posterior surface of middle of ulna and interosseous membrane.	Base of distal phalanx of thumb.	Extends distal phalanx of thumb at interphalangeal joint, first metacarpal of thumb at carpometacarpal joint, and abducts hand at wrist joint.
Extensor indicis (eks-TEN-sor IN-di-kis; *indicis* = index)	Posterior surface of ulna.	Tendon of extensor digitorum of index finger.	Extends distal and middle phalanges of index finger at interphalangeal joints, proximal phalanx of index finger at metacarpophalangeal joint, and hand at wrist joint.

*Reminder: The thumb or pollex is the first digit and has two phalanges: proximal and distal. The remaining digits, the fingers, are numbered 2–5, and each has three phalanges: proximal, middle, and distal.

(continues)

Figure 11.17 Muscles that move the wrist, hand, thumb, and fingers. (See Tortora, *A Photographic Atlas of the Human Body,* Figures 5.12 and 5.13.)

The anterior compartment muscles function as flexors, and the posterior compartment muscles function as extensors.

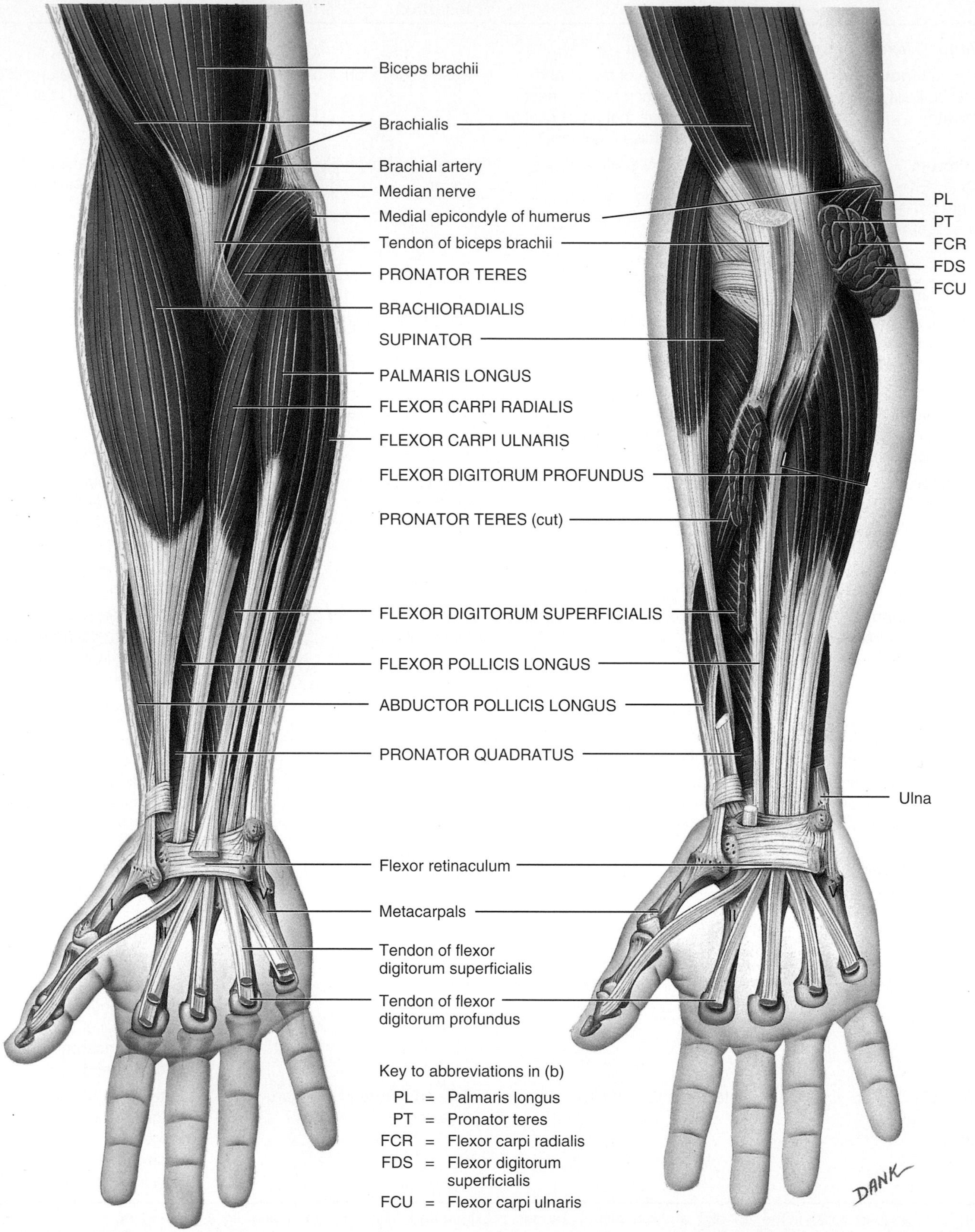

Biceps brachii
Brachialis
Brachial artery
Median nerve
Medial epicondyle of humerus
Tendon of biceps brachii
PRONATOR TERES
BRACHIORADIALIS
SUPINATOR
PALMARIS LONGUS
FLEXOR CARPI RADIALIS
FLEXOR CARPI ULNARIS
FLEXOR DIGITORUM PROFUNDUS
PRONATOR TERES (cut)
FLEXOR DIGITORUM SUPERFICIALIS
FLEXOR POLLICIS LONGUS
ABDUCTOR POLLICIS LONGUS
PRONATOR QUADRATUS
Flexor retinaculum
Metacarpals
Tendon of flexor digitorum superficialis
Tendon of flexor digitorum profundus

PL
PT
FCR
FDS
FCU
Ulna

DANK

Key to abbreviations in (b)
PL = Palmaris longus
PT = Pronator teres
FCR = Flexor carpi radialis
FDS = Flexor digitorum superficialis
FCU = Flexor carpi ulnaris

(a) Anterior superficial view

(b) Anterior deep view

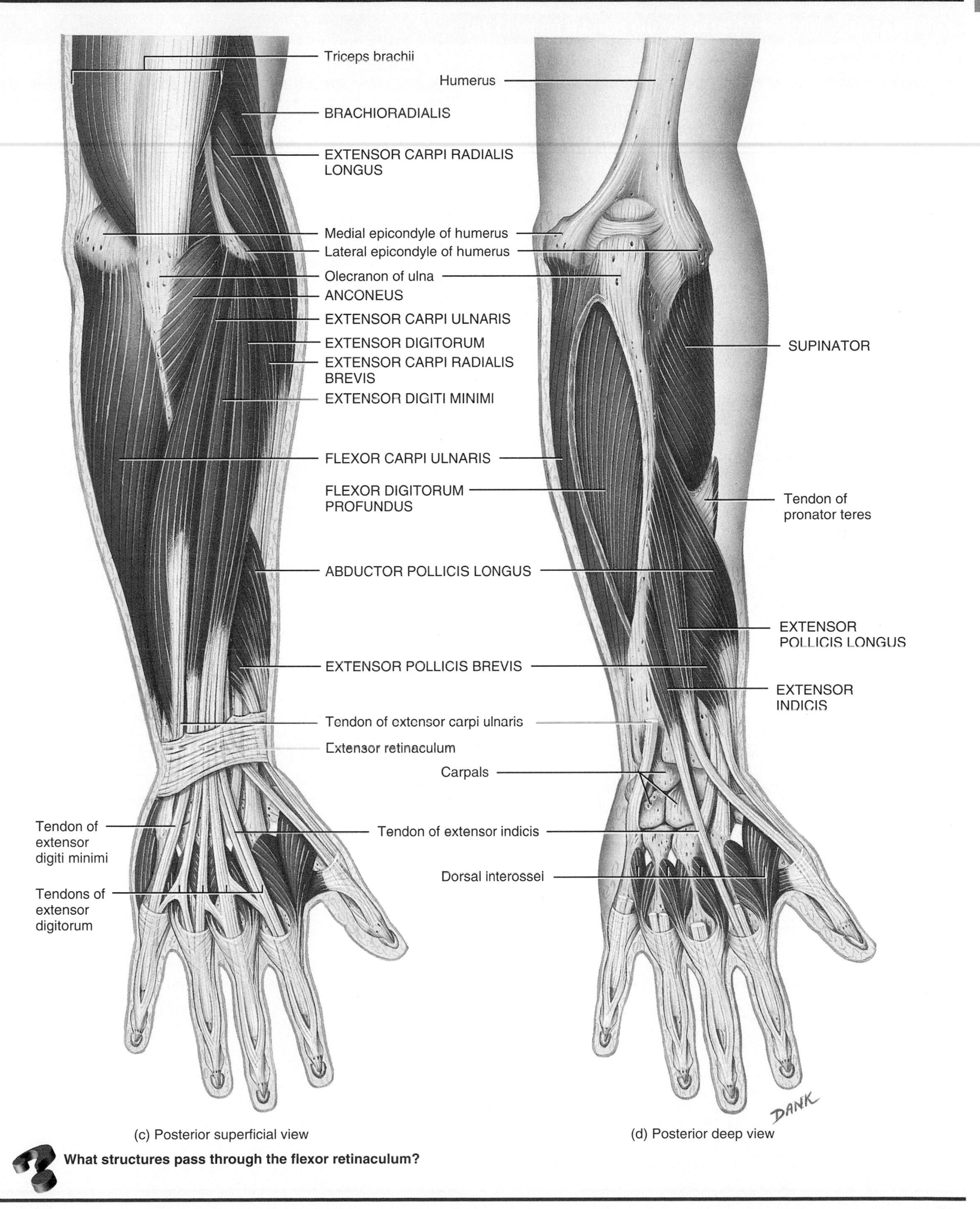

(c) Posterior superficial view

(d) Posterior deep view

What structures pass through the flexor retinaculum?

Exhibit 11.15 Intrinsic Muscles of the Hand (Figure 11.18)

► OBJECTIVE

- Describe the origin, insertion, action, and innervation of the intrinsic muscles of the hand.

Several of the muscles discussed in Exhibit 11.14 are extrinsic muscles of the hand. They originate outside the hand and insert within the hand. The extrinsic muscles of the hand produce powerful but crude movements of the digits. The **intrinsic hand muscles** are located in the palm. They produce the less powerful but more intricate and precise movements of the digits. The muscles in this group are so-named because their origins and insertions are *within* the hand. The functional importance of the hand muscles is readily apparent when one considers that certain hand injuries can cause permanent disability.

The intrinsic muscles of the hand are divided into three groups: (1) **thenar,** (2) **hypothenar,** and (3) **intermediate.** The four thenar muscles act on the thumb and form the **thenar eminence,** the lateral rounded contour on the palm. The thenar muscles include the **abductor pollicis brevis, opponens pollicis, flexor pollicis brevis,** and **adductor pollicis.**

The three hypothenar muscles act on the little finger and form the **hypothenar eminence,** the rounded contour on the medial aspect of the palm. The hypothenar muscles are the **abductor digiti minimi, flexor digiti minimi brevis,** and **opponens digiti minimi.**

The 12 intermediate (midpalmar) muscles act on all the digits except the thumb. The intermediate muscles include the **lumbricals, palmar interossei,** and **dorsal interossei.** Both sets of interossei muscles are located between the metacarpals and are important in abduction, adduction, flexion, and extension of the fingers, and in movements in skilled activities such as writing, typing, and playing a piano.

Besides free motion, the hand performs several special movements. *Power grip* is the forcible movement of the fingers and thumb against the palm, as in hand-shaking. *Precision handling* is a change in position of a handled object that requires exact control of finger and thumb positions, as in winding a watch or threading a needle. *Pinching* is compression between the thumb and index finger or between the thumb and first two fingers.

Movements of the thumb are very important in the precise activities of the hand. They are defined in different planes from comparable movements of the fingers because the thumb can be positioned at a right angle to the fingers. Figure 11.18c shows the five main movements of the thumb. They include *flexion* (movement of the thumb medially across the palm), *extension* (movement of the thumb laterally away from the palm), *abduction* (movement of the thumb in an anteroposterior plane away from the palm), *adduction* (movement of the thumb in an anteroposterior plane toward the palm), and *opposition* (movement of the thumb across the palm so that the tip of the thumb meets the tip of a finger). Opposition is the most distinctive digital movement. It gives humans and other primates the ability to precisely grasp and manipulate objects.

Innervation

Nerves derived from the brachial plexus (shown in Exhibit 13.2 on page 438) innervate intrinsic muscles of the hand.

- Abductor pollicis brevis and opponens pollicis: median nerve
- Adductor pollicis, abductor digiti minimi, flexor digiti minimi brevis, opponens digiti minimi, dorsal interossei, and palmar interossei: ulnar nerve
- Flexor pollicis brevis and lumbricals: median and ulnar nerves

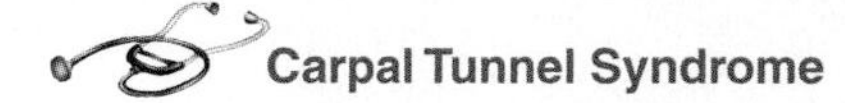

Carpal Tunnel Syndrome

The **carpal tunnel** is a narrow passageway formed anteriorly by the flexor retinaculum and posteriorly by the carpal bones. Through this tunnel pass the median nerve, the most superficial structure, and the long flexor tendons for the digits. Structures within the carpal tunnel, especially the median nerve, are vulnerable to compression, and the resulting condition is called **carpal tunnel syndrome.** Compression of the median nerve leads to sensory changes over the lateral side of the hand and muscle weakness in the thenar eminence. This results in pain, numbness, and tingling of the fingers. The condition may be caused by inflammation of the digital tendon sheaths, fluid retention, excessive exercise, infection, trauma, and repetitive activities that involve flexion of the wrist, such as keyboarding, cutting hair, and playing a piano. ■

Relating Muscles to Movements

Arrange the muscles in this exhibit according to the following actions on the thumb at the carpometacarpal and metaphalangeal joints: (1) abduction, (2) adduction, (3) flexion, and (4) opposition; and the following actions on the fingers at the metacarpophalangeal and interphalangeal joints: (1) abduction, (2) adduction, (3) flexion, and (4) extension. The same muscle may be mentioned more than once.

► CHECKPOINT

How do the actions of the extrinsic and intrinsic muscles of the hand differ?

Muscle	Origin	Insertion	Action
Thenar (lateral aspect of palm)			
Abductor pollicis brevis (ab-DUK-tor POL-li-sis BREV-is; *abductor* = moves part away from middle; *pollic-* = the thumb; *brevis* = short)	Flexor retinaculum, scaphoid, and trapezium.	Lateral side of proximal phalanx of thumb.	Abducts thumb at carpometacarpal joint.
Opponens pollicis (op-PŌ-nenz POL-li-sis; *opponens* = opposes)	Flexor retinaculum and trapezium.	Lateral side of first metacarpal (thumb).	Moves thumb across palm to meet little finger (opposition) at the carpometacarpal joint.
Flexor pollicis brevis (FLEK-sor POL-li-sis BREV-is; *flexor* = decreases angle at joint)	Flexor retinaculum, trapezium, capitate, and trapezoid.	Lateral side of proximal phalanx of thumb.	Flexes thumb at carpometacarpal and metacarpophalangeal joints.
Adductor pollicis (ad-DUK-tor POL-li-sis; *adductor* = moves part toward midline)	Oblique head: capitate and second and third metacarpals; transverse head: third metacarpal.	Medial side of proximal phalanx of thumb by a tendon containing a sesamoid bone.	Adducts thumb at carpometacarpal and metacarpophalangeal joints.
Hypothenar (medial aspect of palm)			
Abductor digiti minimi (ab-DUK-tor DIJ-i-tē MIN-i-mē; *digit* = finger or toe; *minimi* = little)	Pisiform and tendon of flexor carpi ulnaris.	Medial side of proximal phalanx of little finger.	Abducts and flexes little finger at metacarpophalangeal joint.
Flexor digiti minimi brevis (FLEK-sor DIJ-i-tē MIN-i-mē BREV-is)	Flexor retinaculum and hamate.	Medial side of proximal phalanx of little finger.	Flexes little finger at carpometacarpal and metacarpophalangeal joints.
Opponens digiti minimi (op-PŌ-nenz DIJ-i-tē MIN-i-mē)	Flexor retinaculum and hamate.	Medial side of fifth metacarpal (little finger).	Moves little finger across palm to meet thumb (opposition) at the carpometacarpal joint.
Intermediate (Midpalmar)			
Lumbricals (LUM-bri-kals; *lumbric-* = earthworm) (four muscles)	Lateral sides of tendons and flexor digitorum profundus of each finger.	Lateral sides of tendons of extensor digitorum on proximal phalanges of each finger.	Flex each finger at metacarpophalangeal joints and extend each finger at interphalangeal joints.
Palmar interossei (in′-ter-OS-ē-ī (*palmar* = palm; *inter-* = between; *ossei* = bones) (four muscles)	Sides of shafts of metacarpals of all digits (except the middle one)	Sides of bases of proximal phalanges of all digits (except the middle one).	Adduct each finger at metacarpophalangeal joints; flex each finger at metacarpophalangeal joints.
Dorsal interossei (in′-ter-OS-ē-ī; *dorsal* = back surface) (four muscles)	Adjacent sides of metacarpals.	Proximal phalanx of each finger.	Abduct fingers 2–4 at metacarpophalangeal joints; flex fingers 2–4 at metacarpophalangeal joints; and extend each finger at interphalangeal joints.

(continues)

Exhibit 11.15 Intrinsic Muscles of the Hand (Figure 11.18) *(continued)*

Figure 11.18 Intrinsic muscles of the hand. (See Tortora, *A Photographic Atlas of the Human Body,* Figures 5.14 and 5.15.)

The intrinsic muscles of the hand produce intricate and precise movements of the thumb and fingers.

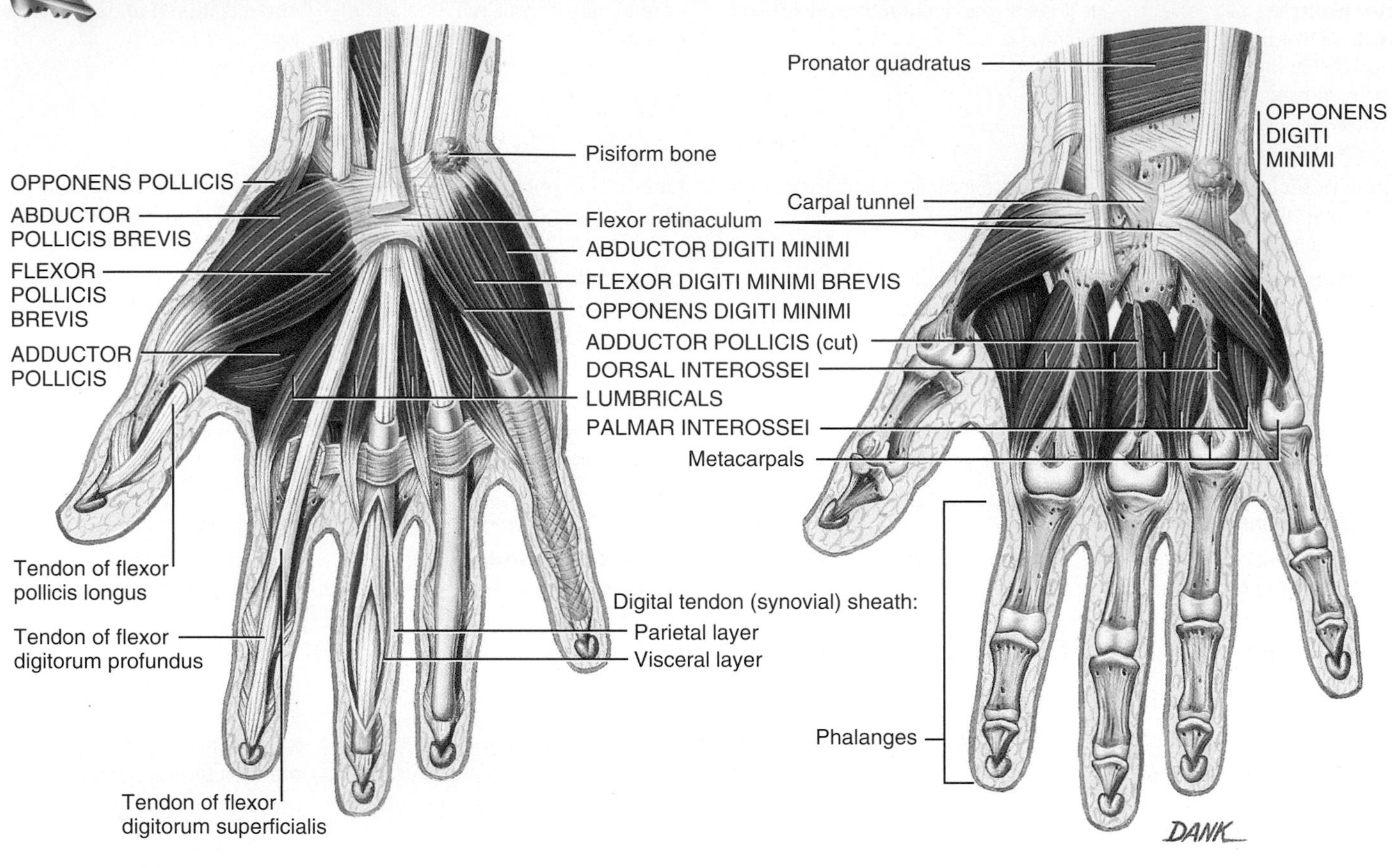

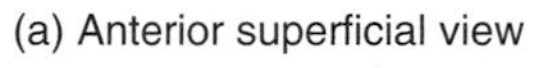

(a) Anterior superficial view

(b) Anterior deep view

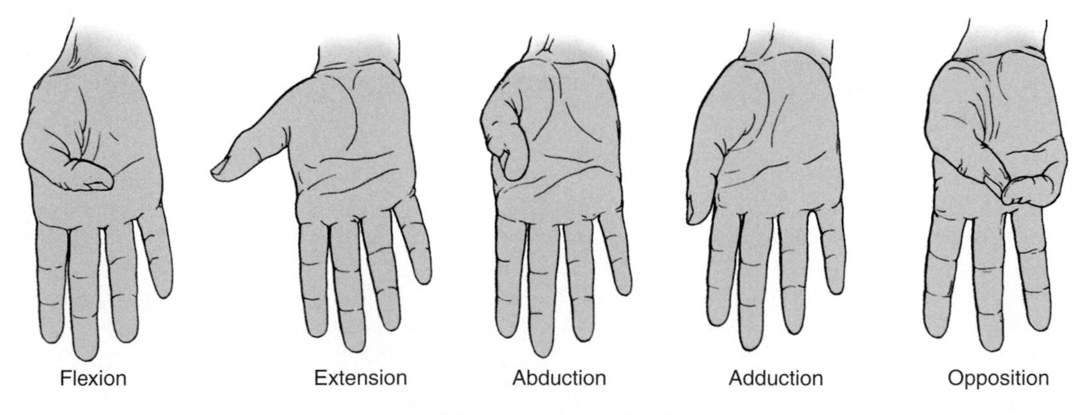

(c) Movements of the thumb

Which structure moves when muscles of the thenar eminence contract?

Exhibit 11.16 Muscles that Move the Vertebral Column (Backbone) (Figure 11.19)

▶ OBJECTIVE

- Describe the origin, insertion, action, and innervation of the muscles that move the vertebral column.

The muscles that move the vertebral column (backbone) are quite complex because they have multiple origins and insertions and there is considerable overlap among them. One way to group the muscles is on the basis of the general direction of the muscle bundles and their approximate lengths. For example, the splenius muscles arise from the midline and extend laterally and superiorly to their insertions (Figure 11.19a). The erector spinae muscle arises from either the midline or more laterally but usually runs almost longitudinally, with neither a significant lateral nor medial direction as it is traced superiorly. The transversospinalis arise laterally but extend toward the midline as they are traced superiorly. Deep to these three muscle groups are small segmental muscles that extend between spinous processes or transverse processes of vertebrae. Because the scalene muscles also assist in moving the vertebral column, they are included in this exhibit. Note in Exhibit 11.17 that the rectus abdominis, external oblique, internal oblique, and quadratus lumborum muscles also play a role in moving the vertebral column.

The bandage-like **splenius** muscles are attached to the sides and back of the neck. The two muscles in this group are named on the basis of their superior attachments (insertions): **splenius capitis** (head region) and **splenius cervicis** (cervical region). They extend the head and laterally flex and rotate the head.

The **erector spinae** is the largest muscle mass of the back, forming a prominent bulge on either side of the vertebral column. It is the chief extensor of the vertebral column. It is also important in controlling flexion, lateral flexion, and rotation of the vertebral column and in maintaining the lumbar curve, because the main mass of the muscle is in the lumbar region. It consists of three groups: iliocostalis (laterally placed), longissimus (intermediately placed), and spinalis (medially placed). These groups, in turn, consist of a series of overlapping muscles, and the muscles within the groups are named according to the regions of the body with which they are associated. The **iliocostalis group** consists of three muscles: the **iliocostalis cervicis** (cervical region), **iliocostalis thoracis** (thoracic region), and **iliocostalis lumborum** (lumbar region). The **longissimus group** resembles a herring bone and consists of three muscles: the **longissimus capitis** (head region), **longissimus cervicis** (cervical region), and **longissimus thoracic** (thoracic region). The **spinalis group** also consists of three muscles: the **spinalis capitis, spinalis cervicis,** and **spinalis thoracis.**

The **transversospinales** are named because their fibers run from the transverse processes to the spinous processes of the vertebrae. The semispinalis muscles in this group are also named according to the region of the body with which they are associated: **semispinalis capitis** (head region), **semispinalis cervicis** (cervical region), and **semispinalis thoracic** (thoracic region). These muscles extend the vertebral column and rotate the head. The **multifidus** muscle in this group, as its name implies, is segmented into several bundles. It extends and laterally flexes the vertebral column and rotates the head. The **rotatores** muscles of this group are short and are found along the entire length of the vertebral column. They extend and rotate the vertebral column.

Within the **segmental** muscle group (Figure 11.17b) are the **interspinales** and **intertransversarii** muscles, which unite the spinous and transverse processes of consecutive vertebrae. They function primarily in stabilizing the vertebral column during its movements.

Within the **scalene** group (Figure 11.17c), the **anterior scalene** muscle is anterior to the middle scalene muscle, the **middle scalene** muscle is intermediate in placement and is the longest and largest of the scalene muscles, and the **posterior scalene** muscle is posterior to the middle scalene muscle and is the smallest of the scalene muscles. These muscles flex, laterally flex, and rotate the head and assist in deep inhalation.

Innervation

- Splenius capitis: middle cervical nerves
- Splenius cervicis: inferior cervical nerves
- Iliocostalis cervicis and semispinalis capitis: cervical and thoracic nerves
- Iliocostalis thoracis: thoracic nerves
- Iliocostalis lumborum: lumbar nerves
- Longissimus capitis: middle and inferior cervical nerves
- Longissimus cervicis, longissimus thoracis, all spinalis muscles, multifidus, rotatores, and interspinales: cervical, thoracic, and lumbar nerves
- Semispinalis cervicis and semispinalis thoracis: cervical and thoracic nerves
- Intertransversarii: cervical, thoracic and lumbar spinal nerves
- Anterior scalene: cervical spinal nerves C5–C6
- Middle scalene: cervical spinal nerves C3–C8
- Posterior scalene: cervical spinal nerves C6–C8.

Back Injuries and Heavy Lifting

Full flexion at the waist, as in touching your toes, overstretches the erector spinae muscles. Because of the length-tension relationship (see page 285), muscles that are overstretched cannot contract effectively. Straightening up from such a position is therefore initiated by the hamstring muscles on the back of the thigh and the gluteus maximus muscles of the buttocks. The erector spinae muscles join in as the degree of flexion decreases. Improperly lifting a heavy weight, however, can strain the erector spinae muscles. The result can be painful muscle spasms, tearing of tendons and ligaments of the lower back, and rupturing of intervertebral discs. The lumbar muscles are adapted for maintaining posture, not for lifting. This is why it is important to kneel and use the powerful extensor muscles of the thighs and buttocks while lifting a heavy load. ■

Relating Muscles to Movements

Arrange the muscles in this exhibit according to the following actions on the head at the atlanto-occipital and intervertebral joints: (1) flexion, (2) extension, (3) lateral flexion, (4) rotation to same side as contracting muscle, and (5) rotation to opposite side as contracting muscle; the following actions on the vertebral column at the intervertebral joints: (1) flexion, (2) extension, (3) lateral flexion, (4) rotation, and (5) stabilization; and the following action on the ribs: elevation during deep inhalation. The same muscle may be mentioned more than once.

▶ CHECKPOINT

What are the four major groups of muscles that move the vertebral column?

(continues)

Exhibit 11.16 Muscles that Move the Vertebral Column (Backbone) (Figure 11.19) *(continued)*

Muscle	Origin	Insertion	Action
Splenius (SPLĒ-nē-us)			
Splenius capitis (KAP-i-tis; *splenium* = bandage; *capit-* = head)	Ligamentum nuchae and spinous processes of seventh cervical vertebra and first three or four thoracic vertebrae.	Occipital bone and mastoid process of temporal bone.	Acting together (bilaterally), extend head; acting singly (unilaterally), laterally flex and rotate head to same side as contracting muscle.
Splenius cervicis (SER-vi-kis; *cervic-* = neck)	Spinous processes of third through sixth thoracic vertebrae.	Transverse processes of first two or four cervical vertebrae.	Acting together, extend head; acting singly, laterally flex and rotate head to same side as contracting muscle.
Erector Spinae (e-REK-tor SPI-nē) Consists of iliocostalis muscles (lateral), longissimus muscles (intermediate), and spinalis muscles (medial).			
Iliocostalis Group (Lateral)			
Iliocostalis cervicis (il′-ē-ō-kos-TAL-is SER-vi-kis; *ilio-* = flank; *costa-* = rib)	Superior six ribs.	Transverse processes of fourth to sixth cervical vertebrae.	Acting together, muscles of each region (cervical, thoracic, and lumbar) extend and maintain erect posture of vertebral column of their respective regions; acting singly, laterally flex vertebral column of their respective regions.
Iliocostalis thoracis (il′-ē-ō-kos-TAL-is thō-RA-sis; *thorac-* = chest)	Inferior six ribs.	Superior six ribs.	
Iliocostalis lumborum (il′-ē-ō-kos-TAL-is lum-BOR-um)	Iliac crest.	Inferior six ribs.	
Longissimus Group (Intermediate)			
Longissimus capitis (lon-JIS-i-mus KAP-i-tis; *longissimus* = longest)	Transverse processes of superior four thoracic vertebrae and articular processes of inferior four cervical vertebrae.	Mastoid process of temporal bone.	Acting together, both longissimus capitis muscles extend head; acting singly, rotate head to same side as contracting muscle. Acting together, longissimus cervicis and both longissimus thoracis muscles extend vertebral column of their respective regions; acting singly, laterally flex vertebral column of their respective regions.
Longissimus cervicis (lon-JIS-i-mus SER-vi-kis)	Transverse processes of fourth and fifth thoracic vertebrae.	Transverse processes of second to sixth cervical vertebrae.	
Longissimus thoracis (lon-JIS-i-mus thō-RA-sis)	Transverse processes of lumbar vertebrae.	Transverse processes of all thoracic and superior lumbar vertebrae and ninth and tenth ribs.	
Spinalis Group (Medial)			
Spinalis capitis (spi-NĀ-lis KAP-i-tis; *spinal-* = vertebral column)	Arises with semispinalis capitis.	Occipital bone.	Acting together, muscles of each region (cervical, thoracic, and lumbar) extend vertebral column of their respective regions.
Spinalis cervicis (spi-NĀ-lis SER-vi-kis)	Ligamentum nuchae and spinous process of seventh cervical vertebra.	Spinous process of axis.	
Spinalis thoracis (spi-NĀ-lis thō-RA-sis)	Spinous processes of superior lumbar and inferior thoracic vertebrae.	Spinous processes of superior thoracic vertebrae.	

Muscle	Origin	Insertion	Action
Transversospinales (trans-ver-sō-spi-NA-lēz)			
Semispinalis capitis (sem′-ē-spi-NĀ-lis KAP-i-tis; *semi-* = partially or one-half)	Transverse processes of first six or seven thoracic vertebrae and seventh cervical vertebra, and articular processes of fourth, fifth, and sixth cervical vertebrae.	Occipital bone.	Acting together, extend head; acting singly, rotate head to side opposite contracting muscle.
Semispinalis cervicis (sem′-ē-spi-NĀ-lis SER-vi-kis)	Transverse processes of superior five or six thoracic vertebrae.	Spinous processes of first to fifth cervical vertebrae.	Acting together, both semispinalis cervicis and both semispinalis thoracis muscles extend vertebral column of their respective regions; acting singly, rotate head to side opposite contracting muscle.
Semispinalis thoracis (sem′-ē-spi-NĀ-lis thō-RA-sis)	Transverse processes of sixth to tenth thoracic vertebrae.	Spinous processes of superior four thoracic and last two cervical vertebrae.	
Multifidus (mul-TIF-i-dus; *multi* = many; *fid-* = segmented)	Sacrum, ilium, transverse processes of lumbar, thoracic, and inferior four cervical vertebrae.	Spinous process of a more superior vertebra.	Acting together, extend vertebral column; acting singly, laterally flex vertebral column and rotate head to side opposite contracting muscle.
Rotatores (rō′-ta-TŌ-rez; singular is **rotatore;** *rotatore* = to rotate)	Transverse processes of all vertebrae.	Spinous process of vertebra superior to the one of origin.	Acting together, extend vertebral column; acting singly, rotate vertebral column to side opposite contracting muscle.
Segmental (seg-MEN-tal)			
Interspinales (in-ter-spī-NĀ-lēz; *inter-* = SPĪ-nāl-ez; *inter-* = between)	Superior surface of all spinous processes.	Inferior surface of spinous process of vertebra superior to the one of origin.	Acting together, extend vertebral column; acting singly, stabilize vertebral column during movement.
Intertransversarii (in′-ter- trans-vers-AR-ē-ī ; singular is **intertransversarius**)	Transverse processes of all vertebrae.	Transverse process of vertebra superior to the one of origin.	Acting together, extend vertebral column; acting singly, laterally flex vertebral column and stabilize it during movements.
Scalenes (SKĀ-lēnz)			
Anterior scalene (SKĀ-lēn; *anterior* = front; *scalene* = uneven)	Transverse processes of third through sixth cervical vertebrae.	First rib.	Acting together, right and left anterior scalene and middle scalene muscles flex head and elevate first ribs during deep inhalation; acting singly, laterally flex head and rotate head to side opposite contracting muscle.
Middle scalene (SKĀ-lēn)	Transverse processes of inferior six cervical vertebrae.	First rib.	
Posterior scalene (SKĀ-lēn)	Transverse processes of fourth through sixth cervical vertebrae.	Second rib.	Acting together, flex head and elevate second ribs during deep inhalation; acting singly, laterally flex head and rotate head to side opposite contracting muscle.

(continues)

Exhibit 11.16 Muscles that Move the Vertebral Column (Backbone) (Figure 11.19) *(continued)*

Figure 11.19 Muscles that move the vertebral column (backbone).

The erector spinae group is the largest muscular mass of the body and is the chief extensor of the vertebral column.

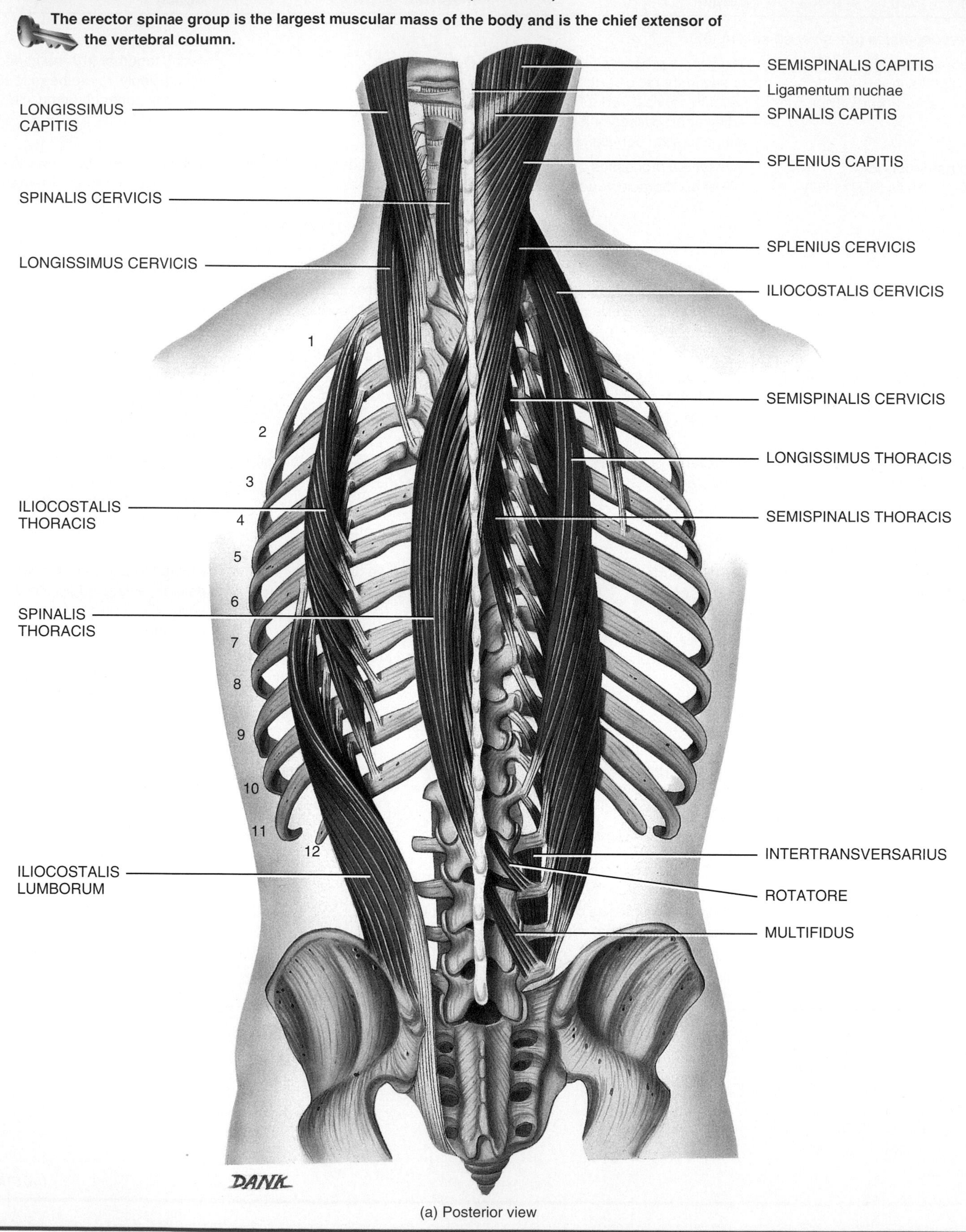

(a) Posterior view

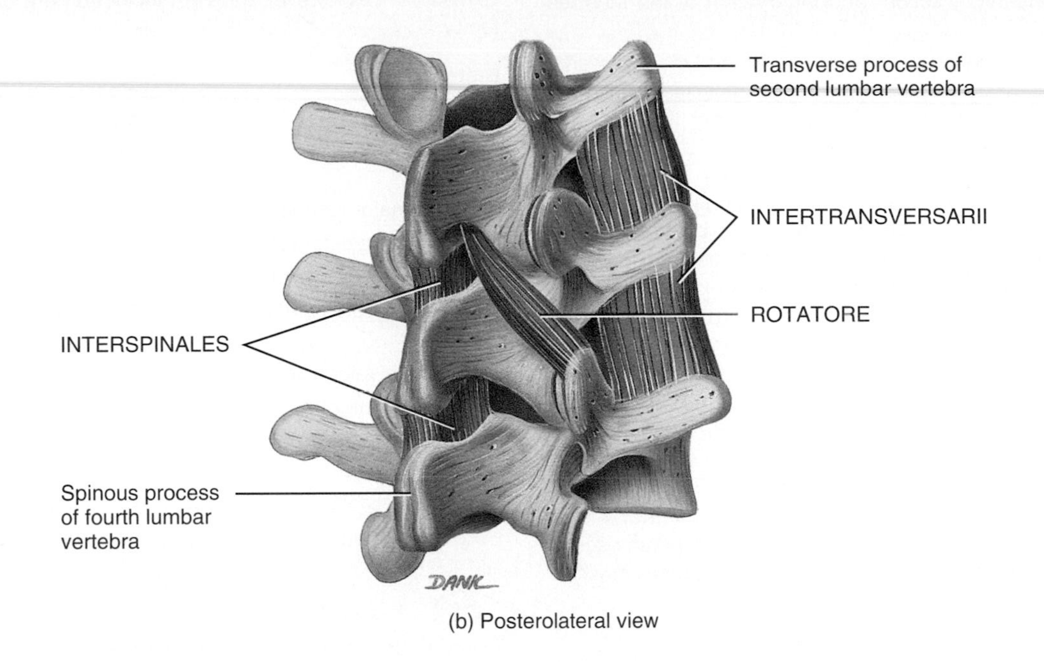

(b) Posterolateral view

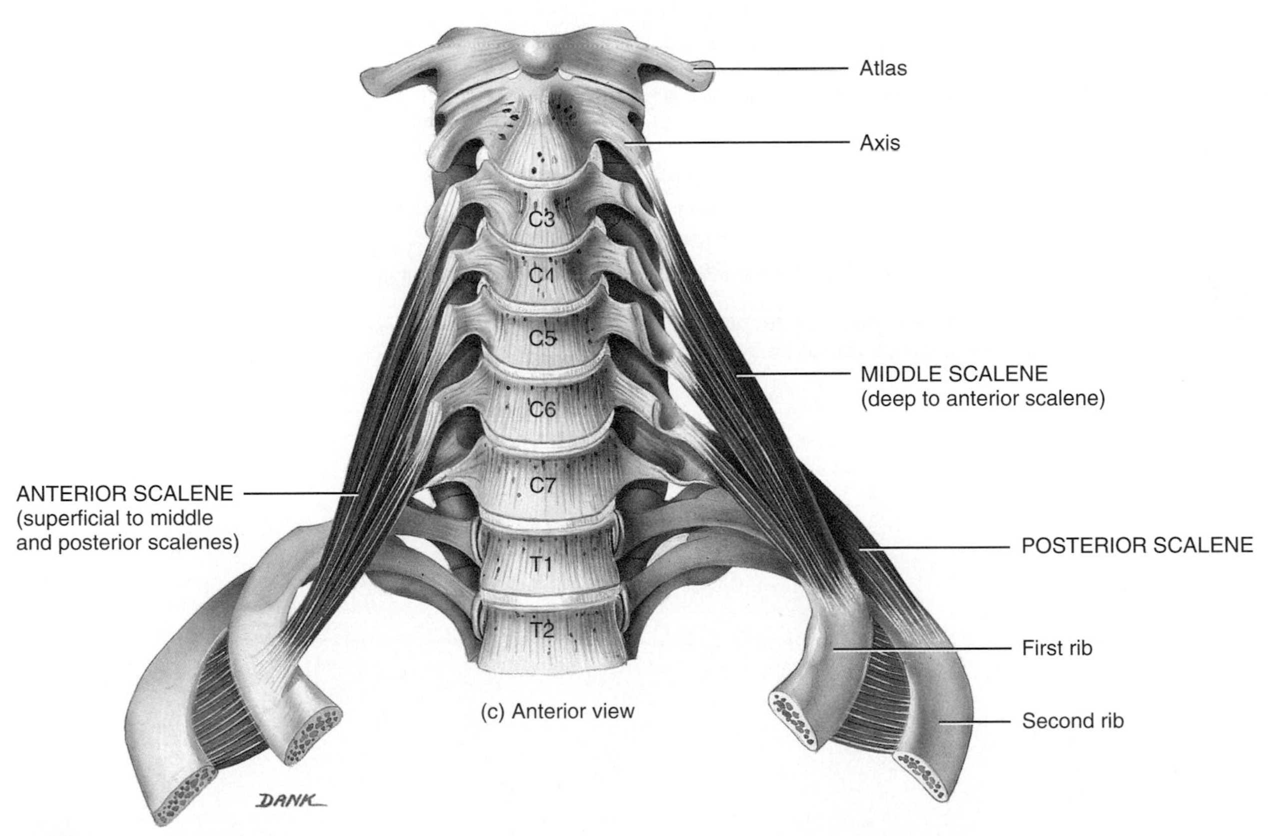

(c) Anterior view

Which muscles originate at the midline and extend laterally and upward to their insertion?

Exhibit 11.17 Muscles that Move the Femur (Thigh Bone) (Figure 11.20)

▶ OBJECTIVE

- Describe the origin, insertion, action, and innervation of the muscles that move the femur.

As you will see, muscles of the lower limbs are larger and more powerful than those of the upper limbs because lower limb muscles function in stability, locomotion, and maintenance of posture. Upper limb muscles are characterized by versatility of movement. In addition, muscles of the lower limbs often cross two joints and act equally on both.

The majority of muscles that move the femur originate on the pelvic girdle and insert on the femur. The **psoas major** and **iliacus** muscles have a common insertion. Together, they are known as the **iliopsoas** (il′-ē-ō-SŌ-as) muscle because they share a common insertion (lesser trochanter of femur). Sometimes people "cheat" when doing situps by using their iliopsoas muscle instead of the abdominal muscles. By doing so, this can potentially produce back pain due to the origin of the iliopsoas on the backbone. There are three gluteal muscles: gluteus maximus, gluteus medius, and gluteus minimus. The **gluteus maximus** is the largest and heaviest of the three muscles and is one of the largest muscles in the body. It is the chief extensor of the femur. The **gluteus medius** is mostly deep to the gluteus maximus and is a powerful abductor of the femur at the hip joint. It is a common site for an intramuscular injection. The **gluteus minimus** is the smallest of the gluteal muscles and lies deep to the gluteus medius.

The **tensor fasciae latae** muscle is located on the lateral surface of the thigh. The *fascia lata* is a layer of deep fascia, composed of dense connective tissue, that encircles the entire thigh. It is well developed laterally where, together with the tendons of the tensor fasciae and gluteus maximus muscles, it forms a structure called the **iliotibial tract.** The tract inserts into the lateral condyle of the tibia.

The **piriformis, obturator internus, obturator externus, superior gemellus, inferior gemellus,** and **quadratus femoris** muscles are all deep to the gluteus maximus muscle and function as lateral rotators of the femur at the hip joint.

Three muscles on the medial aspect of the thigh are the **adductor longus, adductor brevis,** and **adductor magnus.** They originate on the pubic bone and insert on the femur. All three muscles adduct, flex, and medially rotate the femur at the hip joint. The **pectineus** muscle also adducts and flexes the femur at the hip joint.

Technically, the adductor muscles and pectineus muscles are components of the medial compartment of the thigh and could be included in Exhibit 11.18. However, they are included here because they act on the femur.

Innervation

Nerves derived from the lumbar and sacral plexuses (shown in Exhibits 13.3 and 13.4 on pages 442 and 444) innervate the muscles that move the femur.

- Psoas major: lumbar nerves L2–L3
- Iliacus and pectineus: femoral nerve
- Gluteus maximus: inferior gluteal nerve
- Gluteus medius and minimus and tensor fasciae latae: superior gluteal nerve
- Piriformis: sacral spinal nerves S1 or S2, mainly S1
- Obturator internus and superior gemellus: nerve to obturator internus
- Obturator externus, adductor longus, and adductor brevis: obturator nerve
- Inferior gemellus and quadratus femoris: nerve to quadratus femoris
- Adductor magnus: obturator and sciatic nerves

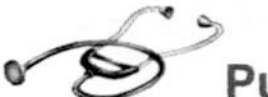

Pulled Groin

Certain athletic and other activities may cause a **pulled groin,** a strain, stretching, or tearing of the distal attachments of the medial muscles of the thigh, iliopsoas, and/or adductor muscles. Generally, a pulled groin results from activities that involve quick sprints, as might occur during soccer, tennis, football, and running. ■

Relating Muscles to Movements

Arrange the muscles in this exhibit according to the following actions on the thigh at the hip joint: (1) flexion, (2) extension, (3) abduction, (4) adduction, (5) medial rotation, and (6) lateral rotation. The same muscle may be mentioned more than once.

▶ CHECKPOINT

What forms the iliotibial tract?

Muscle	Origin	Insertion	Action
Psoas major (SŌ-as; *psoa* = a muscle of the loin)	Transverse processes and bodies of lumbar vertebrae.	With iliacus into lesser trochanter of femur.	Psoas major and iliacus muscles acting together flex thigh at hip joint, rotate thigh laterally, and flex trunk on the hip as in sitting up from the supine position
Iliacus (il′-ē-AK-us; *iliac-* = ilium)	Iliac fossa and sacrum.	With psoas major into lesser trochanter of femur.	
Gluteus maximus (GLOO-tē-us MAK-si-mus; *glute-* = rump or buttock; *maximus* = largest)	Iliac crest, sacrum, coccyx, and aponeurosis of sacrospinalis.	Iliotibial tract of fascia lata and lateral part of linea aspera under greater trochanter (gluteal tuberosity) of femur.	Extends thigh at hip joint and laterally rotates thigh.
Gluteus medius (GLOO-tē-us MĒ-dē-us; *medi-* = middle)	Ilium.	Greater trochanter of femur.	Abducts thigh at hip joint and medially rotates thigh.
Gluteus minimus (GLOO-tē-us MIN-i-mus; *minim-* = smallest)	Ilium.	Greater trochanter of femur.	Abducts thigh at hip joint and medially rotates thigh.
Tensor fasciae latae (TEN-sor FA-shē-ē LĀ-tē; *tensor* = makes tense; *fasciae* = of the band; *lat-* = wide)	Iliac crest.	Tibia by way of the iliotibial tract.	Flexes and abducts thigh at hip joint.
Piriformis (pir-i-FOR-mis; *piri-* = pear; *form-* = shape)	Anterior sacrum.	Superior border of greater trochanter of femur.	Laterally rotates and abducts thigh at hip joint.
Obturator internus (OB-too-rā′-tor in-TER-nus; *obturator* = obturator foramen; *intern-* = inside)	Inner surface of obturator foramen, pubis, and ischium.	Greater trochanter of femur.	Laterally rotates and abducts thigh at hip joint.
Obturator externus (OB-too-rā′-tor ex-TER-nus; *extern-* = outside)	Outer surface of obturator membrane.	Deep depression inferior to greater trochanter (trochanteric fossa) of femur.	Laterally rotates and abducts thigh at hip joint.
Superior gemellus (jem-EL-lus; *superior* = above; *gemell-* = twins)	Ischial spine.	Greater trochanter of femur.	Laterally rotates and abducts thigh at hip joint.
Inferior gemellus (jem-EL-lus; *inferior* = below)	Ischial tuberosity.	Greater trochanter of femur.	Laterally rotates and abducts thigh at hip joint.
Quadratus femoris (kwod-RĀ-tus FEM-or-is; *quad-* = square, four-sided; *femoris* = femur)	Ischial tuberosity.	Elevation superior to mid-portion of intertrochanteric crest (quadrate tubercle) on posterior femur.	Laterally rotates and stabilizes hip joint.
Adductor longus (LONG-us; *adductor* = moves part closer to midline; *longus* = long)	Pubic crest and pubic symphysis.	Linea aspera of femur.	Adducts and flexes thigh at hip joint and medially rotates thigh.
Adductor brevis (BREV-is; *brevis* = short)	Inferior ramus of pubis.	Superior half of linea aspera of femur.	Adducts and flexes thigh at hip joint and medially rotates thigh.
Adductor magnus (MAG-nus; *magnus* = large)	Inferior ramus of pubis and ischium to ischial tuberosity.	Linea aspera of femur.	Adducts thigh at hip joint and medially rotates thigh; anterior part flexes thigh at hip joint, and posterior part extends thigh at hip joint.
Pectineus (pek-TIN-ē-us; *pectin-* = a comb)	Superior ramus of pubis.	Pectineal line of femur, between lesser trochanter and linea aspera.	Flexes and adducts thigh at hip joint.

(continues)

Exhibit 11.17 Muscles that Move the Femur (Thigh Bone) (Figure 11.20) *(continued)*

Figure 11.20 Muscles that move the femur (thigh bone). (See Tortora, *A Photographic Atlas of the Human Body,* Figures 5.16 and 5.17.)

Most muscles that move the femur originate on the pelvic (hip) girdle and insert on the femur.

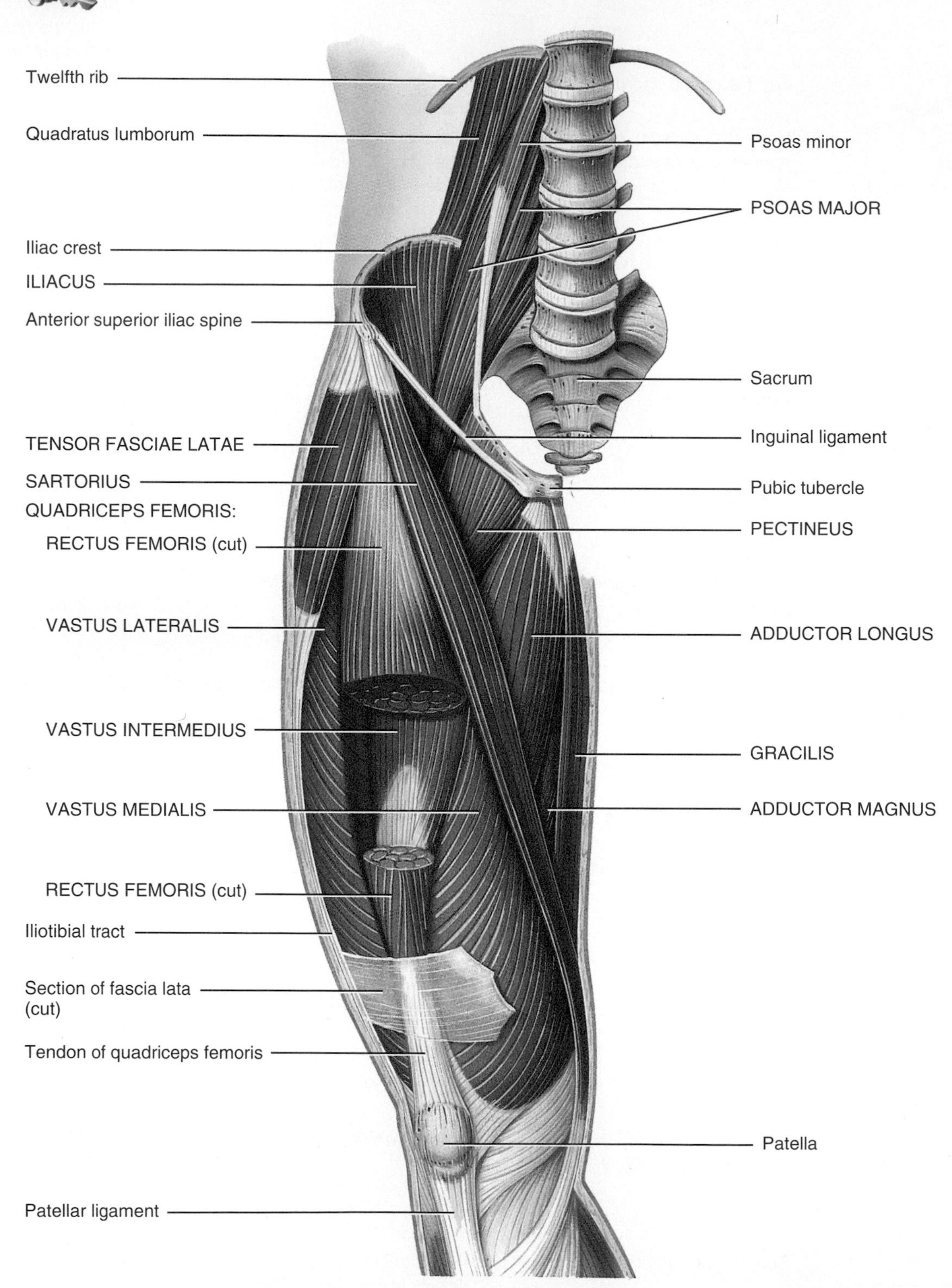

(a) Anterior superficial view

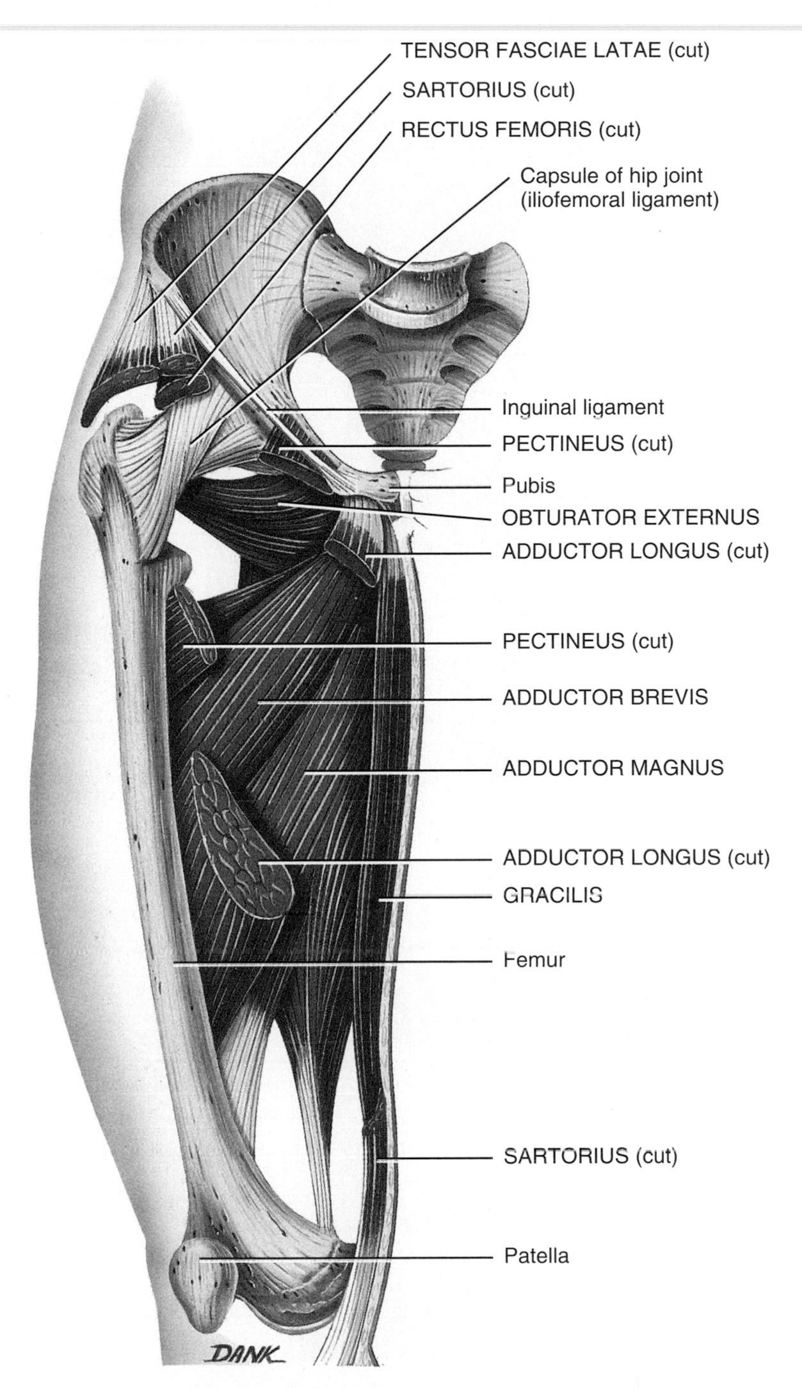

(b) Anterior deep view (femur rotated laterally)

(continues)

Exhibit 11.17 Muscles that Move the Femur (Thigh Bone) (Figure 11.20) *(continued)*

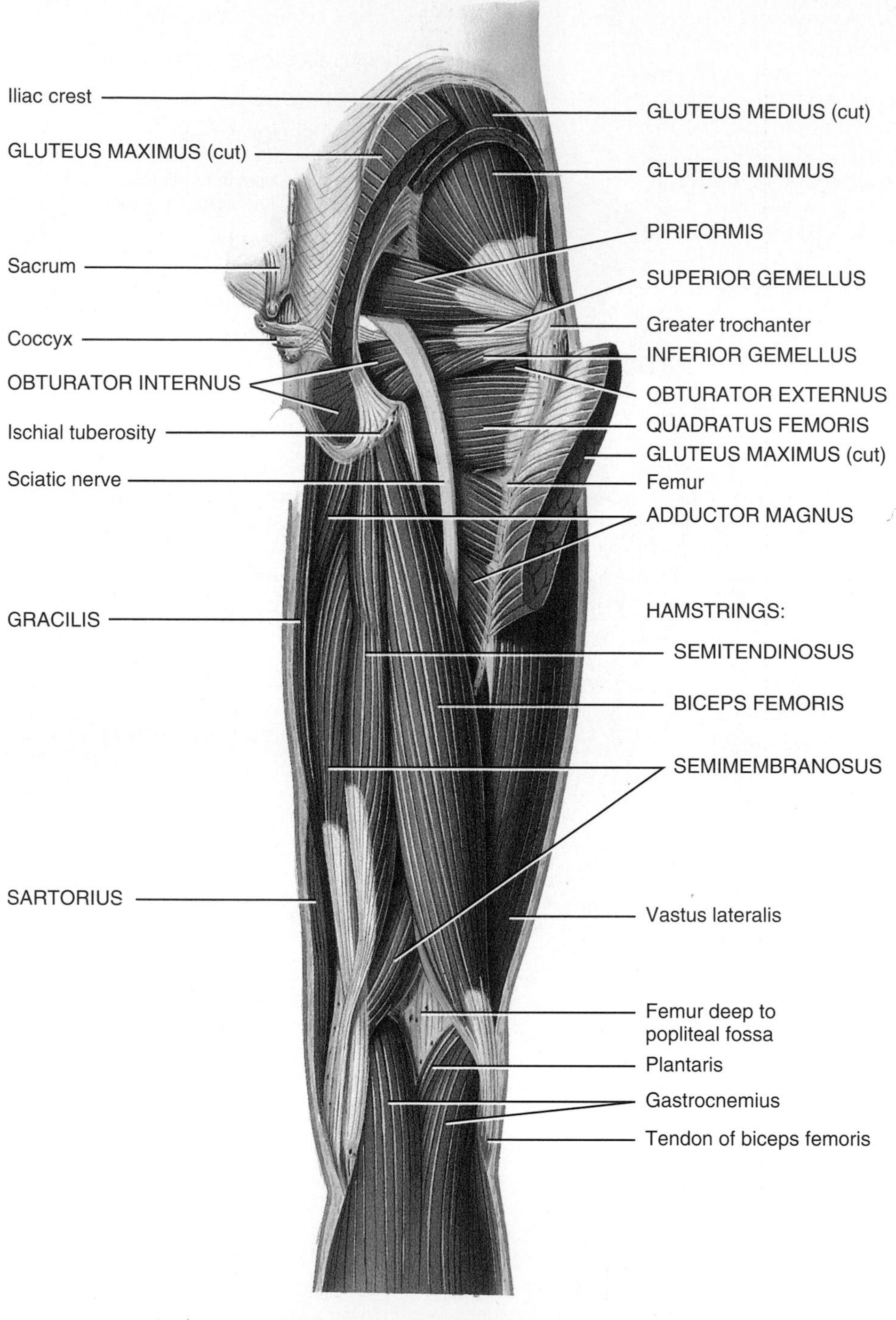

(c) Posterior superficial view

What are the main differences between the muscles of the upper and lower limbs?

Exhibit 11.18 Muscles that Act on the Femur (Thigh Bone) and Tibia and Fibula (Leg Bones) (Figures 11.20 and 11.21)

▶ OBJECTIVE

- Describe the origin, insertion, action, and innervation of the muscles that act on the femur, tibia, and fibula.

Deep fascia separate the muscles that act on the femur (thigh bone) and tibia and fibula (leg bones) into medial, anterior, and posterior compartments. The muscles of the **medial (adductor) compartment of the thigh** adduct the femur at the hip joint. (See the adductor magnus, adductor longus, adductor brevis, and pectineus, which are components of the medial compartment, in Exhibit 11.17.) The **gracilis,** the other muscle in the medial compartment, not only adducts the thigh, but also flexes the leg at the knee joint. For this reason, it is discussed here. The gracilis is a long, straplike muscle on the medial aspect of the thigh and knee.

The muscles of the **anterior (extensor) compartment of the thigh** extend the leg (and flex the thigh). This compartment contains the quadriceps femoris and sartorius muscles. The **quadriceps femoris** muscle is the biggest muscle in the body, covering most of the anterior surface and sides of the thigh. The muscle is actually a composite muscle, usually described as four separate muscles: (1) **rectus femoris,** on the anterior aspect of the thigh; (2) **vastus lateralis,** on the lateral aspect of the thigh; (3) **vastus medialis,** on the medial aspect of the thigh; and (4) **vastus intermedius,** located deep to the rectus femoris between the vastus lateralis and vastus medialis. The common tendon for the four muscles is known as the **quadriceps tendon,** which inserts into the patella. The tendon continues below the patella as the **patellar ligament,** which attaches to the tibial tuberosity. The quadriceps femoris muscle is the great extensor muscle of the leg. The **sartorius** is a long, narrow muscle that forms a band across the thigh from the ilium of the hip bone to the medial side of the tibia. The various movements it produces help effect the cross-legged sitting position in which the heel of one limb is placed on the knee of the opposite limb. It is known as the tailor's muscle because tailors often assume this cross-legged sitting position. (Because the major action of the sartorius muscle is to move the thigh rather than the leg, it could have been included in Exhibit 11.17.)

The muscles of the **posterior (flexor) compartment of the thigh** flex the leg (and extend the thigh). This compartment is composed of three muscles collectively called the **hamstrings:** (1) **biceps femoris,** (2) **semitendinosus,** and (3) **semimembranosus.** The hamstrings are so-named because their tendons are long and stringlike in the popliteal area and from an old practice of butchers in which they hung hams for smoking by these long tendons. Because the hamstrings span two joints (hip and knee), they are both extensors of the thigh and flexors of the leg. The **popliteal fossa** is a diamond-shaped space on the posterior aspect of the knee bordered laterally by the tendons of the biceps femoris muscle and medially by the tendons of the semitendinosus and semimembranosus muscles.

Innervation

The obturator and femoral nerves, derived from the lumbar plexus (shown in Exhibit 13.3 on page 442), and the sciatic nerve, derived from the sacral plexus (shown in Exhibit 13.4 on page 444), innervate muscles that act on the femur, tibia, and fibula.

- Gracilis: obturator nerve
- Quadriceps femoris and sartorius: femoral nerve
- Biceps femoris: tibial and common peroneal nerves from the sciatic nerve
- Semimembranosus and semitendinosus: tibial nerve from the sciatic nerve

Pulled Hamstrings

A strain or partial tear of the proximal hamstring muscles is referred to as **pulled hamstrings** or **hamstring strains.** They are common sports injuries in individuals who run very hard and/or are required to perform quick starts and stops. Sometimes the violent muscular exertion required to perform a feat tears off part of the tendinous origins of the hamstrings, especially the biceps femoris, from the ischial tuberosity. This is usually accompanied by a contusion (bruising), tearing of some of the muscle fibers, and rupture of blood vessels, producing a hematoma (collection of blood) and pain. Adequate training with good balance between the quadriceps femoris and hamstrings and stretching exercises before running or competing are important in preventing this injury. ■

Relating Muscles to Movements

Arrange the muscles in this exhibit according to the following actions on the thigh at the hip joint: (1) abduction, (2) adduction, (3) lateral rotation, (4) flexion, and (5) extension; and according to the following actions on the leg at the knee joint: (1) flexion and (2) extension. The same muscle may be mentioned more than once.

▶ CHECKPOINT

Which muscles are part of the medial, anterior, and posterior compartments of the thigh?

(continues)

Exhibit 11.18 Muscles that Act on the Femur (Thigh Bone) and Tibia and Fibula (Leg Bones) (Figures 11.20 and 11.21)

Muscle	Origin	Insertion	Action
Medial (adductor) compartment of the thigh			
Adductor magnus (MAG-nus)	See Exhibit 11.17.		
Adductor longus (LONG-us)			
Adductor brevis (BREV-is)			
Pectineus (pek-TIN-ē-us)			
Gracilis (gra-SIL-is; *gracilis* = slender)	Pubic symphysis and pubic arch.	Medial surface of body of tibia.	Adducts thigh at hip joint, medially rotates thigh, and flexes leg at knee joint.
Anterior (extensor) compartment of the thigh			
Quadriceps femoris (KWOD-ri-ceps FEM-or-is; *quadriceps* = four heads [of origin]; *femoris* = femur)			
Rectus femoris (REK-tus FEM-or-is; *rectus* = fascicles parallel to midline)	Anterior inferior iliac spine.	Patella via quadriceps tendon and then tibial tuberosity via patellar ligament.	All four heads extend leg at knee joint; rectus femoris muscle acting alone also flexes thigh at hip joint.
Vastus lateralis (VAS-tus lat′-e-RĀ-lis; *vast-* = huge; *lateralis* = lateral)	Greater trochanter and linea aspera of femur.		
Vastus medialis (VAS-tus mē′-dē-A-lis; *medialis* = medial)	Linea aspera of femur.		
Vastus intermedius (VAS-tus in′-ter-MĒ-dē-us; *intermedius* = middle)	Anterior and lateral surfaces of body of femur.		
Sartorius (sar-TOR-ē-us; *sartor* = tailor; longest muscle in body)	Anterior superior iliac spine.	Medial surface of body of tibia.	Flexes leg at knee joint; flexes, abducts, and laterally rotates thigh at hip joint.
Posterior (flexor) compartment of the thigh			
Hamstrings A collective designation for three separate muscles.			
Biceps femoris (BĪ-ceps FEM-or-is; *biceps* = two heads of origin)	Long head arises from ischial tuberosity; short head arises from linea aspera of femur.	Head of fibula and lateral condyle of tibia.	Flexes leg at knee joint and extends thigh at hip joint.
Semitendinosus (sem′-ē-ten-di-NŌ-sus; *semi-* = half; *tendo* = tendon)	Ischial tuberosity.	Proximal part of medial surface of shaft of tibia.	Flexes leg at knee joint and extends thigh at hip joint.
Semimembranosus (sem′-ē-mem-bra-NŌ-sus; *membran-* = membrane)	Ischial tuberosity.	Medial condyle of tibia.	Flexes leg at knee joint and extends thigh at hip joint.

(continued)

Figure 11.21 Muscles that act on the femur (thigh bone) and tibia and fibula (leg bones).

Muscles that act on the leg originate in the hip and thigh; deep fascia separates them into compartments.

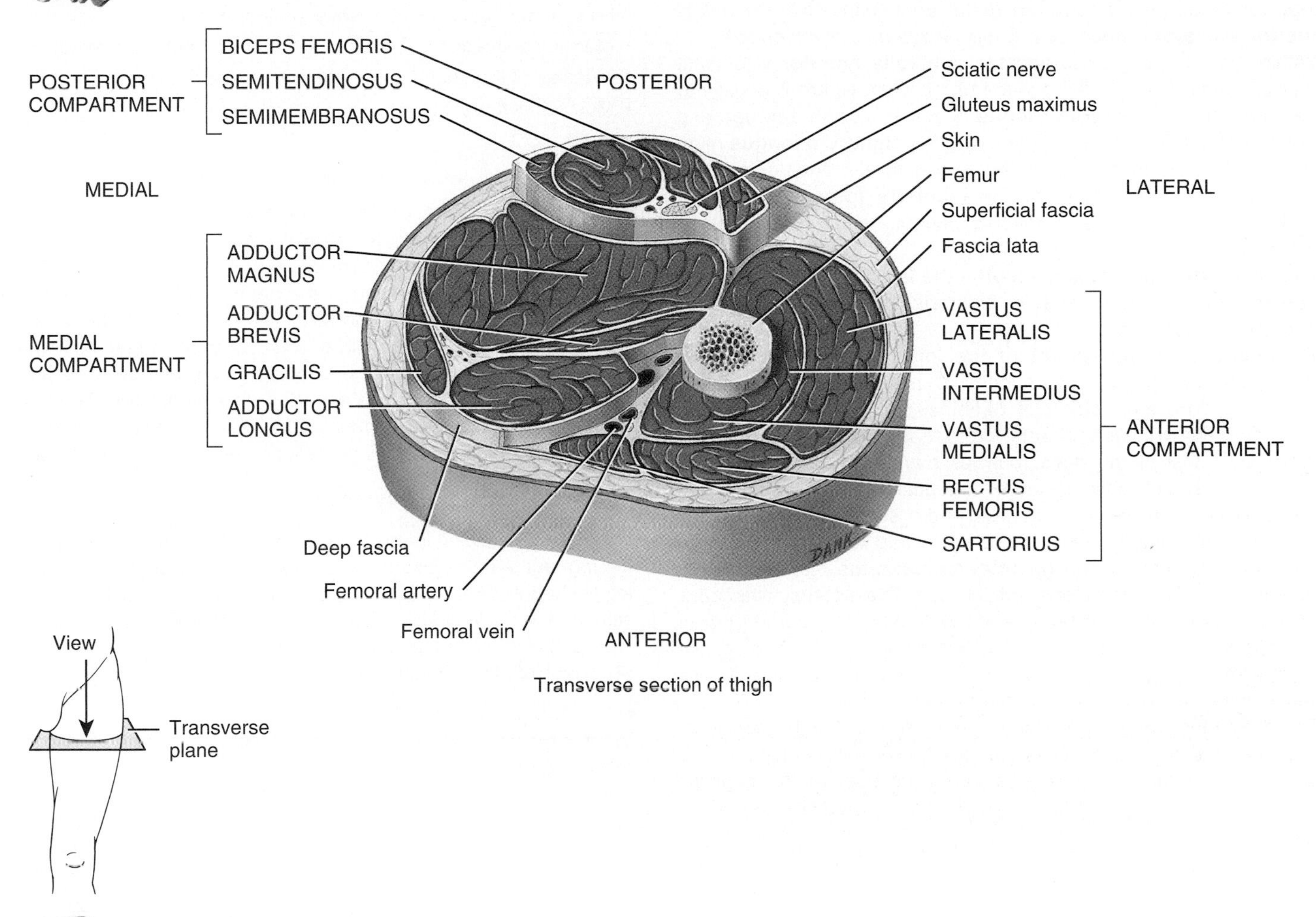

Transverse section of thigh

Which muscles comprise the quadriceps femoris and hamstring muscles?

Exhibit 11.19 Muscles that Move the Foot and Toes (Figure 11.22)

► **OBJECTIVE**

- Describe the origin, insertion, action, and innervation of the muscles that move the foot and toes.

Muscles that move the foot and toes are located in the leg. The muscles of the leg, like those of the thigh, are divided by deep fascia into three compartments: anterior, lateral, and posterior. The **anterior compartment of the leg** consists of muscles that dorsiflex the foot. In a situation analogous to the wrist, the tendons of the muscles of the anterior compartment are held firmly to the ankle by thickenings of deep fascia called the **superior extensor retinaculum** *(transverse ligament of the ankle)* and **inferior extensor retinaculum** *(cruciate ligament of the ankle).*

Within the anterior compartment, the **tibialis anterior** is a long, thick muscle against the lateral surface of the tibia, where it is easy to palpate. The **extensor hallucis longus** is a thin muscle between and partly deep to the tibialis anterior and **extensor digitorum longus** muscles. This featherlike muscle is lateral to the tibialis anterior muscle, where it can easily be palpated. The **fibularis (peroneus) tertius** muscle is part of the extensor digitorum longus, with which it shares a common origin.

The **lateral (fibular) compartment of the leg** contains two muscles that plantar flex and evert the foot: the **fibularis (peroneus) longus** and **fibularis (peroneus) brevis.**

The **posterior compartment of the leg** consists of muscles in superficial and deep groups. The superficial muscles share a common tendon of insertion, the **calcaneal (Achilles) tendon,** the strongest tendon of the body. It inserts into the calcaneal bone of the ankle. The superficial and most of the deep muscles plantar flex the foot at the ankle joint. The superficial muscles of the posterior compartment are the gastrocnemius, soleus, and plantaris—the so-called calf muscles. The large size of these muscles is directly related to the characteristic upright stance of humans. The **gastrocnemius** is the most superficial muscle and forms the prominence of the calf. The **soleus,** which lies deep to the gastrocnemius, is broad and flat. It derives its name from its resemblance to a flat fish (sole). The **plantaris** is a small muscle that may be absent; sometimes there are two of them in each leg. It runs obliquely between the gastrocnemius and soleus muscles.

The deep muscles of the posterior compartment are the popliteus, tibialis posterior, flexor digitorum longus, and flexor hallucis longus. The **popliteus** is a triangular muscle that forms the floor of the popliteal fossa. The **tibialis posterior** is the deepest muscle in the posterior compartment. It lies between the flexor digitorum longus and flexor hallucis longus muscles. The **flexor digitorum longus** is smaller than the **flexor hallucis longus,** even though the former flexes four toes, whereas the latter flexes only the great toe at the interphalangeal joint.

Innervation

Nerves derived from the sacral plexuses (shown in Exhibit 13.4 on page 444) innervate muscles that move the foot and toes.

- Muscles of the anterior compartment: deep fibular (perineal) nerve
- Muscles of the lateral compartment: superficial fibular (perineal) nerve
- Muscles of the posterior compartment: tibial nerve

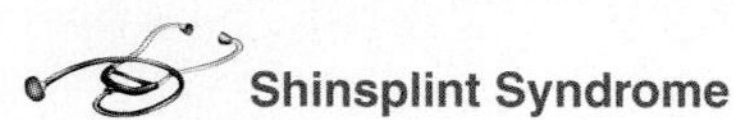

Shinsplint Syndrome

Shinsplint syndrome, or simply **shinsplints,** refers to pain or soreness along the tibia, specifically the medial, distal two-thirds. It may be caused by tendinitis of the anterior compartment muscles, especially the tibialis anterior muscle, inflammation of the periosteum (periostitis) around the tibia, or stress fractures of the tibia. The tendinitis usually occurs when poorly conditioned runners run on hard or banked surfaces with poorly supportive running shoes. The condition may also occur with vigorous activity of the legs following a period of relative inactivity. The muscles in the anterior compartment (mainly the tibialis anterior) can be strengthened to balance the stronger posterior compartment muscles. ■

Relating Muscles to Movements

Arrange the muscles in this exhibit according to the following actions on the foot at the ankle joint: (1) dorsiflexion and (2) plantar flexion; according to the following actions on the foot at the intertarsal joints: (1) inversion and (2) eversion; and according to the following actions on the toes at the metatarsophalangeal and intertarsal joints: (1) flexion and (2) extension. The same muscle may be mentioned more than once.

► **CHECKPOINT**

What are the superior extensor retinaculum and inferior extensor retinaculum?

Muscle	Origin	Insertion	Action
Anterior Compartment of the Leg			
Tibialis anterior (tib′-ē-Ā-lis = tibia; *anterior* = front)	Lateral condyle and body of tibia and interosseous membrane (sheet of fibrous tissue that holds shafts of tibia and fibula together).	First metatarsal and first (medial) cuneiform.	Dorsiflexes foot at ankle joint and inverts foot at intertarsal joints.
Extensor hallucis longus (HAL-ū-sis LON-gus; *extensor* = increases angle at joint; *halluc-* = hallux or great toe; *longus* = long)	Anterior surface of fibula and interosseous membrane.	Distal phalanx of great toe.	Dorsiflexes foot at ankle joint and extends proximal phalanx of great toe at metatarsophalangeal joint.
Extensor digitorum longus (di′-ji-TOR-um LON-gus)	Lateral condyle of tibia, anterior surface of fibula, and interosseous membrane.	Middle and distal phalanges of toes 2–5.*	Dorsiflexes foot at ankle joint and extends distal and middle phalanges of each toe at interphalangeal joints and proximal phalanx of each toe at metatarsophalangeal joint.
Fibularis (Peroneus) tertius (fib-ū-LAR-is TER-shus; *peron-* = fibula; *tertius* = third)	Distal third of fibula and interosseous membrane.	Base of fifth metatarsal.	Dorsiflexes foot at ankle joint and everts foot at intertarsal joints.
Lateral (Fibular) Compartment of the Leg			
Fibularis (Peroneus) longus (fib-ū-LAR-is LON-gus)	Head and body of fibula and lateral condyle of tibia.	First metatarsal and first cuneiform.	Plantar flexes foot at ankle joint and everts foot at intertarsal joints.
Fibularis (Peroneus) brevis (fib-ū-LAR-is BREV-is; *brevis* = short)	Body of fibula.	Base of fifth metatarsal.	Plantar flexes foot at ankle joint and everts foot at intertarsal joints.
Superficial Posterior Compartment of the Leg			
Gastrocnemius (gas′-trok-NĒ-mē-us; *gastro-* = belly; *cnem-* = leg)	Lateral and medial condyles of femur and capsule of knee.	Calcaneus by way of calcaneal (Achilles) tendon.	Plantar flexes foot at ankle joint and flexes leg at knee joint.
Soleus (SŌ-lē-us; *sole* = a type of flat fish)	Head of fibula and medial border of tibia.	Calcaneus by way of calcaneal (Achilles) tendon.	Plantar flexes foot at ankle joint.
Plantaris (plan-TĀR-is; *plantar-* = sole of foot)	Femur superior to lateral condyle.	Calcaneus by way of calcaneal (Achilles) tendon.	Plantar flexes foot at ankle joint and flexes leg at knee joint.
Deep Posterior Compartment of the Leg			
Popliteus (pop-LIT-ē-us; *poplit-* = the back of the knee)	Lateral condyle of femur.	Proximal tibia.	Flexes leg at knee joint and medially rotates tibia to unlock the extended knee.
Tibialis posterior (tib′-ē-Ā-lis; *posterior* = back)	Tibia, fibula, and interosseous membrane.	Second, third, and fourth metatarsals; navicular; all three cuneiforms; and cuboid.	Plantar flexes foot at ankle joint and inverts foot at intertarsal joints.
Flexor digitorum longus (di′-ji-TOR-um LON-gus; *digit* = finger or toe)	Posterior surface of tibia.	Distal phalanges of toes 2–5.	Plantar flexes foot at ankle joint; flexes distal and middle phalanges of each toe at interphalangeal joints and proximal phalanx of each toe at metatarsophalangeal joint.
Flexor hallucis longus (HAL-ū-sis LON-gus; *flexor* = decreases angle at joint)	Inferior two-thirds of fibula.	Distal phalanx of great toe.	Plantar flexes foot at ankle joint; flexes distal phalanx of great toe at interphalangeal joint and proximal phalanx of great toe at metatarsophalangeal joint.

*Reminder: The great toe or hallux is the first toe and has two phalanges: proximal and distal. The remaining toes are numbered 2–5, and each has three phalanges: proximal, middle, and distal.

(continues)

Exhibit 11.19 Muscles that Move the Foot and Toes (Figure 11.22) *(continued)*

Figure 11.22 Muscles that move the foot and toes. (See Tortora, *A Photographic Atlas of the Human Body,* Figures 5.18 and 5.19.)

The superficial muscles of the posterior compartment share a common tendon of insertion, the calcaneal (Achilles) tendon, which inserts into the calcaneal bone of the ankle.

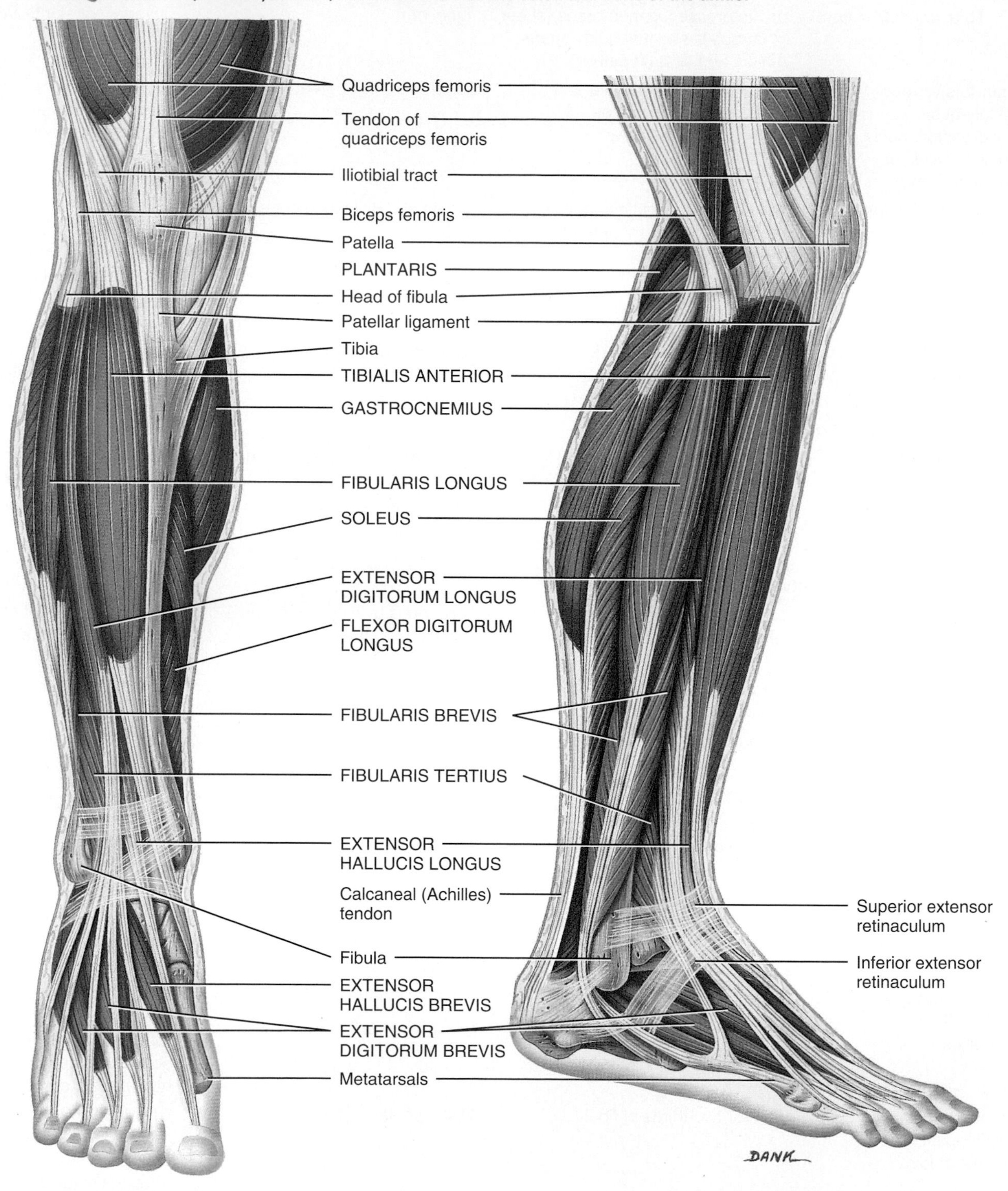

(a) Anterior superficial view

(b) Right lateral superficial view

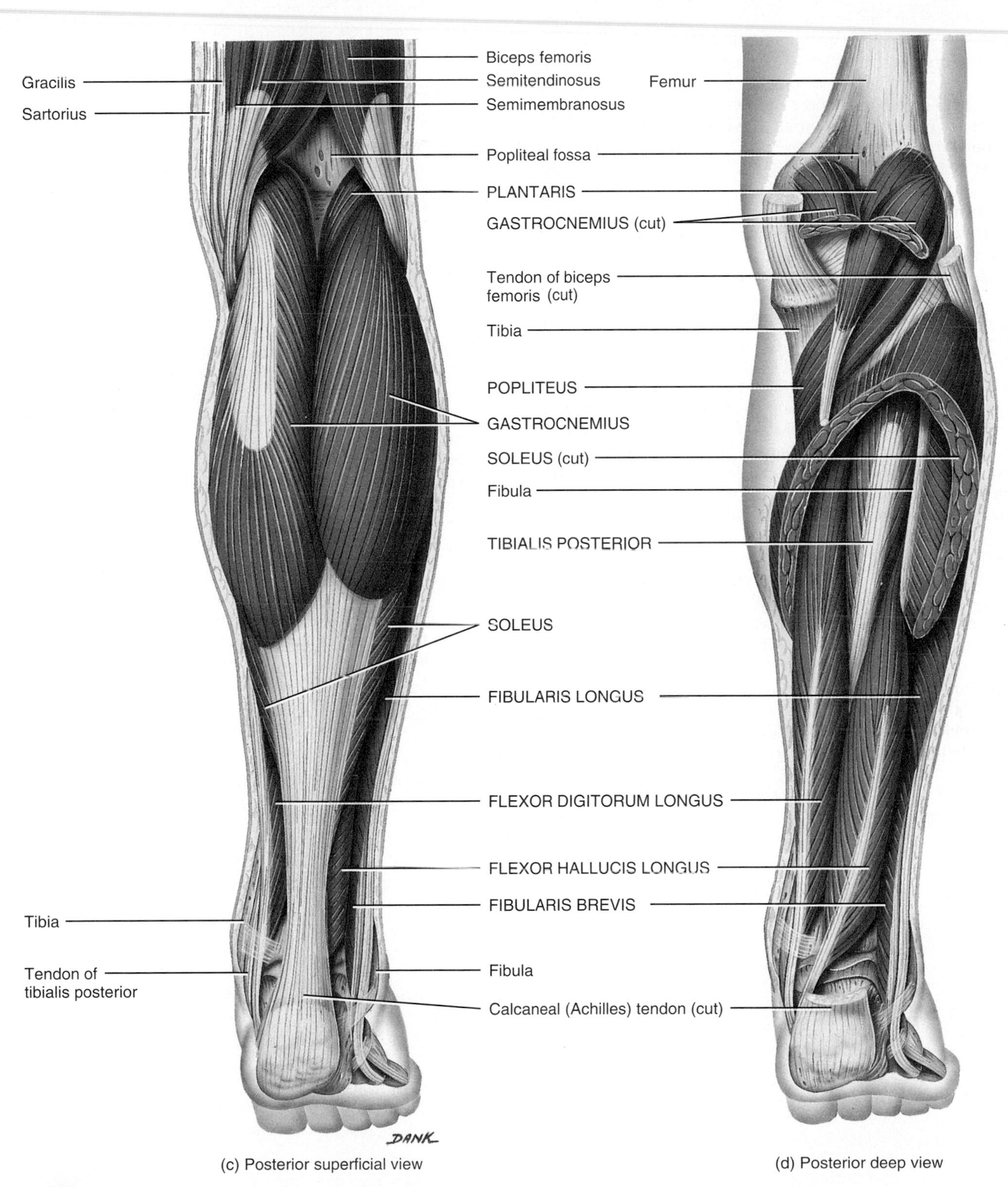

(c) Posterior superficial view

(d) Posterior deep view

What structures firmly hold the tendons of the anterior compartment muscles to the ankle?

Exhibit 11.20 Intrinsic Muscles of the Foot (Figure 11.23)

▶ OBJECTIVE

- Describe the origin, insertion, action, and innervation of the intrinsic muscles of the foot.

The muscles in this exhibit are termed **intrinsic muscles** because they originate and insert *within* the foot. The intrinsic muscles of the foot are, for the most part, comparable to those in the hand. Whereas the muscles of the hand are specialized for precise and intricate movements, those of the foot are limited to support and locomotion. The deep fascia of the foot forms the **plantar aponeurosis (fascia)** that extends from the calcaneus bone to the phalanges of the toes. The aponeurosis supports the longitudinal arch of the foot and encloses the flexor tendons of the foot.

The intrinsic muscles of the foot are divided into two groups: **dorsal** and **plantar.** There is only one dorsal muscle, the **extensor digitorum brevis,** a four-part muscle deep to the tendons of the extensor digitorum longus muscle, which extends toes 2–5 at the metatarsophalangeal joints.

The plantar muscles are arranged in four layers, the most superficial layer being termed the first layer. Three muscles are in the first layer. The **abductor hallucis,** which lies along the medial border of the sole, is comparable to the abductor pollicis brevis in the hand, and abducts the great toe at the metatarsophalangeal joint. The **flexor digitorum brevis,** which lies in the middle of the sole, flexes toes 2–5 at the interphalangeal and metatarsophalangeal joints. The **abductor digiti minimi,** which lies along the lateral border of the sole, is comparable to the same muscle in the hand, and abducts the little toe.

The second layer consists of the **quadratus plantae,** a rectangular muscle that arises by two heads and flexes toes 2–5 at the metatarsophalangeal joints, and the **lumbricals,** four small muscles that are similar to the lumbricals in the hands. They flex the proximal phalanges and extend the distal phalanges of toes 2–5.

Three muscles comprise the third layer. The **flexor hallucis brevis,** which lies adjacent to the plantar surface of the metatarsal of the great toe, is comparable to the same muscle in the hand, and flexes the great toe. The **adductor hallucis,** which has an oblique and transverse head like the adductor pollicis in the hand, adducts the great toe. The **flexor digiti minimi brevis,** which lies superficial to the metatarsal of the little toe, is comparable to the same muscle in the hand, and flexes the little toe.

The fourth layer is the deepest and consists of two muscle groups. The **dorsal interossei** are four muscles that adduct toes 2–4, flex the proximal phalanges, and extend the distal phalanges. The three **plantar interossei** adduct toes 3–5, flex the proximal phalanges, and extend the distal phalanges. The interossei of the feet are similar to those of the hand. However, their actions are relative to the midline of the second digit rather than the third digit as in the hand.

Innervation

Nerves derived from the sacral plexus (shown in Exhibit 13.4 on page 444) innervate intrinsic muscles of the foot.

- Extensor digitorum brevis: deep fibular (perineal) nerve
- Abductor hallucis and flexor digitorum brevis: medial plantar nerve
- Abductor digiti minimi, quadratus plantae, adductor hallucis, flexor digiti minimi brevis, and dorsal and plantar interossei: lateral plantar nerve
- Lumbricals: medial and lateral plantar nerves

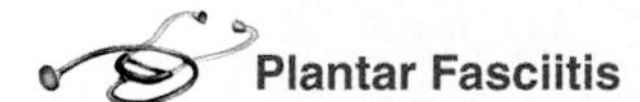

Plantar Fasciitis

Plantar fasciitis (fas-ē-Ī-tis) or **painful heel syndrome** is an inflammatory reaction due to chronic irritation of the plantar aponeurosis (fascia) at its origin on the calcaneus (heel bone). The aponeurosis becomes less elastic with age. This condition is also related to weight-bearing activities (walking, jogging, lifting heavy objects), improperly constructed or fitting shoes, excess weight (puts pressure on the feet), and poor biomechanics (flat feet, high arches, and abnormalities in gait may cause uneven distribution of weight on the feet). Plantar fasciitis is the most common cause of heel pain in runners and arises in response to the repeated impact of running. ■

Relating Muscles to Movements

Arrange the muscles in this exhibit according to the following actions on the great toe at the metatarsophalangeal joint: (1) flexion, (2) extension, (3) abduction, and (4) adduction; and according to the following actions on toes 2–5 at the metatarsophalangeal and interphalangeal joints: (1) flexion, (2) extension, (3) abduction, and (4) adduction. The same muscle may be mentioned more than once.

▶ CHECKPOINT

How do the intrinsic muscles of the hand and foot differ in function?

Muscle	Origin	Insertion	Action
Dorsal			
Extensor digitorum brevis (*extensor* = increases angle at joint; *digit* = finger or toe; *brevis* = short) (see Figure 11.22a, b)	Calcaneus and inferior extensor retinaculum.	Tendons of extensor digitorum longus on toes 2–4 and proximal phalanx of great toe.*	Extensor hallucis brevis extends great toe at metatarsophalangeal joint and extensor digitorum brevis extends toes 2–4 at interphalangeal joints.
Plantar			
First Layer (most superficial)			
Abductor hallucis (*abductor* = moves part away from midline; *hallucis* = hallux or great toe)	Calcaneus, plantar aponeurosis, and flexor retinaculum.	Medial side of proximal phalanx of great toe with the tendon of the flexor hallucis brevis.	Abducts and flexes great toe at metatarsophalangeal joint.
Flexor digitorum brevis (*flexor* = decreases angle at joint)	Calcaneus and plantar aponeurosis.	Sides of middle phalanx of toes 2–5.	Flexes toes 2–5 at proximal interphalangeal and metatarsophalangeal joints.
Abductor digiti minimi (*minimi* = little)	Calcaneus and plantar aponeurosis.	Lateral side of proximal phalanx of little toe with the tendon of the flexor digiti minimi brevis.	Abducts and flexes little toe at metatarsophalangeal joint.
Second layer			
Quadratus plantae (kwod-RĀ-tus; *quad-* = square, four-sided; *planta* = the sole of the foot)	Calcaneus.	Tendon of flexor digitorum longus.	Assists flexor digitorum longus to flex toes 2–5 at interphalangeal and metatarsophalangeal joints.
Lumbricals (LUM-bri-kals; *lumbric-* = earthworm)	Tendons of flexor digitorum longus.	Tendons of extensor digitorum longus on proximal phalanges of toes 2–5.	Extend toes 2–5 at interphalangeal joints and flex toes 2–5 at metatarsophalangeal joints.
Third Layer			
Flexor hallucis brevis	Cuboid and third (lateral) cuneiform.	Medial and lateral sides of proximal phalanx of great toe via a tendon containing a sesamoid bone.	Flexes great toe at metatarsophalangeal joint.
Adductor hallucis	Metatarsals 2–4, ligaments of 3–5 metatarsophalangeal joints, and tendon of peroneus longus.	Lateral side of proximal phalanx of great toe.	Adducts and flexes great toe at metatarsophalangeal joint.
Flexor digiti minimi brevis	Metatarsal 5 and tendon of peroneus longus.	Lateral side of proximal phalanx of little toe.	Flexes little toe at metatarsophalangeal joint.
Fourth Layer (deepest)			
Dorsal interossei (in-ter-OS-ē-ī) (not illustrated)	Adjacent side of metatarsals.	Proximal phalanges: both sides of toe 2 and lateral side of toes 3 and 4.	Abduct and flex toes 2–4 at metatarsophalangeal joints and extend toes at interphalangeal joints.
Plantar interossei	Metatarsals 3–5.	Medial side of proximal phalanges of toes 3–5.	Adduct and flex proximal metatarsophalangeal joints and extend toes at interphalangeal joints.

* The tendon that inserts into the proximal phalanx of the great toe, together with its belly, is often described as a separate muscle, the extensor hallucis brevis.

Exhibit 11.20 Intrinsic Muscles of the Foot (Figure 11.23) (continued)

Figure 11.23 Intrinsic muscles of the foot.

Whereas the intrinsic muscles of the hand are specialized for precise and intricate movements, those of the foot are limited to support and movement.

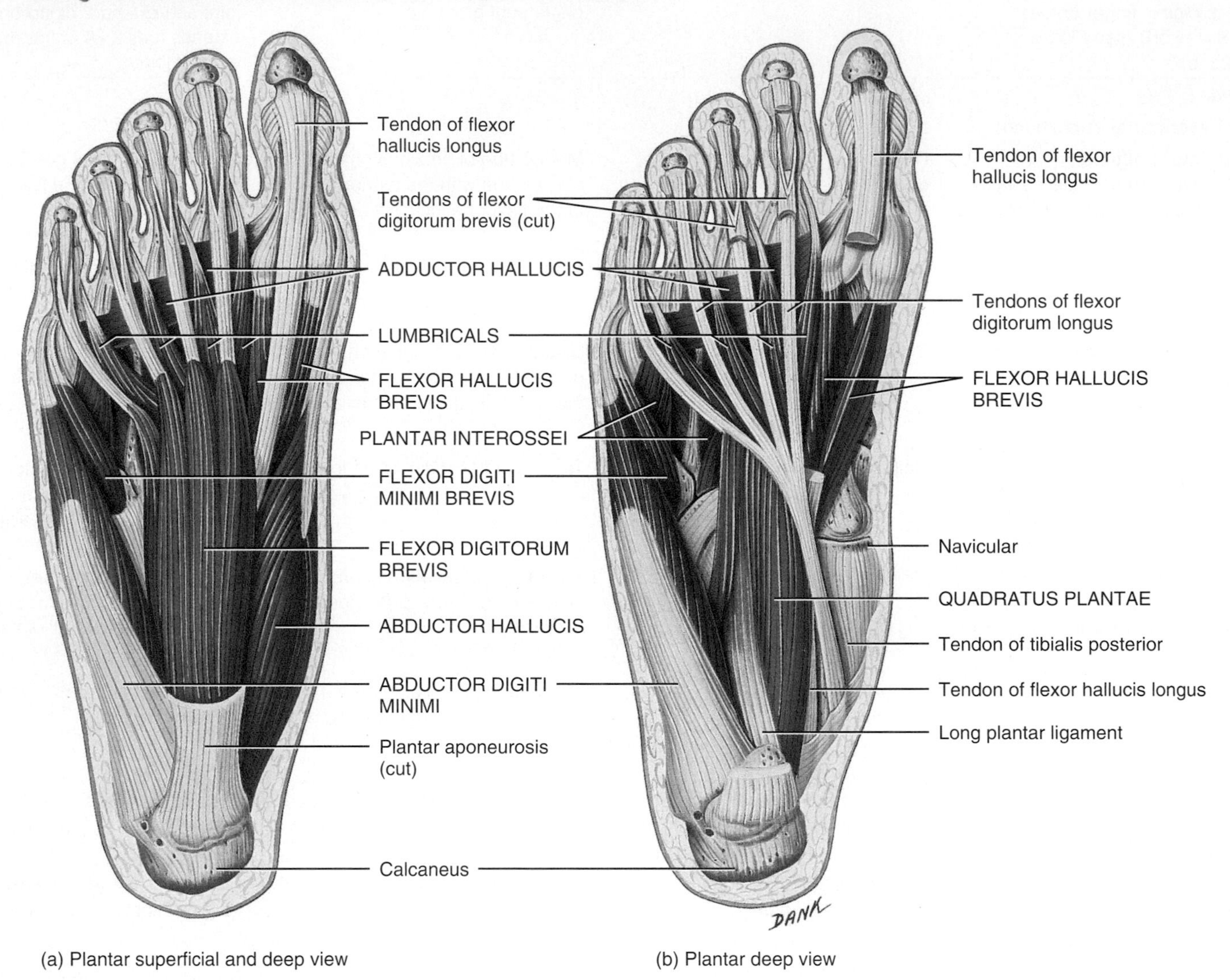

(a) Plantar superficial and deep view

(b) Plantar deep view

What structure supports the longitudinal arch and encloses the flexor tendons of the foot?

Focus on Homeostasis: The Muscular System

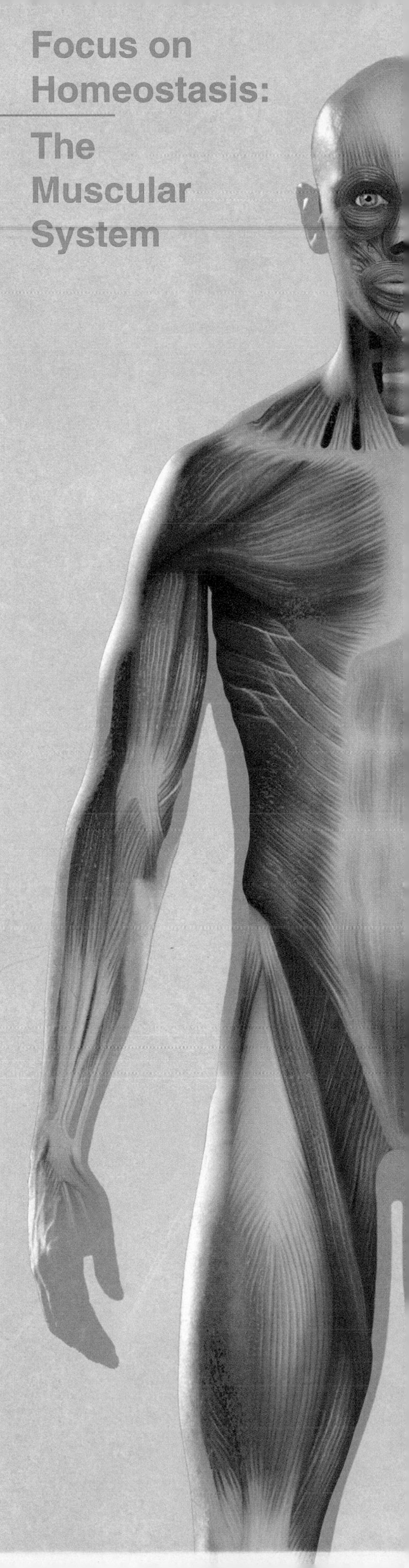

Body System	Contribution of the Muscular System
For all body systems	The muscular system and muscle tissues produce body movements, stabilize body positions, move substances within the body, and produce heat that helps maintain normal body temperature.
Integumentary system 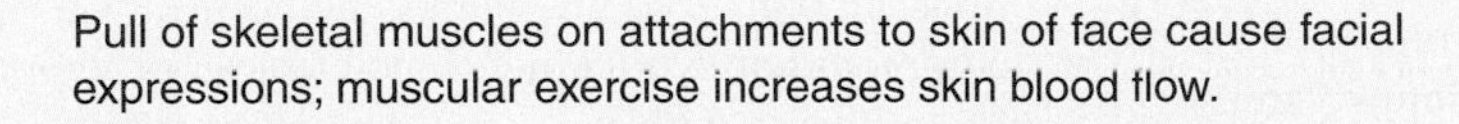	Pull of skeletal muscles on attachments to skin of face cause facial expressions; muscular exercise increases skin blood flow.
Skeletal system 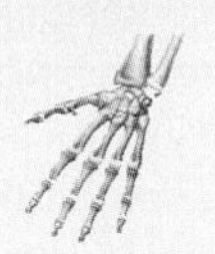	Skeletal muscle causes movement of body parts by pulling on attachments to bones; skeletal muscle provides stability for bones and joints.
Nervous system 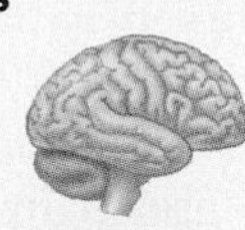	Smooth, cardiac, and skeletal muscles carry out commands for the nervous system; shivering—involuntary contraction of skeletal muscles that is regulated by the brain—generates heat to raise body temperature.
Endocrine system 	Regular activity of skeletal muscles (exercise) improves action of some hormones, such as insulin; muscles protect some endocrine glands.
Cardiovascular system 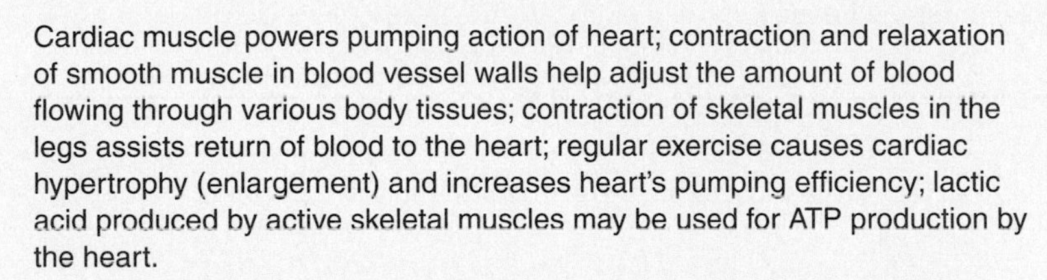	Cardiac muscle powers pumping action of heart; contraction and relaxation of smooth muscle in blood vessel walls help adjust the amount of blood flowing through various body tissues; contraction of skeletal muscles in the legs assists return of blood to the heart; regular exercise causes cardiac hypertrophy (enlargement) and increases heart's pumping efficiency; lactic acid produced by active skeletal muscles may be used for ATP production by the heart.
Lymphatic and immune system 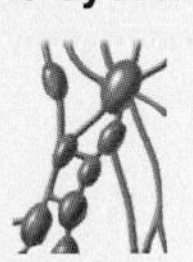	Skeletal muscles protect some lymph nodes and lymphatic vessels and promote flow of lymph inside lymphatic vessels; exercise may increase or decrease some immune responses.
Respiratory system 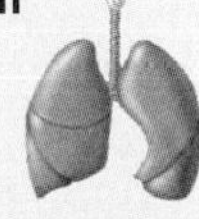	Skeletal muscles involved with breathing cause air to flow into and out of the lungs; smooth muscle fibers adjust size of airways; vibrations in skeletal muscles of larynx control air flowing past vocal cords, regulating voice production; coughing and sneezing, due to skeletal muscle contractions, help clear airways; regular exercise improves efficiency of breathing.
Digestive system	Skeletal muscles protect and support organs in the abdominal cavity; alternating contraction and relaxation of skeletal muscles power chewing and initiate swallowing; smooth muscle sphincters control volume of organs of the gastrointestinal (GI) tract; smooth muscles in walls of GI tract mix and move its contents through the tract.
Urinary system 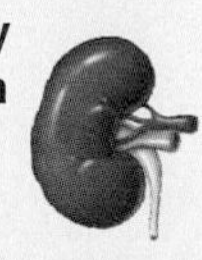	Skeletal and smooth muscle sphincters and smooth muscle in wall of urinary bladder control whether urine is stored in the bladder or voided (urination).
Reproductive systems 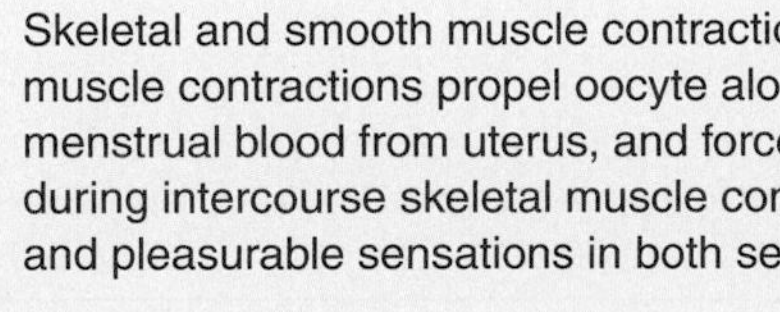	Skeletal and smooth muscle contractions eject semen from male; smooth muscle contractions propel oocyte along uterine tube, help regulate flow of menstrual blood from uterus, and force baby from uterus during childbirth; during intercourse skeletal muscle contractions are associated with orgasm and pleasurable sensations in both sexes.

DISORDERS: HOMEOSTATIC IMBALANCES

Running Injuries

Nearly 70% of those who jog or run sustain some type of running-related injury. Even though most such injuries are minor, some are quite serious. Moreover, untreated or inappropriately treated minor injuries may become chronic. Among runners, common sites of injury include the ankle, knee, calcaneal (Achilles) tendon, hip, groin, foot, and back. Of these, the knee often is the most severely injured area.

Running injuries are frequently related to faulty training techniques. This may involve improper (or lack of) warm-up routines, running too much, or running too soon after an injury. Or it might involve extended running on hard and/or uneven surfaces. Poorly constructed or worn-out running shoes can also contribute to injury, as can any biomechanical problem (such as a fallen arch) aggravated by running.

Most sports injuries should be treated initially with RICE therapy, which stands for Rest, Ice, Compression, and Elevation. Immediately apply ice, and rest and elevate the injured part. Then apply an elastic bandage, if possible, to compress the injured tissue. Continue using RICE for 2 to 3 days, and resist the temptation to apply heat, which may worsen the swelling. Follow-up treatment may include alternating moist heat and ice massage to enhance blood flow in the injured area. Sometimes it is helpful to take nonsteroidal anti-inflammatory drugs (NSAIDs) or to have local injections of corticosteroids. During the recovery period, it is important to keep active using an alternative fitness program that does not worsen the original injury. This activity should be determined in consultation with a physician. Finally, careful exercise is needed to rehabilitate the injured area itself.

Compartment Syndrome

In several exhibits in this chapter, we noted that skeletal muscles in the limbs are organized into units called *compartments.* In a disorder called **compartment syndrome,** some external or internal pressure constricts the structures within a compartment, resulting in damaged blood vessels and subsequent reduction of the blood supply (ischemia) to the structures within the compartment. Common causes of compartment syndrome include crushing and penetrating injuries, contusion (damage to subcutaneous tissues without the skin being broken), muscle strain (overstretching of a muscle), or an improperly fitted cast. As a result of hemorrhage, tissue injury, and edema (buildup of interstitial fluid), the pressure increase in the compartment can have serious consequences. Because the fascia that enclose the compartments are very strong, accumulated blood and interstitial fluid cannot escape, and the increased pressure can literally choke off the blood flow and deprive nearby muscles and nerves of oxygen. One treatment option is **fasciotomy** (fash-ē-OT-ō-mē), a surgical procedure in which muscle fascia is cut to relieve the pressure. Without intervention, nerves suffer damage, and muscles develop scar tissue that results in permanent shortening of the muscles, a condition called *contracture.*

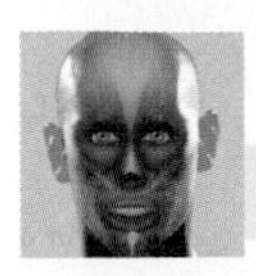

STUDY OUTLINE

HOW SKELETAL MUSCLES PRODUCE MOVEMENT (p. 309)

1. Skeletal muscles that produce movement do so by pulling on bones.
2. The attachment to the more stationary bone is the origin; the attachment to the more movable bone is the insertion.
3. Bones serve as levers, and joints serve as fulcrums. Two different forces act on the lever: load (resistance) and effort.
4. Levers are categorized into three types—first-class, second-class, and third-class (most common)—according to the positions of the fulcrum, the effort, and the load on the lever.
5. Fascicular arrangements include parallel, fusiform, circular, triangular, and pennate. Fascicular arrangement affects a muscle's power and range of motion.
6. A prime mover produces the desired action; an antagonist produces an opposite action. Synergists assist a prime mover by reducing unnecessary movement. Fixators stabilize the origin of a prime mover so that it can act more efficiently.

HOW SKELETAL MUSCLES ARE NAMED (p. 313)

1. Distinctive features of different skeletal muscles include direction of muscle fascicles; size, shape, action, number of origins (or heads), and location of the muscle; and sites of origin and insertion of the muscle.
2. Most skeletal muscles are named based on combinations of these features.

PRINCIPAL SKELETAL MUSCLES (p. 313)

1. Muscles of facial expression move the skin rather than a joint when they contract, and they permit us to express a wide variety of emotions.
2. The extrinsic muscles that move the eyeballs are among the fastest contracting and most precisely controlled skeletal muscles in the body. They permit us to elevate, depress, abduct, adduct, and medially and laterally rotate the eyeballs.
3. Muscles that move the mandible (lower jaw) are also known as the muscles of mastication because they are involved in chewing.
4. The extrinsic muscles that move the tongue are important in chewing, swallowing, and speech.
5. Suprahyoid muscles of the anterior neck are located above the hyoid bone. They elevate the hyoid bone, oral cavity, and tongue during swallowing. Infrahyoid muscles of the anterior neck depress the hyoid bone and some move the larynx during swallowing and speech.
6. Muscles that move the head alter the position of the head and help balance the head on the vertebral column.
7. Muscles that act on the abdominal wall help contain and protect the abdominal viscera, move the vertebral column, compress the abdomen, and produce the force required for defecation, urination, vomiting, and childbirth.
8. Muscles used in breathing alter the size of the thoracic cavity so that inhalation and exhalation can occur.
9. Muscles of the pelvic floor support the pelvic viscera, resist the thrust that accompanies increases in intra-abdominal pressure, and function as sphincters at the anus, urethra, and vagina.
10. Muscles of the perineum assist in urination, erection of the penis and clitoris, ejaculation, and defecation.
11. Muscles that move the pectoral (shoulder) girdle stabilize the scapula so it can function as a stable point of origin for most of the muscles that move the humerus.
12. Muscles that move the humerus (arm bone) originate for the most part on the scapula (scapular muscles); the remaining muscles originate on the axial skeleton (axial muscles).
13. Muscles that move the radius and ulna (forearm bones) are involved in flexion and extension at the elbow joint and are organized into flexor and extensor compartments.
14. Muscles that move the wrist, hand, thumb, and fingers are many and varied. Those muscles that act on the thumb and fingers are called extrinsic muscles.
15. The intrinsic muscles of the hand are important in skilled activities and provide humans with the ability to grasp precisely and manipulate objects.
16. Muscles that move the vertebral column are quite complex because they have multiple origins and insertions and because there is considerable overlap among them.
17. Muscles that move the femur (thigh bone) originate for the most part on the pelvic girdle and insert on the femur. These muscles are larger and more powerful than comparable muscles in the upper limb.
18. Muscles that move the femur (thigh bone) and tibia and fibula (leg bones) are separated into medial (adductor), anterior (extensor), and posterior (flexor) compartments.
19. Muscles that move the foot and toes are divided into anterior, lateral, and posterior compartments.
20. Intrinsic muscles of the foot, unlike those of the hand, normally are limited to the functions of support and locomotion.

SELF-QUIZ QUESTIONS

Fill in the blanks in the following statements.

1. The less movable end of the muscle is the ________; the more movable end is the ________.
2. The muscle that forms the major portion of the cheek is the ________.

Indicate whether the following statements are true or false.

3. Because of their insertions, the muscles of facial expression move the skin rather than a joint when they contract.
4. Opposition is the single most distinctive digital movement that gives humans and other primates the ability to precisely grasp and manipulate objects.

Choose the one best answer to the following questions.

5. Which of the following statements is *false*? (a) Fixators steady the proximal end of a limb whereas movements occur at the distal end. (b) Synergists prevent unwanted movements in intermediate joints. (c) The prime mover, or agonist, causes a desired action. (d) The antagonist supports and aids the agonist's movement. (e) Fascicular arrangement affects a muscle's power and range of motion.
6. Which of the following are important characteristics used in the naming of muscles? (1) muscle shape, (2) muscle size, (3) direction of muscle of fibers, (4) muscle action, (5) muscle location, (6) sites of origin and insertion, (7) number of insertions. (a) 1, 2, 3, 4, 5, and 6, (b) 2, 3, 4, 5, 6, and 7, (c) 2, 4, 6, and 7, (d) 1, 3, 5, and 7, (e) all are correct.
7. The facial nerve (cranial nerve VII) innervates (a) most muscles of facial expression, (b) the muscles that move the head, (c) extrinsic muscles that move the tongue, (d) muscles of mastication, (e) muscles that move the mandible.
8. The muscle that closes the mouth is the (a) risorius, (b) levator labii superioris, (c) platysma, (d) mentalis, (e) masseter.
9. Which of the following are muscle-generated actions on the mandible? (1) elevation, (2) depression, (3) retraction, (4) protraction, (5) abduction, (6) adduction, (7) side-to-side movement. (a) 1, 2, 3, and 5, (b) 2, 3, 4, and 6, (c) 1, 2, 3, 4, and 7, (d) 3, 4, 5, and 7, (e) 1, 3, 5, and 7.
10. Which of the following are functions of the anterolateral abdominal muscle group? (1) helps contain and protect the abdominal viscera, (2) helps contain and protect the lower area of the heart and lungs, (3) flexes, laterally flexes, and rotates the vertebral column, (4) expands the abdomen during forced expiration, (5) produces the force required for defecation, urination, and childbirth. (a) 1, 3, and 5, (b) 2, 4, and 5, (c) 1, 2, and 4, (d) 1 and 5, (e) 1, 2, and 5.
11. In order for movement to occur (a) both the prime mover and antagonist must contract simultaneously, (b) muscles that move a body part cannot lie over the moving part, (c) a muscle must cross at least two joints, (d) the muscle fascicles must have a circular arrangement, (e) a muscle must be innervated by a spinal nerve.
12. Match the following:

____ (a) compression of median nerve resulting in pain and numbness and tingling in the fingers
____ (b) tendinitis of the anterior compartment muscles of the leg; inflammation of the tibial periosteum
____ (c) improperly aligned eyeballs due to lesions in either the oculomotor or abducens nerves
____ (d) stretching or tearing of distal attachments of adductor muscles and/or iliopsoas
____ (e) rupture of a portion of the inguinal area of the abdominal wall resulting in protrusion of part of the small intestine
____ (f) caused by repetitive movement of the arm over the head which results in inflammation of the supraspinatus tendon
____ (g) inflammation due to chronic irritation of the plantar aponeurosis at its origin on the calcaneous; most common cause of heel pain in runners
____ (h) painful inflammation of tendons, tendon sheaths, and the synovial membranes of joints
____ (i) paralysis of facial muscles as a result of damage to the facial nerve
____ (j) common condition in individuals who perform quick starts and stops; tearing away of part of the tendinous origins from the ischial tuberosity
____ (k) may occur as a result of injury to levator ani muscle
____ (l) external or internal pressure constricts structures in a compartment, causing a reduction of blood supply to the structures

(1) tenosynovitis (tendinitis)
(2) Bell's palsy
(3) inguinal hernia
(4) urinary stress incontinence
(5) compartment syndrome
(6) pulled groin
(7) pulled hamstrings
(8) strabismus
(9) shinsplints
(10) plantar fasciitis
(11) impingement syndrome
(12) carpal tunnel syndrome

13. Match the following:

____ (a) most common lever in the body	(1) first-class lever
____ (b) lever formed by the head resting on the vertebral column	(2) second-class lever
____ (c) always produces a mechanical advantage	(3) third-class lever
____ (d) EFL	
____ (e) FLE	
____ (f) FEL	
____ (g) adduction of the thigh	

14. Match the following:

____ (a) rectus femoris, vastus lateralis, vastus medialis, vastus intermedius	(1) breathing muscles
____ (b) biceps femoris, semitendinosus, semimembranosus	(2) constitute flexor compartment of the arm
____ (c) erector spinae; includes iliocostalis, longissimus, and spinalis groups	(3) hamstrings
____ (d) thenar, hypothenar, intermediate	(4) intrinsic muscle groups of the hand
____ (e) biceps brachii, brachialis, coracobrachialis	(5) muscles that strengthen and stabilize the shoulder joint; the rotator cuff
____ (f) latissimus dorsi	(6) quadriceps femoris muscle
____ (g) subscapularis, supraspinatus, infraspinatus, teres minor	(7) largest muscle mass of the back
____ (h) diaphragm, external intercostals, internal intercostals	(8) swimmer's muscle

15. Match the following (some answers may be used more than once):

____ (a) trapezius	(1) muscle of facial expression
____ (b) orbicularis oculi	(2) muscle of mastication
____ (c) levator ani	(3) muscle that moves the eyeballs
____ (d) rectus abdominis	(4) extrinsic muscle that moves the tongue
____ (e) triceps brachii	(5) suprahyoid muscle
____ (f) gastrocnemius	(6) muscle of the perineum
____ (g) temporalis	(7) muscle that moves the head
____ (h) external anal sphincter	(8) abdominal wall muscle
____ (i) external oblique	(9) pelvic floor muscle
____ (j) splenius capitis	(10) pectoral girdle muscle
____ (k) digastric	(11) muscle that moves the humerus
____ (l) styloglossus	(12) muscle that moves the radius and ulna
____ (m) masseter	(13) muscle that moves the wrist, hand, and digits
____ (n) adductor longus	(14) muscle that moves the vertebral column
____ (o) zygomaticus major	(15) muscle that moves the femur
____ (p) latissimus dorsi	(16) muscle that acts on the femur, tibia, and fibula
____ (q) flexor carpi radialis	(17) muscle that moves the foot and toes
____ (r) pronator teres	
____ (s) sternocleidomastoid	
____ (t) quadriceps femoris	
____ (u) deltoid	
____ (v) tibialis anterior	
____ (w) sartorius	
____ (x) gluteus maximus	
____ (y) superior rectus	

CRITICAL THINKING QUESTIONS

1. While you proceed through your daily activities, your mucles take turns being the agonist or antagonist—that is, being contracted or relaxed. Name some antagonistic pairs of muscles in the arm, thigh, torso, and leg.
HINT ***Antagonistic pairs are often located on opposite sides of a limb or body region.***

2. Melody was trying to lift a heavy package using the proper back-saving position that she had been taught at work. She squatted down with her knees bent and her back straight. Then she lifted the package using her leg mucles and keeping her back straight. Identify the resistance, fulcrum, and effort involved in lifting the package. What type of lever is this?
HINT ***Determine the relative order of the effort, the fulcrum, and the resistance.***

3. When Carlos read the first question on his take-home exam, he raised one eyebrow, whistled, then closed his eyes and shook his head. Name the muscles involved in these actions.
HINT ***Try it yourself.***

ANSWERS TO FIGURE QUESTIONS

11.1 The belly of the muscle that extends the forearm, the triceps brachii, is located posterior to the humerus.

11.2 In a third-class lever, the effort is applied between the fulcrum and the load.

11.3 Possible correct responses: direction of fibers: external oblique; shape: deltoid; action: extensor digitorum; size: gluteus maximus; origin and insertion: sternocleidomastoid; location: tibialis anterior; number of tendons of origin: biceps brachii.

11.4 Frowning: frontalis; smiling: zygomaticus major; pouting: platysma; squinting: orbicularis oculi.

11.5 The inferior oblique muscle moves the eyeball superiorly and laterally because it originates at the anteromedial aspect of the floor of the orbit and inserts on the posterolateral aspect of the eyeball.

11.6 The masseter is the strongest of the chewing muscles.

11.7 Functions of the tongue include chewing, perception of taste, swallowing, and speech.

11.8 The suprahyoid and infrahyoid muscles fix the hyoid bone to assist in tongue movements.

11.9 Triangles are important because of the structures that lie within their boundaries.

11.10 The rectus abdominis muscle aids in urination.

11.11 The phrenic nerve innervates the diaphragm.

11.12 The borders of the pelvic diaphragm are the pubic symphysis anteriorly, the coccyx posteriorly, and the walls of the pelvis laterally.

11.13 The borders of the perineum are the pubic symphysis anteriorly, the coccyx posteriorly, and the ischial tuberosities laterally.

11.14 The main action of the muscles that move the pectoral girdle is to stabilize the scapula to assist in movements of the humerus.

11.15 The rotator cuff consists of the flat tendons of the subscapularis, supraspinatus, infraspinatus, and teres minor muscles that form a nearly complete circle around the shoulder joint.

11.16 The most powerful forearm flexor is the brachialis. The most powerful forearm extensor is the triceps brachii.

11.17 Flexor tendons of the digits and wrist and the median nerve pass through the flexor retinaculum.

11.18 Muscles of the thenar eminence act on the thumb.

11.19 The splenius muscles extend laterally and upward from the ligamentum nuchae at the midline.

11.20 Upper limb muscles exhibit diversity of movement; lower limb muscles function in stability, locomotion, and maintenance of posture. Moreover, lower limb muscles usually cross two joints and act equally on both.

11.21 The quadriceps femoris consists of the rectus femoris, vastus lateralis, vastus medialis, and vastus intermedius. The hamstrings consist of the biceps femoris, semitendinosus, and semimembranosus.

11.22 The superior and inferior extensor retinacula firmly hold the tendons of the anterior compartment muscles to the ankle.

11.23 The plantar aponeurosis supports the longitudinal arch and encloses the flexor tendons of the foot.

Appendix A
Measurements

U.S. Customary System

Parameter	Unit	Relation to Other U.S. Units	SI (Metric) Equivalent
Length	inch	1/12 foot	2.54 centimeters
foot	12 inches	0.305 meter	
yard	36 inches	9.144 meters	
mile	5,280 feet	1.609 kilometers	
Mass	grain	1/1000 pound	64.799 milligrams
dram	1/16 ounce	1.772 grams	
ounce	16 drams	28.350 grams	
pound	16 ounces	453.6 grams	
ton	2,000 pounds	907.18 kilograms	
Volume (Liquid)	ounce	1/16 pint	29.574 milliliters
pint	16 ounces	0.473 liter	
quart	2 pints	0.946 liter	
gallon	4 quarts	3.785 liters	
Volume (Dry)	pint	1/2 quart	0.551 liter
quart	2 pints	1.101 liters	
peck	8 quarts	8.810 liters	
bushel	4 pecks	35.239 liters	

International System (SI)

BASE UNITS

Unit	Quantity	Symbol
meter	length	M
kilogram	mass	Kg
second	time	S
liter	volume	L
mole	amount of matter	Mol

PREFIXES

Prefix	Multiplier	Symbol
tera-	10^{12} = 1,000,000,000,000	T
giga-	10^{9} = 1,000,000,000	G
mega-	10^{6} = 1,000,000	M
kilo-	10^{3} = 1,000	k
hecto-	10^{2} = 100	h
deca-	10^{1} = 10	da
deci-	10^{-1} = 0.1	d
centi-	10^{-2} = 0.01	c
milli-	10^{-3} = 0.001	m
micro-	10^{-6} = 0.000,001	μ
nano-	10^{-9} = 0.000,000,001	n
pico-	10^{-12} = 0.000,000,000,001	p

Temperature Conversion
Fahrenheit (F) To Celsius (C)
°C = (°F − 32) ÷ 1.8
Celsius (C) To Fahrenheit (F)
°F = (°C × 1.8) + 32

U.S TO SI (Metric) Conversion

When you know	Multiply by	To find
inches	2.54	centimeters
feet	30.48	centimeters
yards	0.91	meters
miles	1.61	kilometers
ounces	28.35	grams
pounds	0.45	kilograms
tons	0.91	metric tons
fluid ounces	29.57	milliliters
pints	0.47	liters
quarts	0.95	liters
gallons	3.79	liters

SI (Metric) TO U.S. Conversion

When you know	Multiply by	To find
millimeters	0.04	inches
centimeters	0.39	inches
meters	3.28	feet
kilometers	0.62	miles
liters	1.06	quarts
cubic meters	35.32	cubic feet
grams	0.035	ounces
kilograms	2.21	pounds

Appendix B
Periodic Table

The periodic table lists the known **chemical elements,** the basic units of matter. The elements in the table are arranged left-to-right in rows in order of their **atomic number,** the number of protons in the nucleus. Each horizontal row, numbered from 1 to 7, is a **period.** All elements in a given period have the same number of electron shells as their period number. For example, an atom of hydrogen or helium each has one electron shell, while an atom of potassium or calcium each has four electron shells. The elements in each column, or **group,** share chemical properties. For example, the elements in column IA are very chemically reactive, whereas the elements in column VIIIA have full electron shells and thus are chemically inert.

Scientists now recognize 113 different elements; 92 occur naturally on Earth, and the rest are produced from the natural elements using particle accelerators or nuclear reactors. Elements are designated by **chemical symbols,** which are the first one or two letters of the element's name in English, Latin, or another language.

Twenty-six of the 92 naturally occurring elements normally are present in your body. Of these, just four elements—oxygen (O), carbon (C), hydrogen (H), and nitrogen (N) (coded blue)—constitute about 96% of the body's mass. Eight others—calcium (Ca), phosphorus (P), potassium (K), sulfur (S), sodium (Na), chlorine (Cl), magnesium (Mg), and iron (Fe) (coded pink)—contribute 3.8% of the body's mass. An additional 14 elements, called **trace elements** because they are present in tiny amounts, account for the remaining 0.2% of the body's mass. The trace elements are aluminum, boron, chromium, cobalt, copper, fluorine, iodine, manganese, molybdenum, selenium, silicon, tin, vanadium, and zinc (coded yellow). Table 2.1 on page 28 provides information about the main chemical elements in the body.

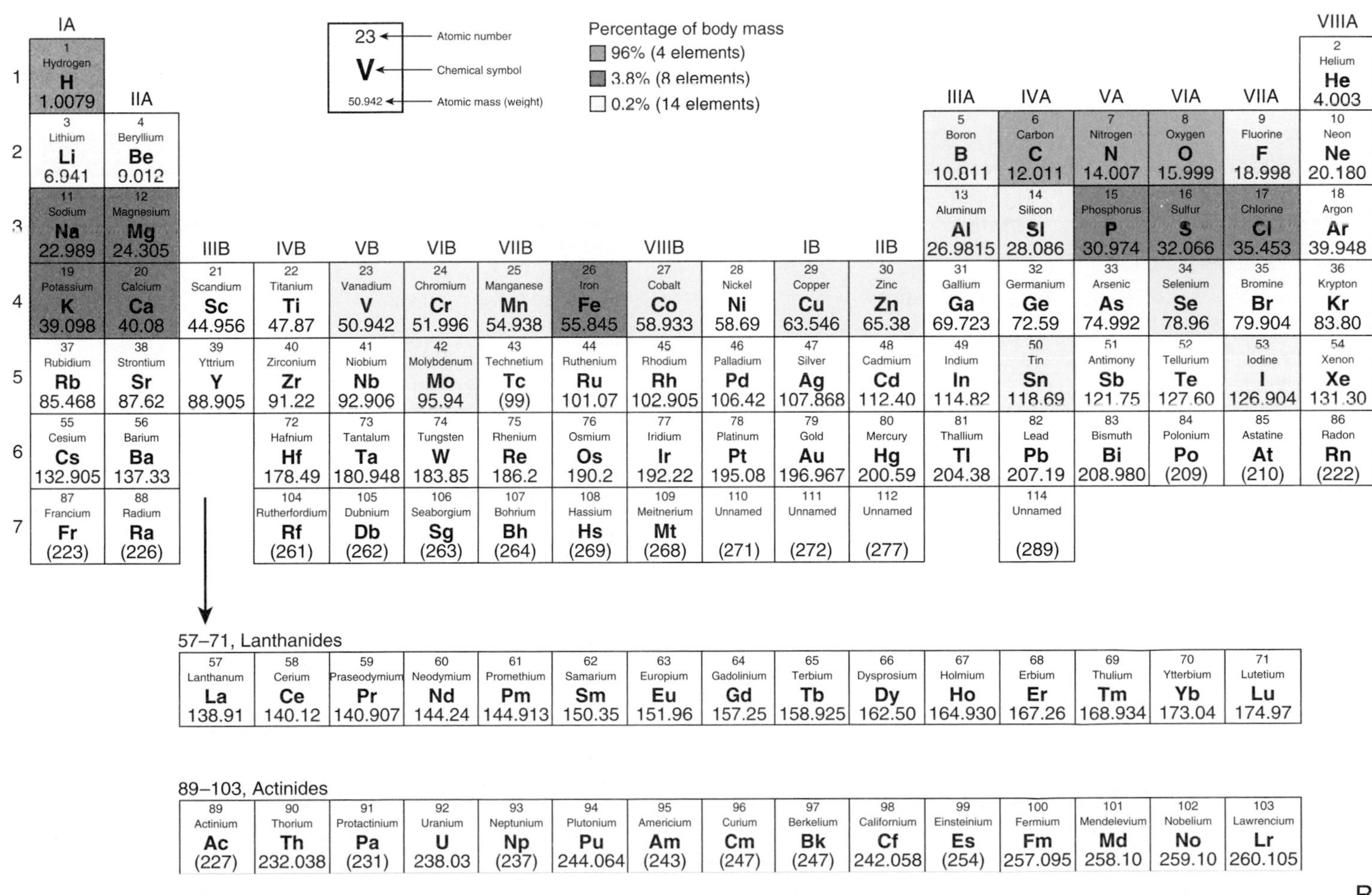

Appendix C
Normal Values For Selected Blood Tests

The system of international (SI) units (Système Internationale d'Unités) is used in most countries and in many medical and scientific journals. Clinical laboratories in the United States, by contrast, usually report values for blood and urine tests in conventional units. The laboratory values in this Appendix give conventional units first, followed by SI equivalents in parentheses. Values listed for various blood tests should be viewed as reference values rather than absolute "normal" values for all well people. Values may vary due to age, gender, diet, and environment of the subject or the equipment, methods, and standards of the lab performing the measurement.

Key To Symbols

g = gram
mg = milligram = 10^{-3} gram
μg = microgram = 10^{-6} gram
U = units
L = liter
dL = deciliter
mL = milliliter
μL = microliter
mEq/L = milliequivalents per liter
mmol/L = millimoles per liter
μmol/L = micromoles per liter
> = greater than; < = less than

Blood Tests

Test (Specimen)	U.S. Reference Values (SI Units)	Values Increase In	Values Decrease In
Aminotransferases (serum)			
Alanine aminotransferase (ALT)	0–35 U/L (same)	Liver disease or liver damage due to toxic drugs.	
Aspartate aminotransferase (AST)	0–35 U/L (same)	Myocardial infarction, liver disease, trauma to skeletal muscles, severe burns.	Beriberi, uncontrolled diabetes mellitus with acidosis, pregnancy.
Ammonia (plasma)	20–120 μg/dL (12–55 μmol/L)	Liver disease, heart failure, emphysema, pneumonia, hemolytic disease of newborn.	Hypertension.
Bilirubin (serum)	Conjugated: <0.5 mg/dL (<5.0 μmol/L)	Conjugated bilirubin: liver dysfunction or gallstones.	
	Unconjugated: 0.2–1.0 mg/dL (18–20 μmol/L) Newborn: 1.0–12.0 mg/dL (<200 μmol/L)	Unconjugated bilirubin: excessive hemolysis of red blood cells.	
Blood urea nitrogen (BUN) (serum)	8–26 mg/dL (2.9–9.3 mmol/L)	Kidney disease, urinary tract obstruction, shock, diabetes, burns, dehydration, myocardial infarction.	Liver failure, malnutrition, overhydration, pregnancy.

Blood Tests

Test (Specimen)	U.S. Reference Values (SI Units)	Values Increase In	Values Decrease In
Carbon dioxide content (bicarbonate + dissolved CO_2) (whole blood)	Arterial: 19–24 mEq/L (19–24 mmol/L) Venous: 22–26 mEq/L (22–26 mmol/L)	Severe diarrhea, severe vomiting, starvation, emphysema, aldosteronism.	Renal failure, diabetic ketoacidosis, shock.
Cholesterol, total (plasma) **HDL cholesterol** (plasma) **LDL cholesterol** (plasma)	$<$200 mg/dL ($<$5.2 mmol/L) is desirable $>$40 mg/dL ($>$1.0 mmol/L) is desirable $<$130 mg/dL ($<$3.2 mmol/L) is desirable	Hypercholesterolemia, uncontrolled diabetes mellitus, hypothyroidism, hypertension, atherosclerosis, nephrosis.	Liver disease, hyperthyroidism, fat malabsorption, pernicious or hemolytic anemia, severe infections.
Creatine (serum)	Males: 0.15–0.5 mg/dL (10–40 μmol/L) Females: 0.35–0.9 mg/dL (30–70 μmol/L)	Muscular dystrophy, damage to muscle tissue, electric shock, chronic alcoholism.	
Creatine Kinase (CK), also known as Creatine phosphokinase (CPK) (serum)	0–130 U/L (same)	Myocardial infarction, progressive muscular dystrophy, hypothyroidism, pulmonary edema.	
Creatinine (serum)	0.5–1.2 mg/dL (45–105 μmol/L)	Impaired renal function, urinary tract obstruction, giantism, acromegaly.	Decreased muscle mass, as occurs in muscular dystrophy or myasthenia gravis.
Electrolytes (plasma)	See Table 27.2 on page 1000.		
Gamma-glutamyl transferase (GGT) (serum)	0–30 U/L (same)	Bile duct obstruction, cirrhosis, alcoholism, metastatic liver cancer, congestive heart failure.	
Glucose (plasma)	70–110 mg/dL (3.9–6.1 mmol/L)	Diabetes mellitus, acute stress, hyperthyroidism, chronic liver disease, Cushing's disease.	Addison's disease, hypothyroidism, hyperinsulinism.
Hemoglobin (whole blood)	Males: 14–18 g/100 mL (140–180 g/L) Females: 12–16 g/100 mL (120–160 g/L) Newborns: 14–20 g/100 mL (140–200 g/L)	Polycythemia, congestive heart failure, chronic obstructive pulmonary disease, living at high altitude.	Anemia, severe hemorrhage, cancer, hemolysis, Hodgkin's disease, nutritional deficiency of vitamin B_{12}, systemic lupus erythematosus, kidney disease.
Iron, total (serum)	Males: 80–180 μg/dL (14–32 μmol/L) Females: 60–160 μg/dL (11–29 μmol/L)	Liver disease, hemolytic anemia, iron poisoning.	Iron-deficiency anemia, chronic blood loss, pregnancy (late), chronic heavy menstruation.
Lactic dehydrogenase (LDH) (serum)	71–207 U/L (same)	Myocardial infarction, liver disease, skeletal muscle necrosis, extensive cancer.	
Lipids (serum) **Total** **Triglycerides**	 400–850 mg/dL (4.0–8.5 g/L) 10–190 mg/dL (0.1–1.9 g/L)	Hyperlipidemia, diabetes mellitus, hypothyroidism.	Fat malabsorption.
Platelet (thrombocyte) count (whole blood)	150,000–400,000/μL	Cancer, trauma, leukemia, cirrhosis.	Anemias, allergic conditions hemorrhage.
Protein (serum) **Total** **Albumin** **Globulin**	 6–8 g/dL (60–80 g/L) 4–6 g/dL (40–60 g/L) 2.3–3.5 g/dL (23–35 g/L)	Dehydration, shock, chronic infections.	Liver disease, poor protein intake, hemorrhage, diarrhea, malabsorption, chronic renal failure, severe burns.

(continues)

Blood Tests

Test (Specimen)	U.S. Reference Values (SI Units)	Values Increase In	Values Decrease In
Red blood cell (erythrocyte) count (whole blood)	Males: 4.5–6.5 million/μL Females: 3.9–5.6 million/μL	Polycythemia, dehydration, living at high altitude.	Hemorrhage, hemolysis, anemias, cancer, overhydration.
Uric acid (urate) (serum)	2.0–7.0 mg/dL (120–420 μmol/L)	Impaired renal function, gout, metastatic cancer, shock, starvation.	
White blood cell (leukocyte) count, total (whole blood)	5,000–10,000/μL (See Table 19.3 on page 646 for relative percentages of different types of WBCs.)	Acute infections, trauma, malignant diseases, cardiovascular diseases. (See also Table 19.2 on page 645.)	Diabetes mellitus, anemia. (See also Table 19.2 on page 645.)

Appendix D
Normal Values For Selected Urine Tests

Urine Tests		
Test (Specimen)	**U.S. Reference Values (SI Units)**	**Clinical Implications**
Amylase (2 hour)	35–260 Somogyi units/hr (6.5–48.1 units/hr)	Values increase in inflammation of the pancreas (pancreatitis) or salivary glands, obstruction of the pancreatic duct, and perforated peptic ulcer.
Bilirubin* (random)	Negative	Values increase in liver disease and obstructive biliary disease.
Blood* (random)	Negative	Values increase in renal disease, extensive burns, transfusion reactions, and hemolytic anemia.
Calcium (Ca^{2+}) (random)	10 mg/dL (2.5 mmol/liter); up to 300 mg/24 hr (7.5 mmol/24 hr)	Amount depends on dietary intake; values increase in hyperparathyroidism, metastatic malignancies, and primary cancer of breasts and lungs; values decrease in hypoparathyroidism and vitamin D deficiency.
Casts (24 hour)		
Epithelial	Occasional	Values increase in nephrosis and heavy metal poisoning.
Granular	Occasional	Values increase in nephritis and pyelonephritis.
Hyaline	Occasional	Values increase in kidney infections.
Red blood cell	Occasional	Values increase in glomerular membrane damage and fever.
White blood cell	Occasional	Values increase in pyelonephritis, kidney stones, and cystitis.
Chloride (Cl^-) (24 hour)	140–250 mEq/24 hr (140–250 mmol/24 hr)	Amount depends on dietary salt intake; values increase in Addison's disease, dehydration, and starvation; values decrease in pyloric obstruction, diarrhea, and emphysema.
Color (random)	Yellow, straw, amber	Varies with many disease states, hydration, and diet.
Creatinine (24 hour)	Males: 1.0–2.0 g/24 hr (9–18 mmol/24 hr) Females: 0.8–1.8 g/24 hr (7–16 mmol/24 hr)	Values increase in infections; values decrease in muscular atrophy, anemia, and kidney diseases.
Glucose*	Negative	Values increase in diabetes mellitus, brain injury, and myocardial infarction.
Hydroxycortico-steroids (17-hydroxysteroids) (24 hour)	Males: 5–15 mg/24 hr (13–41 μmol/24 hr) Females: 2–13 mg/24 hr (5–36 μmol/24 hr)	Values increase in Cushing's syndrome, burns, and infections; values decrease in Addison's disease.
Ketone bodies* (random)	Negative	Values increase in diabetic acidosis, fever, anorexia, fasting, and starvation.
17-ketosteroids (24 hour)	Males: 8–25 mg/24 hr (28–87 μmol/24 hr) Females: 5–15 mg/24 hr (17–53 μmol/24 hr)	Values decrease in surgery, burns, infections, adrenogenital syndrome, and Cushing's syndrome.
Odor (random)	Aromatic	Becomes acetonelike in diabetic ketosis.
Osmolality (24 hour)	500–1400 mOsm/kg water (500–1400 mmol/kg water)	Values increase in cirrhosis, congestive heart failure (CHF), and high-protein diets; values decrease in aldosteronism, diabetes insipidus, and hypokalemia.
pH* (random)	4.6–8.0	Values increase in urinary tract infections and severe alkalosis; values decrease in acidosis, emphysema, starvation, and dehydration.

Test (Specimen)	U.S. Reference Values (SI Units)	Clinical Implications
Phenylpyruvic acid (random)	Negative	Values increase in phenylketonuria (PKU).
Potassium (K^+) (24 hour)	40–80 mEq/24 hr (40–80 mmol/24 hr)	Values increase in chronic renal failure, dehydration, starvation, and Cushing's syndrome; values decrease in diarrhea, malabsorption syndrome, and adrenal cortical insufficiency
Protein* (albumin) (random)	Negative	Values increase in nephritis, fever, severe anemias, trauma, and hyperthyroidism.
Sodium (Na^+) (24 hour)	75–200 mg/24 hr (75–200 mmol/24 hr)	Amount depends on dietary salt intake; values increase in dehydration, starvation, and diabetic acidosis; values decrease in diarrhea, acute renal failure, emphysema, and Cushing's syndrome.
Specific gravity* (random)	1.001–1.035 (same)	Values increase in diabetes mellitus and excessive water loss; values decrease in absence of antidiuretic hormone (ADH) and severe renal damage.
Urea (random)	25–35 g/24 hr (420–580 mmol/24 hr)	Values increase in response to increased protein intake; values decrease in impaired renal function.
Uric acid (24 hour)	0.4–1.0 g/24 hr (1.5–4.0 mmol/24 hr)	Values increase in gout, leukemia, and liver disease; values decrease in kidney disease.
Urobilinogen* (2 hour)	0.3–1.0 Ehrlich units (1.7–6.0 μmol/24 hr)	Values increase in anemias, hepatitis A (infectious), biliary disease, and cirrhosis; values decrease in cholelithiasis and renal insufficiency.
Volume, total (24 hour)	1000–2000 mL/24 hr (1.0–2.0 liters/24 hr)	Varies with many factors.

*Test often performed using a **dipstick,** a plastic strip impregnated with chemicals that is dipped into a urine specimen to detect particular substances. Certain colors indicate the presence or absence of a substance and sometimes give a rough estimate of the amount(s) present.

Appendix E
Answers

Answers to Self-Quiz Questions

Chapter 1

1. cell **2.** catabolism **3.** homeostasis **4.** false **5.** true **6.** true **7.** d **8.** b **9.** b **10.** e **11.** e **12.** (a) 3, (b) 4, (c) 6, (d) 5, (e) 1, (f) 2 **13.** (a) 6, (b) 1, (c) 11, (d) 5, (e) 10, (f) 8, (g) 7, (h) 9, (i) 4, (j) 3, (k) 2 **14.** (a) 4, (b) 6, (c) 8, (d) 1, (e) 5, (f) 2, (g) 7, (h) 3 **15.** (a) 3, (b) 7, (c) 1, (d) 5, (e) 6, (f) 4, (g) 8, (h) 2

Chapter 2

1. protons, electrons, neutrons **2.** 8 **3.** atomic number **4.** concentration, temperature **5.** true **6.** false **7.** b **8.** d **9.** a **10.** c **11.** b **12.** (a) 1, (b) 2 (c) 1, (d) 4 **13.** a **14.** (a) 8, (b) 1, (c) 7, (d) 2, (e) 5, (f) 3, (g) 4, (h) 6 **15.** (a) 7, (b) 5, (c) 1, (d) 2, (e) 6, (f) 3, (g) 8, (h) 4

Chapter 3

1. plasma membrane, cytoplasm, nucleus **2.** apoptosis, necrosis **3.** cytosol **4.** shortening and loss of protective telomeres on chromosomes, cross-link formation between glucose and proteins, free radical formation **5.** true **6.** true **7.** a **8.** d **9.** b **10.** a **11.** c **12.** (a) 2, (b) 3, (c) 5, (d) 7, (e) 6, (f) 8, (g) 1, (h) 4 **13.** (a) 11, (b) 10, (c) 1, (d) 2, (e) 5, (f) 9, (g) 12, (h) 7, (i) 6, (j) 3, (k) 4, (l) 8 **14.** (a) 2, (b) 9, (c) 3, (d) 5, (e) 11, (f) 8, (g) 1, (h) 6, (i) 10, (j) 7, (k) 4 **15.** (a) 3, (b) 9, (c) 1, (d) 5, (e) 11, (f) 4, (g) 8, (h) 7, (i) 2, (j) 10, (k) 6

Chapter 4

1. tissue **2.** endocrine **3.** cells, ground substance, fibers **4.** true **5.** false **6.** a **7.** c **8.** b **9.** c **10.** b **11.** d **12.** (a) 4, (b) 8, (c) 5, (d) 2, (e) 3, (f) 6, (g) 1, (h) 7 **13.** (a) 3, (b) 5, (c) 8, (d) 12, (e) 9, (f) 7, (g) 11, (h) 6, (i) 2, (j) 4, (k) 10, (l) 1 **14.** (a) 2, (b) 5, (c) 3, (d) 4, (e) 1 **15.** (a) 3, (b) 2, (c) 1

Chapter 5

1. palms, soles, fingertips **2.** strength, extensibility, elasticity **3.** ceruminous **4.** melanin, carotene, hemoglobin **5.** a **6.** b **7.** a **8.** c **9.** c **10.** e **11.** (a) 3, (b) 4, (c) 1, (d) 2 **12.** (a) 5, (b) 13, (c) 3, (d) 7, (e) 4, (f) 2, (g) 6, (h) 8, (i) 10, (j) 12, (k) 11, (l) 1, (m) 9 **13.** (a) 3, (b) 5, (c) 4, (d) 1, (e) 2, (f) 6, (g) 7, (h) 8 **14.** (a) 3, (b) 1, (c) 2, (d) 5, (e) 4 **15.** (a) 4, (b) 3, (c) 2, (d) 1, inflammatory, migratory, proliferative, maturation

Chapter 6

1. hardness, tensile strength **2.** osteons (Haversian systems), trabeculae **3.** hyaline cartilage, loose fibrous connective tissue membranes **4.** true **5.** false **6.** b **7.** b **8.** a **9.** d **10.** a **11.** d **12.** (a) 3, (b) 9, (c) 8, (d) 1, (e) 5, (f) 4, (g) 6, (h) 7, (i) 2, (j) 10 **13.** (a) 10, (b) 2, (c) 7, (d) 1, (e) 8, (f) 3, (g) 6, (h) 4, (i) 9, (j) 5 **14.** (a) 9, (b) 7, (c) 6, (d) 1, (e) 4, (f) 2, (g) 5, (h) 3, (i) 8, (j) 10 **15.** (a) 2, (b) 5, (c) 3, (d) 4, (e) 1

Chapter 7

1. fontanels **2.** thoracic **3.** intervertebral discs **4.** pituitary gland **5.** false **6.** false **7.** a **8.** c **9.** b **10.** e **11.** d **12.** (a) 7, (b) 5, (c) 1, (d) 6, (e) 2, (f) 4, (g) 8, (h) 9, (i) 3, (j) 10, (k) 11, (l) 13, (m) 12 **13.** (a) 2, (b) 3, (c) 5, (d) 6, (e) 4, (f) 1, (g) 5, (h) 4, (i) 2, (j) 4, (k) 3 **14.** (a) 4, (b) 7, (c) 6, (d) 5, (e) 3, (f) 1, (g) 2 **15.** (a) 3, (b) 1, (c) 6, (d) 9, (e) 13, (f) 12, (g) 2, (h) 4, (i) 5, (j) 7, (k) 10, (l) 15, (m) 8, (n) 11, (o) 14

Chapter 8

1. protection, movement **2.** metacarpals **3.** true **4.** true **5.** b **6.** b **7.** d **8.** e **9.** e **10.** c **11.** a **12.** b **13.** b **14.** (a) 4, (b) 3, (c) 3, (d) 6, (e) 7, (f) 1, (g) 3, (h) 2, (i) 5, (j) 9, (k) 8, (l) 2, (m) 4, (n) 6, (o) 7 **15.** (a) 2, (b) 6, (c) 9, (d) 7, (e) 4, (f) 5, (g) 8, (h) 10, (i) 1, (j) 3

Chapter 9

1. joint, articulation, or arthrosis **2.** synovial fluid **3.** osteoarthritis **4.** d **5.** b **6.** a **7.** c **8.** e **9.** e **10.** true **11.** false **12.** (a) 5, (b) 3, (c) 6, (d) 1, (e) 4, (f) 7, (g) 2 **13.** (a) 4, (b) 3, (c) 6, (d) 2, (e) 5, (f) 1 **14.** (a) 6, (b) 4, (c) 5, (d) 1, (e) 3, (f) 2 **15.** (a) 6, (b) 7, (c) 9, (d) 4, (e) 8, (f) 3, (g) 10, (h) 2, (i) 11, (j) 1, (k) 5

Chapter 10

1. electrical excitability, contractility, extensibility, elasticity **2.** actin, myosin **3.** motor unit **4.** false **5.** true **6.** e **7.** a **8.** b **9.** e **10.** d **11.** c **12.** (a) 5, (b) 6, (c) 8, (d) 7, (e) 2, (f) 4, (g) 3, (h) 1 **13.** (a) 5, (b) 8, (c) 7, (d) 10, (e) 6, (f) 9, (g) 4, (h) 1, (i) 2, (j) 3 **14.** (a) 2, (b) 3, (c) 1, (d) 1 and 2 (e) 2 and 3, (f) 2, (g) 1, (h) 3, (i) 1 and 2, (j) 3 **15.** (a) 10, (b) 2, (c) 4, (d) 3, (e) 6, (f) 5, (g) 1, (h) 7, (i) 9, (j) 8

Chapter 11

1. origin, insertion **2.** buccinator **3.** true **4.** true **5.** d **6.** e **7.** a **8.** e **9.** c **10.** a **11.** b **12.** (a) 12, (b) 9, (c) 8, (d) 6, (e) 3, (f) 11, (g) 10, (h) 1, (i) 2, (j) 7, (k) 4, (l) 5 **13.** (a) 3, (b) 1, (c) 2, (d) 1, (e) 2, (f) 3, (g) 3 **14.** (a) 6, (b) 3, (c) 7, (d) 4, (e) 2, (f) 8, (g) 5, (h) 1 **15.** (a) 10, (b) 1, (c) 9, (d) 8, (e) 12, (f) 17, (g) 2, (h) 6, (i) 8, (j) 14, (k) 5, (l) 4, (m) 2, (n) 15, (o) 1, (p) 11, (q) 13, (r) 12, (s) 7, (t) 16, (u) 11, (v) 17, (w) 16, (x) 15, (y) 3

Chapter 12

1. somatic nervous system, autonomic nervous system, enteric nervous system **2.** polarized **3.** true **4.** false **5.** c **6.** b **7.** a **8.** e **9.** c **10.** e **11.** b **12.** a **13.** (a) 6, (b) 11, (c) 1, (d) 2, (e) 9, (f) 13, (g) 4, (h) 8, (i) 7, (j) 12, (k) 5, (l) 3, (m) 10 **14.** (a) 2, (b) 1, (c) 9, (d) 8, (e) 6, (f) 3, (g) 4, (h) 5, (i) 10, (j) 7 **15.** (a) 4, (b) 5, (c) 14, (d) 8, (e) 7, (f) 1, (g) 2, (h) 10, (i) 13, (j) 6, (k) 3, (l) 12, (m) 9, (n) 11

Chapter 13

1. reflexes **2.** mixed **3.** true **4.** true **5.** d **6.** e **7.** d **8.** a **9.** c **10.** (a) 1, (b) 8, (c) 4, (d) 2, (e) 10, (f) 1, (g) 6, (h) 5, (i) 3, (j) 9, (k) 1, (l) 11, (m) 7, (n) 2 **11.** a **12.** b **13.** e **14.** (a) 10, (b) 8, (c) 9, (d) 1, (e) 2, (f) 5, (g) 7, (h) 6, (i) 11, (j) 4, (k) 3 **15.** (a) 2, (b) 1, (c) 3, (d) 4, (e) 1, (f) 2, (g) 4, (h) 3, (i) 5, (j) 1

Chapter 14

1. choroid plexuses **2.** longitudinal fissue **3.** true **4.** false **5.** c **6.** d **7.** e **8.** d **9.** e **10.** a **11.** d **12.** (a) 3, (b) 5, (c) 6, (d) 8, (e) 11, (f) 10, (g) 7, (h) 9, (i) 1, (j) 4, (k) 2, (l) 12, (m) 1, (n) 8, (o) 5, (p) 7, (q) 12, (r) 10, (s) 9 **13.** (a) 5, (b) 9, (c) 11, (d) 6, (e) 3, (f) 1, (g) 10, (h) 8, (i) 2, (j) 4, (k) 7 **14.** (a) 9, (b) 2, (c) 6, (d) 10, (e) 4, (f) 11, (g) 1, (h) 2, (i) 5, (j) 8, (k) 12, (l) 7, (m) 3, (n) 6 and 8, (o) 13, (p) 7, (q) 1 **15.** (a) 9, (b) 2, (c) 6, (d) 8, (e) 7, (f) 5, (g) 3, (h) 10, (i) 12, (j) 11, (k) 4, (l) 1

Chapter 15

1. sensation, perception **2.** adaptation **3.** true **4.** true **5.** d **6.** b **7.** a **8.** d **9.** e **10.** e **11.** d **12.** d **13.** (a) 9, (b) 8, (c) 4, (d) 7, (e) 10, (f) 2, (g) 3, (h) 1, (i) 5, (j) 6, (k) 11 **14.** (a) 10, (b) 8, (c) 7, (d) 1, (e) 4, (f) 3, (g) 5, (h) 6, (i) 9, (j) 2 **15.** (a) 3, (b) 2, (c) 5, (d) 7, (e) 1, (f) 3, (g) 8, (h) 4, (i) 6

Chapter 16

1. sweet, sour, salty, bitter, umami **2.** static, dynamic **3.** true **4.** false **5.** a **6.** d **7.** c **8.** c **9.** c **10.** c **11.** e **12.** a **13.** (a) 1, (b) 5, (c) 7, (d) 6, (e) 1, (f) 8, (g) 2, (h) 4, (i) 3 **14.** (a) 3, (b) 6, (c) 8, (d) 12, (e) 1, (f) 5, (g) 9, (h) 11, (i) 7, (j) 13, (k) 2, (l) 10, (m) 4 **15.** (a) 2, (b) 11, (c) 14, (d) 13, (e) 3, (f) 10, (g) 6, (h) 12, (i) 4, (j) 5, (k) 9, (l) 1, (m) 7, (n) 8

Chapter 17

1. acetylcholine, epinephrine, or norepinephrine **2.** thoracolumbar, craniosacral **3.** true **4.** true **5.** b **6.** c **7.** d **8.** a **9.** a **10.** c **11.** d **12.** e **13.** (a) 2, (b) 5, (c) 6, (d) 1, (e) 3, (f) 4 **14.** (a) 3, (b) 2, (c) 1, (d) 1, (e) 2, (f) 3, (g) 3 **15.** (a) 2, (b) 1, (c) 1, (d) 2, (e) 1, (f) 1, (g) 2, (h) 2

Chapter 18

1. fight-or-flight response, resistance reaction, exhaustion **2.** hypothalamus **3.** exocrine, endocrine **4.** false **5.** true **6.** b **7.** c **8.** d **9.** a **10.** e **11.** a **12.** c **13.** (a) 8, (b) 2, (c) 7, (d) 1, (e) 12, (f) 18, (g) 5, (h) 17, (i) 20, (j) 15, (k) 3, (l) 16, (m) 19, (n) 6, (o) 13, (p) 11, (q) 4, (r) 10, (s) 14, (t) 9 **14.** (a) 10, (b) 8, (c) 2, (d) 4, (e) 1, (f) 6, (g) 9, (h) 7, (i) 5, (j) 3 **15.** (a) 12, (b) 1, (c) 11, (d) 7, (e) 3, (f) 10, (g) 2, (h) 9, (i) 4, (j) 8, (k) 5, (l) 6

Chapter 19

1. blood plasma, formed elements, red blood cells, white blood cells, platelets **2.** serum **3.** platelets, thrombocytes **4.** true **5.** false **6.** e **7.** a **8.** e **9.** a **10.** b **11.** c **12.** d **13.** (a) 6, (b) 8, (c) 3, (d) 10, (e) 2, (f) 4, (g) 1, (h) 7, (i) 9, (j) 5 **14.** (a) 5, (b) 1, (c) 2, (d) 3, (e) 6, (f) 8 (g) 9, (h) 4, (i) 7 **15.** (a) 4, (b) 6, (c) 8, (d) 1, (e) 7, (f) 3, (g) 5, (h) 2

Chapter 20

1. left ventricle **2.** systole, diastole **3.** stroke volume **4.** false **5.** false **6.** true **7.** a **8.** c **9.** b **10.** a **11.** b **12.** (a) 3, (b) 2, (c) 9, (d) 8, (e) 7, (f) 4, (g) 1, (h) 5, (i) 6 **13.** (a) 4, (b) 10, (c) 6, (d) 5, (e) 3, (f) 9, (g) 2, (h) 8, (i) 1, (j) 7 **14.** (a) 3, (b) 7, (c) 2, (d) 5, (e) 1, (f) 6, (g) 4 and 7 **15.** (a) 3, (b) 6, (c) 1, (d) 5, (e) 2, (f) 4

Chapter 21

1. carotid sinus, aortic **2.** skeletal muscle pump, respiratory pump **3.** true **4.** true **5.** a **6.** c **7.** e **8.** a **9.** d **10.** e **11.** c **12.** (a) 2, (b) 5, (c) 1, (d) 4, (e) 3 **13.** (a) 11, (b) 1, (c) 4, (d) 9, (e) 3, (f) 8, (g) 6, (h) 2, (i) 7, (j) 5, (k) 10, (l) 12 **14.** (a) 2, (b) 6, (c) 4, (d) 1, (e) 3, (f) 5 **15.** (a) 5, (b) 3, (c) 1, (d) 4, (e) 2, (f) 4, (g) 1, (h) 5

Chapter 22

1. skin, mucous membranes; antimicrobial proteins, NK cells, phagocytes **2.** specificity, memory **3.** true **4.** true **5.** e **6.** e **7.** d **8.** a **9.** c **10.** e **11.** b **12.** c **13.** (a) 3, (b) 1, (c) 4, (d) 2, (e) 5 **14.** (a) 2, (b) 3, (c) 4, (d) 7, (e) 1, (f) 6, (g) 5 **15.** (a) 11, (b) 11, (c) 8, (d) 1, (e) 2, (f) 5, (g) 4, (h) 7, (i) 9, (j) 12, (k) 3, (l) 6, (m) 10

Chapter 23

1. pulmonary ventilation, external respiration, internal respiration **2.** less, greater **3.** oxyhemoglogin; dissolved CO_2, carbamino compounds, and bicarbonate ion **4.** $CO_2 + H_2O \rightarrow H_2CO_3 \rightarrow H^+ + HCO_3^-$ **5.** false **6.** true **7.** b **8.** d **9.** a **10.** c **11.** e **12.** (a) 2, (b) 8, (c) 3, (d) 7, (e) 1, (f) 9, (g) 5, (h) 10, (i) 4, (j) 6 **13.** (a) 7, (b) 8, (c) 1, (d) 5, (e) 6, (f) 2, (g) 3, (h) 4 **14.** (a) 3, (b) 8, (c) 5, (d) 2, (e) 7, (f) 1, (g) 4, (h) 6 **15.** (a) 7, (b) 3, (c) 5, (d) 1, (e) 6, (f) 4, (g) 2

Chapter 24

1. mucosa, submucosa, muscularis, serosa **2.** submucosal plexus or plexus of Meissner; myenteric plexus or plexus of Auerbach **3.** monosaccharides; amino acids; monoglycerides, fatty acids; pentoses, phosphates, nitrogenous bases **4.** false **5.** false **6.** a **7.** c **8.** e **9.** b **10.** a **11.** b **12.** d **13.** (a) 3, (b) 10, (c) 5, (d) 8, (e) 6, (f) 1, (g) 4, (h) 9, (i) 11, (j) 2, (k) 7 **14.** (a) 4, (b) 6, (c) 7, (d) 1, (e) 5, (f) 3, (g) 8, (h) 2, (i) 10, (j) 9 **15.** (a) 4, (b) 8, (c) 2, (d) 10, (e) 7, (f) 11, (g) 1, (h) 6, (i) 3, (j) 9, (k) 5

Chapter 25

1. hypothalamus **2.** glucose 6-phosphate, pyruvic acid, acetyl coenzyme A **3.** hormones **4.** false **5.** true **6.** c **7.** d **8.** a **9.** d **10.** a **11.** a **12.** b **13.** (a) 2, (b) 8, (c) 5, (d) 10, (e) 6, (f) 7, (g) 3, (h) 9, (i) 4, (j) 1 **14.** (a) 9, (b) 12, (c) 11, (d) 10, (e) 4, (f) 7, (g) 3, (h) 5, (i) 8, (j) 1, (k) 6, (l) 2 **15.** (a) 8, (b) 5, (c) 1, (d) 10, (e) 6, (f) 2, (g) 4, (h) 9, (i) 3, (j) 7

Chapter 26

1. transitional epithelial **2.** voluntary, involuntary **3.** glomerular filtration, tubular reabsorption, tubular secretion **4.** true **5.** true **6.** c **7.** a **8.** d **9.** a **10.** e **11.** b **12.** c **13.** (a) 8, (b) 2, (c) 10, (d) 5, (e) 3, (f) 1, (g) 7, (h) 4, (i) 9, (j) 6 **14.** (a) 4, (b) 3, (c) 7, (d) 6, (e) 2, (f) 1, (g) 5 **15.** (a) 5, (b) 4, (c) 6, (d) 1, (e) 2, (f) 7, (g) 3

Chapter 27

1. intracellular fluid, extracellular fluid **2.** bicarbonate ion, carbonic acid **3.** weak acid, weak base **4.** protein buffer, carbonic acid-bicarbonate, phosphate **5.** true **6.** true **7.** a **8.** c **9.** b **10.** e **11.** d **12.** a **13.** b **14.** (a) 8, (b) 9, (c) 7, (d) 1, (e) 6, (f) 2, (g) 4, (h) 5, (i) 3 **15.** (a) 8, (b) 7, (c) 5, (d) 6, (e) 1, (f) 3, (g) 4, (h) 2

Chapter 28

1. puberty, menarche, menopause **2.** semen **3.** true **4.** true **5.** e **6.** b **7.** a **8.** c **9.** b **10.** a **11.** e **12.** (a) 7, (b) 2, (c) 1, (d) 4, (e) 8, (f) 5, (g) 3, (h) 6 **13.** (a) 6, (b) 4, (c) 2, (d) 1, (e) 3, (f) 8, (g) 7, (h) 5 **14.** (a) 6, (b) 4, (c) 1, (d) 12, (e) 8, (f) 5, (g) 7, (h) 13, (i) 11, (j) 3, (k) 2, (l) 10, (m) 9 **15.** (a) 10, (b) 12, (c) 1, (d) 5, (e) 2, (f) 4, (g) 6, (h) 11, (i) 8, (j) 3, (k) 9, (l) 7

Chapter 29

1. dilation, expulsion, placental **2.** corpus luteum **3.** mesoderm; ectoderm; endoderm **4.** true **5.** a **6.** b **7.** e **8.** (a) 3, (b) 4, (c) 5, (d) 1, (e) 2, (f) 6, (g) 7, (h) 8, (i) 9 **9.** e **10.** b **11.** (a) 7, (b) 3, (c) 6, (d) 4, (e) 10, (f) 8, (g) 5, (h) 2, (i) 1, (j) 9, (k) 11, (l) 12 **12.** b **13.** (a) 3, (b) 6, (c) 4, (d) 1, (e) 8, (f) 2, (g) 5, (h) 7 **14.** (a) 5, (b) 2, (c) 8, (d) 4, (e) 7, (f) 1, (g) 3, (h) 6 **15.** d

Answers to Critical Thinking Questions

Chapter 1

1. Homeostasis is the relative constancy of the body's internal environment. Body temperature should vary within a narrow range around normal body temperature (38°C or 98.6° F).
2. Bilateral means two sides or both arms in this case. The carpal region is the wrist area.
3. A x ray provides a good image of dense tissue such as bones. An MRI is used for imaging soft tissues, not bones. An MRI cannot be used when metal is present due to the magnetic field used.

Chapter 2

1. Fatty acids are in all lipids including plant oils and the phospholipids that compose cell membranes. Sugars (monosaccharides) are needed for ATP production, are a component of nucleotides, and are the basic units of disaccharides and polysaccharides including starch, glycogen, and cellulose.
2. DNA = deoxyribonucleic acid. DNA is composed of 4 different nucleotides. The order of the nucleotides is unique in every person. The 20 different amino acids form proteins.
3. One pH unit equals a tenfold change in H^+ concentration. Pure water has a pH of 7, which equals 1×10^{-7} moles of H^+ per liter. Blood has a pH of 7.4, which equals 0.4×10^{-7} moles H^+ per liter. Thus, blood has less than half as many H^+ as water.

Chapter 3

1. The tissues are destroyed due to autolysis of the cells caused by the release of acids and digestive enzymes from the lysosomes.
2. Synthesis of mucin by the ribosomes on rough endoplasmic reticulum, to transport vesicle, to entry face of Golgi, to transfer vesicle, to medial cisternae, where protein is modified, to transfer vesicle, to exit face, to secretory vesicle, to plasma membrane, where it undergoes exocytosis.
3. The sense strand is composed of DNA codons transcribed into mRNA. Only the exons will be found in the mRNA. The introns are clipped out.

Chapter 4

1. Many possible adaptations including: more adipose tissue for insulation, thicker bone for support, more red blood cells for oxygen transport, and so on.
2. The surface layer of the skin, the epidermis, is keratinized, stratified squamous epithelium. It is avascular. If the pin were stuck straight in, the vascularized connective tissue would be pierced and the finger would bleed.
3. The gelatin portion of the salad represents the ground substance of the connective tissue matrix. The grapes are cells such as fibroblasts. The shredded carrots and coconut are fibers embedded in the ground substance.

Chapter 5

1. The dust particles are mostly keratinocytes shed from the stratum corneum of the skin.
2. It would be neither wise nor feasible to remove the exocrine glands. Essential exocrine glands include the sudoriferous glands (sweat helps control body temperature), sebaceous glands (sebum lubricates the skin), and ceruminous glands (provide protective lubricant for ear canal).
3. The superficial layer of the epidermis of the skin is the stratum corneum. Its cells are full of intermediate filaments of keratin, keratohyalin, and lipids from lamellar granules, making this layer a water-repellent barrier. The epidermis also contains an abundance of desmosomes.

Chapter 6

1. In greenstick fracture, which occurs only in children due to the greater flexibility of their bones, the bone breaks on one side but bends on the other side, resembling what happens when one tries to break a green (not dry) stick. Lynne Marie probably broke her fall with her extended arm, breaking the radius or ulna.
2. At Aunt Edith's age, the production of several hormones (such as estrogens and human growth hormone) necessary for bone remodeling is likely to be decreased. Her decreased height results, in part, from compression of her vertebrae due to flattening of the discs between them. She may also have osteoporosis and an increased susceptibility to fractures resulting in damage to the vertebrae and loss of height.
3. Exercise causes mechanical stress on bones, but because there is effectively zero gravity in space, the pull of gravity on bones is missing. The lack of stress from gravity results in bone demineralization and weakness.

Chapter 7

1. Fontanels, the soft spots between cranial bones, are fibrous connective tissue membranes that will be replaced by bone as the baby matures. They allow the infant's head to be molded during its passage through the birth canal and allow for brain and skull growth in the infant.
2. Barbara probably broke her coccyx or tailbone. The coccygeal vertebrae usually fuse by age 30. The coccyx points inferiorly in females.
3. Infants are born with a single concave curve in the vertebral column. Adults have four curves in their spinal cord-at the cervical (convex), thoracic (concave), lumbar (convex), and sacral regions. Buying a mattress from this company and sleeping on it would result in potentially severe back problems.

Chapter 8

1. Clawfoot is an abnormal elevation of the medial longitudinal arch of the foot.
2. There are 14 phalanges in each hand: two bones in the thumb and three in each of the other fingers. Farmer White has lost five phalanges on his left hand so he has nine remaining on his left and 14 remaining on his right for a total of 23.
3. Snakes don't have appendages, so they don't need shoulders and hips. Pelvic and pectoral girdles connect the upper and lower limbs (appendages) to the axial skeleton, and thus are considered part of the appendicular skeleton.

Chapter 9

1. Katie's vertebral column, head, thigh, lower leg, and lower arm are all flexed. Her lower arm and shoulder are medially rotated.
2. The elbow is a hinge-type synovial joint. The trochlea of the humerus articulates with the trochlear notch of the ulna. The elbow has monaxial (opening and closing) movement similar to a door.
3. Cartilaginous joints can be composed of hyaline cartilage, such as at the epiphyseal plate, or fibrocartilage, such as the intervertebral disc. Sutures are an example of fibrous joints, which are composed of dense fibrous tissue. Synovial (diarthrotic) joints are held together by an articular capsule, which has an inner synovial membrane layer.

Chapter 10

1. Smooth muscle contains both thick and thin filaments as well as intermediate filaments attached to dense bodies. The smooth muscle fiber contracts like a corkscrew turns; the fiber twists like a helix as it contracts and shortens.
2. Ming's muscles performed isometric contractions. Peng used primarily isotonic concentric contraction to lift, isometric to hold, and then isotonic eccentric contractions to lower the barbell.
3. Mr. Klopfer's students have muscle fatigue in their hands. Fatigue has many causes including an increase in lactic acid and ADP, and a decrease in Ca^{2+}, oxygen, creatine phosphate, and glycogen.

Chapter 11

1. Some of the antagonistic pairs for the regions are: upper arm-biceps brachii vs. triceps brachii; upper leg-quadriceps femoris vs. hamstrings; torso-rectus abdominus vs. erector spinae; lower leg-gastrocnemius vs. tibialis anterior.
2. The fulcrum is the knee joint, the load (resistance) is the weight of the upper body and the package, and the effort is the thigh muscles. This is a third-class lever.
3. Lifting the eyebrow uses the frontal belly of the occipitofrontalis; whistling uses the buccinator and orbicularis oris; closing the eyes uses the orbicularis oculi; shaking the head can use several muscles, including sternocleidomastoid, semispinalis capitis, splenius capitis, and longissimus capitis.

Chapter 12

1. The motor neuron is multipolar in structure with an axon and several dendrites projecting from the cell body. The converging circuit has several neurons that converge to form synapses with one common neuron. The bipolar neuron has one dendrite and an axon projecting from the cell body. In a simple circuit, each presynaptic neuron makes a single synapse with one postsynaptic neuron.
2. Gray matter appears gray in color due to the absence of myelin. It is composed of cell bodies, and unmyelinated axons and dendrites. White matter contains many myelinated axons. Lipofuscin, a yellowing pigment, collects in neurons with age.
3. Smelling coffee and hearing alarm are somatic sensory, stretching and yawning are somatic motor, salivating is autonomic (parasympathetic) motor, stomach rumble is enteric motor.

Chapter 13

1. A withdrawal/flexor reflex is ipsilateral and polysynaptic. The route is pain receptor of sensory neuron → spinal cord (integrating center) interneurons → motor neurons → flexor muscles in the leg. Also present is a crossed extensor reflex, which is contralateral and polysynaptic. The route diverges at the spinal cord: Interneurons cross to the other side → motor neurons → extensor muscles in the opposite leg.
2. The spinal cord is anchored in place by the filum terminale and the denticulate ligaments.
3. The needle will pierce the epidermis, the dermis, and the subcutaneous layer and then go between the vertebrae through the epidural space, the dura mater, the subdural space, the arachnoid mater, and into the CSF in the subarachnoid space. CSF is produced in the brain, and the spinal meninges are continuous with cranial meninges.

Chapter 14

1. Movement of the right arm is controlled by the left hemisphere's primary motor area, located in the precentral gyrus. Speech is controlled by Broca's area in the left hemisphere's frontal lobe just superior to the lateral cerebral sulcus.
2. The brain is enclosed by the cranial bones and meninges. The temporal bone houses the middle ear and inner ear, separating these from the brain.
3. The dentist has injected anesthetic into the inferior alveolar nerve, a branch of the mandibular nerve, which numbs the lower teeth and the lower lip. The tongue is numbed by blocking the lingual nerve. The upper teeth and lip are anesthetized by injecting the superior alveolar nerve, a branch of the maxillary nerve.

Chapter 15

1. Chemoreceptors in the nose detect odors. Proprioceptors detect body position and are involved in equilibrium. The receptors for smell are rapidly adapting (phasic), whereas proprioceptors are slowly adapting (tonic).
2. The tickle receptors are free nerve endings in the foot. The impulses travel along the first-order sensory neurons to the posterior gray horn, then along the second-order neurons, crossing to the other side of the spinal cord and up the anterior spinothalamic tract to the thalamus. The third-order neurons extend from the thalamus to the "foot" region of the somatosensory area of the cerebral cortex.
3. Yoshio's perception of feeling in his amputated foot is called phantom limb sensation. Impulses from the remaining proximal section of the sensory neuron are perceived by the brain as still coming from the amputated foot.

Chapter 16

1. Because smell and taste have ties to the cortex and limbic areas, Brenna may be recalling a memory of this or similar food. Pathway: olfactory receptors (cranial nerve I) → olfactory bulbs → olfactory tracts → lateral olfactory area in temporal lobe of cerebral cortex.
2. Fred may have cataracts, a loss of transparency in the lens of the eye. Cataracts are often associated with age, smoking, and exposure to UV light.
3. Auricle → external auditory meatus → tympanic membrane → maleus → incus → stapes → oval window → perilymph of scala vestibuli and tympani → vestibular membrane → endolymph of cochlear duct → basilar membrane → spiral organ.

Chapter 17

1. Digestion and relaxation are controlled by increased stimulation of the parasympathetic division of the ANS. The salivary glands, pancreas, and liver will show increased secretion; the stomach and intestines will have increased activity; the gallbladder will have increased contractions; heart contractions will have decreased force and rate.
2. Stretch receptors in the colon → autonomic sensory neurons → sacral region of spinal cord (integrating center) → parasympathetic preganglionic neuron → terminal ganglion → parasympathetic postganglionic neuron → smooth muscle in colon, rectum, and sphincter (effectors).
3. Nicotine from the cigarette smoke binds to nicotinic receptors on skeletal muscles, mimicking the effect of acetylcholine and causing increased contractions ("twitches"). Nicotine also binds to nicotinic receptors on cells in the adrenal medulla, stimulating release of epinephrine and norepinephrine, which mimics fight-or-flight responses.

Chapter 18

1. The beta cells are one of the cell types in the pancreatic islets of the pancreas. In Type 1 diabetes, only the beta cells are destroyed; the rest of the pancreas is not affected. A successful transplantation of

the beta cells would enable the recipient to produce the hormone insulin and would cure the diabetes.
2. Amanda has an enlarged thyroid gland or goiter. The goiter is probably due to hypothyroidism, which is causing the weight gain, fatigue, mental dullness, and other symptoms.
3. The two adrenal glands are located superior to the two kidneys. The adrenals are about 4 cm high, 2 cm wide, and 1 cm thick. The outer adrenal cortex is composed of three layers: zona glomerulosa, zona fasciculata, and zona reticularis. The inner adrenal medulla is composed of chromaffin cells.

Chapter 19
1. Blood is composed of liquid blood plasma and the formed elements: red blood cells (erythrocytes), white blood cells (leukocytes), and platelets. Blood plasma contains water, proteins, electrolytes, gases, and nutrients. There are about 5 million RBCs, 5000 WBCs, and 250,000 platelets per mL of blood. Leukocytes include neutrophils, eosinophils, basophils, monocytes, and lymphocytes.
2. To determine the blood type, a drop of each of three different antibody solutions (antisera) are added to three separate blood drops. The solutions are anti-A, anti-B, and anti-Rh. In Josef's test, anti-B caused agglutination (clump formation) when added to his blood, indicating the presence of antigen B on the RBCs. Anti-A and anti-Rh did not cause agglutination, indicating the absence of these antigens.
3. Hemostasis occurred. The stages involved are vascular spasm (if an arteriole was cut), platelet plug formation, and coagulation (clotting).

Chapter 20
1. With a normal resting CO of about 5.25 liters/min, and a heart rate of 55 beats per minute, Arian's stroke volume would be 95 mL/beat. His CO during strenuous exercise is 6 times his resting CO, about 31,500 mL/min.
2. Rheumatic fever is caused by inflammation of the bicuspid and the aortic valves following a streptococcal infection. Antibodies produced by the immune system to destroy the bacteria may also attack and damage the heart valves. Heart problems later in life may be related to this damage.
3. Mr. Perkins is suffering from angina pectoris and has several risk factors for coronary artery disease such as smoking, obesity, and male gender. Cardiac angiography involves the use of a cardiac catheter to inject a radiopaque medium into the heart and its vessels. The angiogram may reveal blockages such as atherosclerotic plaques in his coronary arteries.

Chapter 21
1. After birth, the foramen ovale and ductus arteriosus close to establish the pulmonary circulation. The umbilical artery and veins close because the placenta is no longer functioning. The ductus venosus closes so that the liver is no longer bypassed.
2. Right hand: left ventricle → ascending aorta → arch of aorta → brachiocephalic trunk → right subclavian artery → right axillary artery → right brachial artery → right radial and ulnar arteries → right superficial palmar arch. Left hand: left ventricle → ascending aorta → arch of aorta → left subclavian artery → left axillary artery → left brachial artery → left radial and ulnar arteries → left superficial palmar arch.
3. A vascular (venous) sinus is a vein that lacks smooth muscle in its tunica media. Dense connective tissue replaces the tunica media and tunica externa.

Chapter 22
1. Tariq had a hypersensitivity reaction to an insect sting; he was probably allergic to the venom. He had a localized anaphylactic or type I reaction.
2. The site of the embedded splinter had probably become infected with bacteria. The red streaks are the lymphatic vessels that drain the infected area; the swollen tender bumps are the axillary lymph nodes, which are swollen due to the immune response to the infection.
3. Influenza vaccination introduces a weakened or killed virus (which will not cause the disease) to the body. The immune system recognizes the antigen and mounts a primary immune response. Upon exposure to the same flu virus that was in the vaccine, the body will produce a secondary response, which will usually prevent a case of the flu. This is artificially acquired active immunity.

Chapter 23
1. Normal average male volumes are 500 mL for tidal volume during quiet breathing, 3100 mL for inspiratory reserve volume, and 1200 mL for expiratory reserve volume. Average female volumes are less than average male volumes, because females are generally smaller than males.
2. The bones of the external nose are the frontal bone, the two nasal bones (the location of the fracture), and the maxillae. The external nose is also composed of the septal, lateral nasal, and alar cartilages, as well as skin, muscle, and mucous membrane.
3. The cerebral cortex can temporarily allow voluntary breath holding. P_{CO_2} and H^+ levels in the blood and CSF increase with breath holding, strongly stimulating the inspiratory area, which stimulates inspiration to begin again. Breathing will begin again even if the person loses consciousness.

Chapter 24
1. Katie lost the two upper central permanent incisors. The remaining deciduous teeth include the lower central incisors, an upper and lower pair of lateral incisors, an upper and lower pair of cuspids, and upper and lower pair of first molars, and upper and lower pair of second molars.
2. CCK promotes ejection of bile, secretion of pancreatic juice, and contraction of the pyloric sphincter. It also promotes pancreatic growth and enhances the effects of secretin. CCK acts on the hypothalamus to induce satiety (the feeling of fullness) and should therefore decrease the appetite.
3. The smaller left lobe of the liver is separated from the larger right lobe by the falciform ligament. The left lobe is inferior to the diaphragm in the epigastric region of the abdominopelvic cavity.

Chapter 25
1. Deb is eating a diet high in carbohydrates in order to store maximum amounts of glycogen in her skeletal muscles and liver. This practice is called carbohydrate loading. Glycogenolysis, the breakdown of stored glycogen, supplies the muscles with the glucose needed for ATP production via cellular respiration.
2. Mr. Hernandez was suffering from heat exhaustion caused by loss of fluids and electrolytes. Lack of NaCl causes muscle cramps, nausea and vomiting, dizziness, and fainting. Low blood pressure may also result.
3. Sara's normal growth may be maintained despite day-to-day fluctuations in food intake. Many factors govern food intake, including neurons in the hypothalamus, blood glucose level, amount of adipose tissue, CCK, distention of the GI tract, and body temperature.

Chapter 26

1. Without reabsorption, initially 105–125 mL of filtrate would be lost per minute, assuming normal glomerular filtration rate. Fluid loss from the blood would cause a decrease in blood pressure, and therefore a decrease in GBHP. When GBHP dropped below 45mmHg, filtration would stop (assuming normal CHP and BCOP) because NFP would be zero.
2. The urinary bladder can stretch considerably due to the presence of transitional epithelium, rugae, and three layers of smooth muscle in the detrusor muscle.
3. In females the urethra is about 4 cm long; in males the urethra is 15–20 cm long, including its passage through the penis, urogenital diaphragm, and prostate.

Chapter 27

1. Gary has water intoxication. The Na^+ concentration of his plasma and interstitial fluid is below normal. Water moved by osmosis into the cells, resulting in hypotonic intracellular fluid and water intoxication. Decreased plasma volume, due to water movement into the interstitial fluid, caused hypovolemic shock.
2. The average female body contains about 55% water, versus 60% water in the average male. Due to the influence of male and female hormones, the average female has relatively more subcutaneous fat (which contains very little water) and relatively less muscle and other tissues (which have higher water content) than the average male.
3. Excessive vomiting causes a loss of hydrochloric acid in gastric juice, and intake of antacids increases the amount of alkali in the body fluids, resulting in metabolic alkalosis. Vomiting also causes a loss of fluid and may cause dehydration.
4. (Step 1) pH = 7.30 indicates slight acidosis, which could be caused by elevated P_{CO_2} or lowered HCO_3^-. (Step 2) The HCO_3^- is lower than normal (20 mEq/liter), so (step 3) the cause is metabolic. (Step 4) the P_{CO_2} is lower than normal (32 mmHg), so hyperventilation is providing some compensation. Diagnosis: Henry has partially compensated metabolic acidosis. A possible cause is kidney damage that resulted from interruption of blood flow during the heart attack.

Chapter 28

1. LH (luteinizing hormone) is an anterior pituitary hormone that acts on both the male and female reproductive systems. In males, LH stimulates the Leydig cells of the testes to secrete testosterone.
2. The germinal epithelium, located on the surface of the ovaries, is a misnomer. The oocytes are located in the ovarian cortex, deep to the tunica albuginea. The tunica vaginalis covers the testes, superficial to the tunica albuginea. It forms from the peritoneum during the descent of the testes from the abdomen into the scrotum during fetal development.
3. No. The first polar body, formed by meiosis I, would contain half the homologous chromosome pairs, the other half being in the secondary oocyte. The second polar body, formed by meiosis II, would contain chromosomes identical to the ovum; however, fertilization would be by two different sperm, so the babies would still not be identical.

Chapter 29

1. Being able to taste PCT is an autosomal dominant trait. Because Kendra is a PCT nontaster, her genotype is homozygous recessive for this trait. Her parents must both be heterozygous for the trait. They exhibit the dominant phenotype for PCT tasting and are carriers of the recessive trait.
2. Because Huntington disease (HD) is a trinucleotide repeat disease, in which the number of repeats expand with each succeeding generation, it can occur in a child whose parents did not suffer HD. Also, because it is a dominant gene, one mutant allele of the *HD* gene is sufficient to cause HD.
3. All arteries carry blood away from the heart. The umbilical arteries, located in the fetus, carry deoxygenated blood away from the fetal heart towards the placenta where the blood will become oxygenated as it passes through the placenta.

Glossary

Pronunciation Key

1. The most strongly accented syllable appears in capital letters, for example, bilateral (bī-LAT-er-al) and diagnosis (dī-ag-NŌ-sis).
2. If there is a secondary accent, it is noted by a prime (′), for example, constitution (kon′-sti-TOO-shun) and physiology (fiz′-ē-OL-ō-jē). Any additional secondary accents are also noted by a prime, for example, decarboxylation (dē′-kar-bok′-si-LĀ-shun).
3. Vowels marked by a line above the letter are pronounced with the long sound, as in the following common words:
   ā as in *māke* — *ō* as in *pōle*
   ē as in *bē* — *ū* as in *cute*
   ī as in *īvy*
4. Vowels not marked by a line above the letter are pronounced with the short sound, as in the following words:
   a as in *above* or *at* — *o* as in *not*
   e as in *bet* — *u* as in *bud*
   i as in *sip*
5. Other vowel sounds are indicated as follows:
   oy as in *oil*
   oo as in *root*
6. Consonant sounds are pronounced as in the following words:
   b as in *bat* — *m* as in *mother*
   ch as in *chair* — *n* as in *no*
   d as in *dog* — *p* as in *pick*
   f as in *father* — *r* as in *rib*
   g as in *get* — *s* as in *so*
   h as in *hat* — *t* as in *tea*
   j as in *jump* — *v* as in *very*
   k as in *can* — *w* as in *welcome*
   ks as in *tax* — *z* as in *zero*
   kw as in *quit* — *zh* as in *lesion*
   l as in *let*

A

Abdomen (ab-DŌ-men *or* AB-dō-men) The area between the diaphragm and pelvis.

Abdominal (ab-DŌM-i-nal) **cavity** Superior portion of the abdominopelvic cavity that contains the stomach, spleen, liver, gallbladder, most of the small intestine, and part of the large intestine.

Abdominal thrust maneuver A first-aid procedure for choking. Employs a quick, upward thrust against the diaphragm that forces air out of the lungs with sufficient force to eject any lodged material. Also called the **Heimlich** (HĪM-lik) **maneuver.**

Abdominopelvic (ab-dom′-i-nō-PEL-vic) **cavity** Inferior component of the ventral body cavity that is subdivided into a superior abdominal cavity and an inferior pelvic cavity.

Abduction (ab-DUK-shun) Movement away from the midline of the body.

Abortion (a-BOR-shun) The premature loss (spontaneous) or removal (induced) of the embryo or nonviable fetus; miscarriage due to a failure in the normal process of developing or maturing.

Abscess (AB-ses) A localized collection of pus and liquefied tissue in a cavity.

Absorption (ab-SORP-shun) Intake of fluids or other substances by cells of the skin or mucous membranes; the passage of digested foods from the gastrointestinal tract into blood or lymph.

Absorptive (fed) state Metabolic state during which ingested nutrients are absorbed into the blood or lymph from the gastrointestinal tract.

Accessory duct A duct of the pancreas that empties into the duodenum about 2.5 cm (1 in.) superior to the ampulla of Vater (hepatopancreatic ampulla). Also called the **duct of Santorini** (san′-tō-RĒ-nē).

Accommodation (a-kom-ō-DĀ-shun) An increase in the curvature of the lens of the eye to adjust for near vision.

Acetabulum (as′-e-TAB-ū-lum) The rounded cavity on the external surface of the hip bone that receives the head of the femur.

Acetylcholine (as′-e-til-KŌ-lēn) **(ACh)** A neurotransmitter liberated by many peripheral nervous system neurons and some central nervous system neurons. It is excitatory at neuromuscular junctions but inhibitory at some other synapses (for example, it slows heart rate).

Achalasia (ak′-a-LĀ-zē-a) A condition, caused by malfunction of the myenteric plexus, in which the lower esophageal sphincter fails to relax normally as food approaches. A whole meal may become lodged in the esophagus and enter the stomach very slowly. Distension of the esophagus results in chest pain that is often confused with pain originating from the heart.

Acid (AS-id) A proton donor, or a substance that dissociates into hydrogen ions (H^+) and anions; characterized by an excess of hydrogen ions and a pH less than 7.

Acidosis (as-i-DŌ-sis) A condition in which blood pH is below 7.35. Also known as **acidemia.**

Acini (AS-i-nē) Groups of cells in the pancreas that secrete digestive enzymes.

Acoustic (a-KOOS-tik) Pertaining to sound or the sense of hearing.

Acquired immunodeficiency syndrome (AIDS) A fatal disease caused by the human immunodeficiency virus (HIV). Characterized by a positive HIV-antibody test, low helper T cell count, and certain indicator diseases (for example Kaposi's sarcoma,

Pneumocystis carinii pneumonia, tuberculosis, fungal diseases). Other symptoms include fever or night sweats, coughing, sore throat, fatigue, body aches, weight loss, and enlarged lymph nodes.

Acrosome (AK-rō-sōm) A lysosomelike organelle in the head of a sperm cell containing enzymes that facilitate the penetration of a sperm cell into a secondary oocyte.

Actin (AK-tin) A contractile protein that is part of thin filaments in muscle fibers.

Action potential An electrical signal that propagates along the membrane of a neuron or muscle fiber (cell); a rapid change in membrane potential that involves a depolarization followed by a repolarization. Also called a **nerve action potential** or **nerve impulse** as it relates to a neuron, and a **muscle action potential** as it relates to a muscle fiber.

Activation (ak′-ti-VĀ-shun) **energy** The minimum amount of energy required for a chemical reaction to occur.

Active transport The movement of substances across cell membranes against a concentration gradient, requiring the expenditure of cellular energy (ATP).

Acute (a-KŪT) Having rapid onset, severe symptoms, and a short course; not chronic.

Adaptation (ad′-ap-TĀ-shun) The adjustment of the pupil of the eye to changes in light intensity. The property by which a sensory neuron relays a decreased frequency of action potentials from a receptor, even though the strength of the stimulus remains constant; the decrease in perception of a sensation over time while the stimulus is still present.

Adduction (ad-DUK-shun) Movement toward the midline of the body.

Adenoids (AD-e-noyds) The pharyngeal tonsils.

Adenosine triphosphate (a-DEN-ō-sēn trī-FOS-fāt) **(ATP)** The main energy currency in living cells; used to transfer the chemical energy needed for metabolic reactions. ATP consists of the purine base *adenine* and the five-carbon sugar *ribose,* to which are added, in linear array, three *phosphate* groups.

Adenylate cyclase (a-DEN-i-lāt SĪ-klās) An enzyme that is activated when certain neurotransmitters or hormones bind to their receptors; the enzyme that converts ATP into cyclic AMP, an important second messenger.

Adhesion (ad-HĒ-zhun) Abnormal joining of parts to each other.

Adipocyte (AD-i-pō-sīt) Fat cell, derived from a fibroblast.

Adipose (AD-i-pōz) **tissue** Tissue composed of adipocytes specialized for triglyceride storage and present in the form of soft pads between various organs for support, protection, and insulation.

Adrenal cortex (a-DRĒ-nal KOR-teks) The outer portion of an adrenal gland, divided into three zones; the zona glomerulosa secretes mineralocorticoids, the zona fasciculata secretes glucocorticoids, and the zona reticularis secretes androgens.

Adrenal glands Two glands located superior to each kidney. Also called the **suprarenal** (soo′-pra-RĒ-nal) **glands.**

Adrenal medulla (me-DUL-a) The inner part of an adrenal gland, consisting of cells that secrete epinephrine, norepinephrine, and a small amount of dopamine in response to stimulation by sympathetic preganglionic neurons.

Adrenergic (ad′-ren-ER-jik) **neuron** A neuron that releases epinephrine (adrenaline) or norepinephrine (noradrenaline) as its neurotransmitter.

Adrenocorticotropic (ad-rē′-nō-kor-ti-kō-TRŌP-ik) **hormone (ACTH)** A hormone produced by the anterior pituitary that influences the production and secretion of certain hormones of the adrenal cortex.

Adventitia (ad-ven-TISH-a) The outermost covering of a structure or organ.

Aerobic (air-Ō-bik) Requiring molecular oxygen.

Afferent arteriole (AF-er-ent ar-TĒ-rē-ōl) A blood vessel of a kidney that divides into the capillary network called a glomerulus; there is one afferent arteriole for each glomerulus.

Agglutination (a-gloo′-ti-NĀ-shun) Clumping of microorganisms or blood cells, typically due to an antigen–antibody reaction.

Aggregated lymphatic follicles Clusters of lymph nodules that are most numerous in the ileum. Also called **Peyer's** (PĪ-erz) **patches.**

Albinism (AL-bin-izm) Abnormal, nonpathological, partial, or total absence of pigment in skin, hair, and eyes.

Albumin (al-BŪ-min) The most abundant (60%) and smallest of the plasma proteins; it is the main contributor to blood colloid osmotic pressure (BCOP).

Aldosterone (al-DOS-ter-ōn) A mineralocorticoid produced by the adrenal cortex that promotes sodium and water reabsorption by the kidneys and potassium excretion in urine.

Alkaline (AL-ka-līn) Containing more hydroxide ions (OH^-) than hydrogen ions (H^+); a pH higher than 7.

Alkalosis (al-ka-LŌ-sis) A condition in which blood pH is higher than 7.45. Also known as **alkalemia.**

Allantois (a-LAN-tō-is) A small, vascularized outpouching of the yolk sac that serves as an early site for blood formation and development of the urinary bladder.

Alleles (a-LĒLZ) Alternate forms of a single gene that control the same inherited trait (such as type A blood) and are located at the same position on homologous chromosomes.

Allergen (AL-er-jen) An antigen that evokes a hypersensitivity reaction.

Alpha (AL-fa) **cell** A type of cell in the pancreatic islets (islets of Langerhans) in the pancreas that secretes the hormone glucagon. Also termed an **A cell.**

Alpha receptor A type of receptor for norepinephrine and epinephrine; present on visceral effectors innervated by sympathetic postganglionic neurons.

Alveolar duct Branch of a respiratory bronchiole around which alveoli and alveolar sacs are arranged.

Alveolar macrophage (MAK-rō-fāj) Highly phagocytic cell found in the alveolar walls of the lungs. Also called a **dust cell.**

Alveolar (al-VĒ-ō-lar) **pressure** Air pressure within the lungs. Also called **intrapulmonic pressure.**

Alveolar sac A cluster of alveoli that share a common opening.

Alveolus (al-VĒ-ō-lus) A small hollow or cavity; an air sac in the lungs; milk-secreting portion of a mammary gland. *Plural is* **alveoli** (al-VĒ-ol-ī).

Alzheimer (ALTZ-hī-mer) **disease (AD)** Disabling neurological disorder characterized by dysfunction and death of specific cerebral neurons, resulting in widespread intellectual impairment, personality changes, and fluctuations in alertness.

Amnesia (am-NE-zē-a) A lack or loss of memory.

Amenorrhea (ā-men-ō-RĒ-a) Absence of menstruation.

Amino (a-MĒ-nō) **acid** An organic acid, containing an acidic carboxyl group (−COOH) and a basic amino group ($-NH_2$); the monomer used to synthesize polypeptides and proteins.

Amnion (AM-nē-on) A thin, protective fetal membrane that develops from the epiblast; holds the fetus suspended in amniotic fluid. Also called the "**bag of waters.**"

Amniotic (am′-nē-OT-ik) **fluid** Fluid in the amniotic cavity, the space between the developing embryo (or fetus) and amnion; the fluid is initially produced as a filtrate from maternal blood and later includes fetal urine. It functions as a shock absorber, helps regulate fetal body temperature, and helps prevent desiccation.

Amphiarthrosis (am′-fē-ar-THRŌ-sis) A slightly movable joint, in which the articulating bony surfaces are separated by fibrous connective tissue or fibrocartilage to which both are attached; types are syndesmosis and symphysis.

Ampulla (am-PUL-la) A saclike dilation of a canal or duct.

Anabolism (a-NAB-ō-lizm) Synthetic, energy-requiring reactions whereby small molecules are built up into larger ones.

Anaerobic (an-ār-Ō-bik) Not requiring oxygen.

Anal (Ā-nal) **canal** The last 2 or 3 cm (1 in.) of the rectum; opens to the exterior through the anus.

Anal column A longitudinal fold in the mucous membrane of the anal canal that contains a network of arteries and veins.

Anal triangle The subdivision of the female or male perineum that contains the anus.

Analgesia (an-al-JĒ-zē-a) Pain relief; absence of the sensation of pain.

Anaphylaxis (an′-a-fi-LAK-sis) A hypersensitivity (allergic) reaction in which IgE antibodies attach to mast cells and basophils, causing them to produce mediators of anaphylaxis (histamine, leukotrienes, kinins, and prostaglandins) that bring about increased blood permeability, increased smooth muscle contraction, and increased mucus production. Examples are hay fever, hives, and anaphylactic shock.

Anaphase (AN-a-fāz) The third stage of mitosis in which the chromatids that have separated at the centromeres move to opposite poles of the cell.

Anastomosis (a-nas-tō-MŌ-sis) An end-to-end union or joining of blood vessels, lymphatic vessels, or nerves.

Anatomical (an′-a-TOM-i-kal) **position** A position of the body universally used in anatomical descriptions in which the body is erect, the head is level, the eyes face forward, the upper limbs are at the sides, the palms face forward, and the feet are flat on the floor.

Anatomic dead space Spaces of the nose, pharynx, larynx, trachea, bronchi, and bronchioles totaling about 150 mL of the 500 mL in a quiet breath (tidal volume); air in the anatomic dead space does not reach the alveoli to participate in gas exchange.

Anatomy (a-NAT-ō-mē) The structure or study of structure of the body and the relation of its parts to each other.

Androgens (AN-drō-jenz) Masculinizing sex hormones produced by the testes in males and the adrenal cortex in both sexes; responsible for libido (sexual desire); the two main androgens are testosterone and dihydrotestosterone.

Anemia (a-NĒ-mē-a) Condition of the blood in which the number of functional red blood cells or their hemoglobin content is below normal.

Anesthesia (an′-es-THĒ-zē-a) A total or partial loss of feeling or sensation; may be general or local.

Aneuploid (an′-ū-PLOYD) A cell that has one or more chromosomes of a set added or deleted.

Aneurysm (AN-ū-rizm) A saclike enlargement of a blood vessel caused by a weakening of its wall.

Angina pectoris (an-JI-na *or* AN-ji-na PEK-tō-ris) A pain in the chest related to reduced coronary circulation due to coronary artery disease (CAD) or spasms of vascular smooth muscle in coronary arteries.

Angiotensin (an-jē-ō-TEN-sin) Either of two forms of a protein associated with regulation of blood pressure. Angiotensin I is produced by the action of renin on angiotensinogen and is converted by the action of ACE (angiotensin-converting enzyme) into angiotensin II, which stimulates aldosterone secretion by the adrenal cortex, stimulates the sensation of thirst, and causes vasoconstriction with resulting increase in systemic vascular resistance.

Anion (AN-ī-on) A negatively charged ion. An example is the chloride ion (Cl^-).

Ankylosis (ang′-ki-LŌ-sis) Severe or complete loss of movement at a joint as the result of a disease process.

Anoxia (an-OK-sē-a) Deficiency of oxygen.

Antagonist (an-TAG-ō-nist) A muscle that has an action opposite that of the prime mover (agonist) and yields to the movement of the prime mover.

Antagonistic (an-tag-ō-NIST-ik) **effect** A hormonal interaction in which the effect of one hormone on a target cell is opposed by another hormone. For example, calcitonin (CT) lowers blood calcium level, whereas parathyroid hormone (PTH) raises it.

Anterior (an-TĒR-ē-or) Nearer to or at the front of the body. Equivalent to **ventral** in bipeds.

Anterior pituitary Anterior lobe of the pituitary gland. Also called the **adenohypophysis** (ad′-e-nō-hī-POF-i-sis).

Anterior root The structure composed of axons of motor (efferent) neurons that emerges from the anterior aspect of the spinal cord and extends laterally to join a posterior root, forming a spinal nerve. Also called a **ventral root.**

Anterolateral (an′-ter-ō-LAT-er-al) **pathway** Sensory pathway that conveys information related to pain, temperature, crude touch, pressure, tickle, and itch.

Antibody (AN-ti-bod′-ē) A protein produced by plasma cells in response to a specific antigen; the antibody combines with that antigen to neutralize, inhibit, or destroy it. Also called an **immunoglobulin** (im-ū-nō-GLOB-ū-lin) or **Ig.**

Antibody-mediated immunity That component of immunity in which B lymphocytes (B cells) develop into plasma cells that produce antibodies that destroy antigens. Also called **humoral** (HŪ-mor-al) **immunity.**

Anticoagulant (an-tī-cō-AG-ū-lant) A substance that can delay, suppress, or prevent the clotting of blood.

Antidiuretic (an′-ti-dī-ū-RET-ik) Substance that inhibits urine formation.

Antidiuretic hormone (ADH) Hormone produced by neurosecretory cells in the paraventricular and supraoptic nuclei of the hypothalamus that stimulates water reabsorption from kidney tubule cells into the blood and vasoconstriction of arterioles. Also called **vasopressin** (vāz-ō-PRES-in).

Antigen (AN-ti-jen) A substance that has immunogenicity (the ability to provoke an immune response) and reactivity (the ability to react with the antibodies or cells that result from the immune response); contraction of *anti*body *gen*erator. Also termed a **complete antigen.**

Antigen-presenting cell (APC) Special class of migratory cell that processes and presents antigens to T cells during an immune response; APCs include macrophages, B cells, and dendritic cells, which are present in the skin, mucous membranes, and lymph nodes.

Antiporter A transmembrane transporter protein that moves two substances, often Na^+ and another substance, in opposite directions across a plasma membrane. Also called a **countertransporter.**

Anulus fibrosus (AN-ū-lus fī-BRŌ-sus) A ring of fibrous tissue and fibrocartilage that encircles the pulpy substance (nucleus pulposus) of an intervertebral disc.

Anuria (an-Ū-rē-a) Absence of urine formation or daily urine output of less than 50 mL.

Anus (Ā-nus) The distal end and outlet of the rectum.

Aorta (ā-OR-ta) The main systemic trunk of the arterial system of the body that emerges from the left ventricle.

Aortic (ā-OR-tik) **body** Cluster of chemoreceptors on or near the arch of the aorta that respond to changes in blood levels of oxygen, carbon dioxide, and hydrogen ions (H^+).

Aortic reflex A reflex that helps maintain normal systemic blood pressure; initiated by baroreceptors in the wall of the ascending aorta and arch of the aorta. Nerve impulses from aortic baroreceptors reach the cardiovascular center via sensory axons of the vagus nerves (cranial nerve X).

Aperture (AP-er-chur) An opening or orifice.

Apex (Ā-peks) The pointed end of a conical structure, such as the apex of the heart.

Aphasia (a-FA-zē-a) Loss of ability to express oneself properly through speech or loss of verbal comprehension.

Apnea (AP-nē-a) Temporary cessation of breathing.

Apneustic (ap-NOO-stik) **area** A part of the respiratory center in the pons that sends stimulatory nerve impulses to the inspiratory area that activate and prolong inhalation and inhibit exhalation.

Apocrine (AP-ō-krin) **gland** A type of gland in which the secretory products gather at the free end of the secreting cell and are pinched off, along with some of the cytoplasm, to become the secretion, as in mammary glands.

Aponeurosis (ap′-ō-noo-RŌ-sis) A sheetlike tendon joining one muscle with another or with bone.

Apoptosis (ap′-ō-TŌ-sis *or* ap′-ōp-TŌ-sis) Programmed cell death; a normal type of cell death that removes unneeded cells during embryological development, regulates the number of cells in tissues, and eliminates many potentially dangerous cells such as cancer cells. During apoptosis, the DNA fragments, the nucleus condenses, mitochondria cease to function, and the cytoplasm shrinks, but the plasma membrane remains intact. Phagocytes engulf and digest the apoptotic cells, and an inflammatory response does not occur.

Appositional (ap′-ō-ZISH-o-nal) **growth** Growth due to surface deposition of material, as in the growth in diameter of cartilage and bone. Also called **exogenous** (eks-OJ-e-nus) **growth.**

Aqueous humor (AK-wē-us HŪ-mer) The watery fluid, similar in composition to cerebrospinal fluid, that fills the anterior cavity of the eye.

Arachnoid (a-RAK-noyd) **mater** The middle of the three meninges (coverings) of the brain and spinal cord. Also termed the **arachnoid.**

Arachnoid villus (VIL-us) Berrylike tuft of the arachnoid mater that protrudes into the superior sagittal sinus and through which cerebrospinal fluid is reabsorbed into the bloodstream.

Arbor vitae (AR-bor VĪ-tē) The white matter tracts of the cerebellum, which have a treelike appearance when seen in midsagittal section.

Arch of the aorta The most superior portion of the aorta, lying between the ascending and descending segments of the aorta.

Areola (a-RĒ-ō-la) Any tiny space in a tissue. The pigmented ring around the nipple of the breast.

Arm The part of the upper limb from the shoulder to the elbow.

Arousal (a-ROW-zal) Awakening from sleep, a response due to stimulation of the reticular activating system (RAS).

Arrector pili (a-REK-tor PI-lē) Smooth muscles attached to hairs; contraction pulls the hairs into a vertical position, resulting in "goose bumps."

Arrhythmia (a-RITH-mē-a) An irregular heart rhythm. Also called a **dysrhythmia.**

Arteriole (ar-TĒ-rē-ōl) A small, almost microscopic, artery that delivers blood to a capillary.

Arteriosclerosis (ar-tē-rē-ō-skle-RŌ-sis) Group of diseases characterized by thickening of the walls of arteries and loss of elasticity.

Artery (AR-ter-ē) A blood vessel that carries blood away from the heart.

Arthritis (ar-THRI-tis) Inflammation of a joint.

Arthrology (ar-THROL-ō-jē) The study or description of joints.

Arthroscopy (ar-THROS-co-pē) A procedure for examining the interior of a joint, usually the knee, by inserting an arthroscope into a small incision; used to determine extent of damage, remove torn cartilage, repair cruciate ligaments, and obtain samples for analysis.

Arthrosis (ar-THRŌ-sis) A joint or articulation.

Articular (ar-TIK-ū-lar) **capsule** Sleevelike structure around a synovial joint composed of a fibrous capsule and a synovial membrane.

Articular cartilage (KAR-ti-lij) Hyaline cartilage attached to articular bone surfaces.

Articular disc Fibrocartilage pad between articular surfaces of bones of some synovial joints. Also called a **meniscus** (men-IS-kus).

Articulation (ar-tik′-ū-LĀ-shun) A joint; a point of contact between bones, cartilage and bones, or teeth and bones.

Arytenoid (ar′-i-TĒ-noyd) **cartilages** A pair of small, pyramidal cartilages of the larynx that attach to the vocal folds and intrinsic pharyngeal muscles and can move the vocal folds.

Ascending colon (KŌ-lon) The part of the large intestine that passes superiorly from the cecum to the inferior border of the liver, where it bends at the right colic (hepatic) flexure to become the transverse colon.

Ascites (as-SĪ-tēz) Abnormal accumulation of serous fluid in the peritoneal cavity.

Association areas Large cortical regions on the lateral surfaces of the occipital, parietal, and temporal lobes and on the frontal lobes anterior to the motor areas connected by many motor and sensory axons to other parts of the cortex. The association areas are concerned with motor patterns, memory, concepts of word-hearing and word-seeing, reasoning, will, judgment, and personality traits.

Asthma (AZ-ma) Usually allergic reaction characterized by smooth muscle spasms in bronchi resulting in wheezing and difficult breathing. Also called **bronchial asthma.**

Astigmatism (a-STIG-ma-tizm) An irregularity of the lens or cornea of the eye causing the image to be out of focus and producing faulty vision.

Astrocyte (AS-trō-sīt) A neuroglial cell having a star shape that participates in brain development and the metabolism of neurotransmitters, helps form the blood–brain barrier, helps maintain the proper balance of K^+ for generation of nerve impulses, and provides a link between neurons and blood vessels.

Ataxia (a-TAK-sē-a) A lack of muscular coordination, lack of precision.

Atherosclerotic (ath′-er-ō-skle-RO-tic) **plaque** (PLAK) A lesion that results from accumulated cholesterol and smooth muscle fibers (cells) of the tunica media of an artery; may become obstructive.

Atom Unit of matter that makes up a chemical element; consists of a nucleus (containing positively charged protons and uncharged neutrons) and negatively charged electrons that orbit the nucleus.

Atomic mass (weight) Average mass of all stable atoms of an element, reflecting the relative proportion of atoms with different mass numbers.

Atomic number Number of protons in an atom.

Atresia (a-TRĒ-zē-a) Degeneration and reabsorption of an ovarian follicle before it fully matures and ruptures; abnormal closure of a passage, or absence of a normal body opening.

Atrial fibrillation (Ā-trē-al fib-ri-LĀ-shun) Asynchronous contraction of cardiac muscle fibers in the atria that results in the cessation of atrial pumping.

Atrial natriuretic (na′-trē-ū-RET-ik) **peptide (ANP)** Peptide hormone, produced by the atria of the heart in response to stretching, that inhibits aldosterone production and thus lowers blood pressure; causes natriuresis, increased urinary excretion of sodium.

Atrioventricular (AV) (ā′-trē-ō-ven-TRIK-ū-lar) **bundle** The part of the conduction system of the heart that begins at the atrioventricular (AV) node, passes through the cardiac skeleton separating the atria and the ventricles, then extends a short distance down the interventricular septum before splitting into right and left bundle branches. Also called the **bundle of His** (HISS).

Atrioventricular (AV) node The part of the conduction system of the heart made up of a compact mass of conducting cells located in the septum between the two atria.

Atrioventricular (AV) valve A heart valve made up of membranous flaps or cusps that allows blood to flow in one direction only, from an atrium into a ventricle.

Atrium (Ā-trē-um) A superior chamber of the heart.

Atrophy (AT-rō-fē) Wasting away or decrease in size of a part, due to a failure, abnormality of nutrition, or lack of use.

Auditory ossicle (AW-di-tō-rē OS-si-kul) One of the three small bones of the middle ear called the **malleus, incus,** and **stapes.**

Auditory tube The tube that connects the middle ear with the nose and nasopharynx region of the throat. Also called the **Eustachian** (ū-STĀ-shun *or* ū-STĀ-kē-an) **tube** or **pharyngotympanic tube.**

Auscultation (aws-kul-TĀ-shun) Examination by listening to sounds in the body.

Autocrine (AW-tō-krin) Local hormone, such as interleukin-2, that acts on the same cell that secreted it.

Autoimmunity An immunological response against a person's own tissues.

Autolysis (aw-TOL-i-sis) Self-destruction of cells by their own lysosomal digestive enzymes after death or in a pathological process.

Autonomic ganglion (aw′-tō-NOM-ik GANG-lē-on) A cluster of cell bodies of sympathetic or parasympathetic neurons located outside the central nervous system.

Autonomic nervous system (ANS) Visceral sensory (afferent) and visceral motor (efferent) neurons. Autonomic motor neurons, both sympathetic and parasympathetic, conduct nerve impulses from the central nervous system to smooth muscle, cardiac muscle, and glands. So named because this part of the nervous system was thought to be self-governing or spontaneous.

Autonomic plexus (PLEK-sus) A network of sympathetic and parasympathetic axons; examples are the cardiac, celiac, and pelvic plexuses, which are located in the thorax, abdomen, and pelvis, respectively.

Autophagy (aw-TOF-a-jē) Process by which worn-out organelles are digested within lysosomes.

Autopsy (AW-top-sē) The examination of the body after death.

Autoregulation (aw-tō-reg-ū-LĀ-shun) A local, automatic adjustment of blood flow in a given region of the body in response to tissue needs.

Autorhythmic cells Cardiac or smooth muscle fibers that are self-excitable (generate impulses without an external stimulus); act as the heart's pacemaker and conduct the pacing impulse through the conduction system of the heart; self-excitable neurons in the central nervous system, as in the inspiratory area of the brain stem.

Autosome (AW-tō-sōm) Any chromosome other than the X and Y chromosomes (sex chromosomes).

Axilla (ak-SIL-a) The small hollow beneath the arm where it joins the body at the shoulders. Also called the **armpit.**

Axon (AK-son) The usually single, long process of a nerve cell that propagates a nerve impulse toward the axon terminals.

Axon terminal Terminal branch of an axon where synaptic vesicles undergo exocytosis to release neurotransmitter molecules.

Azygos (AZ-ī-gos) An anatomical structure that is not paired; occurring singly.

B

B cell A lymphocyte that can develop into a clone of antibody-producing plasma cells or memory cells when properly stimulated by a specific antigen.

Babinski (ba-BIN-skē) **sign** Extension of the great toe, with or without fanning of the other toes, in response to stimulation of the outer margin of the sole; normal up to 18 months of age and indicative of damage to descending motor pathways such as the corticospinal tracts after that.

Back The posterior part of the body; the dorsum.

Ball-and-socket joint A synovial joint in which the rounded surface of one bone moves within a cup-shaped depression or socket of another bone, as in the shoulder or hip joint. Also called a **spheroid** (SFĒ-royd) **joint.**

Baroreceptor (bar′-ō-re-SEP-tor) Neuron capable of responding to changes in blood, air, or fluid pressure. Also called a **pressoreceptor.**

Basal ganglia (GANG-glē-a) Paired clusters of gray matter deep in each cerebral hemisphere including the globus pallidus, putamen, and caudate nucleus. Together, the caudate nucleus and putamen are known as the **corpus striatum.** Nearby structures that are functionally linked to the basal ganglia are the substantia nigra of the midbrain and the subthalamic nuclei of the diencephalon.

Basal metabolic (BĀ-sal met′-a-BOL-ik) **rate (BMR)** The rate of metabolism measured under standard or basal conditions (awake, at rest, fasting).

Base A nonacid or a proton acceptor, characterized by excess of hydroxide ions (OH^-) and a pH greater than 7. A ring-shaped, nitrogen-containing organic molecule that is one of the components of a nucleotide, namely, adenine, guanine, cytosine, thymine, and uracil; also known as a **nitrogenous base.**

Basement membrane Thin, extracellular layer between epithelium and connective tissue consisting of a basal lamina and a reticular lamina.

Basilar (BĀS-i-lar) **membrane** A membrane in the cochlea of the internal ear that separates the cochlear duct from the scala tympani and on which the spiral organ (organ of Corti) rests.

Basophil (BĀ-sō-fil) A type of white blood cell characterized by a pale nucleus and large granules that stain blue-purple with basic dyes.

Belly The abdomen. The gaster or prominent, fleshy part of a skeletal muscle.

Beta (BĀ-ta) **cell** A type of cell in the pancreatic islets (islets of Langerhans) in the pancreas that secretes the hormone insulin. Also termed a **B cell.**

Beta receptor A type of adrenergic receptor for epinephrine and norepinephrine; found on visceral effectors innervated by sympathetic postganglionic neurons.

Bicuspid (bī-KUS-pid) **valve** Atrioventricular (AV) valve on the left side of the heart. Also called the **mitral valve.**

Bilateral (bī-LAT-er-al) Pertaining to two sides of the body.

Bile (BĪL) A secretion of the liver consisting of water, bile salts, bile pigments, cholesterol, lecithin, and several ions; it emulsifies lipids prior to their digestion.

Bilirubin (bil-ē-ROO-bin) An orange pigment that is one of the end products of hemoglobin breakdown in the hepatocytes and is excreted as a waste material in bile.

Blastocele (BLAS-tō-sēl) The fluid-filled cavity within the blastocyst.

Blastocyst (BLAS-tō-sist) In the development of an embryo, a hollow ball of cells that consists of a blastocele (the internal cavity), trophoblast (outer cells), and inner cell mass.

Blastomere (BLAS-tō-mēr) One of the cells resulting from the cleavage of a fertilized ovum.

Blind spot Area in the retina at the end of the optic nerve (cranial nerve II) in which there are no photoreceptors.

Blood The fluid that circulates through the heart, arteries, capillaries, and veins and that constitutes the chief means of transport within the body.

Blood–brain barrier (BBB) A barrier consisting of specialized brain capillaries and astrocytes that prevents the passage of materials from the blood to the cerebrospinal fluid and brain.

Blood island Isolated mass of mesoderm derived from angioblasts and from which blood vessels develop.

Blood pressure (BP) Force exerted by blood against the walls of blood vessels due to contraction of the heart and influenced by the elasticity of the vessel walls; clinically, a measure of the pressure in arteries during ventricular systole and ventricular diastole. *See also* **mean arterial blood pressure.**

Blood reservoir (REZ-er-vwar) Systemic veins that contain large amounts of blood that can be moved quickly to parts of the body requiring the blood.

Blood–testis barrier (BTB) A barrier formed by Sertoli cells that prevents an immune response against antigens produced by spermatogenic cells by isolating the cells from the blood.

Body cavity A space within the body that contains various internal organs.

Body fluid Body water and its dissolved substances; constitutes about 60% of total body mass.

Bohr (BŌR) **effect** In an acidic environment, oxygen unloads more readily from hemoglobin because when hydrogen ions (H^+) bind to hemoglobin, they alter the structure of hemoglobin, thereby reducing its oxygen-carrying capacity.

Bolus (BŌ-lus) A soft, rounded mass, usually food, that is swallowed.

Bony labyrinth (LAB-i-rinth) A series of cavities within the petrous portion of the temporal bone forming the vestibule, cochlea, and semicircular canals of the inner ear.

Brachial plexus (BRĀ-kē-al PLEK-sus) A network of nerve axons of the ventral rami of spinal nerves C5, C6, C7, C8, and T1. The nerves that emerge from the brachial plexus supply the upper limb.

Bradycardia (brād′-i-KAR-dē-a) A slow resting heart or pulse rate (under 60 beats per minute).

Brain The part of the central nervous system contained within the cranial cavity.

Brain stem The portion of the brain immediately superior to the spinal cord, made up of the medulla oblongata, pons, and midbrain.

Brain waves Electrical signals that can be recorded from the skin of the head due to electrical activity of brain neurons.

Broad ligament A double fold of parietal peritoneum attaching the uterus to the side of the pelvic cavity.

Broca's (BRŌ-kaz) **area** Motor area of the brain in the frontal lobe that translates thoughts into speech. Also called the **motor speech area.**

Bohr effect Shifting of the oxygen–hemoglobin dissociation curve to the right when pH decreases; at any given partial pressure of oxygen, hemoglobin is less saturated with O_2 at lower pH.

Bronchi (BRONG-kē) Branches of the respiratory passageway including primary bronchi (the two divisions of the trachea), secondary or lobar bronchi (divisions of the primary bronchi that are distributed to the lobes of the lung), and tertiary or segmental bronchi (divisions of the secondary bronchi that are distributed to bronchopulmonary segments of the lung). *Singular is* **bronchus.**

Bronchial tree The trachea, bronchi, and their branching structures up to and including the terminal bronchioles.

Bronchiole (BRONG-kē-ōl) Branch of a tertiary bronchus further dividing into terminal bronchioles (distributed to lobules of the lung), which divide into respiratory bronchioles (distributed to alveolar sacs).

Bronchitis (brong-KĪ-tis) Inflammation of the mucous membrane of the bronchial tree; characterized by hypertrophy and hyperplasia of seromucous glands and goblet cells that line the bronchi and which results in a productive cough.

Bronchopulmonary (brong′-kō-PUL-mō-ner-ē) **segment** One of the smaller divisions of a lobe of a lung supplied by its own branches of a bronchus.

Buccal (BUK-al) Pertaining to the cheek or mouth.

Buffer (BUF-er) **system** A pair of chemicals—one a weak acid and the other the salt of the weak acid, which functions as a weak base—that resists changes in pH.

Bulbourethral (bul′-bō-ū-RĒ-thral) **gland** One of a pair of glands located inferior to the prostate on either side of the urethra that secretes an alkaline fluid into the cavernous urethra. Also called a **Cowper's** (KOW-perz) **gland.**

Bulimia (boo-LIM-ē-a *or* boo-LĒ-mē-a) A disorder characterized by overeating at least twice a week followed by purging by self-induced vomiting, strict dieting or fasting, vigorous exercise, or use of laxatives or diuretics. Also called **binge–purge syndrome.**

Bulk flow The movement of large numbers of ions, molecules, or particles in the same direction due to pressure differences (osmotic, hydrostatic, or air pressure).

Bundle branch One of the two branches of the atrioventricular (AV) bundle made up of specialized muscle fibers (cells) that transmit electrical impulses to the ventricles.

Bursa (BUR-sa) A sac or pouch of synovial fluid located at friction points, especially about joints.

Buttocks (BUT-oks) The two fleshy masses on the posterior aspect of the inferior trunk, formed by the gluteal muscles.

C

Calcaneal (kal-KĀ-nē-al) **tendon** The tendon of the soleus, gastrocnemius, and plantaris muscles at the back of the heel. Also called the **Achilles** (a-KIL-ēz) **tendon.**

Calcification (kal-si-fi-KĀ-shun) Deposition of mineral salts, primarily hydroxyapatite, in a framework formed by collagen fibers in which the tissue hardens. Also called **mineralization** (min′-e-ral-i-ZĀ-shun).

Calcitonin (kal-si-TŌ-nin) **(CT)** A hormone produced by the parafollicular cells of the thyroid gland that can lower the amount of blood calcium and phosphates by inhibiting bone resorption (breakdown of bone matrix) and by accelerating uptake of calcium and phosphates into bone matrix.

Calculus (KAL-kū-lus) A stone, or insoluble mass of crystallized salts or other material, formed within the body, as in the gallbladder, kidney, or urinary bladder.

Callus (KAL-lus) A growth of new bone tissue in and around a fractured area, ultimately replaced by mature bone. An acquired, localized thickening.

Calorie (KAL-ō-rē) A unit of heat. A calorie (cal) is the standard unit and is the amount of heat needed to raise the temperature of 1 g of water from 14°C to 15°C. The **kilocalorie (kcal)** or **Calorie** (spelled with an uppercase C), used to express the caloric value of foods and to measure metabolic rate, is equal to 1000 cal.

Calyx (KĀL-iks) Any cuplike division of the kidney pelvis. *Plural is* **calyces** (KĀ-li-sēz).

Canal (ka-NAL) A narrow tube, channel, or passageway.

Canaliculus (kan′-a-LIK-ū-lus) A small channel or canal, as in bones, where they connect lacunae. *Plural is* **canaliculi** (kan′-a-LIK-ū-lī).

Cancellous (KAN-sel-us) Having a reticular or latticework structure, as in spongy tissue of bone.

Capacitation (ka′-pas-i-TĀ-shun) The functional changes that sperm undergo in the female reproductive tract that allow them to fertilize a secondary oocyte.

Capillary (KAP-i-lar′-ē) A microscopic blood vessel located between an arteriole and venule through which materials are exchanged between blood and interstitial fluid.

Carbohydrate (kar′-bō-HĪ-drāt) An organic compound containing carbon, hydrogen, and oxygen in a particular amount and arrangement and composed of monosaccharide subunits; usually has the general formula $(CH_2O)n$.

Carcinogen (kar-SIN-ō-jen) A chemical substance or radiation that causes cancer.

Cardiac (KAR-dē-ak) **arrest** Cessation of an effective heartbeat in which the heart is completely stopped or in ventricular fibrillation.

Cardiac cycle A complete heartbeat consisting of systole (contraction) and diastole (relaxation) of both atria plus systole and diastole of both ventricles.

Cardiac muscle Striated muscle fibers (cells) that form the wall of the heart; stimulated by an intrinsic conduction system and autonomic motor neurons.

Cardiac notch An angular notch in the anterior border of the left lung into which part of the heart fits.

Cardiac output (CO) The volume of blood pumped from one ventricle of the heart (usually measured from the left ventricle) in 1 min; normally about 5.2 liters/min in an adult at rest.

Cardiac reserve The maximum percentage that cardiac output can increase above normal.

Cardinal ligament A ligament of the uterus, extending laterally from the cervix and vagina as a continuation of the broad ligament.

Cardiology (kar-dē-OL-ō-jē) The study of the heart and diseases associated with it.

Cardiovascular (kar-dē-ō-VAS-kū-lar) **center** Groups of neurons scattered within the medulla oblongata that regulate heart rate, force of contraction, and blood vessel diameter.

Carotene (KAR-o-tēn) Antioxidant precursor of vitamin A, which is needed for synthesis of photopigments; yellow-orange pigment present in the stratum corneum of the epidermis. Accounts for the yellowish coloration of skin. Also termed **beta-carotene.**

Carotid (ka-ROT-id) **body** Cluster of chemoreceptors on or near the carotid sinus that respond to changes in blood levels of oxygen, carbon dioxide, and hydrogen ions.

Carotid sinus A dilated region of the internal carotid artery just above the point where it branches from the common carotid artery; it contains baroreceptors that monitor blood pressure.

Carotid sinus reflex A reflex that helps maintain normal blood pressure in the brain. Nerve impulses propagate from the carotid sinus baroreceptors over sensory axons in the glossopharyngeal nerves (cranial nerve IX) to the cardiovascular center in the medulla oblongata.

Carpal bones The eight bones of the wrist. Also called **carpals.**

Carpus (KAR-pus) A collective term for the eight bones of the wrist.

Cartilage (KAR-ti-lij) A type of connective tissue consisting of chondrocytes in lacunae embedded in a dense network of collagen and elastic fibers and a matrix of chondroitin sulfate.

Cartilaginous (kar′-ti-LAJ-i-nus) **joint** A joint without a synovial (joint) cavity where the articulating bones are held tightly together by cartilage, allowing little or no movement.

Cast A small mass of hardened material formed within a cavity in the body and then discharged from the body; can originate in different areas and can be composed of various materials.

Catabolism (ka-TAB-ō-lizm) Chemical reactions that break down complex organic compounds into simple ones, with the net release of energy.

Catalyst (KAT-a-list) A substance that speeds up a chemical reaction without itself being altered; an enzyme.

Cataract (KAT-a-rakt) Loss of transparency of the lens of the eye or its capsule or both.

Cation (KAT-ī-on) A positively charged ion. An example is a sodium ion (Na^+).

Cauda equina (KAW-da ē-KWĪ-na) A tail-like array of roots of spinal nerves at the inferior end of the spinal cord.

Caudal (KAW-dal) Pertaining to any tail-like structure; inferior in position.

Cecum (SĒ-kum) A blind pouch at the proximal end of the large intestine that attaches to the ileum.

Celiac plexus (PLEK-sus) A large mass of autonomic ganglia and axons located at the level of the superior part of the first lumbar vertebra. Also called the **solar plexus.**

Cell The basic structural and functional unit of all organisms; the smallest structure capable of performing all the activities vital to life.

Cell cycle Growth and division of a single cell into two daughter cells; consists of interphase and cell division.

Cell division Process by which a cell reproduces itself that consists of a nuclear division (mitosis) and a cytoplasmic division (cytokinesis); types include somatic and reproductive cell division.

Cell-mediated immunity That component of immunity in which specially sensitized T lymphocytes (T cells) attach to antigens to destroy them. Also called **cellular immunity.**

Cementum (se-MEN-tum) Calcified tissue covering the root of a tooth.

Center of ossification (os′-i-fi-KĀ-shun) An area in the cartilage model of a future bone where the cartilage cells hypertrophy and then secrete enzymes that result in the calcification of their matrix, resulting in the death of the cartilage cells, followed by the invasion of the area by osteoblasts that then lay down bone.

Central canal A microscopic tube running the length of the spinal cord in the gray commissure. A circular channel running longitudinally in the center of an osteon (Haversian system) of mature compact bone, containing blood and lymphatic vessels and nerves. Also called a **Haversian** (ha-VER-shan) **canal.**

Central fovea (FŌ-vē-a) A depression in the center of the macula lutea of the retina, containing cones only and lacking blood vessels; the area of highest visual acuity (sharpness of vision).

Central nervous system (CNS) That portion of the nervous system that consists of the brain and spinal cord.

Centrioles (SEN-trē-ōlz) Paired, cylindrical structures of a centrosome, each consisting of a ring of microtubules and arranged at right angles to each other.

Centromere (SEN-trō-mēr) The constricted portion of a chromosome where the two chromatids are joined; serves as the point of attachment for the microtubules that pull chromatids during anaphase of cell division.

Centrosome (SEN-trō-sōm) A dense network of small protein fibers near the nucleus of a cell, containing a pair of centrioles and pericentriolar material.

Cephalic (se-FAL-ik) Pertaining to the head; superior in position.

Cerebellar peduncle (ser-e-BEL-ar pe-DUNG-kul) A bundle of nerve axons connecting the cerebellum with the brain stem.

Cerebellum (ser-e-BEL-um) The part of the brain lying posterior to the medulla oblongata and pons; governs balance and coordinates skilled movements.

Cerebral aqueduct (SER-ē-bral AK-we-dukt) A channel through the midbrain connecting the third and fourth ventricles and containing cerebrospinal fluid. Also termed the **aqueduct of Sylvius.**

Cerebral arterial circle A ring of arteries forming an anastomosis at the base of the brain between the internal carotid and basilar arteries and arteries supplying the cerebral cortex. Also called the **circle of Willis.**

Cerebral cortex The surface of the cerebral hemispheres, 2–4 mm thick, consisting of gray matter; arranged in six layers of neuronal cell bodies in most areas.

Cerebral peduncle One of a pair of nerve axon bundles located on the anterior surface of the midbrain, conducting nerve impulses between the pons and the cerebral hemispheres.

Cerebrospinal (se-rē′-brō-SPĪ-nal) **fluid (CSF)** A fluid produced by ependymal cells that cover choroid plexuses in the ventricles of the brain; the fluid circulates in the ventricles, the central canal, and the subarachnoid space around the brain and spinal cord.

Cerebrovascular (se rē′-brō-VAS-kū-lar) **accident (CVA)** Destruction of brain tissue (infarction) resulting from obstruction or rupture of blood vessels that supply the brain. Also called a **stroke** or **brain attack.**

Cerebrum (SER-e-brum *or* se-RĒ-brum) The two hemispheres of the forebrain (derived from the telencephalon), making up the largest part of the brain.

Cerumen (se-ROO-men) Waxlike secretion produced by ceruminous glands in the external auditory meatus (ear canal). Also termed **ear wax.**

Ceruminous (se-ROO-mi-nus) **gland** A modified sudoriferous (sweat) gland in the external auditory meatus that secretes cerumen (ear wax).

Cervical ganglion (SER-vi-kul GANG-glē-on) A cluster of cell bodies of postganglionic sympathetic neurons located in the neck, near the vertebral column.

Cervical plexus (PLEK-sus) A network formed by nerve axons from the ventral rami of the first four cervical nerves and receiving gray rami communicates from the superior cervical ganglion.

Cervix (SER-viks) Neck; any constricted portion of an organ, such as the inferior cylindrical part of the uterus.

Chemical bond Force of attraction in a molecule or compound that holds its atoms together. Examples include ionic and covalent bonds.

Chemical element Unit of matter that cannot be decomposed into a simpler substance by ordinary chemical reactions. Examples include hydrogen (H), carbon (C), and oxygen (O).

Chemically gated channel A channel in a membrane that opens and closes in response to a specific chemical stimulus, such as a neurotransmitter, hormone, or specific type of ion.

Chemical reaction The combination or separation of atoms in which chemical bonds are formed or broken and new products with different properties are produced.

Chemiosmosis (kem′-ē-oz-MŌ-sis) Mechanism for ATP generation that links chemical reactions (electrons passing along the electron transport chain) with pumping of H^+ out of the mitochondrial matrix. ATP synthesis occurs as H^+ diffuse back into the mitochondrial matrix through special H^+ channels in the membrane.

Chemoreceptor (kē′-mō-rē-SEP-tor) Sensory receptor that detects the presence of a specific chemical.

Chemotaxis (kē-mō-TAK-sis) Attraction of phagocytes to microbes by a chemical stimulus.

Chiasm (KĪ-azm) A crossing; especially the crossing of axons in the optic nerve (cranial nerve II).

Chief cell The secreting cell of a gastric gland that produces pepsinogen, the precursor of the enzyme pepsin, and the enzyme gastric lipase. Also called a **zymogenic** (zī′-mō-JEN-ik) **cell.** Cell in the parathyroid glands that secretes parathyroid hormone (PTH). Also called a **principal cell.**

Chiropractic (kī-rō-PRAK-tik) A system of treating disease by using one's hands to manipulate body parts, mostly the vertebral column.

Chloride shift Exchange of bicarbonate ions (HCO_3^-) for chloride ions (Cl^-) between red blood cells and plasma; maintains electrical balance inside red blood cells as bicarbonate ions are produced or eliminated during respiration.

Cholecystectomy (kō′-lē-sis-TEK-tō-mē) Surgical removal of the gallbladder.

Cholecystitis (kō′-lē-sis-TĪ-tis) Inflammation of the gallbladder.

Cholesterol (kō-LES-te-rol) Classified as a lipid, the most abundant steroid in animal tissues; located in cell membranes and used for the synthesis of steroid hormones and bile salts.

Cholinergic (kō′-lin-ER-jik) **neuron** A neuron that liberates acetylcholine as its neurotransmitter.

Chondrocyte (KON-drō-sīt) Cell of mature cartilage.

Chondroitin (kon-DROY-tin) **sulfate** An amorphous matrix material found outside connective tissue cells.

Chordae tendineae (KOR-dē TEN-di-nē-ē) Tendonlike, fibrous cords that connect atrioventricular valves of the heart with papillary muscles.

Chorion (KŌ-rē-on) The most superficial fetal membrane that becomes the principal embryonic portion of the placenta; serves a protective and nutritive function.

Chorionic villi (kō-rē-ON-ik VIL-lī) Fingerlike projections of the chorion that grow into the decidua basalis of the endometrium and contain fetal blood vessels.

Choroid (KŌ-royd) One of the vascular coats of the eyeball.

Choroid plexus (PLEK-sus) A network of capillaries located in the roof of each of the four ventricles of the brain; ependymal cells around choroid plexuses produce cerebrospinal fluid.

Chromaffin (KRŌ-maf-in) **cell** Cell that has an affinity for chrome salts, due in part to the presence of the precursors of the neurotransmitter epinephrine; found, among other places, in the adrenal medulla.

Chromatid (KRŌ-ma-tid) One of a pair of identical connected nucleoprotein strands that are joined at the centromere and separate during cell division, each becoming a chromosome of one of the two daughter cells.

Chromatin (KRŌ-ma-tin) The threadlike mass of genetic material, consisting of DNA and histone proteins, that is present in the nucleus of a nondividing or interphase cell.

Chromatolysis (krō-ma-TOL-i-sis) The breakdown of Nissl bodies into finely granular masses in the cell body of a neuron whose axon has been damaged.

Chromosome (KRŌ-mō-sōm) One of the small, threadlike structures in the nucleus of a cell, normally 46 in a human diploid cell, that bears the genetic material; composed of DNA and proteins (histones) that form a delicate chromatin thread during interphase; becomes packaged into compact rodlike structures that are visible under the light microscope during cell division.

Chronic (KRON-ik) Long term or frequently recurring; applied to a disease that is not acute.

Chronic obstructive pulmonary disease (COPD) A disease, such as bronchitis or emphysema, in which there is some degree of obstruction of airways and consequent increase in airway resistance.

Chyle (KĪL) The milky-appearing fluid found in the lacteals of the small intestine after absorption of lipids in food.

Chylomicron (kī-lō-MĪ-kron) Protein-coated spherical structure that contains triglycerides, phospholipids, and cholesterol and is absorbed into the lacteal of a villus in the small intestine.

Chyme (KĪM) The semifluid mixture of partly digested food and digestive secretions found in the stomach and small intestine during digestion of a meal.

Ciliary (SIL-ē-ar′-ē) **body** One of the three parts of the vascular tunic of the eyeball, the others being the choroid and the iris; includes the ciliary muscle and the ciliary processes.

Ciliary ganglion (GANG-glē-on) A very small parasympathetic ganglion whose preganglionic axons come from the oculomotor nerve (cranial nerve III) and whose postganglionic axons carry nerve impulses to the ciliary muscle and the sphincter muscle of the iris.

Cilium (SIL-ē-um) A hair or hairlike process projecting from a cell that may be used to move the entire cell or to move substances along the surface of the cell. *Plural is* **cilia.**

Circadian (ser-KĀ-dē-an) **rhythm** A cycle of active and nonactive periods in organisms determined by internal mechanisms and repeating about every 24 hours.

Circular folds Permanent, deep, transverse folds in the mucosa and submucosa of the small intestine that increase the surface area for absorption. Also called **plicae circulares** (PLĪ-kē SER-kū-lar-ēs).

Circulation time Time required for blood to pass from the right atrium, through pulmonary circulation, back to the left ventricle, through systemic circulation to the foot, and back again to the right atrium; normally about 1 min.

Circumduction (ser′-kum-DUK-shun) A movement at a synovial joint in which the distal end of a bone moves in a circle while the proximal end remains relatively stable.

Cirrhosis (si-RŌ-sis) A liver disorder in which the parenchymal cells are destroyed and replaced by connective tissue.

Cisterna chyli (sis-TER-na-KĪ-lē) The origin of the thoracic duct.

Cleavage The rapid mitotic divisions following the fertilization of a secondary oocyte, resulting in an increased number of progressively smaller cells, called blastomeres.

Climacteric (klī-mak-TER-ik) Cessation of the reproductive function in the female or diminution of testicular activity in the male.

Climax The peak period or moments of greatest intensity during sexual excitement.

Clitoris (KLI-to-ris) An erectile organ of the female, located at the anterior junction of the labia minora, that is homologous to the male penis.

Clone (KLŌN) A population of identical cells.

Clot The end result of a series of biochemical reactions that changes liquid plasma into a gelatinous mass; specifically, the conversion of fibrinogen into a tangle of polymerized fibrin molecules.

Clot retraction (rē-TRAK-shun) The consolidation of a fibrin clot to pull damaged tissue together.

Clotting Process by which a blood clot is formed. Also known as **coagulation** (cō-ag-ū-LĀ-shun).

Coccyx (KOK-six) The fused bones at the inferior end of the vertebral column.

Cochlea (KŌK-lē-a) A winding, cone-shaped tube forming a portion of the inner ear and containing the spiral organ (organ of Corti).

Cochlear duct The membranous cochlea consisting of a spirally arranged tube enclosed in the bony cochlea and lying along its outer wall. Also called the **scala media** (SCA-la MĒ-dē-a).

Coenzyme A nonprotein organic molecule that is associated with and activates an enzyme; many are derived from vitamins. An example is nicotinamide adenine dinucleotide (NAD), derived from the B vitamin niacin.

Coitus (KŌ-i-tus) Sexual intercourse.

Collagen (KOL-a-jen) A protein that is the main organic constituent of connective tissue.

Collateral circulation The alternate route taken by blood through an anastomosis.

Colliculus (ko-LIK-ū-lus) A small elevation.

Colloid (KOL-loyd) The material that accumulates in the center of thyroid follicles, consisting of thyroglobulin and stored thyroid hormones.

Colon The portion of the large intestine consisting of ascending, transverse, descending, and sigmoid portions.

Colony-stimulating factor (CSF) One of a group of molecules that stimulates development of white blood cells. Examples are macrophage CSF and granulocyte CSF.

Colostrum (kō-LOS-trum) A thin, cloudy fluid secreted by the mammary glands a few days prior to or after delivery before true milk is produced.

Column (KOL-um) Group of white matter tracts in the spinal cord.

Common bile duct A tube formed by the union of the common hepatic duct and the cystic duct that empties bile into the duodenum at the hepatopancreatic ampulla (ampulla of Vater).

Compact (dense) bone tissue Bone tissue that contains few spaces between osteons (Haversian systems); forms the external portion of all bones and the bulk of the diaphysis (shaft) of long bones; is found immediately deep to the periosteum and external to spongy bone.

Complement (KOM-ple-ment) A group of at least 20 normally inactive proteins found in plasma that forms a component of nonspecific resistance and immunity by bringing about cytolysis, inflammation, and opsonization.

Compliance The ease with which the lungs and thoracic wall or blood vessels can be expanded.

Compound A substance that can be broken down into two or more other substances by chemical means.

Concha (KONG-ka) A scroll-like bone found in the skull. *Plural is* **conchae** (KONG-kē).

Concussion (kon-KUSH-un) Traumatic injury to the brain that produces no visible bruising but may result in abrupt, temporary loss of consciousness.

Conduction system A group of autorhythmic cardiac muscle fibers that generates and distributes electrical impulses to stimulate coordinated contraction of the heart chambers; includes the sinoatrial (SA) node, the atrioventricular (AV) node, the atrioventricular (AV) bundle, the right and left bundle branches, and the Purkinje fibers.

Conductivity (kon′-duk-TIV-i-tē) The ability of a cell to propagate (conduct) action potentials along its plasma membrane; characteristic of neurons and muscle fibers (cells).

Condyloid (KON-di-loyd) **joint** A synovial joint structured so that an oval-shaped condyle of one bone fits into an elliptical cavity of another bone, permitting side-to-side and back-and-forth movements, such as the joint at the wrist between the radius and carpals. Also called an **ellipsoidal** (ē-lip-SOYD-al) **joint.**

Cone (KŌN) The type of photoreceptor in the retina that is specialized for highly acute color vision in bright light.

Congenital (kon-JEN-i-tal) Present at the time of birth.

Conjunctiva (kon′-junk-TĪ-va) The delicate membrane covering the eyeball and lining the eyes.

Connective tissue One of the most abundant of the four basic tissue types in the body, performing the functions of binding and supporting; consists of relatively few cells in a generous matrix (the ground substance and fibers between the cells).

Consciousness (KON-shus-nes) A state of wakefulness in which an individual is fully alert, aware, and oriented, partly as a result of feedback between the cerebral cortex and reticular activating system.

Continuous conduction (kon-DUK-shun) Propagation of an action potential (nerve impulse) in a step-by-step depolarization of each adjacent area of an axon membrane.

Contraception (kon′-tra-SEP-shun) The prevention of fertilization or impregnation without destroying fertility.

Contractility (kon′-trak-TIL-i-tē) The ability of cells or parts of cells to actively generate force to undergo shortening for movements. Muscle fibers (cells) exhibit a high degree of contractility.

Contralateral (kon′-tra-LAT-er-al) On the opposite side; affecting the opposite side of the body.

Control center The component of a feedback system, such as the brain, that determines the point at which a controlled condition, such as body temperature, is maintained.

Conus medullaris (KŌ-nus med-ū-LAR-is) The tapered portion of the spinal cord inferior to the lumbar enlargement.

Convergence (con-VER-jens) A synaptic arrangement in which the synaptic end bulbs of several presynaptic neurons terminate on one postsynaptic neuron. The medial movement of the two eyeballs so that both are directed toward a near object being viewed in order to produce a single image.

Convulsion (con-VUL-shun) Violent, involuntary contractions or spasms of an entire group of muscles.

Cornea (KOR-nē-a) The nonvascular, transparent fibrous coat through which the iris of the eye can be seen.

Corona radiata The innermost layer of granulosa cells that is firmly attached to the zona pellucida around a secondary oocyte.

Coronary artery disease (CAD) A condition such as atherosclerosis that causes narrowing of coronary arteries so that blood flow to the heart is reduced. The result is **coronary heart disease (CHD),** in which the heart muscle receives inadequate blood flow due to an interruption of its blood supply.

Coronary circulation The pathway followed by the blood from the ascending aorta through the blood vessels supplying the heart and returning to the right atrium. Also called **cardiac circulation.**

Coronary sinus (SĪ-nus) A wide venous channel on the posterior surface of the heart that collects the blood from the coronary circulation and returns it to the right atrium.

Corpus (KOR-pus) The principal part of any organ; any mass or body.

Corpus albicans (KOR-pus AL-bi-kanz) A white fibrous patch in the ovary that forms after the corpus luteum regresses.

Corpus callosum (kal-LŌ-sum) The great commissure of the brain between the cerebral hemispheres.

Corpus luteum (LOO-tē-um) A yellowish body in the ovary formed when a follicle has discharged its secondary oocyte; secretes estrogens, progesterone, relaxin, and inhibin.

Corpuscle of touch The sensory receptor for the sensation of touch; found in the dermal papillae, especially in palms and soles. Also called a **Meissner** (MĪZ-ner) **corpuscle.**

Cortex (KOR-teks) An outer layer of an organ. The convoluted layer of gray matter covering each cerebral hemisphere.

Costal (KOS-tal) Pertaining to a rib.

Costal cartilage (KAR-ti-lij) Hyaline cartilage that attaches a rib to the sternum.

Countercurrent mechanism One mechanism involved in the ability of the kidneys to produce hypertonic urine.

Cramp A spasmodic, usually painful contraction of a muscle.

Cranial (KRĀ-ne-al) **cavity** A subdivision of the dorsal body cavity formed by the cranial bones and containing the brain.

Cranial nerve One of 12 pairs of nerves that leave the brain; pass through foramina in the skull; and supply sensory and motor neurons to the head, neck, part of the trunk, and viscera of the thorax and abdomen. Each is designated by a Roman numeral and a name.

Craniosacral (krā-nē-ō-SĀ-kral) **outflow** The axons of parasympathetic preganglionic neurons, which have their cell bodies located in nuclei in the brain stem and in the lateral gray matter of the sacral portion of the spinal cord.

Cranium (KRĀ-nē-um) The skeleton of the skull that protects the brain and the organs of sight, hearing, and balance; includes the frontal, parietal, temporal, occipital, sphenoid, and ethmoid bones.

Creatine phosphate (KRĒ-a-tin FOS-fāt) Molecule in striated muscle fibers that contains high-energy phosphate bonds; used to generate ATP rapidly from ADP by transfer of a phosphate group. Also called **phosphocreatine** (fos′-fō-KRĒ-a-tin).

Crenation (krē-NĀ-shun) The shrinkage of red blood cells into knobbed, starry forms when they are placed in a hypertonic solution.

Crista (KRIS-ta) A crest or ridged structure. A small elevation in the ampulla of each semicircular duct that contains receptors for dynamic equilibrium.

Crossed extensor reflex A reflex in which extension of the joints in one limb occurs together with contraction of the flexor muscles of the opposite limb.

Crossing-over The exchange of a portion of one chromatid with another during meiosis. It permits an exchange of genes among chromatids and is one factor that results in genetic variation of progeny.

Crus (KRUS) **of penis** Separated, tapered portion of the corpora cavernosa penis. *Plural is* **crura** (KROO-ra).

Cryptorchidism (krip-TOR-ki-dizm) The condition of undescended testes.

Cuneate (KŪ-nē-āt) **nucleus** A group of neurons in the inferior part of the medulla oblongata in which axons of the cuneate fasciculus terminate.

Cupula (KUP-ū-la) A mass of gelatinous material covering the hair cells of a crista; a sensory receptor in the ampulla of a semicircular canal stimulated when the head moves.

Cushing's syndrome Condition caused by a hypersecretion of glucocorticoids characterized by spindly legs, "moon face," "buffalo hump," pendulous abdomen, flushed facial skin, poor wound healing, hyperglycemia, osteoporosis, hypertension, and increased susceptibility to disease.

Cutaneous (kū-TĀ-nē-us) Pertaining to the skin.

Cyanosis (sī-a-NŌ-sis) A blue or dark purple discoloration, most easily seen in nail beds and mucous membranes, that results from an increased concentration of deoxygenated (reduced) hemoglobin (more than 5 gm/dL).

Cyclic AMP (cyclic adenosine-3′, 5′-monophosphate) Molecule formed from ATP by the action of the enzyme adenylate cyclase; serves as second messenger for some hormones and neurotransmitters.

Cyst (SIST) A sac with a distinct connective tissue wall, containing a fluid or other material.

Cystic (SIS-tik) **duct** The duct that carries bile from the gallbladder to the common bile duct.

Cystitis (sis-TĪ-tis) Inflammation of the urinary bladder.

Cytosol (SĪ-tō-sol) Fluid located within cells. Also called **intracellular** (in′-tra-SEL-ū-lar) **fluid (ICF).**

Cytolysis (sī-TOL-i-sis) The rupture of living cells in which the contents leak out.

Cytochrome (SĪ-tō-krōm) A protein with an iron-containing group (heme) capable of alternating between a reduced form (Fe^{2+}) and an oxidized form (Fe^{3+}).

Cytokines (SĪ-to-kīns) Small protein hormones produced by lymphocytes, fibroblasts, endothelial cells, and antigen-presenting cells that act as autocrine or paracrine substances to stimulate or inhibit cell growth and differentiation, regulate immune responses, or aid nonspecific defenses.

Cytokinesis (sī-tō-ki-NĒ-sis) Distribution of the cytoplasm into two separate cells during cell division; coordinated with nuclear division (mitosis).

Cytoplasm (SĪ-tō-plazm) Cytosol plus all organelles except the nucleus.

Cytoskeleton Complex internal structure of cytoplasm consisting of microfilaments, microtubules, and intermediate filaments.

Cytosol (SĪ-tō-sol) Semifluid portion of cytoplasm in which organelles and inclusions are suspended and solutes are dissolved. Also called **intracellular fluid.**

D

Dartos (DAR-tōs) The contractile tissue deep to the skin of the scrotum.

Decibel (DES-i-bel) (**dB**) A unit for expressing the relative intensity (loudness) of sound.

Decidua (dē-SID-ū-a) That portion of the endometrium of the uterus (all but the deepest layer) that is modified during pregnancy and shed after childbirth.

Deciduous (dē-SID-ū-us) Falling off or being shed seasonally or at a particular stage of development. In the body, referring to the first set of teeth.

Decussation (dē-ku-SĀ-shun) A crossing-over to the opposite (contralateral) side; an example is the crossing of 90% of the axons in the large motor tracts to opposite sides in the medullary pyramids.

Deep Away from the surface of the body or an organ.

Deep fascia (FASH-ē-a) A sheet of connective tissue wrapped around a muscle to hold it in place.

Deep inguinal (IN-gwi-nal) **ring** A slitlike opening in the aponeurosis of the transversus abdominis muscle that represents the origin of the inguinal canal.

Defecation (def-e-KĀ-shun) The discharge of feces from the rectum.

Deglutition (dē-gloo-TISH-un) The act of swallowing.

Dehydration (dē-hī-DRĀ-shun) Excessive loss of water from the body or its parts.

Delta cell A cell in the pancreatic islets (islets of Langerhans) in the pancreas that secretes somatostatin. Also termed a **D cell.**

Demineralization (de-min′-er-al-ı-ZĀ-shun) Loss of calcium and phosphorus from bones.

Denaturation (de-nā-chur-A-shun) Disruption of the tertiary structure of a protein by heat, changes in pH, or other physical or chemical methods, in which the protein loses its physical properties and biological activity.

Dendrite (DEN-drīt) A neuronal process that carries electrical signals, usually graded potentials, toward the cell body.

Dendritic (den-DRIT-ik) **cell** One type of antigen-presenting cell with long branchlike projections that commonly is present in mucosal linings such as the vagina, in the skin (Langerhans cells in the epidermis), and in lymph nodes (follicular dendritic cells).

Dental caries (KA-rēz) Gradual demineralization of the enamel and dentin of a tooth that may invade the pulp and alveolar bone. Also called **tooth decay.**

Denticulate (den-TIK-ū-lāt) Finely toothed or serrated; characterized by a series of small, pointed projections.

Dentin (DEN-tin) The bony tissues of a tooth enclosing the pulp cavity.

Dentition (den-TI-shun) The eruption of teeth. The number, shape, and arrangement of teeth.

Deoxyribonucleic (dē-ok′-sē-rī′-bō-noo-KLĒ-ik) **acid (DNA)** A nucleic acid constructed of nucleotides consisting of one of four bases (adenine, cytosine, guanine, or thymine), deoxyribose, and a phosphate group; encoded in the nucleotides is genetic information.

Depolarization (dē-pō-lar-i-ZĀ-shun) A reduction of voltage across a plasma membrane; expressed as a change toward less negative (more positive) voltages on the interior surface of the plasma membrane.

Depression (de-PRESH-un) Movement in which a part of the body moves inferiorly.

Dermal papilla (pa-PILL-a) Fingerlike projection of the papillary region of the dermis that may contain blood capillaries or corpuscles of touch (Meissner corpuscles).

Dermatology (der-ma-TOL-ō-jē) The medical specialty dealing with diseases of the skin.

Dermatome (DER-ma-tōm) The cutaneous area developed from one embryonic spinal cord segment and receiving most of its sensory innervation from one spinal nerve. An instrument for incising the skin or cutting thin transplants of skin.

Dermis (DER-mis) A layer of dense irregular connective tissue lying deep to the epidermis.

Descending colon (KŌ-lon) The part of the large intestine descending from the left colic (splenic) flexure to the level of the left iliac crest.

Detritus (de-TRĪ-tus) Particulate matter produced by or remaining after the wearing away or disintegration of a substance or tissue; scales, crusts, or loosened skin.

Detrusor (de-TROO-ser) **muscle** Smooth muscle that forms the wall of the urinary bladder.

Developmental biology The study of development from the fertilized egg to the adult form.

Diagnosis (dī-ag-NŌ-sis) Distinguishing one disease from another or determining the nature of a disease from signs and symptoms by inspection, palpation, laboratory tests, and other means.

Dialysis (dī-AL-i-sis) The removal of waste products from blood by diffusion through a selectively permeable membrane.

Diaphragm (DĪ-a-fram) Any partition that separates one area from another, especially the dome-shaped skeletal muscle between the thoracic and abdominal cavities. Also a dome-shaped device that is placed over the cervix, usually with a spermicide, to prevent conception.

Diaphysis (dī-AF-i-sis) The shaft of a long bone.

Diarrhea (dī-a-RE-a) Frequent defecation of liquid feces caused by increased motility of the intestines.

Diarthrosis (dī-ar-THRŌ-sis) A freely movable joint; types are gliding, hinge, pivot, condyloid, saddle, and ball-and-socket.

Diastole (dī-AS-tō-lē) In the cardiac cycle, the phase of relaxation or dilation of the heart muscle, especially of the ventricles.

Diastolic (dī-as-TOL-ik) **blood pressure** The force exerted by blood on arterial walls during ventricular relaxation; the lowest blood pressure measured in the large arteries, normally about 80 mmHg in a young adult.

Diencephalon (dī′-en-SEF-a-lon) A part of the brain consisting of the thalamus, hypothalamus, epithalamus, and subthalamus.

Diffusion (dif-Ū-zhun) A passive process in which there is a net or greater movement of molecules or ions from a region of high concentration to a region of low concentration until equilibrium is reached.

Digestion (dī-JES-chun) The mechanical and chemical breakdown of food to simple molecules that can be absorbed and used by body cells.

Dilate (DĪ-lāt) To expand or swell.

Diploid (DIP-loyd) Having the number of chromosomes characteristically found in the somatic cells of an organism; having two haploid sets of chromosomes, one each from the mother and father. Symbolized 2*n*.

Direct motor pathways Collections of upper motor neurons with cell bodies in the motor cortex that project axons into the spinal cord, where they synapse with lower motor neurons or interneurons in the anterior horns. Also called the **pyramidal pathways.**

Disease Any change from a state of health.

Dislocation (dis-lō-KĀ-shun) Displacement of a bone from a joint with tearing of ligaments, tendons, and articular capsules. Also called **luxation** (luks-Ā-shun).

Dissect (di-SEKT) To separate tissues and parts of a cadaver or an organ for anatomical study.

Distal (DIS-tal) Farther from the attachment of a limb to the trunk; farther from the point of origin or attachment.

Diuretic (dī-ū-RET-ik) A chemical that increases urine volume by decreasing reabsorption of water, usually by inhibiting sodium reabsorption.

Divergence (dī-VER-jens) A synaptic arrangement in which the synaptic end bulbs of one presynaptic neuron terminate on several postsynaptic neurons.

Diverticulum (dī-ver-TIK-ū-lum) A sac or pouch in the wall of a canal or organ, especially in the colon.

Dominant allele An allele that overrides the influence of an alternate allele on the homologous chromosome; the allele that is expressed.

Dorsal body cavity Cavity near the dorsal (posterior) surface of the body that consists of a cranial cavity and vertebral canal.

Dorsal ramus (RĀ-mus) A branch of a spinal nerve containing motor and sensory axons supplying the muscles, skin, and bones of the posterior part of the head, neck, and trunk.

Dorsiflexion (dor′-si-FLEK-shun) Bending the foot in the direction of the dorsum (upper surface).

Down-regulation Phenomenon in which there is a decrease in the number of receptors in response to an excess of a hormone or neurotransmitter.

Ductus arteriosus (DUK-tus ar-tē-rē-Ō-sus) A small vessel connecting the pulmonary trunk with the aorta; found only in the fetus.

Ductus (vas) deferens (DEF-er-ens) The duct that carries sperm from the epididymis to the ejaculatory duct. Also called the **seminal duct.**

Ductus epididymis (ep′-i-DID-i-mis) A tightly coiled tube inside the epididymis, distinguished into a head, body, and tail, in which sperm undergo maturation.

Ductus venosus (ve-NŌ-sus) A small vessel in the fetus that helps the circulation bypass the liver.

Duodenal (doo-ō-DĒ-nal) **gland** Gland in the submucosa of the duodenum that secretes an alkaline mucus to protect the lining of the small intestine from the action of enzymes and to help neutralize the acid in chyme. Also called **Brunner's** (BRUN-erz) **gland.**

Duodenum (doo′-ō-DĒ-num *or* doo-OD-e-num) The first 25 cm (10 in.) of the small intestine, which connects the stomach and the ileum.

Dura mater (DOO-ra MĀ-ter) The outermost of the three meninges (coverings) of the brain and spinal cord.

Dynamic equilibrium (ē-kwi-LIB-rē-um) The maintenance of body position, mainly the head, in response to sudden movements such as rotation.

Dysfunction (dis-FUNK-shun) Absence of completely normal function.

Dysmenorrhea (dis′-men-ō-RĒ-a) Painful menstruation.

Dysplasia (dis-PLĀ-zē-a) Change in the size, shape, and organization of cells due to chronic irritation or inflammation; may either revert to normal if stress is removed or progress to neoplasia.

Dyspnea (DISP-nē-a) Shortness of breath.

E

Eardrum A thin, semitransparent partition of fibrous connective tissue between the external auditory meatus and the middle ear. Also called the **tympanic membrane.**

Ectoderm The primary germ layer that gives rise to the nervous system and the epidermis of skin and its derivatives.

Ectopic (ek-TOP-ik) Out of the normal location, as in ectopic pregnancy.

Edema (e-DĒ-ma) An abnormal accumulation of interstitial fluid.

Effector (e-FEK-tor) An organ of the body, either a muscle or a gland, that is innervated by somatic or autonomic motor neurons.

Efferent arteriole (EF-er-ent ar-TĒ-rē-ōl) A vessel of the renal vascular system that carries blood from a glomerulus to a peritubular capillary.

Efferent (EF-er-ent) **ducts** A series of coiled tubes that transport sperm from the rete testis to the epididymis.

Eicosanoids (ī-KŌ-sa-noyds) Local hormones derived from a 20-carbon fatty acid (arachidonic acid); two important types are prostaglandins and leukotrienes.

Ejaculation (e-jak-ū-LĀ-shun) The reflex ejection or expulsion of semen from the penis.

Ejaculatory (e-JAK-ū-la-tō-rē) **duct** A tube that transports sperm from the ductus (vas) deferens to the prostatic urethra.

Elasticity (e-las-TIS-i-tē) The ability of tissue to return to its original shape after contraction or extension.

Electrocardiogram (e-lek′-trō-KAR-dē-ō-gram) (**ECG** or **EKG**) A recording of the electrical changes that accompany the cardiac cycle that can be detected at the surface of the body; may be resting, stress, or ambulatory.

Electroencephalogram (e-lek′-trō-en-SEF-a-lō-gram) (**EEG**) A recording of the electrical activity of the brain from the scalp surface; used to diagnose certain diseases (such as epilepsy), furnish information regarding sleep and wakefulness, and confirm brain death.

Electrolyte (ē-LEK-trō-līt) Any compound that separates into ions when dissolved in water and that conducts electricity.

Electromyography (e-lek′-trō-mī-OG-ra-fē) Evaluation of the electrical activity of resting and contracting muscle to ascertain causes of muscular weakness, paralysis, involuntary twitching, and abnormal levels of muscle enzymes; also used as part of biofeedback studies.

Electron transport chain A sequence of electron carrier molecules on the inner mitochondrial membrane that undergo oxidation and reduction as they pump hydrogen ions (H^+) through the membrane. ATP synthesis then occurs as H^+ diffuse back into the mitochondrial matrix through special H^+ channels. *See also* **Chemiosmosis.**

Elevation (el-e-VĀ-shun) Movement in which a part of the body moves superiorly.

Embolism (EM-bō-lizm) Obstruction or closure of a vessel by an embolus.

Embolus (EM-bō-lus) A blood clot, bubble of air or fat from broken bones, mass of bacteria, or other debris or foreign material transported by the blood.

Embryo (EM-brē-ō) The young of any organism in an early stage of development; in humans, the developing organism from fertilization to the end of the eighth week of development.

Embryology (em′-brē-OL-ō-jē) The study of development from the fertilized egg to the end of the eighth week of development.

Emesis (EM-e-sis) Vomiting.

Emigration (em′-e-GRĀ-shun) Process whereby white blood cells (WBCs) leave the bloodstream by rolling along the endothelium, sticking to it, and squeezing between the endothelial cells. Adhesion molecules help WBCs stick to the endothelium. Also known as **migration** or **extravasation.**

Emission (ē-MISH-un) Propulsion of sperm into the urethra due to peristaltic contractions of the ducts of the testes, epididymides, and ductus (vas) deferens as a result of sympathetic stimulation.

Emmetropia (em′-e-TRŌ-pē-a) Normal vision in which light rays are focused exactly on the retina.

Emphysema (em′-fi′-SĒ-ma) A lung disorder in which alveolar walls disintegrate, producing abnormally large air spaces and loss of elasticity in the lungs; typically caused by exposure to cigarette smoke.

Emulsification (ē-mul′-si-fi-KĀ-shun) The dispersion of large lipid globules into smaller, uniformly distributed particles in the presence of bile.

Enamel (e-NAM-el) The hard, white substance covering the crown of a tooth.

End-diastolic (dī-as-TO-lik) **volume (EDV)** The volume of blood, about 130 mL, remaining in a ventricle at the end of its diastole (relaxation).

Endergonic (end′-er-GON-ik) **reaction** Type of chemical reaction in which the energy released as new bonds form is less than the energy needed to break apart old bonds; an energy-requiring reaction.

Endocardium (en-dō-KAR-dē-um) The layer of the heart wall, composed of endothelium and smooth muscle, that lines the inside of the heart and covers the valves and tendons that hold the valves open.

Endochondral ossification (en′-dō-KON-dral os′-i-fi-KĀ-shun) The replacement of cartilage by bone. Also called **intracartilaginous** (in′-tra-kar′-ti-LAJ-i-nus) **ossification.**

Endocrine (EN-dō-krin) **gland** A gland that secretes hormones into interstitial fluid and then the blood; a ductless gland.

Endocrinology (en′-dō-kri-NOL-ō-jē) The science concerned with the structure and functions of endocrine glands and the diagnosis and treatment of disorders of the endocrine system.

Endocytosis (en′-dō-sī-TŌ-sis) The uptake into a cell of large molecules and particles in which a segment of plasma membrane surrounds the substance, encloses it, and brings it in; includes phagocytosis, pinocytosis, and receptor-mediated endocytosis.

Endoderm (EN-dō-derm) A primary germ layer of the developing embryo; gives rise to the gastrointestinal tract, urinary bladder, urethra, and respiratory tract.

Endodontics (en′-dō-DON-tiks) The branch of dentistry concerned with the prevention, diagnosis, and treatment of diseases that affect the pulp, root, periodontal ligament, and alveolar bone.

Endogenous (en-DOJ-e-nus) Growing from or beginning within the organism.

Endolymph (EN-dō-limf′) The fluid within the membranous labyrinth of the internal ear.

Endometrium (en′-do-MĒ-trē-um) The mucous membrane lining the uterus.

Endomysium (en′-dō-MĪZ-ē-um) Invagination of the perimysium separating each individual muscle fiber (cell).

Endoneurium (en′-dō-NOO-rē-um) Connective tissue wrapping around individual nerve axons (cells).

Endoplasmic reticulum (en′-do-PLAZ-mik re-TIK-ū-lum) **(ER)** A network of channels running through the cytoplasm of a cell that serves in intracellular transportation, support, storage, synthesis, and packaging of molecules. Portions of ER where ribosomes are attached to the outer surface are called **rough ER;** portions that have no ribosomes are called **smooth ER.**

Endorphin (en-DOR-fin) A neuropeptide in the central nervous system that acts as a painkiller.

Endosteum (en-DOS-tē-um) The membrane that lines the medullary (marrow) cavity of bones, consisting of osteogenic cells and scattered osteoclasts.

Endothelium (en′-dō-THĒ-lē-um) The layer of simple squamous epithelium that lines the cavities of the heart, blood vessels, and lymphatic vessels.

End-systolic (sis-TO-lik) **volume (ESV)** The volume of blood, about 60 mL, remaining in a ventricle after its systole (contraction).

Energy The capacity to do work.

Enkephalin (en-KEF-a-lin) A peptide found in the central nervous system that acts as a painkiller.

Enteric (EN-ter-ik) **nervous system** The part of the nervous system that is embedded in the submucosa and muscularis of the gastrointestinal (GI) tract; governs motility and secretions of the GI tract.

Enteroendocrine (en-ter-ō-EN-dō-krin) **cell** A cell of the mucosa of the gastrointestinal tract that secretes a hormone that governs function of the GI tract; hormones secreted include gastrin, cholecystokinin, glucose-dependent insulinotropic peptide (GIP), and secretin.

Enterogastric (en-te-rō-GAS-trik) **reflex** A reflex that inhibits gastric secretion; initiated by food in the small intestine.

Enzyme (EN-zīm) A substance that accelerates chemical reactions; an organic catalyst, usually a protein.

Eosinophil (ē′-ō-SIN-ō-fil) A type of white blood cell characterized by granules that stain red or pink with acid dyes.

Ependymal (e-PEN-de-mal) **cells** Neuroglial cells that cover choroid plexuses and produce cerebrospinal fluid (CSF); they also line the ventricles of the brain and probably assist in the circulation of CSF.

Epicardium (ep′-i-KAR-dē-um) The thin outer layer of the heart wall, composed of serous tissue and mesothelium. Also called the **visceral pericardium.**

Epidemiology (ep′-i-dē-mē-OL-ō-jē) Study of the occurrence and distribution of diseases and disorders in human populations.

Epidermis (ep-i-DERM-is) The superficial, thinner layer of skin, composed of keratinized stratified squamous epithelium.

Epididymis (ep′-i-DID-i-mis) A comma-shaped organ that lies along the posterior border of the testis and contains the ductus epididymis, in which sperm undergo maturation. *Plural is* **epididymides** (ep′-i-DID-i-mi-dēz).

Epidural (ep′-i-DOO-ral) **space** A space between the spinal dura mater and the vertebral canal, containing areolar connective tissue and a plexus of veins.

Epiglottis (ep′-i-GLOT-is) A large, leaf-shaped piece of cartilage lying on top of the larynx, attached to the thyroid cartilage and its unattached portion is free to move up and down to cover the glottis (vocal folds and rima glottidis) during swallowing.

Epimysium (ep′-i-MĪZ-ē-um) Fibrous connective tissue around muscles.

Epinephrine (ep-ē-NEF-rin) Hormone secreted by the adrenal medulla that produces actions similar to those that result from sympathetic stimulation. Also called **adrenaline** (a-DREN-a-lin).

Epineurium (ep′-i-NOO-rē-um) The superficial connective tissue covering around an entire nerve.

Epiphyseal (ep′-i-FIZ-ē-al) **line** The remnant of the epiphyseal plate in the metaphysis of a long bone.

Epiphyseal (ep′-i-FIZ-ē-al) **plate** The hyaline cartilage plate in the metaphysis of a long bone; site of lengthwise growth of long bones.

Epiphysis (ē-PIF-i-sis) The end of a long bone, usually larger in diameter than the shaft (diaphysis).

Episiotomy (e-piz′-ē-OT-ō-mē) A cut made with surgical scissors to avoid tearing of the perineum at the end of the second stage of labor.

Epistaxis (ep′-i-STAK-sis) Loss of blood from the nose due to trauma, infection, allergy, neoplasm, and bleeding disorders. Also called **nosebleed.**

Epithelial (ep′-i-THĒ-lē-al) **tissue** The tissue that forms innermost and outermost surfaces of body structures and forms glands.

Eponychium (ep′-o-NIK-ē-um) Narrow band of stratum corneum at the proximal border of a nail that extends from the margin of the nail wall. Also called the **cuticle.**

Erectile dysfunction Failure to maintain an erection long enough for sexual intercourse. Also known as **impotence** (IM-pō-tens).

Erection (ē-REK-shun) The enlarged and stiff state of the penis or clitoris resulting from the engorgement of the spongy erectile tissue with blood.

Eructation (e-ruk′-TĀ-shun) The forceful expulsion of gas from the stomach. Also called **belching.**

Erythema (er′-i-THĒ-ma) Skin redness usually caused by dilation of the capillaries.

Erythrocyte (e-RITH-rō-sīt) A mature red blood cell.

Erythropoiesis (e-rith′-rō-poy-Ē-sis) The process by which red blood cells are formed.

Erythropoietin (e-rith′-rō-POY-e-tin) A hormone released by the juxtaglomerular cells of the kidneys that stimulates red blood cell production.

Esophagus (e-SOF-a-gus) The hollow muscular tube that connects the pharynx and the stomach.

Essential amino acids Those 10 amino acids that cannot be synthesized by the human body at an adequate rate to meet its needs and therefore must be obtained from the diet.

Estrogens (ES-tro-jenz) Feminizing sex hormones produced by the ovaries; govern development of oocytes, maintenance of female reproductive structures, and appearance of secondary sex characteristics; also affect fluid and electrolyte balance, and protein anabolism. Examples are β-estradiol, estrone, and estriol.

Etiology (ē′-tē-OL-ō-jē) The study of the causes of disease, including theories of the origin and organisms (if any) involved.

Eupnea (ŪP-nē-a) Normal quiet breathing.

Eversion (ē-VER-zhun) The movement of the sole laterally at the ankle joint or of an atrioventricular valve into an atrium during ventricular contraction.

Excitability (ek-sīt′-a-BIL-i-tē) The ability of muscle fibers to receive and respond to stimuli; the ability of neurons to respond to stimuli and generate nerve impulses.

Excitatory postsynaptic potential (EPSP) A small depolarization of a postsynaptic membrane when it is stimulated by an excitatory neurotransmitter. An EPSP is both graded (variable size) and localized (decreases in size away from the point of excitation).

Excrement (EKS-kre-ment) Material eliminated from the body as waste, especially fecal matter.

Excretion (eks-KRĒ-shun) The process of eliminating waste products from the body; also the products excreted.

Exergonic (eks′-er-GON-ik) **reaction** Type of chemical reaction in which the energy released as new bonds form is greater than the energy needed to break apart old bonds; an energy-releasing reaction.

Exocrine (EK-sō-krin) **gland** A gland that secretes its products into ducts that carry the secretions into body cavities, into the lumen of an organ, or to the outer surface of the body.

Exocytosis (ex′-ō-sī-TŌ-sis) A process in which membrane-enclosed secretory vesicles form inside the cell, fuse with the plasma membrane, and release their contents into the interstitial fluid; achieves secretion of materials from a cell.

Exogenous (ex-SOJ-e-nus) Originating outside an organ or part.

Exhalation (eks-ha-LĀ-shun) Breathing out; expelling air from the lungs into the atmosphere. Also called **expiration.**

Exon (EX-on) A region of DNA that codes for synthesis of a protein.

Expiratory (eks-PĪ-ra-tō-rē) **reserve volume** The volume of air in excess of tidal volume that can be exhaled forcibly; about 1200 mL.

Extensibility (ek-sten′-si-BIL-i-tē) The ability of muscle tissue to stretch when it is pulled.

Extension (ek-STEN-shun) An increase in the angle between two bones; restoring a body part to its anatomical position after flexion.

External Located on or near the surface.

External auditory (AW-di-tōr-ē) **canal** or **meatus** (mē-Ā-tus) A curved tube in the temporal bone that leads to the middle ear.

External ear The outer ear, consisting of the pinna, external auditory canal, and tympanic membrane (eardrum).

External nares (NĀ-rez) The external nostrils, or the openings into the nasal cavity on the exterior of the body.

External respiration The exchange of respiratory gases between the lungs and blood. Also called **pulmonary respiration.**

Exteroceptor (eks′-ter-ō-SEP-tor) A sensory receptor adapted for the reception of stimuli from outside the body.

Extracellular fluid (ECF) Fluid outside body cells, such as interstitial fluid and plasma.

Extravasation (eks-trav-a-SĀ-shun) The escape of fluid, especially blood, lymph, or serum, from a vessel into the tissues.

Extrinsic (ek-STRIN-sik) Of external origin.

Extrinsic pathway (of blood clotting) Sequence of reactions leading to blood clotting that is initiated by the release of tissue factor (TF), also known as thromboplastin, that leaks into the blood from damaged cells outside the blood vessels.

Exudate (EKS-oo-dāt) Escaping fluid or semifluid material that oozes from a space and that may contain serum, pus, and cellular debris.

Eyebrow The hairy ridge superior to the eye.

F

Face The anterior aspect of the head.

Facilitated diffusion (fa-SIL-i-tā-ted dif-Ū-zhun) Diffusion in which a substance not soluble by itself in lipids diffuses across a selectively permeable membrane with the help of a transporter protein.

Falciform ligament (FAL-si-form LIG-a-ment) A sheet of parietal peritoneum between the two principal lobes of the liver. The ligamentum teres, or remnant of the umbilical vein, lies within its fold.

Falx cerebelli (FALKS ser′-e-BEL-lē) A small triangular process of the dura mater attached to the occipital bone in the posterior cranial fossa and projecting inward between the two cerebellar hemispheres.

Falx cerebri (FALKS SER-e-brē) A fold of the dura mater extending deep into the longitudinal fissure between the two cerebral hemispheres.

Fascia (FASH-ē-a) A fibrous membrane covering, supporting, and separating muscles.

Fascicle (FAS-i-kul) A small bundle or cluster, especially of nerve or muscle fibers (cells). Also called a **fasciculus** (fa-SIK-ū-lus). *Plural is* **fasciculi** (fa-SIK-yoo-lī).

Fasciculation (fa-sik′-ū-LĀ-shun) Abnormal, spontaneous twitch of all skeletal muscle fibers in one motor unit that is visible at the skin surface; not associated with movement of the affected muscle; present in progressive diseases of motor neurons, for example, poliomyelitis.

Fauces (FAW-sēz) The opening from the mouth into the pharynx.

F cell A cell in the pancreatic islets (islets of Langerhans) that secretes pancreatic polypeptide.

Feces (FĒ-sēz) Material discharged from the rectum and made up of bacteria, excretions, and food residue. Also called **stool.**

Feedback system A sequence of events in which information about the status of a situation is continually reported (fed back) to a control center.

Female reproductive cycle General term for the ovarian and uterine cycles, the hormonal changes that accompany them, and cyclic changes in the breasts and cervix; includes changes in the endometrium of a nonpregnant female that prepares the lining of the uterus to receive a fertilized ovum. Less correctly termed the **menstrual cycle.**

Fertilization (fer′-ti-li-ZĀ-shun) Penetration of a secondary oocyte by a sperm cell, meiotic division of secondary oocyte to form an ovum, and subsequent union of the nuclei of the gametes.

Fetal circulation The cardiovascular system of the fetus, including the placenta and special blood vessels involved in the exchange of materials between fetus and mother.

Fetus (FĒ-tus) In humans, the developing organism *in utero* from the beginning of the third month to birth.

Fever An elevation in body temperature above the normal temperature of 37°C (98.6°F) due to a resetting of the hypothalamic thermostat.

Fibrillation (fi-bri-LĀ-shun) Abnormal, spontaneous twitch of a single skeletal muscle fiber (cell) that can be detected with electromyography but is not visible at the skin surface; not associated with movement of the affected muscle; present in certain disorders of motor neurons, for example, amyotrophic lateral sclerosis (ALS). With reference to cardiac muscle, *see* **Atrial fibrillation** and **Ventricular fibrillation.**

Fibrin (FĪ-brin) An insoluble protein that is essential to blood clotting; formed from fibrinogen by the action of thrombin.

Fibrinogen (fī-BRIN-ō-jen) A clotting factor in blood plasma that by the action of thrombin is converted to fibrin.

Fibrinolysis (fī-bri-NOL-i-sis) Dissolution of a blood clot by the action of a proteolytic enzyme, such as plasmin (fibrinolysin), that dissolves fibrin threads and inactivates fibrinogen and other blood-clotting factors.

Fibroblast (FĪ-brō-blast) A large, flat cell that secretes most of the matrix (extracellular) material of areolar and dense connective tissues.

Fibrous (FĪ-brus) **joint** A joint that allows little or no movement, such as a suture or a syndesmosis.

Fibrous tunic (TOO-nik) The superficial coat of the eyeball, made up of the posterior sclera and the anterior cornea.

Fight-or-flight response The effects produced upon stimulation of the sympathetic division of the autonomic nervous system.

Filiform papilla (FIL-i-form pa-PIL-a) One of the conical projections that are distributed in parallel rows over the anterior two-thirds of the tongue and lack taste buds.

Filtration (fil-TRĀ-shun) The flow of a liquid through a filter (or membrane that acts like a filter) due to a hydrostatic pressure; occurs in capillaries due to blood pressure.

Filtration fraction The percentage of plasma entering the kidneys that becomes glomerular filtrate.

Filtration membrane Site of blood filtration in nephrons of the kidneys, consisting of the endothelium and basement membrane of the glomerulus and the epithelium of the visceral layer of the glomerular (Bowman's) capsule.

Filum terminale (FĪ-lum ter-mi-NAL-ē) Non-nervous fibrous tissue of the spinal cord that extends inferiorly from the conus medullaris to the coccyx.

Fimbriae (FIM-brē-ē) Fingerlike structures, especially the lateral ends of the uterine (Fallopian) tubes.

Fissure (FISH-ur) A groove, fold, or slit that may be normal or abnormal.

Fistula (FIS-tū-la) An abnormal passage between two organs or between an organ cavity and the outside.

Fixator A muscle that stabilizes the origin of the prime mover so that the prime mover can act more efficiently.

Fixed macrophage (MAK-rō-fāj) Stationary phagocytic cell found in the liver, lungs, brain, spleen, lymph nodes, subcutaneous tissue, and red bone marrow. Also called a **histiocyte** (HIS-tē-ō-sīt).

Flaccid (FLAS-sid) Relaxed, flabby, or soft; lacking muscle tone.

Flagellum (fla-JEL-um) A hairlike, motile process on the extremity of a bacterium, protozoan, or sperm cell. *Plural is* **flagella** (fla-JEL-a).

Flatus (FLĀ-tus) Gas in the stomach or intestines, commonly used to denote expulsion of gas through the anus.

Flexion (FLEK-shun) Movement in which there is a decrease in the angle between two bones.

Flexor reflex A protective reflex in which flexor muscles are stimulated while extensor muscles are inhibited.

Follicle (FOL-i-kul) A small secretory sac or cavity; the group of cells that contains a developing oocyte in the ovaries.

Follicle-stimulating hormone (FSH) Hormone secreted by the anterior pituitary that initiates development of ova and stimulates the ovaries to secrete estrogens in females, and initiates sperm production in males.

Fontanel (fon′-ta-NEL) A fibrous connective tissue membrane-filled space where bone formation is not yet complete, especially between the cranial bones of an infant's skull.

Foot The terminal part of the lower limb, from the ankle to the toes.

Foramen (fō-RĀ-men) A passage or opening; a communication between two cavities of an organ, or a hole in a bone for passage of vessels or nerves. *Plural is* **foramina** (fō-RAM-i-na).

Foramen ovale (fō-RĀ-men-ō-VAL-ē) An opening in the fetal heart in the septum between the right and left atria. A hole in the greater wing of the sphenoid bone that transmits the mandibular branch of the trigeminal nerve (cranial nerve V).

Forearm (FOR-arm) The part of the upper limb between the elbow and the wrist.

Fornix (FOR-niks) An arch or fold; a tract in the brain made up of association fibers, connecting the hippocampus with the mammillary bodies; a recess around the cervix of the uterus where it protrudes into the vagina.

Fossa (FOS-a) A furrow or shallow depression.

Fourth ventricle (VEN-tri-kul) A cavity filled with cerebrospinal fluid within the brain lying between the cerebellum and the medulla oblongata and pons.

Fracture (FRAK-choor) Any break in a bone.

Frenulum (FREN-ū-lum) A small fold of mucous membrane that connects two parts and limits movement.

Frontal plane A plane at a right angle to a midsagittal plane that divides the body or organs into anterior and posterior portions. Also called a **coronal** (kō-RŌ-nal) **plane.**

Functional residual (re-ZID-ū-al) **capacity** The sum of residual volume plus expiratory reserve volume; about 2400 mL.

Fundus (FUN-dus) The part of a hollow organ farthest from the opening.

Fungiform papilla (FUN-ji-form pa-PIL-a) A mushroomlike elevation on the upper surface of the tongue appearing as a red dot; most contain taste buds.

G

Gallbladder A small pouch, located inferior to the liver, that stores bile and empties by means of the cystic duct.

Gallstone A solid mass, usually containing cholesterol, in the gallbladder or a bile-containing duct; formed anywhere between bile canaliculi in the liver and the hepatopancreatic ampulla (ampulla of Vater), where bile enters the duodenum. Also called a **biliary calculus.**

Gamete (GAM-ēt) A male or female reproductive cell; a sperm cell or secondary oocyte.

Ganglion (GANG-glē-on) Usually, a group of neuronal cell bodies lying outside the central nervous system (CNS). *Plural is* **ganglia** (GANG-glē-a).

Gastric (GAS-trik) **glands** Glands in the mucosa of the stomach composed of cells that empty their secretions into narrow channels called gastric pits. Types of cells are chief cells (secrete pepsinogen), parietal cells (secrete hydrochloric acid and intrinsic factor), surface mucous and mucous neck cells (secrete mucus), and G cells (secrete gastrin).

Gastroenterology (gas′-trō-en′-ter-OL-ō-jē) The medical specialty that deals with the structure, function, diagnosis, and treatment of diseases of the stomach and intestines.

Gastrointestinal (gas-trō-in-TES-ti-nal) **(GI) tract** A continuous tube running through the ventral body cavity extending from the mouth to the anus. Also called the **alimentary** (al′-i-MEN-tar-ē) **canal.**

Gastrulation (gas′-troo-LĀ-shun) The migration of groups of cells from the epiblast that transform a bilaminar embryonic disc into a trilaminar embryonic disc that consists of the three primary germ layers; transformation of the blastula into the gastrula.

Gene (JĒN) Biological unit of heredity; a segment of DNA located in a definite position on a particular chromosome; a sequence of DNA that codes for a particular mRNA, rRNA, or tRNA.

Generator potential The graded depolarization that results in a change in the resting membrane potential in a receptor (specialized neuronal ending); may trigger a nerve action potential (nerve impulse) if depolarization reaches threshold.

Genetic engineering The manufacture and manipulation of genetic material.

Genetics The study of genes and heredity.

Genitalia (jen′-i-TĀ-lē-a) Reproductive organs.

Genome (JĒ-nōm) The complete set of genes of an organism.

Genotype (JĒ-nō-tīp) The genetic makeup of an individual; the combination of alleles present at one or more chromosomal locations, as distinguished from the appearance, or phenotype, that results from those alleles.

Geriatrics (jer′-e-AT-riks) The branch of medicine devoted to the medical problems and care of elderly persons.

Gestation (jes-TĀ-shun) The period of development from fertilization to birth.

Gingivae (jin-JI-vē) Gums. They cover the alveolar processes of the mandible and maxilla and extend slightly into each socket.

Gland Specialized epithelial cell or cells that secrete substances; may be exocrine or endocrine.

Glans penis (glanz PĒ-nis) The slightly enlarged region at the distal end of the penis.

Glaucoma (glaw-KŌ-ma) An eye disorder in which there is increased intraocular pressure due to an excess of aqueous humor.

Gliding joint A synovial joint having articulating surfaces that are usually flat, permitting only side-to-side and back-and-forth movements, as between carpal bones, tarsal bones, and the scapula and clavicle. Also called an **arthrodial** (ar-THRŌ-dē-al) **joint.**

Glomerular (glō-MER-ū-lar) **capsule** A double-walled globe at the proximal end of a nephron that encloses the glomerular capillaries. Also called **Bowman's** (BŌ-manz) **capsule.**

Glomerular filtrate (glō-MER-ū-lar FIL-trāt) The fluid produced when blood is filtered by the filtration membrane in the glomeruli of the kidneys.

Glomerular filtration The first step in urine formation in which substances in blood pass through the filtration membrane and the filtrate enters the proximal convoluted tubule of a nephron.

Glomerular filtration rate (GFR) The total volume of fluid that enters all the glomerular (Bowman's) capsules of the kidneys in 1 min; about 100–125 mL/min.

Glomerulus (glō-MER-ū-lus) A rounded mass of nerves or blood vessels, especially the microscopic tuft of capillaries that is surrounded by the glomerular (Bowman's) capsule of each kidney tubule. *Plural is* **glomeruli.**

Glottis (GLOT-is) The vocal folds (true vocal cords) in the larynx plus the space between them (rima glottidis).

Glucagon (GLOO-ka-gon) A hormone produced by the alpha cells of the pancreatic islets (islets of Langerhans) that increases blood glucose level.

Glucocorticoids (gloo-kō-KOR-ti-koyds) Hormones secreted by the cortex of the adrenal gland, especially cortisol, that influence glucose metabolism.

Gluconeogenesis (gloo′-kō-nē-ō-JEN-e-sis) The synthesis of glucose from certain amino acids or lactic acid.

Glucose (GLOO-kōs) A hexose (six-carbon sugar), $C_6H_{12}O_6$, that is a major energy source for the production of ATP by body cells.

Glucosuria (gloo′-kō-SOO-rē-a) The presence of glucose in the urine; may be temporary or pathological. Also called **glycosuria.**

Glycogen (GLĪ-kō-jen) A highly branched polymer of glucose containing thousands of subunits; functions as a compact store of glucose molecules in liver and muscle fibers (cells).

Glycogenesis (glī′-kō-JEN-e-sis) The chemical reactions by which many molecules of glucose are used to synthesize glycogen.

Glycogenolysis (glī-kō-je-NOL-i-sis) The breakdown of glycogen into glucose.

Glycolysis (glī-KOL-i-sis) Series of chemical reactions in the cytosol of a cell in which a molecule of glucose is split into two molecules of pyruvic acid with net production of two ATPs.

Goblet cell A goblet-shaped unicellular gland that secretes mucus; present in epithelium of the airways and intestines.

Goiter (GOY-ter) An enlarged thyroid gland.

Golgi (GOL-jē) **complex** An organelle in the cytoplasm of cells consisting of four to six flattened sacs (cisternae), stacked on one another, with expanded areas at their ends; functions in processing, sorting, packaging, and delivering proteins and lipids to the plasma membrane, lysosomes, and secretory vesicles.

Gomphosis (gom-FŌ-sis) A fibrous joint in which a cone-shaped peg fits into a socket.

Gonad (GŌ-nad) A gland that produces gametes and hormones; the ovary in the female and the testis in the male.

Gonadotropic hormone Anterior pituitary hormone that affects the gonads.

Gracile (GRAS-īl) **nucleus** A group of nerve cells in the inferior part of the medulla oblongata in which axons of the gracile fasciculus terminate.

Gray commissure (KOM-i-shur) A narrow strip of gray matter connecting the two lateral gray masses within the spinal cord.

Gray matter Areas in the central nervous system and ganglia containing neuronal cell bodies, dendrites, unmyelinated axons, axon terminals, and neuroglia; Nissl bodies impart a gray color and there is little or no myelin in gray matter.

Gray ramus communicans (RĀ-mus kō-MŪ-ni-kans) A short nerve containing axons of sympathetic postganglionic neurons; the cell bodies of the neurons are in a sympathetic chain ganglion, and the unmyelinated axons extend via the gray ramus to a spinal nerve and then to the periphery to supply smooth muscle in blood vessels, arrector pili muscles, and sweat glands. *Plural is* **rami communicantes** (RĀ-mē kō-mū-ni-KAN-tēz).

Greater omentum (ō-MEN-tum) A large fold in the serosa of the stomach that hangs down like an apron anterior to the intestines.

Greater vestibular (ves-TIB-ū-lar) **glands** A pair of glands on either side of the vaginal orifice that open by a duct into the space between the hymen and the labia minora. Also called **Bartholin's** (BAR-to-linz) **glands.**

Groin (GROYN) The depression between the thigh and the trunk; the inguinal region.

Gross anatomy The branch of anatomy that deals with structures that can be studied without using a microscope. Also called **macroscopic anatomy.**

Growth An increase in size due to an increase in (1) the number of cells, (2) the size of existing cells as internal components increase in size, or (3) the size of intercellular substances.

Gustatory (GUS-ta-tō′-rē) Pertaining to taste.

Gynecology (gī′-ne-KOL-ō-jē) The branch of medicine dealing with the study and treatment of disorders of the female reproductive system.

Gynecomastia (gīn′-e-kō-MAS-tē-a) Excessive growth (benign) of the male mammary glands due to secretion of estrogens by an adrenal gland tumor (feminizing adenoma).

Gyrus (JĪ-rus) One of the folds of the cerebral cortex of the brain. *Plural is* **gyri** (JĪ-rī). Also called a **convolution.**

H

Hair A threadlike structure produced by hair follicles that develops in the dermis. Also called a **pilus** (PĪ-lus).

Hair follicle (FOL-li-kul) Structure, composed of epithelium and surrounding the root of a hair, from which hair develops.

Hair root plexus (PLEK-sus) A network of dendrites arranged around the root of a hair as free or naked nerve endings that are stimulated when a hair shaft is moved.

Haldane effect Less carbon dioxide binds to hemoglobin when the partial pressure of oxygen is higher.

Hand The terminal portion of an upper limb, including the carpus, metacarpus, and phalanges.

Haploid (HAP-loyd) Having half the number of chromosomes characteristically found in the somatic cells of an organism; characteristic of mature gametes. Symbolized *n*.

Hard palate (PAL-at) The anterior portion of the roof of the mouth, formed by the maxillae and palatine bones and lined by mucous membrane.

Haustra (HAWS-tra) A series of pouches that characterize the colon; caused by tonic contractions of the teniae coli. *Singular is* **haustrum.**

Head The superior part of a human, cephalic to the neck. The superior or proximal part of a structure.

Heart A hollow muscular organ lying slightly to the left of the midline of the chest that pumps the blood through the cardiovascular system.

Heart block An arrhythmia (dysrhythmia) of the heart in which the atria and ventricles contract independently because of a blocking of electrical impulses through the heart at some point in the conduction system.

Heart murmur (MER-mer) An abnormal sound that consists of a flow noise that is heard before, between, or after the normal heart sounds, or that may mask normal heart sounds.

Heat exhaustion Condition characterized by cool, clammy skin, profuse perspiration, and fluid and electrolyte (especially sodium and chloride) loss that results in muscle cramps, dizziness, vomiting, and fainting. Also called **heat prostration.**

Heat stroke Condition produced when the body cannot easily lose heat and characterized by reduced perspiration and elevated body temperature. Also called **sunstroke.**

Hematocrit (hē-MAT-ō-krit) **(Hct)** The percentage of blood made up of red blood cells. Usually measured by centrifuging a blood sample in a graduated tube and then reading the volume of red blood cells and dividing it by the total volume of blood in the sample.

Hematology (hē′-ma-TOL-ō-jē) The study of blood.

Hemiplegia (hem-i-PLĒ-jē-a) Paralysis of the upper limb, trunk, and lower limb on one side of the body.

Hemodynamics (hē-mō-dī-NA-miks) The study of factors and forces that govern the flow of blood through blood vessels.

Hemoglobin (hē′-mō-GLŌ-bin) **(Hb)** A substance in red blood cells consisting of the protein globin and the iron-containing red pigment heme that transports most of the oxygen and some carbon dioxide in blood.

Hemolysis (hē-MOL-i-sis) The escape of hemoglobin from the interior of a red blood cell into the surrounding medium; results from disruption of the cell membrane by toxins or drugs, freezing or thawing, or hypotonic solutions.

Hemolytic disease of the newborn A hemolytic anemia of a newborn child that results from the destruction of the infant's erythrocytes (red blood cells) by antibodies produced by the mother; usually the antibodies are due to an Rh blood type incompatibility. Also called **erythroblastosis fetalis** (e-rith′-rō-blas-TŌ-sis fe-TAL-is).

Hemophilia (hē′-mō-FIL-ē-a) A hereditary blood disorder where there is a deficient production of certain factors involved in blood clotting, resulting in excessive bleeding into joints, deep tissues, and elsewhere.

Hemopoiesis (hē-mō-poy-Ē-sis) Blood cell production, which occurs in red bone marrow after birth. Also called **hematopoiesis** (hem′-a-tō-poy-Ē-sis).

Hemorrhage (HEM-or-rij) Bleeding; the escape of blood from blood vessels, especially when the loss is profuse.

Hemorrhoids (HEM-ō-royds) Dilated or varicosed blood vessels (usually veins) in the anal region. Also called **piles.**

Hemostasis (hē-MŌS-tā-sis) The stoppage of bleeding.

Heparin (HEP-a-rin) An anticoagulant given to slow the conversion of prothrombin to thrombin, thus reducing the risk of blood clot formation; found in basophils, mast cells, and various other tissues, especially the liver and lungs.

Hepatic (he-PAT-ik) Refers to the liver.

Hepatic duct A duct that receives bile from the bile capillaries. Small hepatic ducts merge to form the larger right and left hepatic ducts that unite to leave the liver as the common hepatic duct.

Hepatic portal circulation The flow of blood from the gastrointestinal organs to the liver before returning to the heart.

Hepatocyte (he-PAT-ō-cyte) A liver cell.

Hepatopancreatic (hep′-a-tō-pan′-krē-A-tik) **ampulla** A small, raised area in the duodenum where the combined common bile duct and main pancreatic duct empty into the duodenum. Also called the **ampulla of Vater** (VA-ter).

Hernia (HER-nē-a) The protrusion or projection of an organ or part of an organ through a membrane or cavity wall, usually the abdominal cavity.

Herniated (HER-nē-Ā′-ted) **disc** A rupture of an intervertebral disc so that the nucleus pulposus protrudes into the vertebral cavity. Also called a **slipped disc.**

Heterozygous (he-ter-ō-ZĪ-gus) Possessing different alleles on homologous chromosomes for a particular hereditary trait.

Hiatus (hī-Ā-tus) An opening; a foramen.

Hilus (HĪ-lus) An area, depression, or pit where blood vessels and nerves enter or leave an organ. Also called a **hilum.**

Hinge joint A synovial joint in which a convex surface of one bone fits into a concave surface of another bone, such as the elbow, knee, ankle, and interphalangeal joints. Also called a **ginglymus** (JIN-gli-mus) **joint.**

Hirsutism (HER-soot-izm) An excessive growth of hair in females and children, with a distribution similar to that in adult males, due to the conversion of vellus hairs into large terminal hairs in response to higher-than-normal levels of androgens.

Histamine (HISS-ta-mēn) Substance found in many cells, especially mast cells, basophils, and platelets, released when the cells are injured; results in vasodilation, increased permeability of blood vessels, and constriction of bronchioles.

Histology (hiss-TOL-ō-jē) Microscopic study of the structure of tissues.

Holocrine (HŌL-ō-krin) **gland** A type of gland in which entire secretory cells, along with their accumulated secretions, make up the secretory product of the gland, as in the sebaceous (oil) glands.

Homeostasis (hō′-mē-ō-STĀ-sis) The condition in which the body's internal environment remains relatively constant, within physiological limits.

Homologous chromosomes Two chromosomes that belong to a pair. Also called **homologues.**

Homozygous (hō-mō-ZĪ-gus) Possessing the same alleles on homologous chromosomes for a particular hereditary trait.

Hormone (HOR-mōn) A secretion of endocrine cells that alters the physiological activity of target cells of the body.

Horn An area of gray matter (anterior, lateral, or posterior) in the spinal cord.

Human chorionic gonadotropin (kō-rē-ON-ik gō-nad-ō-TRŌ-pin) **(hCG)** A hormone produced by the developing placenta that maintains the corpus luteum.

Human chorionic somatomammotropin (sō-mat-ō-mam-ō-TRŌ-pin) **(hCS)** Hormone produced by the chorion of the placenta that stimulates breast tissue for lactation, enhances body growth, and regulates metabolism. Also called **human placental lactogen (hPL).**

Human growth hormone (hGH) Hormone secreted by the anterior pituitary that stimulates growth of body tissues, especially skeletal and muscular tissues. Also known as **somatotropin** and **somatotropic hormone (STH).**

Hyaluronic (hī′-a-loo-RON-ik) **acid** A viscous, amorphous extracellular material that binds cells together, lubricates joints, and maintains the shape of the eyeballs.

Hyaluronidase (hī′-a-loo-RON-i-dās) An enzyme that breaks down hyaluronic acid, increasing the permeability of connective tissues by dissolving the substances that hold body cells together.

Hydride ion (HĪ-drīd Ī-on) A hydrogen nucleus with two orbiting electrons (H^-), in contrast to the more common **hydrogen ion,** which has no orbiting electrons (H^+).

Hymen (HĪ-men) A thin fold of vascularized mucous membrane at the vaginal orifice.

Hypercalcemia (hī′-per-kal-SĒ-mē-a) An excess of calcium in the blood.

Hypercapnia (hī′-per-KAP-nē-a) An abnormal increase in the amount of carbon dioxide in the blood.

Hyperextension (hī′-per-ek-STEN-shun) Continuation of extension beyond the anatomical position, as in bending the head backward.

Hyperglycemia (hī′-per-glī-SĒ-mē-a) An elevated blood glucose level.

Hyperkalemia (hī′-per-kā-LĒ-mē-a) An excess of potassium ions in the blood.

Hypermagnesemia (hī′-per-mag′-ne-SĒ-mē-a) An excess of magnesium ions in the blood.

Hypermetropia (hī′-per-mē-TRŌ-pē-a) A condition in which visual images are focused behind the retina, with resulting defective vision of near objects; farsightedness.

Hyperphosphatemia (hī-per-fos′-fa-TĒ-mē-a) An abnormally high level of phosphates in the blood.

Hyperplasia (hī′-per-PLĀ-zē-a) An abnormal increase in the number of normal cells in a tissue or organ, increasing its size.

Hyperpolarization (hī′-per-PŌL-a-ri-zā′-shun) Increase in the internal negativity across a cell membrane, thus increasing the voltage and moving it farther away from the threshold value.

Hypersecretion (hī′-per-se-KRĒ-shun) Overactivity of glands resulting in excessive secretion.

Hypersensitivity (hī′-per-sen-si-TI-vi-tē) Overreaction to an allergen that results in pathological changes in tissues. Also called **allergy.**

Hypertension (hī′-per-TEN-shun) High blood pressure.

Hyperthermia (hī′-per-THERM-ē-a) An elevated body temperature.

Hypertonia (hī′-per-TŌ-nē-a) Increased muscle tone that is expressed as spasticity or rigidity.

Hypertonic (hī′-per-TON-ik) Solution that causes cells to shrink due to loss of water by osmosis.

Hypertrophy (hī-PER-trō-fē) An excessive enlargement or overgrowth of tissue without cell division.

Hyperventilation (hī′-per-ven-ti-LĀ-shun) A rate of respiration higher than that required to maintain a normal partial pressure of carbon dioxide in the blood.

Hypocalcemia (hī′-pō-kal-SĒ-mē-a) A below-normal level of calcium in the blood.

Hypochloremia (hī′-pō-klō-RĒ-mē-a) Deficiency of chloride ions in the blood.

Hypoglycemia (hī′-pō-glī-SĒ-mē-a) An abnormally low concentration of glucose in the blood; can result from excess insulin (injected or secreted).

Hypokalemia (hī′-pō-ka-LĒ-mē-a) Deficiency of potassium ions in the blood.

Hypomagnesemia (hī′-pō-mag′-ne-SĒ-mē-a) Deficiency of magnesium ions in the blood.

Hyponatremia (hī′-pō-na-TRĒ-mē-a) Deficiency of sodium ions in the blood.

Hyponychium (hī′-pō-NIK-ē-um) Free edge of the fingernail.

Hypophosphatemia (hī-pō-fos′-fa-TĒ-mē-a) An abnormally low level of phosphates in the blood.

Hypophyseal fossa (hī′-pō-FIZ-ē-al FOS-a) A depression on the superior surface of the sphenoid bone that houses the pituitary gland.

Hypophyseal (hī′-pō-FIZ-ē-al) **pouch** An outgrowth of ectoderm from the roof of the mouth from which the anterior pituitary develops.

Hypophysis (hī-POF-i-sis) Pituitary gland.

Hyposecretion (hī′-pō-se-KRĒ-shun) Underactivity of glands resulting in diminished secretion.

Hypothalamohypophyseal (hī-pō-tha-lam′-ō-hī-pō-FIZ-ē-al) **tract** A bundle of axons containing secretory vesicles filled with oxytocin or antidiuretic hormone that extend from the hypothalamus to the posterior pituitary.

Hypothalamus (hī′-pō-THAL-a-mus) A portion of the diencephalon, lying beneath the thalamus and forming the floor and part of the wall of the third ventricle.

Hypothermia (hī′-pō-THER-mē-a) Lowering of body temperature below 35°C (95°F); in surgical procedures, it refers to deliberate cooling of the body to slow down metabolism and reduce oxygen needs of tissues.

Hypotonia (hī′-pō-TŌ-nē-a) Decreased or lost muscle tone in which muscles appear flaccid.

Hypotonic (hī′-pō-TON-ik) Solution that causes cells to swell and perhaps rupture due to gain of water by osmosis.

Hypoventilation (hī-pō-ven-ti-LĀ-shun) A rate of respiration lower than that required to maintain a normal partial pressure of carbon dioxide in plasma.

Hypovolemic (hī-pō-vō-LĒ-mik) **shock** A type of shock characterized by decreased blood volume; may be caused by acute hemorrhage or excessive loss of other body fluids, for example, by vomiting, diarrhea, or excessive sweating.

Hypoxia (hī-POKS-ē-a) Lack of adequate oxygen at the tissue level.

Hysterectomy (hiss-te-REK-tō-mē) The surgical removal of the uterus.

I

Ileocecal (il-ē-ō-SĒ-kal) **sphincter** A fold of mucous membrane that guards the opening from the ileum into the large intestine. Also called the **ileocecal valve.**

Ileum (IL-ē-um) The terminal part of the small intestine.

Immunity (im-Ū-ni-tē) The state of being resistant to injury, particularly by poisons, foreign proteins, and invading pathogens.

Immunogenicity (im′-ū-nō-je-NIS-i-tē) Ability of an antigen to provoke an immune response.

Immunoglobulin (im-ū-nō-GLOB-ū-lin) **(Ig)** An antibody synthesized by plasma cells derived from B lymphocytes in response to the introduction of an antigen. Immunoglobulins are divided into five kinds (IgG, IgM, IgA, IgD, IgE).

Immunology (im′-ū-NOL-ō-jē) The study of the responses of the body when challenged by antigens.

Implantation (im-plan-TĀ-shun) The insertion of a tissue or a part into the body. The attachment of the blastocyst to the stratum basalis of the endometrium about 6 days after fertilization.

Incontinence (in-KON-ti-nens) Inability to retain urine, semen, or feces through loss of sphincter control.

Indirect motor pathways Motor tracts that convey information from the brain down the spinal cord for automatic movements, coordination of body movements with visual stimuli, skeletal muscle tone and posture, and balance. Also known as **extrapyramidal pathways.**

Infarction (in-FARK-shun) A localized area of necrotic tissue, produced by inadequate oxygenation of the tissue.

Infection (in-FEK-shun) Invasion and multiplication of microorganisms in body tissues, which may be inapparent or characterized by cellular injury.

Inferior (in-FĒR-ē-or) Away from the head or toward the lower part of a structure. Also called **caudad** (KAW-dad).

Inferior vena cava (VĒ-na CĀ-va) **(IVC)** Large vein that collects blood from parts of the body inferior to the heart and returns it to the right atrium.

Infertility Inability to conceive or to cause conception. Also called **sterility.**

Inflammation (in′-fla-MĀ-shun) Localized, protective response to tissue injury designed to destroy, dilute, or wall off the infecting agent or injured tissue; characterized by redness, pain, heat, swelling, and sometimes loss of function.

Inflation reflex Reflex that prevents overinflation of the lungs. Also called **Hering–Breuer reflex.**

Infundibulum (in′-fun-DIB-ū-lum) The stalklike structure that attaches the pituitary gland to the hypothalamus of the brain. The funnel-shaped, open, distal end of the uterine (Fallopian) tube.

Ingestion (in-JES-chun) The taking in of food, liquids, or drugs, by mouth.

Inguinal (IN-gwi-nal) Pertaining to the groin.

Inguinal canal An oblique passageway in the anterior abdominal wall just superior and parallel to the medial half of the inguinal ligament that transmits the spermatic cord and ilioinguinal nerve in the male and round ligament of the uterus and ilioinguinal nerve in the female.

Inhalation (in-ha-LĀ-shun) The act of drawing air into the lungs. Also termed **inspiration.**

Inheritance The acquisition of body traits by transmission of genetic information from parents to offspring.

Inhibin A hormone secreted by the gonads that inhibits release of follicle-stimulating hormone (FSH) by the anterior pituitary.

Inhibiting hormone Hormone secreted by the hypothalamus that can suppress secretion of hormones by the anterior pituitary.

Inhibitory postsynaptic potential (IPSP) A small hyperpolarization in a neuron; caused by an inhibitory neurotransmitter at a synapse.

Inner cell mass A region of cells of a blastocyst that differentiates into the three primary germ layers—ectoderm, mesoderm, and endoderm—from which all tissues and organs develop; also called an **embryoblast.**

Inorganic (in′-or-GAN-ik) **compound** Compound that usually lacks carbon, usually is small, and often contains ionic bonds. Examples include water and many acids, bases, and salts.

Insertion (in-SER-shun) The attachment of a muscle tendon to a movable bone or the end opposite the origin.

Inspiratory (in-SPĪ-ra-tor-ē) **capacity** Total inspiratory capacity of the lungs; the total of tidal volume plus inspiratory reserve volume; averages 3600 mL.

Inspiratory (in-SPĪ-ra-tor-ē) **reserve volume** Additional inspired air over and above tidal volume; averages 3100 mL.

Insula (IN-soo-la) A triangular area of the cerebral cortex that lies deep within the lateral cerebral fissue, under the parietal, frontal, and temporal lobes.

Insulin (IN-soo-lin) A hormone produced by the beta cells of a pancreatic islet (islet of Langerhans) that decreases the blood glucose level.

Insulinlike growth factor (IGF) Small protein, produced by the liver and other tissues in response to stimulation by human growth hormone (hGH), that mediates most of the effects of human growth hormone. Previously called **somatomedin** (sō′-ma-tō-MĒ-din).

Integrins (IN-te-grinz) A family of transmembrane glycoproteins in plasma membranes that function in cell adhesion; they are present in hemidesmosomes, which anchor cells to a basement membrane, and they mediate adhesion of neutrophils to endothelial cells during emigration.

Integumentary (in-teg′-ū-MEN-tar-e) Relating to the skin.

Intercalated (in-TER-ka-lāt-ed) **disc** An irregular transverse thickening of sarcolemma that contains desmosomes, which hold cardiac muscle fibers (cells) together, and gap junctions, which aid in conduction of muscle action potentials from one fiber to the next.

Intercostal (in′-ter-KOS-tal) **nerve** A nerve supplying a muscle located between the ribs.

Interferons (in′-ter-FĒR-ons) **(IFNs)** Antiviral proteins produced by virus-infected host cells; induce uninfected host cells to synthesize proteins that inhibit viral replication and enhance phagocytic activity of macrophages; types include alpha interferon, beta interferon, and gamma interferon.

Intermediate Between two structures, one of which is medial and one of which is lateral.

Intermediate filament Protein filament, ranging from 8 to 12 nm in diameter, that may provide structural reinforcement, hold organelles in place, and give shape to a cell.

Internal Away from the surface of the body.

Internal capsule A large tract of projection fibers lateral to the thalamus that is the major connection between the cerebral cortex and the brain stem and spinal cord; contains axons of sensory neurons carrying auditory, visual, and somatic sensory signals to the cerebral cortex plus axons of motor neurons descending from the cerebral cortex to the thalamus, subthalamus, brain stem, and spinal cord.

Internal ear The inner ear or labyrinth, lying inside the temporal bone, containing the organs of hearing and balance.

Internal nares (NĀ-rez) The two openings posterior to the nasal cavities opening into the nasopharynx. Also called the **choanae** (kō-Ā-nē).

Internal respiration The exchange of respiratory gases between blood and body cells. Also called **tissue respiration.**

Interneurons (in′-ter-NOO-ronz) Neurons whose axons extend only for a short distance and contact nearby neurons in the brain, spinal cord, or a ganglion; they comprise the vast majority of neurons in the body.

Interoceptor (in′-ter-ō-SEP-tor) Sensory receptor located in blood vessels and viscera that provides information about the body's internal environment.

Interphase (IN-ter-fāz) The period of the cell cycle between cell divisions, consisting of the G_1-(gap or growth) phase, when the cell is engaged in growth, metabolism, and production of substances required for division; S-(synthesis) phase, during which chromosomes are replicated; and G_2-phase.

Interstitial (in′-ter-STISH-al) **fluid** The portion of extracellular fluid that fills the microscopic spaces between the cells of tissues; the internal environment of the body. Also called **intercellular** or **tissue fluid.**

Interstitial growth Growth from within, as in the growth of cartilage. Also called **endogenous** (en-DOJ-e-nus) **growth.**

Interventricular (in′-ter-ven-TRIK-ū-lar) **foramen** A narrow, oval opening through which the lateral ventricles of the brain communicate with the third ventricle. Also called the **foramen of Monro.**

Intervertebral (in′-ter-VER-te-bral) **disc** A pad of fibrocartilage located between the bodies of two vertebrae.

Intestinal gland A gland that opens onto the surface of the intestinal mucosa and secretes digestive enzymes. Also called a **crypt of Lieberkühn** (LĒ-ber-kūn).

Intrafusal (in′-tra-FŪ-sal) **fibers** Three to ten specialized muscle fibers (cells), partially enclosed in a spindle-shaped connective tissue capsule, that make up a muscle spindle.

Intramembranous ossification (in′-tra-MEM-bra-nus os′-i′-fi-KĀ-shun) The method of bone formation in which the bone is formed directly in membranous tissue.

Intraocular (in′-tra-OK-ū-lar) **pressure (IOP)** Pressure in the eyeball, produced mainly by aqueous humor.

Intrapleural pressure Air pressure between the two pleurae of the lungs, usually subatmospheric. Also called **intrathoracic pressure.**

Intrinsic (in-TRIN-sik) Of internal origin.

Intrinsic pathway (of blood clotting) Sequence of reactions leading to blood clotting that is initiated by damage to blood vessel endothelium or platelets; activators of this pathway are contained within blood itself or are in direct contact with blood.

Intrinsic factor (IF) A glycoprotein, synthesized and secreted by the parietal cells of the gastric mucosa, that facilitates vitamin B_{12} absorption in the small intestine.

Intron (IN-tron) A region of DNA that does not code for the synthesis of a protein.

In utero (Ū-ter-ō) Within the uterus.

Invagination (in-vaj′-i-NĀ-shun) The pushing of the wall of a cavity into the cavity itself.

Inversion (in-VER-zhun) The movement of the sole medially at the ankle joint.

In vitro (VĒ-trō) Literally, in glass; outside the living body and in an artificial environment such as a laboratory test tube.

In vivo (VĒ-vō) In the living body.

Ion (Ī-on) Any charged particle or group of particles; usually formed when a substance, such as a salt, dissolves and dissociates.

Ionization (ī′-on-i-ZĀ-shun) Separation of inorganic acids, bases, and salts into ions when dissolved in water. Also called **dissociation.**

Ipsilateral (ip′-si-LAT-er-al) On the same side, affecting the same side of the body.

Iris The colored portion of the vascular tunic of the eyeball seen through the cornea that contains circular and radial smooth muscle; the hole in the center of the iris is the pupil.

Irritable bowel syndrome (IBS) Disease of the entire gastrointestinal tract in which a person reacts to stress by developing symptoms (such as cramping and abdominal pain) associated with alternating patterns of diarrhea and constipation. Excessive amounts of mucus may appear in feces, and other symptoms include flatulence, nausea, and loss of appetite. Also known as **irritable colon** or **spastic colitis.**

Ischemia (is-KĒ-mē-a) A lack of sufficient blood to a body part due to obstruction or constriction of a blood vessel.

Isoantibody A specific antibody in blood plasma that reacts with specific isoantigens and causes the clumping of bacteria, red blood cells, or particles. Also called an **agglutinin.**

Isoantigen A genetically determined antigen located on the surface of red blood cells; basis for the ABO grouping and Rh system of blood classification. Also called an **agglutinogen.**

Isometric contraction A muscle contraction in which tension on the muscle increases, but there is only minimal muscle shortening so that no visible movement is produced.

Isotonic (Ī′-sō-TON-ik) Having equal tension or tone. A solution having the same concentration of impermeable solutes as cytosol.

Isotonic (i′-so-TON-ik) **contraction** Contraction in which the tension remains the same; occurs when a constant load is moved through the range of motions possible at a joint.

Isotopes (Ī-sō-tōps′) Chemical elements that have the same number of protons but different numbers of neutrons. Radioactive isotopes change into other elements with the emission of alpha or beta particles or gamma rays.

Isovolumetric (ī-sō-vol-ū′-MET-rik) **contraction** The period of time, about 0.05 sec, between the start of ventricular systole and the opening of the semilunar valves; the ventricles contract and ventricular pressure rises rapidly, but ventricular volume does not change.

Isovolumetric relaxation The period of time, about 0.05 sec, between the closing of the semilunar valves and the opening of the atrioventricular (AV) valves; there is a drastic decrease in ventricular pressure without a change in ventricular volume.

Isthmus (IS-mus) A narrow strip of tissue or narrow passage connecting two larger parts.

J

Jaundice (JAWN-dis) A condition characterized by yellowness of the skin, the white of the eyes, mucous membranes, and body fluids because of a buildup of bilirubin.

Jejunum (je-JOO-num) The middle part of the small intestine.

Joint kinesthetic (kin′-es-THET-ik) **receptor** A proprioceptive receptor located in a joint, stimulated by joint movement.

Juxtaglomerular (juks-ta-glō-MER-ū-lar) **apparatus (JGA)** Consists of the macula densa (cells of the distal convoluted tubule adjacent to the afferent and efferent arteriole) and juxtaglomerular cells (modified cells of the afferent and sometimes efferent arteriole); secretes renin when blood pressure starts to fall.

K

Keratin (KER-a-tin) An insoluble protein found in the hair, nails, and other keratinized tissues of the epidermis.

Keratinocyte (ke-RAT-in′-ō-sīt) The most numerous of the epidermal cells; produces keratin.

Ketone (KĒ-tōn) **bodies** Substances produced primarily during excessive triglyceride catabolism, such as acetone, acetoacetic acid, and β-hydroxybutyric acid.

Ketosis (kē-TŌ-sis) Abnormal condition marked by excessive production of ketone bodies.

Kidney (KID-nē) One of the paired reddish organs located in the lumbar region that regulates the composition, volume, and pressure of blood and produces urine.

Kidney stone A solid mass, usually consisting of calcium oxalate, uric acid, or calcium phosphate crystals, that may form in any portion of the urinary tract. Also called a **renal calculus.**

Kinesiology (ki-nē′-sē-OL-ō-jē) The study of the movement of body parts.

Kinesthesia (kin-es-THĒ-zē-a) The perception of the extent and direction of movement of body parts; this sense is possible due to nerve impulses generated by proprioceptors.

Korotkoff (kō-ROT-kof) **sounds** The various sounds that are heard while taking blood pressure.

Krebs cycle A series of biochemical reactions that occurs in the matrix of mitochondria in which electrons are transferred to coenzymes and carbon dioxide is formed. The electrons carried by the coenzymes then enter the electron transport chain, which generates a large quantity of ATP. Also called the **citric acid cycle** or **tricarboxylic acid (TCA) cycle.**

Kyphosis (kī-FŌ-sis) An exaggeration of the thoracic curve of the vertebral column, resulting in a "round-shouldered" appearance. Also called **hunchback.**

L

Labia majora (LĀ-bē-a ma-JŌ-ra) Two longitudinal folds of skin extending downward and backward from the mons pubis of the female.

Labia minora (min-OR-a) Two small folds of mucous membrane lying medial to the labia majora of the female.

Labial frenulum (LĀ-bē-al FREN-ū-lum) A medial fold of mucous membrane between the inner surface of the lip and the gums.

Labium (LĀ-bē-um) A lip. A liplike structure. *Plural is* **labia** (LĀ-bē-a).

Labor The process of giving birth in which a fetus is expelled from the uterus through the vagina.

Lacrimal canal A duct, one on each eyelid, beginning at the punctum at the medial margin of an eyelid and conveying tears medially into the nasolacrimal sac.

Lacrimal gland Secretory cells, located at the superior anterolateral portion of each orbit, that secrete tears into excretory ducts that open onto the surface of the conjunctiva.

Lacrimal sac The superior expanded portion of the nasolacrimal duct that receives the tears from a lacrimal canal.

Lactation (lak-TĀ-shun) The secretion and ejection of milk by the mammary glands.

Lacteal (LAK-tē-al) One of many lymphatic vessels in villi of the intestines that absorb triglycerides and other lipids from digested food.

Lacuna (la-KOO-na) A small, hollow space, such as that found in bones in which the osteocytes lie. *Plural is* **lacunae** (la-KOO-nē).

Lambdoid (lam-DOYD) **suture** The joint in the skull between the parietal bones and the occipital bone; sometimes contains sutural (Wormian) bones.

Lamellae (la-MEL-ē) Concentric rings of hard, calcified matrix found in compact bone.

Lamellated corpuscle Oval-shaped pressure receptor located in the dermis or subcutaneous tissue and consisting of concentric layers of connective tissue wrapped around the dendrites of a sensory neuron. Also called a **Pacinian** (pa-SIN-ē-an) **corpuscle.**

Lamina (LAM-i-na) A thin, flat layer or membrane, as the flattened part of either side of the arch of a vertebra. *Plural is* **laminae** (LAM-i-nē).

Lamina propria (PRŌ-prē-a) The connective tissue layer of a mucosa.

Langerhans (LANG-er-hans) **cell** Epidermal dendritic cell that functions as an antigen-presenting cell (APC) during an immune response.

Lanugo (la-NOO-gō) Fine downy hairs that cover the fetus.

Large intestine The portion of the gastrointestinal tract extending from the ileum of the small intestine to the anus, divided structurally into the cecum, colon, rectum, and anal canal.

Laryngopharynx (la-rin′-gō-FAR-inks) The inferior portion of the pharynx, extending downward from the level of the hyoid bone that divides posteriorly into the esophagus and anteriorly into the larynx. Also called the **hypopharynx.**

Laryngotracheal (la-rin′-gō-TRĀ-ke-al) **bud** An outgrowth of endoderm of the foregut from which the respiratory system develops.

Larynx (LAR-inks) The voice box, a short passageway that connects the pharynx with the trachea.

Lateral (LAT-er-al) Farther from the midline of the body or a structure.

Lateral ventricle (VEN-tri-kul) A cavity within a cerebral hemisphere that communicates with the lateral ventricle in the other cerebral hemisphere and with the third ventricle by way of the interventricular foramen.

Leg The part of the lower limb between the knee and the ankle.

Lens A transparent organ constructed of proteins (crystallins) lying posterior to the pupil and iris of the eyeball and anterior to the vitreous body.

Lesion (LĒ-zhun) Any localized, abnormal change in a body tissue.

Lesser omentum (ō-MEN-tum) A fold of the peritoneum that extends from the liver to the lesser curvature of the stomach and the first part of the duodenum.

Lesser vestibular (ves-TIB-ū-lar) **gland** One of the paired mucus-secreting glands with ducts that open on either side of the urethral orifice in the vestibule of the female.

Leukemia (loo-KĒ-mē-a) A malignant disease of the blood-forming tissues characterized by either uncontrolled production and accumulation of immature leukocytes in which many cells fail to reach maturity (acute) or an accumulation of mature leukocytes in the blood because they do not die at the end of their normal life span (chronic).

Leukocyte (LOO-kō-sīt) A white blood cell.

Leukocytosis (loo′-kō-sī-TŌ-sis) An increase in the number of white blood cells, above 10,000 per μL, characteristic of many infections and other disorders.

Leukopenia (loo-kō-PĒ-nē-a) A decrease in the number of white blood cells below 5000 cells per μL.

Leukotriene (loo′-kō-TRĪ-ēn) A type of eicosanoid produced by basophils and mast cells; acts as a local hormone; produces increased vascular permeability and acts as a chemotactic agent for phagocytes in tissue inflammation.

Leydig (LĪ-dig) **cell** A type of cell that secretes testosterone; located in the connective tissue between seminiferous tubules in a mature testis. Also known as **interstitial cell of Leydig** or **interstitial endocrinocyte**.

Libido (li-BĒ-dō) Sexual desire.

Ligament (LIG-a-ment) Dense regular connective tissue that attaches bone to bone.

Ligand (LĪ-gand) A chemical substance that binds to a specific receptor.

Limbic system A part of the forebrain, sometimes termed the visceral brain, concerned with various aspects of emotion and behavior; includes the limbic lobe, dentate gyrus, amygdala, septal nuclei, mammillary bodies, anterior thalamic nucleus, olfactory bulbs, and bundles of myelinated axons.

Lingual frenulum (LIN-gwal FREN-ū-lum) A fold of mucous membrane that connects the tongue to the floor of the mouth.

Lipase An enzyme that splits fatty acids from triglycerides and phospholipids.

Lipid (LIP-id) An organic compound composed of carbon, hydrogen, and oxygen that is usually insoluble in water, but soluble in alcohol, ether, and chloroform; examples include triglycerides (fats and oils), phospholipids, steroids, and eicosanoids.

Lipid bilayer Arrangement of phospholipid, glycolipid, and cholesterol molecules in two parallel sheets in which the hydrophilic "heads" face outward and the hydrophobic "tails" face inward; found in cellular membranes.

Lipogenesis (li-pō-GEN-e-sis) The synthesis of triglycerides.

Lipolysis (lip-OL-i-sis) The splitting of fatty acids from a triglyceride or phospholipid.

Lipoprotein (lip′-ō-PRŌ-tēn) One of several types of particles containing lipids (cholesterol and triglycerides) and proteins that make it water soluble for transport in the blood; high levels of **low-density lipoproteins (LDLs)** are associated with increased risk of atherosclerosis, whereas high levels of **high-density lipoproteins (HDLs)** are associated with decreased risk of atherosclerosis.

Liver Large organ under the diaphragm that occupies most of the right hypochondriac region and part of the epigastric region. Functionally, it produces bile and synthesizes most plasma proteins; interconverts nutrients; detoxifies substances; stores glycogen, iron, and vitamins; carries on phagocytosis of worn-out blood cells and bacteria; and helps synthesize the active form of vitamin D.

Long-term potentiation (LTP) Prolonged, enhanced synaptic transmission that occurs at certain synapses within the hippocampus of the brain; believed to underlie some aspects of memory.

Lordosis (lor-DŌ-sis) An exaggeration of the lumbar curve of the vertebral column. Also called **swayback.**

Lower limb The appendage attached at the pelvic (hip) girdle, consisting of the thigh, knee, leg, ankle, foot, and toes. Also called **lower extremity.**

Lumbar (LUM-bar) Region of the back and side between the ribs and pelvis; loin.

Lumbar plexus (PLEK-sus) A network formed by the anterior (ventral) branches of spinal nerves L1 through L4.

Lumen (LOO-men) The space within an artery, vein, intestine, renal tubule, or other tubular structure.

Lungs Main organs of respiration that lie on either side of the heart in the thoracic cavity.

Lunula (LOO-noo-la) The moon-shaped white area at the base of a nail.

Luteinizing (LOO-tē-in′-īz-ing) **hormone (LH)** A hormone secreted by the anterior pituitary that stimulates ovulation, stimulates progesterone secretion by the corpus luteum, and readies the mammary glands for milk secretion in females; stimulates testosterone secretion by the testes in males.

Lymph (LIMF) Fluid confined in lymphatic vessels and flowing through the lymphatic system until it is returned to the blood.

Lymph node An oval or bean-shaped structure located along lymphatic vessels.

Lymphatic (lim-FAT-ik) **capillary** Closed-ended microscopic lymphatic vessel that begins in spaces between cells and converges with other lymphatic capillaries to form lymphatic vessels.

Lymphatic tissue A specialized form of reticular tissue that contains large numbers of lymphocytes.

Lymphatic vessel A large vessel that collects lymph from lymphatic capillaries and converges with other lymphatic vessels to form the thoracic and right lymphatic ducts.

Lymphocyte (LIM-fō-sīt) A type of white blood cell that helps carry out cell-mediated and antibody-mediated immune responses; found in blood and in lymphatic tissues.

Lysosome (LĪ-sō-sōm) An organelle in the cytoplasm of a cell, enclosed by a single membrane and containing powerful digestive enzymes.

Lysozyme (LĪ-sō-zīm) A bactericidal enzyme found in tears, saliva, and perspiration.

M

Macrophage (MAK-rō-fāj) Phagocytic cell derived from a monocyte; may be fixed or wandering.

Macula (MAK-ū-la) A discolored spot or a colored area. A small, thickened region on the wall of the utricle and saccule that contains receptors for static equilibrium.

Macula lutea (MAK-ū-la LOO-tē-a) The yellow spot in the center of the retina.

Major histocompatibility (MHC) antigens Surface proteins on white blood cells and other nucleated cells that are unique for each person (except for identical siblings); used to type tissues and help prevent rejection of transplanted tissues. Also known as **human leukocyte antigens (HLA).**

Malignant (ma-LIG-nant) Referring to diseases that tend to become worse and cause death, especially the invasion and spreading of cancer.

Mammary (MAM-ar-ē) **gland** Modified sudoriferous (sweat) gland of the female that produces milk for the nourishment of the young.

Mammillary (MAM-i-ler-ē) **bodies** Two small rounded bodies on the inferior aspect of the hypothalamus that are involved in reflexes related to the sense of smell.

Marrow (MAR-ō) Soft, spongelike material in the cavities of bone. Red bone marrow produces blood cells; yellow bone marrow contains adipose tissue that stores triglycerides.

Mast cell A cell found in areolar connective tissue that releases histamine, a dilator of small blood vessels, during inflammation.

Mastication (mas′-ti-KĀ-shun) Chewing.

Matrix (MĀ-triks) The ground substance and fibers between cells in a connective tissue.

Matter Anything that occupies space and has mass.

Mature follicle A large, fluid-filled follicle containing a secondary oocyte and surrounding granulosa cells that secrete estrogens. Also called a **Graafian** (GRAF-ē-an) **follicle.**

Maximal oxygen uptake Maximum rate of oxygen consumption during aerobic catabolism of pyruvic acid that is determined by age, sex, body size, and aerobic training.

Mean arterial blood pressure (MABP) The average force of blood pressure exerted against the walls of arteries; approximately equal to diastolic pressure plus one-third of pulse pressure; for example, 93 mmHg when systolic pressure is 120 mmHg and diastolic pressure is 80 mmHg.

Meatus (mē-Ā-tus) A passage or opening, especially the external portion of a canal.

Mechanoreceptor (me-KAN-ō-rē-sep-tor) Sensory receptor that detects mechanical deformation of the receptor itself or adjacent cells; stimuli so detected include those related to touch, pressure, vibration, proprioception, hearing, equilibrium, and blood pressure.

Medial (MĒ-dē-al) Nearer the midline of the body or a structure.

Medial lemniscus (lem-NIS-kus) A white matter tract that originates in the gracile and cuneate nuclei of the medulla oblongata and extends to the thalamus on the same side; sensory axons in this tract conduct nerve impulses for the sensations of proprioception, fine touch, vibration, hearing, and equilibrium.

Median aperture (AP-er-choor) One of the three openings in the roof of the fourth ventricle through which cerebrospinal fluid enters the subarachnoid space of the brain and cord. Also called the **foramen of Magendie.**

Median plane A vertical plane dividing the body into right and left halves. Situated in the middle.

Mediastinum (mē′-dē-as-TĪ-num) The broad, median partition between the pleurae of the lungs, that extends from the sternum to the vertebral column in the thoracic cavity.

Medulla (me-DUL-la) An inner layer of an organ, such as the medulla of the kidneys.

Medulla oblongata (me-DUL-la ob′-long-GA-ta) The most inferior part of the brain stem. Also termed the **medulla.**

Medullary (MED-ū-lar′-ē) **cavity** The space within the diaphysis of a bone that contains yellow bone marrow. Also called the **marrow cavity.**

Medullary rhythmicity (rith-MIS-i-tē) **area** The neurons of the respiratory center in the medulla oblongata that control the basic rhythm of respiration.

Meiosis (mē-Ō-sis) A type of cell division that occurs during production of gametes, involving two successive nuclear divisions that result in daughter cells with the haploid *(n)* number of chromosomes.

Melanin (MEL-a-nin) A dark black, brown, or yellow pigment found in some parts of the body such as the skin, hair, and pigmented layer of the retina.

Melanocyte (MEL-a-nō-sīt′) A pigmented cell, located between or beneath cells of the deepest layer of the epidermis, that synthesizes melanin.

Melanocyte-stimulating hormone (MSH) A hormone secreted by the anterior pituitary that stimulates the dispersion of melanin granules in melanocytes in amphibians; continued administration produces darkening of skin in humans.

Melatonin (mel-a-TŌN-in) A hormone secreted by the pineal gland that helps set the timing of the body's biological clock.

Membrane A thin, flexible sheet of tissue composed of an epithelial layer and an underlying connective tissue layer, as in an epithelial membrane, or of areolar connective tissue only, as in a synovial membrane.

Membranous labyrinth (mem-BRA-nus LAB-i-rinth) The part of the labyrinth of the internal ear that is located inside the bony labyrinth and separated from it by the perilymph; made up of the semicircular ducts, the saccule and utricle, and the cochlear duct.

Menarche (me-NAR-kē) The first menses (menstrual flow) and beginning of ovarian and uterine cycles.

Meninges (me-NIN-jēz) Three membranes covering the brain and spinal cord, called the dura mater, arachnoid mater, and pia mater. *Singular is* **meninx** (MEN-inks).

Menopause (MEN-ō-pawz) The termination of the menstrual cycles.

Menstruation (men′-stroo-Ā-shun) Periodic discharge of blood, tissue fluid, mucus, and epithelial cells that usually lasts for 5 days; caused by a sudden reduction in estrogens and progesterone. Also called the **menstrual phase** or **menses.**

Merkel (MER-kel) **cell** Type of cell in the epidermis of hairless skin that makes contact with a tactile (Merkel) disc, which functions in touch.

Merocrine (MER-ō-krin) **gland** Gland made up of secretory cells that remain intact throughout the process of formation and discharge of the secretory product, as in the salivary and pancreatic glands.

Mesenchyme (MEZ-en-kīm) An embryonic connective tissue from which all other connective tissues arise.

Mesentery (MEZ-en-ter′-ē) A fold of peritoneum attaching the small intestine to the posterior abdominal wall.

Mesocolon (mez′-ō-KŌ-lon) A fold of peritoneum attaching the colon to the posterior abdominal wall.

Mesoderm The middle primary germ layer that gives rise to connective tissues, blood and blood vessels, and muscles.

Mesothelium (mez′-ō-THĒ-lē-um) The layer of simple squamous epithelium that lines serous membranes.

Mesovarium (mez′-ō-VAR-ē-um) A short fold of peritoneum that attaches an ovary to the broad ligament of the uterus.

Metabolism (me-TAB-ō-lizm) All the biochemical reactions that occur within an organism, including the synthetic (anabolic) reactions and decomposition (catabolic) reactions.

Metacarpus (met′-a-KAR-pus) A collective term for the five bones that make up the palm.

Metaphase (MET-a-phāz) The second stage of mitosis, in which chromatid pairs line up on the metaphase plate of the cell.

Metaphysis (me-TAF-i-sis) Region of a long bone between the diaphysis and epiphysis that contains the epiphyseal plate in a growing bone.

Metarteriole (met′-ar-TĒ-rē-ōl) A blood vessel that emerges from an arteriole, traverses a capillary network, and empties into a venule.

Metastasis (me-TAS-ta-sis) The spread of cancer to surrounding tissues (local) or to other body sites (distant).

Metatarsus (met′-a-TAR-sus) A collective term for the five bones located in the foot between the tarsals and the phalanges.

Micelle (mī-SEL) A spherical aggregate of bile salts that dissolves fatty acids and monoglycerides so that they can be absorbed into small intestinal epithelial cells.

Microfilament (mī-krō-FIL-a-ment) Rodlike protein filament about 6 nm in diameter; constitutes contractile units in muscle fibers (cells) and provides support, shape, and movement in nonmuscle cells.

Microglia (mī-krō-GLĒ-a) Neuroglial cells that carry on phagocytosis.

Microtubule (mī-krō-TOO-būl′) Cylindrical protein filament, ranging in diameter from 18 to 30 nm, consisting of the protein tubulin; provides support, structure, and transportation.

Microvilli (mī′-krō-VIL-ē) Microscopic, fingerlike projections of the plasma membranes of cells that increase surface area for absorption, especially in the small intestine and proximal convoluted tubules of the kidneys.

Micturition (mik′-choo-RISH-un) The act of expelling urine from the urinary bladder. Also called **urination** (yoo-ri-NĀ-shun).

Midbrain The part of the brain between the pons and the diencephalon. Also called the **mesencephalon** (mes′-en-SEF-a-lon).

Middle ear A small, epithelial-lined cavity hollowed out of the temporal bone, separated from the external ear by the eardrum and from the internal ear by a thin bony partition containing the oval and round windows; extending across the middle ear are the three auditory ossicles. Also called the **tympanic** (tim-PAN-ik) **cavity.**

Midline An imaginary vertical line that divides the body into equal left and right sides.

Midsagittal plane A vertical plane through the midline of the body that divides the body or organs into *equal* right and left sides. Also called a **median plane.**

Milk ejection reflex Contraction of alveolar cells to force milk into ducts of mammary glands, stimulated by oxytocin (TO), which is released from the posterior pituitary in response to suckling action. Also called the **milk letdown reflex.**

Mineral Inorganic, homogeneous solid substance that may perform a function vital to life; examples include calcium and phosphorus.

Mineralocorticoids (min′-er-al-ō-KOR-ti-koyds) A group of hormones of the adrenal cortex that help regulate sodium and potassium balance.

Minimal volume The volume of air in the lungs even after the thoracic cavity has been opened, forcing out some of the residual volume.

Minute ventilation (MV) Total volume of air inhaled and exhaled per minute; about 6000 mL.

Miosis Constriction of the pupil of the eye.

Mitochondrion (mī′-tō-KON-drē-on) A double-membraned organelle that plays a central role in the production of ATP; known as the "powerhouse" of the cell.

Mitosis (mī-TŌ-sis) The orderly division of the nucleus of a cell that ensures that each new daughter nucleus has the same number and kind of chromosomes as the original parent nucleus. The process includes the replication of chromosomes and the distribution of the two sets of chromosomes into two separate and equal nuclei.

Mitotic spindle Collective term for a football-shaped assembly of microtubules (nonkinetochore, kinetochore, and aster) that is responsible for the movement of chromosomes during cell division.

Modality (mō-DAL-i-tē) Any of the specific sensory entities, such as vision, smell, taste, or touch.

Modiolus (mō-DĪ-ō′-lus) The central pillar or column of the cochlea.

Mole The weight, in grams, of the combined atomic weights of the atoms that make up a molecule of a substance.

Molecule (MOL-e-kūl) The chemical combination of two or more atoms covalently bonded together.

Monocyte (MON-ō-sīt′) The largest type of white blood cell, characterized by agranular cytoplasm.

Monounsaturated fat A fatty acid that contains one double covalent bond between its carbon atoms; it is not completely saturated with hydrogen atoms. Plentiful in triglycerides of olive and peanut oils.

Mons pubis (monz PŪ-bis) The rounded, fatty prominence over the pubic symphysis, covered by coarse pubic hair.

Morula (MOR-ū-la) A solid sphere of cells produced by successive cleavages of a fertilized ovum about four days after fertilization.

Motor end plate Region of the sarcolemma of a muscle fiber (cell) that includes acetylcholine (ACh) receptors, which bind ACh released by synaptic end bulbs of somatic motor neurons.

Motor neurons (NOO-ronz) Neurons that conduct impulses from the brain toward the spinal cord or out of the brain and spinal cord into cranial or spinal nerves to effectors that may be either muscles or glands. Also called **efferent neurons.**

Motor unit A motor neuron together with the muscle fibers (cells) it stimulates.

Mucin (MŪ-sin) A protein found in mucus.

Mucosa-associated lymphatic tissue (MALT) Lymphatic nodules scattered throughout the lamina propria (connective tissue) of mucous membranes lining the gastrointestinal tract, respiratory airways, urinary tract, and reproductive tract.

Mucous (MŪ-kus) **cell** A unicellular gland that secretes mucus. Two types are mucous neck cells and surface mucous cells in the stomach.

Mucous membrane A membrane that lines a body cavity that opens to the exterior. Also called the **mucosa** (mū-KŌ-sa).

Mucus The thick fluid secretion of goblet cells, mucous cells, mucous glands, and mucous membranes.

Muscarinic (mus′-ka-RIN-ik) **receptor** Receptor for the neurotransmitter acetylcholine found on all effectors innervated by parasympathetic postganglionic axons and on sweat glands innervated by cholinergic sympathetic postganglionic axons; so named because muscarine activates these receptors but does not activate nicotinic receptors for acetylcholine.

Muscle An organ composed of one of three types of muscle tissue (skeletal, cardiac, or smooth), specialized for contraction to produce voluntary or involuntary movement of parts of the body.

Muscle action potential A stimulating impulse that propagates along the sarcolemma and transverse tubules; in skeletal muscle, it is generated by acetylcholine, which increases the permeability of the sarcolemma to cations, especially sodium ions (Na^+).

Muscle fatigue (fa-TĒG) Inability of a muscle to maintain its strength of contraction or tension; may be related to insufficient oxygen, depletion of glycogen, and/or lactic acid buildup.

Muscle spindle An encapsulated proprioceptor in a skeletal muscle, consisting of specialized intrafusal muscle fibers and nerve endings; stimulated by changes in length or tension of muscle fibers.

Muscle tissue A tissue specialized to produce motion in response to muscle action potentials by its qualities of contractility, extensibility, elasticity, and excitability; types include skeletal, cardiac, and smooth.

Muscle tone A sustained, partial contraction of portions of a skeletal or smooth muscle in response to activation of stretch receptors or a baseline level of action potentials in the innervating motor neurons.

Muscular dystrophies (DIS-trō-fēz′) Inherited muscle-destroying diseases, characterized by degeneration of muscle fibers (cells), which causes progressive atrophy of the skeletal muscle.

Muscularis (MUS-kū-la′-ris) A muscular layer (coat or tunic) of an organ.

Muscularis mucosae (mū-KŌ-sē) A thin layer of smooth muscle fibers that underlie the lamina propria of the mucosa of the gastrointestinal tract.

Mutation (mū-TĀ-shun) Any change in the sequence of bases in a DNA molecule resulting in a permanent alteration in some inheritable trait.

Myasthenia (mī-as-THĒ-nē-a) **gravis** Weakness and fatigue of skeletal muscles caused by antibodies directed against acetylcholine receptors.

Myelin (MĪ-e-lin) **sheath** Multilayered lipid and protein covering, formed by Schwann cells and oligodendrocytes, around axons of many peripheral and central nervous system neurons.

Myenteric plexus A network of autonomic axons and postganglionic cell bodies located in the muscularis of the gastrointestinal tract. Also called the **plexus of Auerbach** (OW-er-bak).

Myocardial infarction (mī′-ō-KAR-dē-al in-FARK-shun) **(MI)** Gross necrosis of myocardial tissue due to interrupted blood supply. Also called a **heart attack.**

Myocardium (mī′-ō-KAR-dē-um) The middle layer of the heart wall, made up of cardiac muscle tissue, lying between the epicardium and the endocardium and constituting the bulk of the heart.

Myofibril (mī-ō-FĪ-bril) A threadlike structure, extending longitudinally through a muscle fiber (cell) consisting mainly of thick filaments (myosin) and thin filaments (actin, troponin, and tropomyosin).

Myoglobin (mī-ō-GLŌ-bin) The oxygen-binding, iron-containing protein present in the sarcoplasm of muscle fibers (cells); contributes the red color to muscle.

Myogram (MĪ-ō-gram) The record or tracing produced by a myograph, an apparatus that measures and records the force of muscular contractions.

Myology (mī-OL-ō-jē) The study of muscles.

Myometrium (mī′-ō-MĒ-trē-um) The smooth muscle layer of the uterus.

Myopathy (mī-OP-a-thē) Any abnormal condition or disease of muscle tissue.

Myopia (mī-Ō-pē-a) Defect in vision in which objects can be seen distinctly only when very close to the eyes; nearsightedness.

Myosin (MĪ-ō-sin) The contractile protein that makes up the thick filaments of muscle fibers.

Myotome (MĪ-ō-tōm) A group of muscles innervated by the motor neurons of a single spinal segment. In an embryo, the portion of a somite that develops into some skeletal muscles.

N

Nail A hard plate, composed largely of keratin, that develops from the epidermis of the skin to form a protective covering on the dorsal surface of the distal phalanges of the fingers and toes.

Nail matrix (MĀ-triks) The part of the nail beneath the body and root from which the nail is produced.

Nasal (NĀ-zal) **cavity** A mucosa-lined cavity on either side of the nasal septum that opens onto the face at the external nares and into the nasopharynx at the internal nares.

Nasal septum (SEP-tum) A vertical partition composed of bone (perpendicular plate of ethmoid and vomer) and cartilage, covered with a mucous membrane, separating the nasal cavity into left and right sides.

Nasolacrimal (nā′-zō-LAK-ri-mal) **duct** A canal that transports the lacrimal secretion (tears) from the nasolacrimal sac into the nose.

Nasopharynx (nā′-zō-FAR-inks) The superior portion of the pharynx, lying posterior to the nose and extending inferiorly to the soft palate.

Neck The part of the body connecting the head and the trunk. A constricted portion of an organ such as the neck of the femur or uterus.

Necrosis (ne-KRŌ-sis) A pathological type of cell death that results from disease, injury, or lack of blood supply in which many adjacent cells swell, burst, and spill their contents into the interstitial fluid, triggering an inflammatory response.

Negative feedback The principle governing most control systems; a mechanism of response in which a stimulus initiates actions that reverse or reduce the stimulus.

Neonatal (nē-ō-NĀ-tal) Pertaining to the first four weeks after birth.

Neoplasm (NĒ-ō-plazm) A new growth that may be benign or malignant.

Nephron (NEF-ron) The functional unit of the kidney.

Nerve A cordlike bundle of neuronal axons and/or dendrites and associated connective tissue coursing together outside the central nervous system.

Nerve fiber General term for any process (axon or dendrite) projecting from the cell body of a neuron.

Nerve impulse A wave of depolarization and repolarization that self-propagates along the plasma membrane of a neuron; also called a **nerve action potential.**

Nervous tissue Tissue containing neurons that initiate and conduct nerve impulses to coordinate homeostasis, and neuroglia that provide support and nourishment to neurons.

Net filtration pressure (NFP) Net pressure that promotes fluid outflow at the arterial end of a capillary, and fluid inflow at the venous end of a capillary; net pressure that promotes glomerular filtration in the kidneys.

Neural plate A thickening of ectoderm, induced by the notochord, that forms early in the third week of development and represents the beginning of the development of the nervous system.

Neuralgia (noo-RAL-jē-a) Attacks of pain along the entire course or branch of a peripheral sensory nerve.

Neuritis (noo-RĪ-tis) Inflammation of one or more nerves.

Neuroglia (noo-RŌG-lē-a) Cells of the nervous system that perform various supportive functions. The neuroglia of the central nervous system are the astrocytes, oligodendrocytes, microglia, and ependymal cells; neuroglia of the peripheral nervous system include Schwann cells and satellite cells. Also called **glial** (GLĒ-al) **cells.**

Neurohypophyseal (noo′-rō-hī′-pō-FIZ-ē-al) **bud** An outgrowth of ectoderm located on the floor of the hypothalamus that gives rise to the posterior pituitary.

Neurolemma (noo-rō-LEM-ma) The peripheral, nucleated cytoplasmic layer of the Schwann cell. Also called **sheath of Schwann** (SCHVON).

Neurology (noo-ROL-ō-jē) The study of the normal functioning and disorders of the nervous system.

Neuromuscular (noo-rō-MUS-kū-lar) **junction** A synapse between the axon terminals of a motor neuron and the sarcolemma of a muscle fiber (cell).

Neuron (NOO-ron) A nerve cell, consisting of a cell body, dendrites, and an axon.

Neuropeptide (noo-rō-PEP-tīd) Chain of three to about 40 amino acids that occurs naturally in the nervous system, and that acts primarily to modulate the response of or to a neurotransmitter. Examples are enkephalins and endorphins.

Neurosecretory (noo-rō-SĒC-re-tō-rē) **cell** A neuron that secretes a hypothalamic releasing hormone or inhibiting hormone into blood capillaries of the hypothalmus; a neuron that secretes oxytocin or antidiuretic hormone into blood capillaries of the posterior pituitary.

Neurotransmitter One of a variety of molecules within axon terminals that are released into the synaptic cleft in response to a nerve impulse, and that change the membrane potential of the postsynaptic neuron.

Neutrophil (NOO-trō-fil) A type of white blood cell characterized by granules that stain pale lilac with a combination of acidic and basic dyes.

Nicotinic (nik′-ō-TIN-ik) **receptor** Receptor for the neurotransmitter acetylcholine found on both sympathetic and parasympathetic postganglionic neurons and on skeletal muscle in the motor end plate; so named because nicotine activates these receptors but does not activate muscarinic receptors for acetylcholine.

Nipple A pigmented, wrinkled projection on the surface of the breast that is the location of the openings of the lactiferous ducts for milk release.

Nociceptor (nō′-sē-SEP-tor) A free (naked) nerve ending that detects painful stimuli.

Node of Ranvier (ron-vē-Ā) A space, along a myelinated axon, between the individual Schwann cells that form the myelin sheath and the neurolemma. Also called a **neurofibral node.**

Nondisjunction (non′-dis-JUNGK-shun) Failure of sister chromatids to separate properly during anaphase of mitosis (or meiosis II) or failure of homologous chromosomes to separate properly during meiosis I in which chromatids or chromosomes pass into the same daughter cell; the result is too many copies of that chromosome in the daughter cell or gamete.

Norepinephrine (nor′-ep-ē-NEF-rin) **(NE)** A hormone secreted by the adrenal medulla that produces actions similar to those that result from sympathetic stimulation. Also called **noradrenaline** (nor-a-DREN-a-lin).

Notochord (NŌ-tō-cord) A flexible rod of mesodermal tissue that lies where the future vertebral column will develop and plays a role in induction.

Nuclear medicine The branch of medicine concerned with the use of radioisotopes in the diagnosis and therapy of disease.

Nucleic (noo-KLĒ-ic) **acid** An organic compound that is a long polymer of nucleotides, with each nucleotide containing a pentose sugar, a phosphate group, and one of four possible nitrogenous bases (adenine, cytosine, guanine, and thymine or uracil).

Nucleolus (noo-KLĒ-ō-lus) Spherical body within a cell nucleus composed of protein, DNA, and RNA that is the site of the assembly of small and large ribosomal subunits.

Nucleosome (NOO-klē-ō-sōm) Structural subunit of a chromosome consisting of histones and DNA.

Nucleus (NOO-klē-us) A spherical or oval organelle of a cell that contains the hereditary factors of the cell, called genes. A cluster of unmyelinated nerve cell bodies in the central nervous system. The central part of an atom made up of protons and neutrons.

Nucleus pulposus (pul-PŌ-sus) A soft, pulpy, highly elastic substance in the center of an intervertebral disc; a remnant of the notochord.

Nutrient A chemical substance in food that provides energy, forms new body components, or assists in various body functions.

O

Obesity (ō-BĒS-i-tē) Body weight more than 20% above a desirable standard due to excessive accumulation of fat.

Oblique (ō-BLĒK) **plane** A plane that passes through the body or an organ at an angle between the transverse plane and either the midsagittal, parasagittal, or frontal plane.

Obstetrics (ob-STET-riks) The specialized branch of medicine that deals with pregnancy, labor, and the period of time immediately after delivery (about 6 weeks).

Olfactory (ōl-FAK-tō-rē) Pertaining to smell.

Olfactory bulb A mass of gray matter containing cell bodies of neurons that form synapses with neurons of the olfactory nerve (cranial nerve I), lying inferior to the frontal lobe of the cerebrum on either side of the crista galli of the ethmoid bone.

Olfactory receptor A bipolar neuron with its cell body lying between supporting cells located in the mucous membrane lining the superior portion of each nasal cavity; transduces odors into neural signals.

Olfactory tract A bundle of axons that extends from the olfactory bulb posteriorly to olfactory regions of the cerebral cortex.

Oligodendrocyte (ol′-i-gō-DEN-drō-sīt) A neuroglial cell that supports neurons and produces a myelin sheath around axons of neurons of the central nervous system.

Oligospermia (ol′-i-gō-SPER-mē-a) A deficiency of sperm cells in the semen.

Olive A prominent oval mass on each lateral surface of the superior part of the medulla oblongata.

Oncogenes (ONG-kō-jēnz) Cancer-causing genes; they derive from normal genes, termed **proto-oncogenes,** that encode proteins involved in cell growth or cell regulation but have the ability to transform a normal cell into a cancerous cell when they are mutated or inappropriately activated. One example is *p53*.

Oncology (ong-KOL-ō-jē) The study of tumors.

Oogenesis (ō′-ō-JEN-e-sis) Formation and development of female gametes (oocytes).

Oophorectomy (ō′-of-ō-REK-tō-me) Surgical removal of the ovaries.

Ophthalmic (of-THAL-mik) Pertaining to the eye.

Ophthalmologist (of′-thal-MOL-ō-jist) A physician who specializes in the diagnosis and treatment of eye disorders using drugs, surgery, and corrective lenses.

Ophthalmology (of′-thal-MOL-ō-jē) The study of the structure, function, and diseases of the eye.

Opsin (OP-sin) The glycoprotein portion of a photopigment.

Opsonization (op-sō-ni-ZĀ-shun) The action of some antibodies that renders bacteria and other foreign cells more susceptible to phagocytosis.

Optic (OP-tik) Refers to the eye, vision, or properties of light.

Optic chiasm (KĪ-azm) A crossing point of the optic nerves (cranial nerve II), anterior to the pituitary gland. Also called **optic chiasma.**

Optic disc A small area of the retina containing openings through which the axons of the ganglion cells emerge as the optic nerve (cranial nerve II). Also called the **blind spot.**

Optician (op-TISH-an) A technician who fits, adjusts, and dispenses corrective lenses on prescription of an ophthalmologist or optometrist.

Optic tract A bundle of axons that carry nerve impulses from the retina of the eye between the optic chiasm and the thalamus.

Optometrist (op-TOM-e-trist) Specialist with a doctorate degree in optometry who is licensed to examine and test the eyes and treat visual defects by prescribing corrective lenses.

Ora serrata (Ō-ra ser-RĀ-ta) The irregular margin of the retina lying internal and slightly posterior to the junction of the choroid and ciliary body.

Orbit (OR-bit) The bony, pyramidal-shaped cavity of the skull that holds the eyeball.

Organ A structure composed of two or more different kinds of tissues with a specific function and usually a recognizable shape.

Organelle (or-gan-EL) A permanent structure within a cell with characteristic morphology that is specialized to serve a specific function in cellular activities.

Organic (or-GAN-ik) **compound** Compound that always contains carbon in which the atoms are held together by covalent bonds. Examples include carbohydrates, lipids, proteins, and nucleic acids (DNA and RNA).

Organism (OR-ga-nizm) A total living form; one individual.

Orgasm (OR-gazm) Sensory and motor events involved in ejaculation for the male and involuntary contraction of the perineal muscles in the female at the climax of sexual intercourse.

Orifice (OR-i-fis) Any aperture or opening.

Origin (OR-i-jin) The attachment of a muscle tendon to a stationary bone or the end opposite the insertion.

Oropharynx (or′-ō-FAR-inks) The intermediate portion of the pharynx, lying posterior to the mouth and extending from the soft palate to the hyoid bone.

Orthopedics (or′-thō-PĒ-diks) The branch of medicine that deals with the preservation and restoration of the skeletal system, articulations, and associated structures.

Osmoreceptor (oz′-mō-re-CEP-tor) Receptor in the hypothalamus that is sensitive to changes in blood osmolarity and, in response to high osmolarity (low water concentration), stimulates synthesis and release of antidiuretic hormone (ADH).

Osmosis (os-MŌ-sis) The net movement of water molecules through a selectively permeable membrane from an area of higher water concentration to an area of lower water concentration until equilibrium is reached.

Osmotic pressure The pressure required to prevent the movement of pure water into a solution containing solutes when the solutions are separated by a selectively permeable membrane.

Osseous (OS-ē-us) Bony.

Ossicle (OS-si-kul) One of the small bones of the middle ear (malleus, incus, stapes).

Ossification (os′-i-fi-KĀ-shun) Formation of bone. Also called **osteogenesis.**

Osteoblast (OS-tē-ō-blast) Cell formed from an osteogenic cell that participates in bone formation by secreting some organic components and inorganic salts.

Osteoclast (OS-tē-ō-clast′) A large, multinuclear cell that resorbs (destroys) bone matrix.

Osteocyte (OS-tē-ō-sīt′) A mature bone cell that maintains the daily activities of bone tissue.

Osteogenic (os′-tē-ō-prō-JEN-i-tor) **cell** Stem cell derived from mesenchyme that has mitotic potential and the ability to differentiate into an osteoblast.

Osteogenic (os′-tē-ō-JEN-ik) **layer** The inner layer of the periosteum that contains cells responsible for forming new bone during growth and repair.

Osteology (os′-tē-OL-ō-jē) The study of bones.

Osteon (OS-tē-on) The basic unit of structure in adult compact bone, consisting of a central (Haversian) canal with its concentrically arranged lamellae, lacunae, osteocytes, and canaliculi. Also called a **Haversian** (ha-VER-shan) **system.**

Osteoporosis (os′-tē-ō-pō-RŌ-sis) Age-related disorder characterized by decreased bone mass and increased susceptibility to fractures, often as a result of decreased levels of estrogens.

Otic (Ō-tik) Pertaining to the ear.

Otolith (Ō-tō-lith) A particle of calcium carbonate embedded in the otolithic membrane that functions in maintaining static equilibrium.

Otolithic (ō-tō-LITH-ik) **membrane** Thick, gelatinous, glycoprotein layer located directly over hair cells of the macula in the saccule and utricle of the internal ear.

Otorhinolaryngology (ō′-tō-rī-nō-lar′-in-GOL-ō-jē) The branch of medicine that deals with the diagnosis and treatment of diseases of the ears, nose, and throat.

Oval window A small, membrane-covered opening between the middle ear and inner ear into which the footplate of the stapes fits.

Ovarian (ō-VAR-ē-an) **cycle** A monthly series of events in the ovary associated with the maturation of a secondary oocyte.

Ovarian follicle (FOL-i-kul) A general name for oocytes (immature ova) in any stage of development, along with their surrounding epithelial cells.

Ovarian ligament (LIG-a-ment) A rounded cord of connective tissue that attaches the ovary to the uterus.

Ovary (Ō-var-ē) Female gonad that produces oocytes and the estrogens, progesterone, inhibin, and relaxin hormones.

Ovulation (ov-ū-LĀ-shun) The rupture of a mature ovarian (Graafian) follicle with discharge of a secondary oocyte into the pelvic cavity.

Ovum (Ō-vum) The female reproductive or germ cell; an egg cell; arises through completion of meiosis in a secondary oocyte after penetration by a sperm.

Oxidation (ok-si-DĀ-shun) The removal of electrons from a molecule or, less commonly, the addition of oxygen to a molecule that results in a decrease in the energy content of the molecule. The oxidation of glucose in the body is called **cellular respiration.**

Oxyhemoglobin (ok′-sē-HĒ-mō-glō-bin) **($Hb-O_2$)** Hemoglobin combined with oxygen.

Oxytocin (ok′-sē-TŌ-sin) **(OT)** A hormone secreted by neurosecretory cells in the paraventricular and supraoptic nuclei of the hypothalamus that stimulates contraction of smooth muscle in the pregnant uterus and myoepithelial cells around the ducts of mammary glands.

P

P wave The deflection wave of an electrocardiogram that signifies atrial depolarization.

Palate (PAL-at) The horizontal structure separating the oral and the nasal cavities; the roof of the mouth.

Palpate (PAL-pāt) To examine by touch; to feel.

Pancreas (PAN-krē-as) A soft, oblong organ lying along the greater curvature of the stomach and connected by a duct to the duodenum. It is both an exocrine gland (secreting pancreatic juice) and an endocrine gland (secreting insulin, glucagon, somatostatin, and pancreatic polypeptide).

Pancreatic (pan′-krē-AT-ik) **duct** A single large tube that unites with the common bile duct from the liver and gallbladder and drains pancreatic juice into the duodenum at the hepatopancreatic ampulla (ampulla of Vater). Also called the **duct of Wirsung.**

Pancreatic islet A cluster of endocrine gland cells in the pancreas that secretes insulin, glucagon, somatostatin, and pancreatic polypeptide. Also called an **islet of Langerhans** (LANG-er-hanz).

Pancreatic polypeptide Hormone secreted by the F cells of pancreatic islets (islets of Langerhans) that regulates release of pancreatic digestive enzymes.

Papanicolaou (pap′-a-NIK-ō-la-oo) **test** A cytological staining test for the detection and diagnosis of premalignant and malignant conditions of the female genital tract. Cells scraped from the epithelium of the cervix of the uterus are examined microscopically. Also called a **Pap test** or **Pap smear.**

Papilla (pa-PIL-a) A small nipple-shaped projection or elevation.

Paracrine (PAR-a-krin) Local hormone, such as histamine, that acts on neighboring cells without entering the bloodstream.

Paralysis (pa-RAL-a-sis) Loss or impairment of motor function due to a lesion of nervous or muscular origin.

Paranasal sinus (par′-a-NĀ-zal SĪ-nus) A mucus-lined air cavity in a skull bone that communicates with the nasal cavity. Paranasal sinuses are located in the frontal, maxillary, ethmoid, and sphenoid bones.

Paraplegia (par-a-PLĒ-jē-a) Paralysis of both lower limbs.

Parasagittal plane (par-a-SAJ-i-tal) A vertical plane that does not pass through the midline and that divides the body or organs into *unequal* left and right portions.

Parasympathetic (par′-a-sim-pa-THET-ik) **division** One of the two subdivisions of the autonomic nervous system, having cell bodies of preganglionic neurons in nuclei in the brain stem and in the lateral gray horn of the sacral portion of the spinal cord; primarily concerned with activities that conserve and restore body energy.

Parathyroid (par′-a-THĪ-royd) **gland** One of usually four small endocrine glands embedded in the posterior surfaces of the lateral lobes of the thyroid gland.

Parathyroid hormone (PTH) A hormone secreted by the chief (principal) cells of the parathyroid glands that increases blood calcium level and decreases blood phosphate level.

Paraurethral (par′-a-ū-RĒ-thral) **gland** Gland embedded in the wall of the urethra whose duct opens on either side of the urethral orifice and secretes mucus. Also called **Skene's** (SKĒNZ) **gland.**

Parenchyma (par-EN-ki-ma) The functional parts of any organ, as opposed to tissue that forms its stroma or framework.

Parietal (pa-RĪ-e-tal) Pertaining to or forming the outer wall of a body cavity.

Parietal cell A type of secretory cell in gastric glands that produces hydrochloric acid and intrinsic factor. Also called an **oxyntic cell.**

Parkinson disease (PD) Progressive degeneration of the basal ganglia and substantia nigra of the cerebrum resulting in decreased production of dopamine (DA) that leads to tremor, slowing of voluntary movements, and muscle weakness.

Parotid (pa-ROT-id) **gland** One of the paired salivary glands located inferior and anterior to the ears and connected to the oral cavity via a duct (Stensen's) that opens into the inside of the cheek opposite the maxillary (upper) second molar tooth.

Parturition (par′-too-RISH-un) Act of giving birth to young; childbirth, delivery.

Patellar (pa-TELL-ar) **reflex** Extension of the leg by contraction of the quadriceps femoris muscle in response to tapping the patellar ligament. Also called the **knee jerk reflex.**

Patent ductus arteriosus Congenital anatomical heart defect in which the fetal connection between the aorta and pulmonary trunk remains open instead of closing completely after birth.

Pathogen (PATH-ō-jen) A disease-producing microbe.

Pathological (path′-ō-LOJ-i-kal) **anatomy** The study of structural changes caused by disease.

Pectoral (PEK-tō-ral) Pertaining to the chest or breast.

Pediatrician (pē′-dē-a-TRISH-un) A physician who specializes in the care and treatment of children.

Pedicel (PED-i-sel) Footlike structure, as on podocytes of a glomerulus.

Pelvic (PEL-vik) **cavity** Inferior portion of the abdominopelvic cavity that contains the urinary bladder, sigmoid colon, rectum, and internal female and male reproductive structures.

Pelvic splanchnic (PEL-vik SPLANGK-nik) **nerves** Consist of preganglionic parasympathetic axons from the levels of S2, S3, and S4 that supply the urinary bladder, reproductive organs, and the descending and sigmoid colon and rectum.

Pelvis The basinlike structure formed by the two hip bones, the sacrum, and the coccyx. The expanded, proximal portion of the ureter, lying within the kidney and into which the major calyces open.

Penis (PĒ-nis) The organ of urination and copulation in males; used to deposit semen into the female vagina.

Pepsin Protein-digesting enzyme secreted by chief cells of the stomach in the inactive form pepsinogen, which is converted to active pepsin by hydrochloric acid.

Peptic ulcer An ulcer that develops in areas of the gastrointestinal tract exposed to hydrochloric acid; classified as a gastric ulcer if in the lesser curvature of the stomach and as a duodenal ulcer if in the first part of the duodenum.

Percussion (per-KUSH-un) The act of striking (percussing) an underlying part of the body with short, sharp blows as an aid in diagnosing the part by the quality of the sound produced.

Perforating canal A minute passageway by means of which blood vessels and nerves from the periosteum penetrate into compact bone. Also called **Volkmann's** (FŌLK-manz) **canal.**

Pericardial (per′-i-KAR-dē-al) **cavity** Small potential space between the visceral and parietal layers of the serous pericardium that contains pericardial fluid.

Pericardium (per′-i-KAR-dē-um) A loose-fitting membrane that encloses the heart, consisting of a superficial fibrous layer and a deep serous layer.

Perichondrium (per′-i-KON-drē-um) The membrane that covers cartilage.

Perilymph (PER-i-limf) The fluid contained between the bony and membranous labyrinths of the inner ear.

Perimetrium (per′-i-MĒ-trē-um) The serosa of the uterus.

Perimysium (per′-i-MĪZ-ē-um) Invagination of the epimysium that divides muscles into bundles.

Perineum (per′-i-NĒ-um) The pelvic floor; the space between the anus and the scrotum in the male and between the anus and the vulva in the female.

Perineurium (per′-i-NOO-rē-um) Connective tissue wrapping around fascicles in a nerve.

Periodontal (per-ē-ō-DON-tal) **disease** A collective term for conditions characterized by degeneration of gingivae, alveolar bone, periodontal ligament, and cementum.

Periodontal ligament The periosteum lining the alveoli (sockets) for the teeth in the alveolar processes of the mandible and maxillae.

Periosteum (per′-ē-OS-tē-um) The membrane that covers bone and consists of connective tissue, osteogenic cells, and osteoblasts; is essential for bone growth, repair, and nutrition.

Peripheral (pe-RIF-er-al) Located on the outer part or a surface of the body.

Peripheral nervous system (PNS) The part of the nervous system that lies outside the central nervous system, consisting of nerves and ganglia.

Peristalsis (per′-i-STAL-sis) Successive muscular contractions along the wall of a hollow muscular structure.

Peritoneum (per′-i-tō-NĒ-um) The largest serous membrane of the body that lines the abdominal cavity and covers the viscera.

Peritonitis (per′-i-tō-NĪ-tis) Inflammation of the peritoneum.

Permissive (per-MIS-sive) **effect** A hormonal interaction in which the effect of one hormone on a target cell requires previous or simultaneous exposure to another hormone(s) to enhance the response of a target cell or increase the activity of another hormone.

Peroxisome (per-OK-si-sōm) Organelle similar in structure to a lysosome that contains enzymes that use molecular oxygen to oxidize various organic compounds; such reactions produce hydrogen peroxide; abundant in liver cells.

Perspiration Sweat; produced by sudoriferous (sweat) glands and containing water, salts, urea, uric acid, amino acids, ammonia, sugar, lactic acid, and ascorbic acid. Helps maintain body temperature and eliminate wastes.

pH A measure of the concentration of hydrogen ions (H^+) in a solution. The pH scale extends from 0 to 14, with a value of 7 expressing neutrality, values lower than 7 expressing increasing acidity, and values higher than 7 expressing increasing alkalinity.

Phagocytosis (fag′-ō-sī-TŌ-sis) The process by which phagocytes ingest particulate matter; the ingestion and destruction of microbes, cell debris, and other foreign matter.

Phalanx (FĀ-lanks) The bone of a finger or toe. *Plural is* **phalanges** (fa-LAN-jēz).

Pharmacology (far′-ma-KOL-ō-jē) The science of the effects and uses of drugs in the treatment of disease.

Pharynx (FAR-inks) The throat; a tube that starts at the internal nares and runs partway down the neck, where it opens into the esophagus posteriorly and the larynx anteriorly.

Phenotype (FĒ-nō-tīp) The observable expression of genotype; physical characteristics of an organism determined by genetic makeup and influenced by interaction between genes and internal and external environmental factors.

Phlebitis (fle-BĪ-tis) Inflammation of a vein, usually in a lower limb.

Phosphorylation (fos′-for-i-LĀ-shun) The addition of a phosphate group to a chemical compound; types include substrate-level phosphorylation, oxidative phosphorylation, and photo-phosphorylation.

Photopigment A substance that can absorb light and undergo structural changes that can lead to the development of a receptor potential. An example is rhodopsin. In the eye, also called **visual pigment.**

Photoreceptor Receptor that detects light shining on the retina of the eye.

Physiology (fiz′-ē-OL-ō-jē) Science that deals with the functions of an organism or its parts.

Pia mater (PĪ-a-MĀ-ter *or* PĒ-a MA-ter) The innermost of the three meninges (coverings) of the brain and spinal cord.

Pineal (PĪN-ē-al) **gland** A cone-shaped gland located in the roof of the third ventricle that secretes melatonin. Also called the **epiphysis cerebri** (ē-PIF-i-sis se-RĒ-brē).

Pinealocyte (pin-ē-AL-ō-sīt) Secretory cell of the pineal gland that releases melatonin.

Pinna (PIN-na) The projecting part of the external ear composed of elastic cartilage and covered by skin and shaped like the flared end of a trumpet. Also called the **auricle** (OR-i-kul).

Pinocytosis (pi′-nō-sī-TŌ-sis) A process by which most body cells can ingest membrane-surrounded droplets of interstitial fluid.

Pituicyte (pi-TOO-i-sīt) Supporting cell of the posterior pituitary.

Pituitary (pi-TOO-i-tār-ē) **gland** A small endocrine gland occupying the hypophyseal fossa above the sella turcica of the sphenoid bone and attached to the hypothalamus by the infundibulum. Also called the **hypophysis** (hī-POF-i-sis).

Pivot joint A synovial joint in which a rounded, pointed, or conical surface of one bone articulates with a ring formed partly by another bone and partly by a ligament, as in the joint between the atlas and axis and between the proximal ends of the radius and ulna. Also called a **trochoid** (TRŌ-koyd) **joint.**

Placenta (pla-SEN-ta) The special structure through which the exchange of materials between fetal and maternal circulations occurs. Also called the **afterbirth.**

Plantar flexion (PLAN-tar FLEK-shun) Bending the foot in the direction of the plantar surface (sole).

Plaque (PLAK) A layer of dense proteins on the inside of a plasma membrane in adherens junctions and desmosomes. A mass of bacterial cells, dextran (polysaccharide), and other debris that adheres to teeth (dental plaque). See also atherosclerotic plaque.

Plasma (PLAZ-ma) The extracellular fluid found in blood vessels; blood minus the formed elements.

Plasma cell Cell that develops from a B cell (lymphocyte) and produces antibodies.

Plasma (cell) membrane Outer, limiting membrane that separates the cell's internal parts from extracellular fluid or the external environment.

Platelet (PLĀT-let) A fragment of cytoplasm enclosed in a cell membrane and lacking a nucleus; found in the circulating blood; plays a role in hemostasis. Also called a **thrombocyte** (THROM-bō-sīt).

Platelet plug Aggregation of platelets (thrombocytes) at a site where a blood vessel is damaged that helps stop or slow blood loss.

Pleura (PLOOR-a) The serous membrane that covers the lungs and lines the walls of the chest and the diaphragm.

Pleural cavity Small potential space between the visceral and parietal pleurae.

Plexus (PLEK-sus) A network of nerves, veins, or lymphatic vessels.

Pluripotent stem cell Immature stem cell in red bone marrow that gives rise to precursors of all the different mature blood cells.

Pneumotaxic (noo-mō-TAK-sik) **area** A part of the respiratory center in the pons that continually sends inhibitory nerve impulses to the inspiratory area, limiting inhalation and facilitating exhalation.

Podiatry (pō-DĪ-a-trē) The diagnosis and treatment of foot disorders.

Polar body The smaller cell resulting from the unequal division of primary and secondary oocytes during meiosis. The polar body has no function and degenerates.

Polycythemia (pol′-ē-sī-THĒ-mē-a) Disorder characterized by an above-normal hematocrit (above 55%) in which hypertension, thrombosis, and hemorrhage can occur.

Polysaccharide (pol′-ē-SAK-a-rīd) A carbohydrate in which three or more monosaccharides are joined chemically.

Polyunsaturated fat A fatty acid that contains more than one double covalent bond between its carbon atoms; abundant in triglycerides of corn oil, safflower oil, and cottonseed oil.

Polyuria (pol′-ē-Ū-rē-a) An excessive production of urine.

Pons (PONZ) The part of the brain stem that forms a "bridge" between the medulla oblongata and the midbrain, anterior to the cerebellum.

Positive feedback A feedback mechanism in which the response enhances the original stimulus.

Postabsorptive (fasting) state Metabolic state after absorption of food is complete and energy needs of the body must be satisfied using molecules stored previously.

Postcentral gyrus Gyrus of cerebral cortex located immediately posterior to the central sulcus; contains the primary somatosensory area.

Posterior (pos-TĒR-ē-or) Nearer to or at the back of the body. Equivalent to **dorsal** in bipeds.

Posterior column–medial lemniscus pathways Sensory pathways that carry information related to proprioception, fine touch, two-point discrimination, pressure, and vibration. First-order neurons project from the spinal cord to the ipsilateral medulla in the posterior columns (gracile fasciculus and cuneate fasciculus). Second-order neurons project from the medulla to the contralateral thalamus in the medial lemniscus. Third-order neurons project from the thalamus to the somatosensory cortex (postcentral gyrus) on the same side.

Posterior pituitary Posterior lobe of the pituitary gland. Also called the **neurohypophysis** (noo-rō-hī-POF-i-sis).

Posterior root The structure composed of sensory axons lying between a spinal nerve and the dorsolateral aspect of the spinal cord. Also called the **dorsal (sensory) root.**

Posterior root ganglion (GANG-glē-on) A group of cell bodies of sensory neurons and their supporting cells located along the posterior root of a spinal nerve. Also called a **dorsal (sensory) root ganglion.**

Postganglionic neuron (pōst′-gang-lē-ON-ik NOO-ron) The second autonomic motor neuron in an autonomic pathway, having its cell body and dendrites located in an autonomic ganglion and its unmyelinated axon ending at cardiac muscle, smooth muscle, or a gland.

Postsynaptic (pōst-sin-AP-tik) **neuron** The nerve cell that is activated by the release of a neurotransmitter from another neuron and carries nerve impulses away from the synapse.

Precapillary sphincter (SFINGK-ter) A ring of smooth muscle fibers (cells) at the site of origin of true capillaries that regulate blood flow into true capillaries.

Precentral gyrus Gyrus of cerebral cortex located immediately anterior to the central sulcus; contains the primary motor area.

Preganglionic (prē′ gang-lē-ON-ik) **neuron** The first autonomic motor neuron in an autonomic pathway, with its cell body and dendrites in the brain or spinal cord and its myelinated axon ending at an autonomic ganglion, where it synapses with a postganglionic neuron.

Pregnancy Sequence of events that normally includes fertilization, implantation, embryonic growth, and fetal growth and terminates in birth.

Premenstrual syndrome (PMS) Severe physical and emotional stress ocurring late in the postovulatory phase of the menstrual cycle and sometimes overlapping with menstruation.

Prepuce (PRĒ-poos) The loose-fitting skin covering the glans of the penis and clitoris. Also called the **foreskin.**

Presbyopia (prez-bē-Ō-pē-a) A loss of elasticity of the lens of the eye due to advancing age with resulting inability to focus clearly on near objects.

Presynaptic (prē-sin-AP-tik) **inhibition** Type of inhibition in which neurotransmitter released by an inhibitory neuron depresses the release of neurotransmitter by a presynaptic neuron.

Presynaptic (prē-sin-AP-tik) **neuron** A neuron that propagates nerve impulses toward a synapse.

Prevertebral ganglion (prē-VER-te-bral GANG-lē-on) A cluster of cell bodies of postganglionic sympathetic neurons anterior to the spinal column and close to large abdominal arteries. Also called a **collateral ganglion.**

Primary germ layer One of three layers of embryonic tissue, called ectoderm, mesoderm, and endoderm, that give rise to all tissues and organs of the body.

Primary motor area A region of the cerebral cortex in the precentral gyrus of the frontal lobe of the cerebrum that controls specific muscles or groups of muscles.

Primary somatosensory area A region of the cerebral cortex posterior to the central sulcus in the postcentral gyrus of the parietal lobe of the cerebrum that localizes exactly the points of the body where somatic sensations originate.

Prime mover The muscle directly responsible for producing a desired motion. Also called an **agonist** (AG-ō-nist).

Primitive gut Embryonic structure formed from the dorsal part of the yolk sac that gives rise to most of the gastrointestinal tract.

Principal cell Cell type in the distal convoluted tubules and collecting ducts of the kidneys that is stimulated by aldosterone and antidiuretic hormone.

Proctology (prok-TOL-ō-jē) The branch of medicine concerned with the rectum and its disorders.

Progeny (PROJ-e-nē) Offspring or descendants.

Progesterone (prō-JES-te-rōn) A female sex hormone produced by the ovaries that helps prepare the endometrium of the uterus for implantation of a fertilized ovum and the mammary glands for milk secretion.

Prognosis (prog-NŌ-sis) A forecast of the probable results of a disorder; the outlook for recovery.

Prolactin (prō-LAK-tin) **(PRL)** A hormone secreted by the anterior pituitary that initiates and maintains milk secretion by the mammary glands.

Prolapse (PRŌ-laps) A dropping or falling down of an organ, especially the uterus or rectum.

Proliferation (prō-lif′-er-Ā-shun) Rapid and repeated reproduction of new parts, especially cells.

Pronation (prō-NĀ-shun) A movement of the forearm in which the palm is turned posteriorly or inferiorly.

Prophase (PRŌ-fāz) The first stage of mitosis during which chromatid pairs are formed and aggregate around the metaphase plate of the cell.

Proprioception (prō-prē-ō-SEP-shun) The perception of the position of body parts, especially the limbs, independent of vision; this sense is possible due to nerve impulses generated by proprioceptors.

Proprioceptor (prō′-prē-ō-SEP-tor) A receptor located in muscles, tendons, joints, or the internal ear (muscle spindles, tendon organs, joint kinesthetic receptors, and hair cells of the vestibular apparatus) that provides information about body position and movements.

Prostaglandin (pros′-ta-GLAN-din) **(PG)** A membrane-associated lipid; released in small quantities and acts as a local hormone.

Prostate (PROS-tāt) A doughnut-shaped gland inferior to the urinary bladder that surrounds the superior portion of the male urethra and secretes a slightly acidic solution that contributes to sperm motility and viability.

Protein An organic compound consisting of carbon, hydrogen, oxygen, nitrogen, and sometimes sulfur and phosphorus; synthesized on ribosomes and made up of amino acids linked by peptide bonds.

Proteasome Tiny cellular organelle in cytosol and nucleus containing proteases that destroy unneeded, damaged, or faulty proteins.

Prothrombin (prō-THROM-bin) An inactive blood-clotting factor synthesized by the liver, released into the blood, and converted to active thrombin in the process of blood clotting by the activated enzyme prothrombinase.

Protraction (prō-TRAK-shun) The movement of the mandible or shoulder girdle forward on a plane parallel with the ground.

Proximal (PROK-si-mal) Nearer the attachment of a limb to the trunk; nearer to the point of origin or attachment.

Pseudopods (SOO-dō-pods) Temporary protrusions of the leading edge of a migrating cell; cellular projections that surround a particle undergoing phagocytosis.

Pterygopalatine ganglion (ter′-i-gō-PAL-a-tīn GANG-glē-on) A cluster of cell bodies of parasympathetic postganglionic neurons ending at the lacrimal and nasal glands.

Ptosis (TŌ-sis) Drooping, as of the eyelid or the kidney.

Puberty (PŪ-ber-tē) The time of life during which the secondary sex characteristics begin to appear and the capability for sexual reproduction is possible; usually occurs between the ages of 10 and 17.

Pubic symphysis A slightly movable cartilaginous joint between the anterior surfaces of the hip bones.

Puerperium (pū′-er-PER-ē-um) The period immediately after childbirth, usually 4–6 weeks.

Pulmonary (PUL-mo-ner′-ē) Concerning or affected by the lungs.

Pulmonary circulation The flow of deoxygenated blood from the right ventricle to the lungs and the return of oxygenated blood from the lungs to the left atrium.

Pulmonary edema (e-DĒ-ma) An abnormal accumulation of interstitial fluid in the tissue spaces and alveoli of the lungs due to increased pulmonary capillary permeability or increased pulmonary capillary pressure.

Pulmonary embolism (EM-bō-lizm) **(PE)** The presence of a blood clot or a foreign substance in a pulmonary arterial blood vessel that obstructs circulation to lung tissue.

Pulmonary ventilation The inflow (inhalation) and outflow (exhalation) of air between the atmosphere and the lungs. Also called **breathing**.

Pulp cavity A cavity within the crown and neck of a tooth, which is filled with pulp, a connective tissue containing blood vessels, nerves, and lymphatic vessels.

Pulse (PULS) The rhythmic expansion and elastic recoil of a systemic artery after each contraction of the left ventricle.

Pulse pressure The difference between the maximum (systolic) and minimum (diastolic) pressures; normally about 40 mmHg.

Pupil The hole in the center of the iris, the area through which light enters the posterior cavity of the eyeball.

Purkinje (pur-KIN-jē) **fiber** Muscle fiber (cell) in the ventricular tissue of the heart specialized for conducting an action potential to the myocardium; part of the conduction system of the heart.

Pus The liquid product of inflammation containing leukocytes or their remains and debris of dead cells.

Pyloric (pī-LOR-ik) **sphincter** A thickened ring of smooth muscle through which the pylorus of the stomach communicates with the duodenum. Also called the **pyloric valve.**

Pyogenesis (pi′-ō-JEN-e-sis) Formation of pus.

Pyramid (PIR-a-mid) A pointed or cone-shaped structure. One of two roughly triangular structures on the anterior aspect of the medulla oblongata composed of the largest motor tracts that run from the cerebral cortex to the spinal cord. A triangular structure in the renal medulla.

Pyramidal (pi-RAM-i-dal) **tracts (pathways)** *See* **Direct motor pathways**.

Q

QRS wave The deflection waves of an electrocardiogram that represent onset of ventricular depolarization.

Quadriplegia (kwod′-ri-PLĒ-jē-a) Paralysis of four limbs: two upper and two lower.

R

Radiographic (rā′-dē-ō-GRAF-ic) **anatomy** Diagnostic branch of anatomy that includes the use of x rays.

Rami communicantes (RĀ-mē kō-mū-ni-KAN-tēz) Branches of a spinal nerve. *Singular is* **ramus communicans** (RĀ-mus kō-MŪ-ni-kans).

Rapid eye movement (REM) sleep Stage of sleep in which dreaming occurs, lasting for 5 to 10 minutes several times during a sleep cycle; characterized by rapid movements of the eyes beneath the eyelids.

Reactivity (rē-ak-TI-vi-tē) Ability of an antigen to react specifically with the antibody whose formation it induced.

Receptor A specialized cell or a distal portion of a neuron that responds to a specific sensory modality, such as touch, pressure, cold, light, or sound, and converts it to an electrical signal (generator or receptor potential). A specific molecule or cluster of molecules that recognizes and binds a particular ligand.

Receptor-mediated endocytosis A highly selective process whereby cells take up specific ligands, which usually are large molecules or particles, by enveloping them within a sac of plasma membrane. Ligands are eventually broken down by enzymes in lysosomes.

Receptor potential Depolarization or hyperpolarization of the plasma membrane of a receptor that alters release of neurotransmitter from the cell; if the neuron that synapses with the receptor cell becomes depolarized to threshold, a nerve impulse is triggered.

Recessive allele An allele whose presence is masked in the presence of a dominant allele on the homologous chromosome.

Reciprocal innervation (re-SIP-rō-kal in-ner-VĀ-shun) The phenomenon by which action potentials stimulate contraction of one muscle and simultaneously inhibit contraction of antagonistic muscles.

Recombinant DNA Synthetic DNA, formed by joining a fragment of DNA from one source to a portion of DNA from another.

Recovery oxygen consumption Elevated oxygen use after exercise ends due to metabolic changes that start during exercise and continue after exercise. Previously called **oxygen debt.**

Recruitment (rē-KROOT-ment) The process of increasing the number of active motor units. Also called **motor unit summation.**

Rectouterine pouch A pocket formed by the parietal peritoneum as it moves posteriorly from the surface of the uterus and is reflected onto the rectum; the most inferior point in the pelvic cavity. Also called the **pouch** or **cul de sac of Douglas.**

Rectum (REK-tum) The last 20 cm (8 in.) of the gastrointestinal tract, from the sigmoid colon to the anus.

Red nucleus A cluster of cell bodies in the midbrain, occupying a large part of the tectum from which axons extend into the rubroreticular and rubrospinal tracts.

Red pulp That portion of the spleen that consists of venous sinuses filled with blood and thin plates of splenic tissue called splenic (Billroth's) cords.

Reduction The addition of electrons to a molecule or, less commonly, the removal of oxygen from a molecule that results in an increase in the energy content of the molecule.

Referred pain Pain that is felt at a site remote from the place of origin.

Reflex Fast response to a change (stimulus) in the internal or external environment that attempts to restore homeostasis.

Reflex arc The most basic conduction pathway through the nervous system, connecting a receptor and an effector and consisting of a receptor, a sensory neuron, an integrating center in the central nervous system, a motor neuron, and an effector.

Refraction (rē-FRAK-shun) The bending of light as it passes from one medium to another.

Refractory (re-FRAK-to-rē) **period** A time period during which an excitable cell (neuron or muscle fiber) cannot respond to a stimulus that is usually adequate to evoke an action potential.

Regional anatomy The division of anatomy dealing with a specific region of the body, such as the head, neck, chest, or abdomen.

Regurgitation (rē-gur′-ji-TĀ-shun) Return of solids or fluids to the mouth from the stomach; backward flow of blood through incompletely closed heart valves.

Relaxin (RLX) A female hormone produced by the ovaries and placenta that increases flexibility of the pubic symphysis and helps dilate the uterine cervix to ease delivery of a baby.

Releasing hormone Hormone secreted by the hypothalamus that can stimulate secretion of hormones of the anterior pituitary.

Remodeling Replacement of old bone by new bone tissue.

Renal (RĒ-nal) Pertaining to the kidneys.

Renal corpuscle (KOR-pus-l) A glomerular (Bowman's) capsule and its enclosed glomerulus.

Renal pelvis A cavity in the center of the kidney formed by the expanded, proximal portion of the ureter, lying within the kidney, and into which the major calyces open.

Renal pyramid A triangular structure in the renal medulla containing the straight segments of renal tubules and the vasa recta.

Renin (RĒ-nin) An enzyme released by the kidney into the plasma, where it converts angiotensinogen into angiotensin I.

Renin–angiotensin–aldosterone (RAA) pathway A mechanism for the control of blood pressure, initiated by the secretion of renin by juxtaglomerular cells of the kidney in response to low blood pressure; renin catalyzes formation of angiotensin I, which is converted to angiotensin II by angiotensin-converting enzyme (ACE), and angiotensin II stimulates secretion of aldosterone.

Repolarization (rē-pō-lar-i-ZĀ-shun) Restoration of a resting membrane potential after depolarization.

Reproduction (rē-prō-DUK-shun) The formation of new cells for growth, repair, or replacement; the production of a new individual.

Reproductive cell division Type of cell division in which gametes (sperm and oocytes) are produced; consists of meiosis and cytokinesis.

Residual (re-ZID-ū-al) **volume** The volume of air still contained in the lungs after a maximal exhalation; about 1200 mL.

Resistance (re-ZIS-tans) Hindrance (impedance) to blood flow as a result of higher viscosity, longer total blood vessel length, and smaller blood vessel radius. Ability to ward off disease. The hindrance encountered by electrical charges as they move from one point to another. The hindrance encountered by air as it moves through the respiratory passageways.

Respiration (res-pi-RĀ-shun) Overall exchange of gases between the atmosphere, blood, and body cells consisting of pulmonary ventilation, external respiration, and internal respiration.

Respiratory center Neurons in the pons and medulla oblongata of the brain stem that regulate the rate and depth of pulmonary ventilation.

Respiratory membrane Structure in the lungs consisting of the alveolar wall and its basement membrane and a capillary endothelium and its basement membrane through which the diffusion of respiratory gases occurs.

Resting membrane potential The voltage difference between the inside and outside of a cell membrane when the cell is not responding to a stimulus; in many neurons and muscle fibers it is −70 to −90 mV, with the inside of the cell negative relative to the outside.

Retention (rē-TEN-shun) A failure to void urine due to obstruction, nervous contraction of the urethra, or absence of sensation of desire to urinate.

Rete (RĒ-tē) **testis** The network of ducts in the testes.

Reticular (re-TIK-ū-lar) **activating system (RAS)** A portion of the reticular formation that has many ascending connections with the cerebral cortex; when this area of the brain stem is active, nerve impulses pass to the thalamus and widespread areas of the cerebral cortex, resulting in generalized alertness or arousal from sleep.

Reticular formation A network of small groups of neuronal cell bodies scattered among bundles of axons (mixed gray and white matter) beginning in the medulla oblongata and extending superiorly through the central part of the brain stem.

Reticulocyte (re-TIK-ū-lō-sīt) An immature red blood cell.

Reticulum (re-TIK-ū-lum) A network.

Retina (RET-i-na) The deep coat of the posterior portion of the eyeball consisting of nervous tissue (where the process of vision begins) and a pigmented layer of epithelial cells that contact the choroid.

Retinal (RE-ti-nal) A derivative of vitamin A that functions as the light-absorbing portion of the photopigment rhodopsin.

Retraction (rē-TRAK-shun) The movement of a protracted part of the body posteriorly on a plane parallel to the ground, as in pulling the lower jaw back in line with the upper jaw.

Retroflexion (re-trō-FLEK-shun) A malposition of the uterus in which it is tilted posteriorly.

Retrograde degeneration (RE-trō-grād dē-jen-er-Ā-shun) Changes that occur in the proximal portion of a damaged axon only as far as the first node of Ranvier; similar to changes that occur during Wallerian degeneration.

Retroperitoneal (re′-trō-per-i-tō-NĒ-al) External to the peritoneal lining of the abdominal cavity.

Rh factor An inherited antigen on the surface of red blood cells in Rh^+ individuals; not present in Rh^- individuals.

Rhinology (rī-NOL-ō-jē) The study of the nose and its disorders.

Rhodopsin (rō-DOP-sin) The photopigment in rods of the retina, consisting of a glycoprotein called opsin and a derivative of vitamin A called retinal.

Ribonucleic (rī-bō-noo-KLĒ-ik) **acid (RNA)** A single-stranded nucleic acid made up of nucleotides, each consisting of a nitrogenous base (adenine, cytosine, guanine, or uracil), ribose, and a phosphate group; three types are messenger RNA (mRNA), transfer RNA (tRNA), and ribosomal RNA (rRNA), each of which has a specific role during protein synthesis.

Ribosome (RĪ-bō-sōm) An organelle in the cytoplasm of cells, composed of a small subunit and a large subunit that contain ribosomal RNA and ribosomal proteins; the site of protein synthesis.

Right heart (atrial) reflex A reflex concerned with maintaining normal venous blood pressure.

Right lymphatic (lim-FAT-ik) **duct** A vessel of the lymphatic system that drains lymph from the upper right side of the body and empties it into the right subclavian vein.

Rigidity (ri-JID-i-tē) Hypertonia characterized by increased muscle tone, but reflexes are not affected.

Rigor mortis State of partial contraction of muscles after death due to lack of ATP; myosin heads (cross bridges) remain attached to actin, thus preventing relaxation.

Rod One of two types of photoreceptor in the retina of the eye; specialized for vision in dim light.

Root canal A narrow extension of the pulp cavity lying within the root of a tooth.

Root of penis Attached portion of penis that consists of the bulb and crura.

Rotation (rō-TĀ-shun) Moving a bone around its own axis, with no other movement.

Round ligament (LIG-a-ment) A band of fibrous connective tissue enclosed between the folds of the broad ligament of the uterus, emerging from the uterus just inferior to the uterine tube, extending laterally along the pelvic wall and through the deep inguinal ring to end in the labia majora.

Round window A small opening between the middle and internal ear, directly inferior to the oval window, covered by the secondary tympanic membrane.

Rugae (ROO-gē) Large folds in the mucosa of an empty hollow organ, such as the stomach and vagina.

S

Saccule (SAK-ūl) The inferior and smaller of the two chambers in the membranous labyrinth inside the vestibule of the internal ear containing a receptor organ for static equilibrium.

Sacral plexus (SĀ-kral PLEK-sus) A network formed by the ventral branches of spinal nerves L4 through S3.

Sacral promontory (PROM-on-tor′-ē) The superior surface of the body of the first sacral vertebra that projects anteriorly into the pelvic cavity; a line from the sacral promontory to the superior border of the pubic symphysis divides the abdominal and pelvic cavities.

Saddle joint A synovial joint in which the articular surface of one bone is saddle shaped and the articular surface of the other bone is shaped like the legs of the rider sitting in the saddle, as in the joint between the trapezium and the metacarpal of the thumb.

Sagittal (SAJ-i-tal) **plane** A plane that divides the body or organs into left and right portions. Such a plane may be **midsagittal (median),** in which the divisions are equal, or **parasagittal,** in which the divisions are unequal.

Saliva (sa-LĪ-va) A clear, alkaline, somewhat viscous secretion produced mostly by the three pairs of salivary glands; contains various salts, mucin, lysozyme, salivary amylase, and lingual lipase (produced by glands in the tongue).

Salivary amylase (SAL-i-ver-ē AM-i-lās) An enzyme in saliva that initiates the chemical breakdown of starch.

Salivary gland One of three pairs of glands that lie external to the mouth and pour their secretory product (saliva) into ducts that empty into the oral cavity; the parotid, submandibular, and sublingual glands.

Salt A substance that, when dissolved in water, ionizes into cations and anions, neither of which are hydrogen ions (H^+) or hydroxide ions (OH^-).

Saltatory (sal-ta-TŌ-rē) **conduction** The propagation of an action potential (nerve impulse) along the exposed parts of a myelinated axon. The action potential appears at successive nodes of Ranvier and therefore seems to leap from node to node.

Sarcolemma (sar′-kō-LEM-ma) The cell membrane of a muscle fiber (cell), especially of a skeletal muscle fiber.

Sarcomere (SAR-kō-mēr) A contractile unit in a striated muscle fiber (cell) extending from one Z disc to the next Z disc.

Sarcoplasm (SAR-kō-plazm) The cytoplasm of a muscle fiber (cell).

Sarcoplasmic reticulum (sar′-kō-PLAZ-mik re-TIK-ū-lum) A network of saccules and tubes surrounding myofibrils of a muscle fiber (cell), comparable to endoplasmic reticulum; functions to reabsorb calcium ions during relaxation and to release them to cause contraction.

Saturated fat A fatty acid that contains only single bonds (no double bonds) between its carbon atoms; all carbon atoms are bonded to the maximum number of hydrogen atoms; prevalent in triglycerides of animal products such as meat, milk, milk products, and eggs.

Scala tympani (SKA-la TIM-pan-ē) The inferior spiral-shaped channel of the bony cochlea, filled with perilymph.

Scala vestibuli (ves-TIB-ū-lē) The superior spiral-shaped channel of the bony cochlea, filled with perilymph.

Schwann (SCHVON) **cell** A neuroglial cell of the peripheral nervous system that forms the myelin sheath and neurolemma around a nerve axon by wrapping around the axon in a jelly-roll fashion.

Sciatica (sī-AT-i-ka) Inflammation and pain along the sciatic nerve; felt along the posterior aspect of the thigh extending down the inside of the leg.

Sclera (SKLE-ra) The white coat of fibrous tissue that forms the superficial protective covering over the eyeball except in the most anterior portion; the posterior portion of the fibrous tunic.

Scleral venous sinus A circular venous sinus located at the junction of the sclera and the cornea through which aqueous humor drains from the anterior chamber of the eyeball into the blood. Also called the **canal of Schlemm** (SHLEM).

Sclerosis (skle-RŌ-sis) A hardening with loss of elasticity of tissues.

Scoliosis (skō′-lē-Ō-sis) An abnormal lateral curvature from the normal vertical line of the backbone.

Scrotum (SKRŌ-tum) A skin-covered pouch that contains the testes and their accessory structures.

Sebaceous (se-BĀ-shus) **gland** An exocrine gland in the dermis of the skin, almost always associated with a hair follicle, that secretes sebum. Also called an **oil gland.**

Sebum (SĒ-bum) Secretion of sebaceous (oil) glands.

Secondary response Accelerated, more intense cell-mediated or antibody-mediated immune response upon a subsequent exposure to an antigen after the primary response.

Secondary sex characteristic A characteristic of the male or female body that develops at puberty under the influence of sex hormones but is not directly involved in sexual reproduction; examples are distribution of body hair, voice pitch, body shape, and muscle development.

Second messenger An intracellular mediator molecule that is produced in response to a first messenger (hormone or neurotransmitter) binding to its receptor in the plasma membrane of a target cell. Initiates a cascade of chemical reactions that produce characteristic effects for that particular target cell.

Secretion (se-KRĒ-shun) Production and release from a cell or a gland of a physiologically active substance.

Selective permeability (per′-mē-a-BIL-i-tē) The property of a membrane by which it permits the passage of certain substances but restricts the passage of others.

Semen (SĒ-men) A fluid discharged at ejaculation by a male that consists of a mixture of sperm and the secretions of the seminiferous tubules, seminal vesicles, prostate, and bulbourethral (Cowper's) glands.

Semicircular canals Three bony channels (anterior, posterior, lateral), filled with perilymph, in which lie the membranous semicircular canals filled with endolymph. They contain receptors for equilibrium.

Semicircular ducts The membranous semicircular canals filled with endolymph and floating in the perilymph of the bony semicircular canals; they contain cristae that are concerned with dynamic equilibrium.

Semilunar (sem′-ē-LOO-nar) **valve** A valve between the aorta or the pulmonary trunk and a ventricle of the heart.

Seminal vesicle (SEM-i-nal VES-i-kul) One of a pair of convoluted, pouchlike structures, lying posterior and inferior to the urinary bladder and anterior to the rectum, that secrete a component of semen into the ejaculatory ducts. Also termed **seminal gland.**

Seminiferous tubule (sem′-i-NI-fer-us TOO-būl) A tightly coiled duct, located in the testis, where sperm are produced.

Senescence (se-NES-ens) The process of growing old.

Sensation A state of awareness of external or internal conditions of the body.

Sensory neurons (NOO-ronz) Neurons that carry sensory information from cranial and spinal nerves into the brain and spinal cord or from a lower to a higher level in the spinal cord and brain. Also called **afferent neurons.**

Septal defect An opening in the septum (interatrial or interventricular) between the left and right sides of the heart.

Septum (SEP-tum) A wall dividing two cavities.

Serous (SIR-us) **membrane** A membrane that lines a body cavity that does not open to the exterior. The external layer of an organ formed by a serous membrane. The membrane that lines the pleural, pericardial, and peritoneal cavities. Also called a **serosa** (se-RŌ-sa).

Sertoli (ser-TŌ-lē) **cell** A supporting cell in the seminiferous tubules that secretes fluid for supplying nutrients to sperm and the hormone inhibin, removes excess cytoplasm from spermatogenic cells, and mediates the effects of FSH and testosterone on spermatogenesis. Also called a **sustentacular** (sus′-ten-TAK-ū-lar) **cell.**

Serum Blood plasma minus its clotting proteins.

Sesamoid (SES-a-moyd) **bones** Small bones usually found in tendons.

Sex chromosomes The twenty-third pair of chromosomes, designated X and Y, which determine the genetic sex of an individual; in males, the pair is XY; in females, XX.

Sexual intercourse The insertion of the erect penis of a male into the vagina of a female. Also called **coitus** (KŌ-i-tus).

Shivering Involuntary contraction of skeletal muscles that generates heat. Also called **involuntary thermogenesis.**

Shock Failure of the cardiovascular system to deliver adequate amounts of oxygen and nutrients to meet the metabolic needs of the body due to inadequate cardiac output. It is characterized by hypotension; clammy, cool, and pale skin; sweating; reduced urine formation; altered mental state; acidosis; tachycardia; weak, rapid pulse; and thirst. Types include hypovolemic, cardiogenic, vascular, and obstructive.

Shoulder joint A synovial joint where the humerus articulates with the scapula.

Sigmoid colon (SIG-moyd KŌ-lon) The S-shaped part of the large intestine that begins at the level of the left iliac crest, projects medially, and terminates at the rectum at about the level of the third sacral vertebra.

Sign Any objective evidence of disease that can be observed or measured such as a lesion, swelling, or fever.

Sinoatrial (si-nō-Ā-trē-al) **(SA) node** A small mass of cardiac muscle fibers (cells) located in the right atrium inferior to the opening of the superior vena cava that spontaneously depolarize and generate a cardiac action potential about 100 times per minute. Also called the **pacemaker.**

Sinus (SĪ-nus) A hollow in a bone (paranasal sinus) or other tissue; a channel for blood (vascular sinus); any cavity having a narrow opening.

Sinusoid (SĪ-nū-soyd) A large, thin-walled, and leaky type of capillary, having large intercellular clefts that may allow proteins and blood cells to pass from a tissue into the bloodstream; present in the liver, spleen, anterior pituitary, parathyroid glands, and red bone marrow.

Skeletal muscle An organ specialized for contraction, composed of striated muscle fibers (cells), supported by connective tissue, attached to a bone by a tendon or an aponeurosis, and stimulated by somatic motor neurons.

Skin The external covering of the body that consists of a superficial, thinner epidermis (epithelial tissue) and a deep, thicker dermis (connective tissue) that is anchored to the subcutaneous layer.

Skull The skeleton of the head consisting of the cranial and facial bones.

Sleep A state of partial unconsciousness from which a person can be aroused; associated with a low level of activity in the reticular activating system.

Sliding-filament mechanism The explanation of how thick and thin filaments slide relative to one another during striated muscle contraction to decrease sarcomere length.

Small intestine A long tube of the gastrointestinal tract that begins at the pyloric sphincter of the stomach, coils through the central and inferior part of the abdominal cavity, and ends at the large intestine; divided into three segments: duodenum, jejunum, and ileum.

Smooth muscle A tissue specialized for contraction, composed of smooth muscle fibers (cells), located in the walls of hollow internal organs, and innervated by autonomic motor neurons.

Sodium-potassium ATPase An active transport pump located in the plasma membrane that transports sodium ions out of the cell and potassium ions into the cell at the expense of cellular ATP. It functions to keep the ionic concentrations of these ions at physiological levels. Also called the **sodium-potassium pump.**

Soft palate (PAL-at) The posterior portion of the roof of the mouth, extending from the palatine bones to the uvula. It is a muscular partition lined with mucous membrane.

Solution A homogeneous molecular or ionic dispersion of one or more substances (solutes) in a dissolving medium (solvent) that is usually liquid.

Somatic (sō-MAT-ik) **cell division** Type of cell division in which a single starting parent cell duplicates itself to produce two identical cells; consists of mitosis and cytokinesis.

Somatic nervous system (SNS) The portion of the peripheral nervous system consisting of somatic sensory (afferent) neurons and somatic motor (efferent) neurons.

Somite (SŌ-mīt) Block of mesodermal cells in a developing embryo that is distinguished into a myotome (which forms most of the skeletal muscles), dermatome (which forms connective tissues), and sclerotome (which forms the vertebrae).

Spasm (SPAZM) A sudden, involuntary contraction of large groups of muscles.

Spasticity (spas-TIS-i-tē) Hypertonia characterized by increased muscle tone, increased tendon reflexes, and pathological reflexes (Babinski sign).

Spermatic (sper-MAT-ik) **cord** A supporting structure of the male reproductive system, extending from a testis to the deep inguinal ring, that includes the ductus (vas) deferens, arteries, veins, lymphatic vessels, nerves, cremaster muscle, and connective tissue.

Spermatogenesis (sper′-ma-tō-JEN-e-sis) The formation and development of sperm in the seminiferous tubules of the testes.

Sperm cell A mature male gamete. Also termed **spermatozoon** (sper′-ma-tō-ZŌ-on).

Spermiogenesis (sper′-mē-ō-JEN-e-sis) The maturation of spermatids into sperm.

Sphincter (SFINGK-ter) A circular muscle that constricts an opening.

Sphincter of the hepatopancreatic ampulla A circular muscle at the opening of the common bile and main pancreatic ducts in the duodenum. Also called the **sphincter of Oddi** (OD-ē).

Sphygmomanometer (sfig′-mō-ma-NOM-e-ter) An instrument for measuring arterial blood pressure.

Spinal (SPĪ-nal) **cord** A mass of nerve tissue located in the vertebral canal from which 31 pairs of spinal nerves originate.

Spinal nerve One of the 31 pairs of nerves that originate on the spinal cord from posterior and anterior roots.

Spinal shock A period from several days to several weeks following transection of the spinal cord and characterized by the abolition of all reflex activity.

Spinothalamic (spī-nō-tha-LAM-ik) **tracts** Sensory (ascending) tracts that convey information up the spinal cord to the thalamus for sensations of pain, temperature, crude touch, and deep pressure.

Spinous (SPĪ-nus) **process** A sharp or thornlike process or projection. Also called a **spine.** A sharp ridge running diagonally across the posterior surface of the scapula.

Spiral organ The organ of hearing, consisting of supporting cells and hair cells that rest on the basilar membrane and extend into the endolymph of the cochlear duct. Also called the **organ of Corti** (KOR-tē).

Spirometer (spī-ROM-e-ter) An apparatus used to measure lung volumes and capacities.

Splanchnic (SPLANK-nik) Pertaining to the viscera.

Spleen (SPLĒN) Large mass of lymphatic tissue between the fundus of the stomach and the diaphragm that functions in formation of blood cells during early fetal development, phagocytosis of ruptured blood cells, and proliferation of B cells during immune responses.

Spongy (cancellous) bone tissue Bone tissue that consists of an irregular latticework of thin plates of bone called trabeculae; spaces between trabeculae of some bones are filled with red bone marrow; found inside short, flat, and irregular bones and in the epiphyses (ends) of long bones.

Sprain Forcible wrenching or twisting of a joint with partial rupture or other injury to its attachments without dislocation.

Squamous (SKWĀ-mus) Flat or scalelike.

Starling's law of the capillaries The movement of fluid between plasma and interstitial fluid is in a state of near equilibrium at the arterial and venous ends of a capillary; that is, filtered fluid and absorbed fluid plus that returned to the lymphatic system are nearly equal.

Starling's law of the heart The force of muscular contraction is determined by the length of the cardiac muscle fibers just before they contract; within limits, the greater the length of stretched fibers, the stronger the contraction.

Starvation (star-VĀ-shun) The loss of energy stores in the form of glycogen, triglycerides, and proteins due to inadequate intake of nutrients or inability to digest, absorb, or metabolize ingested nutrients.

Stasis (STĀ-sis) Stagnation or halt of normal flow of fluids, as blood or urine, or of the intestinal contents.

Static equilibrium (ē-kwi-LIB-rē-um) The maintenance of posture in response to changes in the orientation of the body, mainly the head, relative to the ground.

Stellate reticuloendothelial (STEL-āt re-tik′-ū-lō-en′-dō-THĒ-lē-al) **cell** Phagocytic cell bordering a sinusoid of the liver. Also called a **Kupffer** (KOOP-fer) **cell.**

Stenosis (sten-Ō-sis) An abnormal narrowing or constriction of a duct or opening.

Stereocilia (ste′-rē-ō-SIL-ē-a) Groups of extremely long, slender, nonmotile microvilli projecting from epithelial cells lining the epididymis.

Stereognosis (ste′-rē-og-NŌ-sis) The ability to recognize the size, shape, and texture of an object by touching it.

Sterile (STE-ril) Free from any living microorganisms. Unable to conceive or produce offspring.

Sterilization (ster′-i-li-ZĀ-shun) Elimination of all living microorganisms. Any procedure that renders an individual incapable of reproduction (for example, castration, vasectomy, hysterectomy, or oophorectomy).

Stimulus Any stress that changes a controlled condition; any change in the internal or external environment that excites a sensory receptor, a neuron, or a muscle fiber.

Stomach The J-shaped enlargement of the gastrointestinal tract directly inferior to the diaphragm in the epigastric, umbilical, and left hypochondriac regions of the abdomen, between the esophagus and small intestine.

Straight tubule (TOO-būl) A duct in a testis leading from a convoluted seminiferous tubule to the rete testis.

Stratum (STRĀ-tum) A layer.

Stratum basalis (ba-SAL-is) The layer of the endometrium next to the myometrium that is maintained during menstruation and gestation and produces a new stratum functionalis following menstruation or parturition.

Stratum functionalis (funk′-shun-AL-is) The layer of the endometrium next to the uterine cavity that is shed during menstruation and that forms the maternal portion of the placenta during gestation.

Stressor A stress that is extreme, unusual, or long-lasting and triggers the stress response.

Stress response Wide-ranging set of bodily changes, triggered by a stressor, that gears the body to meet an emergency. Also known as **general adaptation syndrome (GAS).**

Stretch receptor Receptor in the walls of blood vessels, airways, or organs that monitors the amount of stretching. Also termed **baroreceptor.**

Stretch reflex A monosynaptic reflex triggered by sudden stretching of muscle spindles within a muscle that elicits contraction of that same muscle. Also called a **tendon jerk.**

Stroke volume The volume of blood ejected by either ventricle in one systole; about 70 mL in an adult at rest.

Stroma (STRŌ-ma) The tissue that forms the ground substance, foundation, or framework of an organ, as opposed to its functional parts (parenchyma).

Subarachnoid (sub′-a-RAK-noyd) **space** A space between the arachnoid mater and the pia mater that surrounds the brain and spinal cord and through which cerebrospinal fluid circulates.

Subcutaneous (sub′-kū-TĀ-nē-us) Beneath the skin. Also called **hypodermic** (hi-pō-DER-mik).

Subcutaneous layer A continuous sheet of areolar connective tissue and adipose tissue between the dermis of the skin and the deep fascia of the muscles. Also called the **superficial fascia** (FASH-ē-a).

Subdural (sub-DOO-ral) **space** A space between the dura mater and the arachnoid mater of the brain and spinal cord that contains a small amount of fluid.

Sublingual (sub-LING-gwal) **gland** One of a pair of salivary glands situated in the floor of the mouth deep to the mucous membrane and to the side of the lingual frenulum, with a duct (Rivinus') that opens into the floor of the mouth.

Submandibular (sub′-man-DIB-ū-lar) **gland** One of a pair of salivary glands found inferior to the base of the tongue deep to the mucous membrane in the posterior part of the floor of the mouth, posterior to the sublingual glands, with a duct (Wharton's) situated to the side of the lingual frenulum. Also called the **submaxillary** (sub′-MAK-si-ler-ē) **gland.**

Submucosa (sub-mū-KO-sa) A layer of connective tissue located deep to a mucous membrane, as in the gastrointestinal tract or the urinary bladder; the submucosa connects the mucosa to the muscularis layer.

Submucosal plexus A network of autonomic nerve fibers located in the superficial part of the submucous layer of the small intestine. Also called the **plexus of Meissner** (MIZ-ner).

Substrate A molecule upon which an enzyme acts.

Subthreshold stimulus A stimulus of such weak intensity that it cannot initiate an action potential (nerve impulse).

Sudoriferous (soo′-dor-IF-er-us) **gland** An apocrine or eccrine exocrine gland in the dermis or subcutaneous layer that produces perspiration. Also called a **sweat gland.**

Sulcus (SUL-kus) A groove or depression between parts, especially between the convolutions of the brain. *Plural is* **sulci** (SUL-sī).

Summation (sum-MĀ-shun) The addition of the excitatory and inhibitory effects of many stimuli applied to a neuron. The increased strength of muscle contraction that results when stimuli follow one another in rapid succession.

Superficial (soo′-per-FISH-al) Located on or near the surface of the body or an organ.

Superficial fascia (FASH-ē-a) A continuous sheet of fibrous connective tissue between the dermis of the skin and the deep fascia of the muscles. Also called **subcutaneous** (sub′-kū-TĀ-nē-us) **layer.**

Superficial inguinal (IN-gwi-nal) **ring** A triangular opening in the aponeurosis of the external oblique muscle that represents the termination of the inguinal canal.

Superior (soo-PĒR-ē-or) Toward the head or upper part of a structure. Also called **cephalad** (SEF-a-lad) or **craniad.**

Superior vena cava (VĒ-na CĀ-va) **(SVC)** Large vein that collects blood from parts of the body superior to the heart and returns it to the right atrium.

Supination (soo-pi-NĀ-shun) A movement of the forearm in which the palm is turned anteriorly or superiorly.

Surface anatomy The study of the structures that can be identified from the outside of the body.

Surfactant (sur-FAK-tant) Complex mixture of phospholipids and lipoproteins, produced by type II alveolar (septal) cells in the lungs, that decreases surface tension.

Susceptibility (sus-sep′-ti-BIL-i-tē) Lack of resistance to the damaging effects of an agent such as a pathogen.

Suspensory ligament (sus-PEN-so-rē LIG-a-ment) A fold of peritoneum extending laterally from the surface of the ovary to the pelvic wall.

Sutural (SOO-cher-al) **bone** A small bone located within a suture between certain cranial bones. Also called **Wormian** (WER-mē-an) **bone.**

Suture (SOO-cher) An immovable fibrous joint that joins skull bones.

Sympathetic (sim′-pa-THET-ik) **division** One of the two subdivisions of the autonomic nervous system, having cell bodies of preganglionic neurons in the lateral gray columns of the thoracic segment and the first two or three lumbar segments of the spinal cord; primarily concerned with process involving the expenditure of energy.

Sympathetic trunk ganglion (GANG-glē-on) A cluster of cell bodies of sympathetic postganglionic neurons lateral to the vertebral column, close to the body of a vertebra. These ganglia extend inferiorly through the neck, thorax, and abdomen to the coccyx on both sides of the vertebral column and are connected to one another to form a chain on each side of the vertebral column. Also called **sympathetic chain** or **vertebral chain ganglia.**

Sympathomimetic (sim′-pa-thō-mi-MET-ik) Producing effects that mimic those brought about by the sympathetic division of the autonomic nervous system.

Symphysis (SIM-fi-sis) A line of union. A slightly movable cartilaginous joint such as the pubic symphysis.

Symporter A transmembrane transporter protein that moves two substances, often Na^+ and another substance, in the same direction across a plasma membrane. Also called a **cotransporter.**

Symptom (SIMP-tum) A subjective change in body function not apparent to an observer, such as pain or nausea, that indicates the presence of a disease or disorder of the body.

Synapse (SYN-aps) The functional junction between two neurons or between a neuron and an effector, such as a muscle or gland; may be electrical or chemical.

Synapsis (sin-AP-sis) The pairing of homologous chromosomes during prophase I of meiosis.

Synaptic (sin-AP-tik) **cleft** The narrow gap at a chemical synapse that separates the axon terminal of one neuron from another neuron or muscle fiber (cell) and across which a neurotransmitter diffuses to affect the postsynaptic cell.

Synaptic delay The length of time between the arrival of the action potential at a presynaptic axon terminal and the membrane potential

(IPSP or EPSP) change on the postsynaptic membrane; usually about 0.5 msec.

Synaptic end bulb Expanded distal end of an axon terminal that contains synaptic vesicles. Also called a **synaptic knob.**

Synaptic vesicle Membrane-enclosed sac in a synaptic end bulb that stores neurotransmitters.

Synarthrosis (sin′-ar-THRŌ-sis) An immovable joint such as a suture, gomphosis, and synchondrosis.

Synchondrosis (sin′-kon-DRŌ-sis) A cartilaginous joint in which the connecting material is hyaline cartilage.

Syndesmosis (sin′-dez-MŌ-sis) A slightly movable joint in which articulating bones are united by fibrous connective tissue.

Syndrome (SIN-drōm) A group of signs and symptoms that occur together in a pattern that is characteristic of a particular disease or abnormal condition.

Synergist (SIN-er-jist) A muscle that assists the prime mover by reducing undesired action or unnecessary movement.

Synergistic (syn-er-JIS-tik) **effect** A hormonal interaction in which the effects of two or more hormones acting together is greater or more extensive than the sum of each hormone acting alone.

Synostosis (sin′-os-TŌ-sis) A joint in which the dense fibrous connective tissue that unites bones at a suture has been replaced by bone, resulting in a complete fusion across the suture line.

Synovial (si-NŌ-vē-al) **cavity** The space between the articulating bones of a synovial joint, filled with synovial fluid. Also called a **joint cavity.**

Synovial fluid Secretion of synovial membranes that lubricates joints and nourishes articular cartilage.

Synovial joint A fully movable or diarthrotic joint in which a synovial (joint) cavity is present between the two articulating bones.

Synovial membrane The deeper of the two layers of the articular capsule of a synovial joint, composed of areolar connective tissue that secretes synovial fluid into the synovial (joint) cavity.

System An association of organs that have a common function.

Systemic (sis-TEM-ik) Affecting the whole body; generalized.

Systemic anatomy The anatomic study of particular systems of the body, such as the skeletal, muscular, nervous, cardiovascular, or urinary systems.

Systemic circulation The routes through which oxygenated blood flows from the left ventricle through the aorta to all the organs of the body and deoxygenated blood returns to the right atrium.

Systemic vascular resistance (SVR) All the vascular resistance offered by systemic blood vessels. Also called **total peripheral resistance.**

Systole (SIS-tō-lē) In the cardiac cycle, the phase of contraction of the heart muscle, especially of the ventricles.

Systolic (sis-TOL-ik) **blood pressure** The force exerted by blood on arterial walls during ventricular contraction; the highest pressure measured in the large arteries, about 120 mmHg under normal conditions for a young adult.

T

T cell A lymphocyte that becomes immunocompetent in the thymus and can differentiate into a helper T cell or a cytotoxic T cell, both of which function in cell-mediated immunity.

T wave The deflection wave of an electrocardiogram that represents ventricular repolarization.

Tachycardia (tak′-i-KAR-dē-a) An abnormally rapid resting heartbeat or pulse rate (over 100 beats per minute).

Tactile (TAK-tīl) Pertaining to the sense of touch.

Tactile disc Modified epidermal cell in the stratum basale of hairless skin that functions as a cutaneous receptor for discriminative touch. Also called a **Merkel** (MER-kel) **disc.**

Teniae coli (TĒ-nē-ē KŌ-lī) The three flat bands of thickened, longitudinal smooth muscle running the length of the large intestine, except in the rectum. *Singular is* **tenia coli.**

Target cell A cell whose activity is affected by a particular hormone.

Tarsal bones The seven bones of the ankle. Also called **tarsals.**

Tarsal gland Sebaceous (oil) gland that opens on the edge of each eyelid. Also called a **Meibomian** (mī-BŌ-mē-an) **gland.**

Tarsal plate A thin, elongated sheet of connective tissue, one in each eyelid, giving the eyelid form and support. The aponeurosis of the levator palpebrae superioris is attached to the tarsal plate of the superior eyelid.

Tarsus (TAR-sus) A collective term for the seven bones of the ankle.

Tectorial (tek-TŌ-rē-al) **membrane** A gelatinous membrane projecting over and in contact with the hair cells of the spiral organ (organ of Corti) in the cochlear duct.

Teeth (TĒTH) Accessory structures of digestion, composed of calcified connective tissue and embedded in bony sockets of the mandible and maxilla, that cut, shred, crush, and grind food. Also called **dentes** (DEN-tēz).

Telophase (TEL-ō-fāz) The final stage of mitosis in which the daughter nuclei become established.

Tendon (TEN-don) A white fibrous cord of dense regular connective tissue that attaches muscle to bone.

Tendon organ A proprioceptive receptor, sensitive to changes in muscle tension and force of contraction, found chiefly near the junctions of tendons and muscles. Also called a **Golgi** (GOL-jē) **tendon organ.**

Tendon reflex A polysynaptic, ipsilateral reflex that protects tendons and their associated muscles from damage that might be brought about by excessive tension. The receptors involved are called tendon organs (Golgi tendon organs).

Tentorium cerebelli (ten-TŌ-rē-um ser′-e-BEL-ē) A transverse shelf of dura mater that forms a partition between the occipital lobe of the cerebral hemispheres and the cerebellum and that covers the cerebellum.

Teratogen (TER-a-tō-jen) Any agent or factor that causes physical defects in a developing embryo.

Terminal ganglion (TER-min-al GANG-glē-on) A cluster of cell bodies of parasympathetic postganglionic neurons either lying very close to the visceral effectors or located within the walls of the visceral effectors supplied by the postganglionic neurons.

Testis (TES-tis) Male gonad that produces sperm and the hormones testosterone and inhibin. Also called a **testicle.**

Testosterone (tes-TOS-te-rōn) A male sex hormone (androgen) secreted by interstitial endocrinocytes (Leydig cells) of a mature testis; needed for development of sperm; together with a second androgen termed **dihydrotestosterone (DHT),** controls the growth and development of male reproductive organs, secondary sex characteristics, and body growth.

Tetany (TET-a-nē) Hyperexcitability of neurons and muscle fibers (cells) caused by hypocalcemia and characterized by intermittent or continuous tonic muscular contractions; may be due to hypoparathyroidism.

Thalamus (THAL-a-mus) A large, oval structure located bilaterally on either side of the third ventricle, consisting of two masses of gray matter organized into nuclei; main relay center for sensory impulses ascending to the cerebral cortex.

Thermoreceptor (THER-mō-rē-sep-tor) Sensory receptor that detects changes in temperature.

Thigh The portion of the lower limb between the hip and the knee.

Third ventricle (VEN-tri-kul) A slitlike cavity between the right and left halves of the thalamus and between the lateral ventricles of the brain.

Thirst center A cluster of neurons in the hypothalamus that is sensitive to the osmotic pressure of extracellular fluid and brings about the sensation of thirst.

Thoracic (thō-RAS-ik) **cavity** Superior portion of the ventral body cavity that contains two pleural cavities, the mediastinum, and the pericardial cavity.

Thoracic duct A lymphatic vessel that begins as a dilation called the cisterna chyli, receives lymph from the left side of the head, neck, and chest, the left arm, and the entire body below the ribs, and empties into the left subclavian vein. Also called the **left lymphatic** (lim-FAT-ik) **duct.**

Thoracolumbar (thō′-ra-kō-LUM-bar) **outflow** The axons of sympathetic preganglionic neurons, which have their cell bodies in the lateral gray columns of the thoracic segments and first two or three lumbar segments of the spinal cord.

Thorax (THŌ-raks) The chest.

Threshold potential The membrane voltage that must be reached to trigger an action potential.

Threshold stimulus Any stimulus strong enough to initiate an action potential or activate a sensory receptor.

Thrombin (THROM-bin) The active enzyme formed from prothrombin that converts fibrinogen to fibrin during formation of a blood clot.

Thrombolytic (throm-bō-LIT-ik) **agent** Chemical substance injected into the body to dissolve blood clots and restore circulation; mechanism of action is direct or indirect activation of plasminogen; examples include tissue plasminogen activator (t-PA), streptokinase, and urokinase.

Thrombosis (throm-BŌ-sis) The formation of a clot in an unbroken blood vessel, usually a vein.

Thrombus A stationary clot formed in an unbroken blood vessel, usually a vein.

Thymus (THĪ-mus) A bilobed organ, located in the superior mediastinum posterior to the sternum and between the lungs, in which T cells develop immunocompetence.

Thyroglobulin (thī-rō-GLŌ-bū-lin) **(TGB)** A large glycoprotein molecule produced by follicular cells of the thyroid gland in which some tyrosines are iodinated and coupled to form thyroid hormones.

Thyroid cartilage (THĪ-royd KAR-ti-lij) The largest single cartilage of the larynx, consisting of two fused plates that form the anterior wall of the larynx.

Thyroid follicle (FOL-i-kul) Spherical sac that forms the parenchyma of the thyroid gland and consists of follicular cells that produce thyroxine (T_4) and triiodothyronine (T_3).

Thyroid gland An endocrine gland with right and left lateral lobes on either side of the trachea connected by an isthmus; located anterior to the trachea just inferior to the cricoid cartilage; secretes thyroxine (T_4), triiodothyronine (T_3), and calcitonin.

Thyroid-stimulating hormone (TSH) A hormone secreted by the anterior pituitary that stimulates the synthesis and secretion of thyroxine (T_4) and triiodothyronine (T_3).

Thyroxine (thī-ROK-sēn) **(T_4)** A hormone secreted by the thyroid gland that regulates metabolism, growth and development, and the activity of the nervous system.

Tic Spasmodic, involuntary twitching of muscles that are normally under voluntary control.

Tidal volume The volume of air breathed in and out in any one breath; about 500 mL in quiet, resting conditions.

Tissue A group of similar cells and their intercellular substance joined together to perform a specific function.

Tissue factor (TF) A factor, or collection of factors, whose appearance initiates the blood clotting process. Also called **thromboplastin** (throm-bō-PLAS-tin).

Tissue plasminogen activator (t-PA) An enzyme that dissolves small blood clots by initiating a process that converts plasminogen to plasmin, which degrades the fibrin of a clot.

Tongue A large skeletal muscle covered by a mucous membrane located on the floor of the oral cavity.

Tonicity (tō-NIS-i-tē) A measure of the concentration of impermeable solute particles in a solution relative to cytosol. When cells are bathed in an **isotonic solution,** they neither shrink nor swell.

Tonsil (TON-sil) An aggregation of large lymphatic nodules embedded in the mucous membrane of the throat.

Torn cartilage A tearing of an articular disc (meniscus) in the knee.

Total lung capacity The sum of tidal volume, inspiratory reserve volume, expiratory reserve volume, and residual volume; about 6000 mL in an average adult.

Trabecula (tra-BEK-ū-la) Irregular latticework of thin plates of spongy bone. Fibrous cord of connective tissue serving as supporting fiber by forming a septum extending into an organ from its wall or capsule. *Plural is* **trabeculae** (tra-BEK-ū-lē).

Trabeculae carneae (KAR-nē-ē) Ridges and folds of the myocardium in the ventricles.

Trachea (TRĀ-kē-a) Tubular air passageway extending from the larynx to the fifth thoracic vertebra. Also called the **windpipe.**

Tract A bundle of nerve axons in the central nervous system.

Transcription (trans-KRIP-shun) The first step in the expression of genetic information in which a single strand of DNA serves as a template for the formation of an RNA molecule.

Translation (trans-LĀ-shun) The synthesis of a new protein on a ribosome as dictated by the sequence of codons in messenger RNA.

Transverse colon (trans-VERS KŌ-lon) The portion of the large intestine extending across the abdomen from the right colic (hepatic) flexure to the left colic (splenic) flexure.

Transverse fissure (FISH-er) The deep cleft that separates the cerebrum from the cerebellum.

Transverse plane A plane that divides the body or organs into superior and inferior portions. Also called a **horizontal plane.**

Transverse tubules (TOO-būls) **(T tubules)** Small, cylindrical invaginations of the sarcolemma of striated muscle fibers (cells) that conduct muscle action potentials toward the center of the muscle fiber.

Trauma (TRAW-ma) An injury, either a physical wound or psychic disorder, caused by an external agent or force, such as a physical blow or emotional shock; the agent or force that causes the injury.

Tremor (TREM-or) Rhythmic, involuntary, purposeless contraction of opposing muscle groups.

Triad (TRĪ-ad) A complex of three units in a muscle fiber composed of a transverse tubule and the sarcoplasmic reticulum terminal cisterns on both sides of it.

Tricuspid (trī-KUS-pid) **valve** Atrioventricular (AV) valve on the right side of the heart.

Triglyceride (trī-GLI-cer-īd) A lipid formed from one molecule of glycerol and three molecules of fatty acids that may be either solid (fats) or liquid (oils) at room temperature; the body's most highly concentrated source of chemical potential energy. Found mainly within adipocytes. Also called a **neutral fat** or a **triacylglycerol.**

Trigone (TRĪ-gon) A triangular region at the base of the urinary bladder.

Triiodothyronine (trī-ī-ō-dō-THĪ-rō-nēn) **(T_3)** A hormone produced by the thyroid gland that regulates metabolism, growth and development, and the activity of the nervous system.

Trophoblast (TRŌF-ō-blast) The superficial covering of cells of the blastocyst.

Tropic (TRŌ-pik) **hormone** A hormone whose target is another endocrine gland.

Trunk The part of the body to which the upper and lower limbs are attached.

Tubal ligation (lī-GĀ-shun) A sterilization procedure in which the uterine (Fallopian) tubes are tied and cut.

Tubular reabsorption The movement of filtrate from renal tubules back into blood in response to the body's specific needs.

Tubular secretion The movement of substances in blood into renal tubular fluid in response to the body's specific needs.

Tumor suppressor gene A gene coding for a protein that normally inhibits cell division; loss or alteration of a tumor suppressor gene called *p53* is the most common genetic change in a wide variety of cancer cells.

Tunica albuginea (TOO-ni-ka al′-bū-JIN-ē-a) A dense white fibrous capsule covering a testis or deep to the surface of an ovary.

Tunica externa (eks-TER-na) The superficial coat of an artery or vein, composed mostly of elastic and collagen fibers. Also called the **adventitia.**

Tunica interna (in-TER-na) The deep coat of an artery or vein, consisting of a lining of endothelium, basement membrane, and internal elastic lamina. Also called the **tunica intima** (IN-ti-ma).

Tunica media (MĒ-dē-a) The intermediate coat of an artery or vein, composed of smooth muscle and elastic fibers.

Twitch contraction Brief contraction of all muscle fibers (cells) in a motor unit triggered by a single action potential in its motor neuron.

Tympanic antrum (tim-PAN-ik AN-trum) An air space in the middle ear that leads into the mastoid air cells or sinus.

Type II cutaneous mechanoreceptor A sensory receptor embedded deeply in the dermis and deeper tissues that detects stretching of skin. Also called a **Ruffini corpuscle.**

U

Umbilical cord The long, ropelike structure containing the umbilical arteries and vein that connect the fetus to the placenta.

Umbilicus (um-BIL-i-kus *or* um-bil-Ī-kus) A small scar on the abdomen that marks the former attachment of the umbilical cord to the fetus. Also called the **navel.**

Upper limb The appendage attached at the shoulder girdle, consisting of the arm, forearm, wrist, hand, and fingers. Also called **upper extremity.**

Up-regulation Phenomenon in which there is an increase in the number of receptors in response to a deficiency of a hormone or neurotransmitter.

Uremia (ū-RĒ-mē-a) Accumulation of toxic levels of urea and other nitrogenous waste products in the blood, usually resulting from severe kidney malfunction.

Ureter (Ū-rē-ter) One of two tubes that connect the kidney with the urinary bladder.

Urethra (ū-RĒ-thra) The duct from the urinary bladder to the exterior of the body that conveys urine in females and urine and semen in males.

Urinary (Ū-ri-ner-ē) **bladder** A hollow, muscular organ situated in the pelvic cavity posterior to the pubic symphysis; receives urine via two ureters and stores urine until it is excreted through the urethra.

Urine The fluid produced by the kidneys that contains wastes and excess materials; excreted from the body through the urethra.

Urogenital (ū′-rō-JEN-i-tal) **triangle** The region of the pelvic floor inferior to the pubic symphysis, bounded by the pubic symphysis and the ischial tuberosities, and containing the external genitalia.

Urology (ū-ROL-ō-jē) The specialized branch of medicine that deals with the structure, function, and diseases of the male and female urinary systems and the male reproductive system.

Uterine (Ū-ter-in) **tube** Duct that transports ova from the ovary to the uterus. Also called the **Fallopian** (fal-LŌ-pē-an) **tube** or **oviduct.**

Uterosacral ligament (ū′-ter-ō-SĀ-kral LIG-a-ment) A fibrous band of tissue extending from the cervix of the uterus laterally to the sacrum.

Uterus (Ū-te-rus) The hollow, muscular organ in females that is the site of menstruation, implantation, development of the fetus, and labor. Also called the **womb.**

Utricle (Ū-tri-kul) The larger of the two divisions of the membranous labyrinth located inside the vestibule of the inner ear, containing a receptor organ for static equilibrium.

Uvea (Ū-vē-a) The three structures that together make up the vascular tunic of the eye.

Uvula (Ū-vū-la) A soft, fleshy mass, especially the V-shaped pendant part, descending from the soft palate.

V

Vagina (va-JĪ-na) A muscular, tubular organ that leads from the uterus to the vestibule, situated between the urinary bladder and the rectum of the female.

Vallate papilla (VAL-āt pa-PIL-a) One of the circular projections that is arranged in an inverted V-shaped row at the back of the tongue; the largest of the elevations on the upper surface of the tongue containing taste buds. Also called **circumvallate papilla.**

Valence (VĀ-lens) The combining capacity of an atom; the number of deficit or extra electrons in the outermost electron shell of an atom.

Varicocele (VAR-i-kō-sēl) A twisted vein; especially, the accumulation of blood in the veins of the spermatic cord.

Varicose (VAR-i-kōs) Pertaining to an unnatural swelling, as in the case of a varicose vein.

Vasa recta (VĀ-sa REK-ta) Extensions of the efferent arteriole of a juxtamedullary nephron that run alongside the loop of the nephron (Henle) in the medullary region of the kidney.

Vasa vasorum (va-SŌ-rum) Blood vessels that supply nutrients to the larger arteries and veins.

Vascular (VAS-kū-lar) Pertaining to or containing many blood vessels.

Vascular spasm Contraction of the smooth muscle in the wall of a damaged blood vessel to prevent blood loss.

Vascular (venous) sinus A vein with a thin endothelial wall that lacks a tunica media and externa and is supported by surrounding tissue.

Vascular tunic (TOO-nik) The middle layer of the eyeball, composed of the choroid, ciliary body, and iris. Also called the **uvea** (Ū-ve-a).
Vasectomy (va-SEK-to-me) A means of sterilization of males in which a portion of each ductus (vas) deferens is removed.
Vasoconstriction (vāz-ō-kon-STRIK-shun) A decrease in the size of the lumen of a blood vessel caused by contraction of the smooth muscle in the wall of the vessel.
Vasodilation (vāz′-ō-DĪ-lā-shun) An increase in the size of the lumen of a blood vessel caused by relaxation of the smooth muscle in the wall of the vessel.
Vasomotion (vāz-ō-MŌ-shun) Intermittent contraction and relaxation of the smooth muscle of the metarterioles and precapillary sphincters that result in an intermittent blood flow.
Vein A blood vessel that conveys blood from tissues back to the heart.
Vena cava (VĒ-na KĀ-va) One of two large veins that open into the right atrium, returning to the heart all of the deoxygenated blood from the systemic circulation except from the coronary circulation.
Ventral (VEN-tral) Pertaining to the anterior or front side of the body; opposite of dorsal.
Ventral body cavity Cavity near the ventral aspect of the body that contains viscera and consists of a superior thoracic cavity and an inferior abdominopelvic cavity.
Ventral ramus (RĀ-mus) The anterior branch of a spinal nerve, containing sensory and motor fibers to the muscles and skin of the anterior surface of the head, neck, trunk, and the limbs.
Ventricle (VEN-tri-kul) A cavity in the brain filled with cerebrospinal fluid. An inferior chamber of the heart.
Ventricular fibrillation (ven-TRIK-ū-lar fib-ri-LĀ-shun) Asynchronous ventricular contractions; unless reversed by defibrillation, results in heart failure.
Venule (VEN-ūl) A small vein that collects blood from capillaries and delivers it to a vein.
Vermiform appendix (VER-mi-form a-PEN-diks) A twisted, coiled tube attached to the cecum.
Vermilion (ver-MIL-yon) The area of the mouth where the skin on the outside meets the mucous membrane on the inside.
Vermis (VER-mis) The central constricted area of the cerebellum that separates the two cerebellar hemispheres.
Vertebral (VER-te-bral) **canal** A cavity within the vertebral column formed by the vertebral foramina of all the vertebrae and containing the spinal cord. Also called the **spinal canal.**
Vertebral column The 26 vertebrae of an adult and 33 vertebrae of a child; encloses and protects the spinal cord and serves as a point of attachment for the ribs and back muscles. Also called the **backbone, spine,** or **spinal column.**
Vesicle (VES-i-kul) A small bladder or sac containing liquid.
Vesicouterine (ves′-ik-ō-Ū-ter-in) **pouch** A shallow pouch formed by the reflection of the peritoneum from the anterior surface of the uterus, at the junction of the cervix and the body, to the posterior surface of the urinary bladder.
Vestibular (ves-TIB-ū-lar) **apparatus** Collective term for the organs of equilibrium, which includes the saccule, utricle, and semicircular ducts.
Vestibular membrane The membrane that separates the cochlear duct from the scala vestibuli.
Vestibule (VES-ti-būl) A small space or cavity at the beginning of a canal, especially the inner ear, larynx, mouth, nose, and vagina.
Villus (VIL-lus) A projection of the intestinal mucosal cells containing connective tissue, blood vessels, and a lymphatic vessel; functions in the absorption of the end products of digestion. *Plural is* **villi** (VIL-ī).
Viscera (VIS-er-a) The organs inside the ventral body cavity. *Singular is* **viscus** (VIS-kus).
Visceral (VIS-er-al) Pertaining to the organs or to the covering of an organ.
Visceral effectors (e-FEK-torz) Organs of the ventral body cavity that respond to neural stimulation, including cardiac muscle, smooth muscle, and glands.
Vital capacity The sum of inspiratory reserve volume, tidal volume, and expiratory reserve volume; about 4800 mL.
Vital signs Signs necessary to life that include temperature (T), pulse (P), respiratory rate (RR), and blood pressure (BP).
Vitamin An organic molecule necessary in trace amounts that acts as a catalyst in normal metabolic processes in the body.
Vitreous (VIT-rē-us) **body** A soft, jellylike substance that fills the vitreous chamber of the eyeball, lying between the lens and the retina.
Vocal folds Pair of mucous membrane folds below the ventricular folds that function in voice production. Also called **true vocal cords.**
Voltage-gated channel An ion channel in a plasma membrane composed of integral proteins that functions like a gate to permit or restrict the movement of ions across the membrane in response to changes in the voltage.
Vulva (VUL-va) Collective designation for the external genitalia of the female. Also called the **pudendum** (poo-DEN-dum).

W

Wallerian (wal-LE-rē-an) **degeneration** Degeneration of the portion of the axon and myelin sheath of a neuron distal to the site of injury.
Wandering macrophage (MAK-rō-fāj) Phagocytic cell that develops from a monocyte, leaves the blood, and migrates to infected tissues.
Wave summation (sum-MĀ-shun) The increased strength of muscle contraction that results when muscle action potentials occur one after another in rapid succession.
White matter Aggregations or bundles of myelinated and unmyelinated axons located in the brain and spinal cord.
White pulp The regions of the spleen composed of lymphatic tissue, mostly B lymphocytes.
White ramus communicans (RĀ-mus kō-MŪ-ni-kans) The portion of a preganglionic sympathetic axon that branches from the anterior ramus of a spinal nerve to enter the nearest sympathetic trunk ganglion.

X

X-chromosome inactivation The random and permanent inactivation of one X chromosome in each cell of a developing female embryo. Also called **lyonization.**
Xiphoid (ZĪ-foyd) Sword-shaped. The inferior portion of the sternum is the **xiphoid process.**

Y

Yolk sac An extraembryonic membrane composed of the exocoelomic membrane and hypoblast. It transfers nutrients to the embryo, is a source of blood cells, contains primordial germ cells that migrate into the gonads to form primitive germ cells, forms part of the gut, and helps prevent desiccation of the embryo.

Z

Zona fasciculata (ZŌ-na fa-sik′-ū-LA-ta) The middle zone of the adrenal cortex consisting of cells arranged in long, straight cords that secrete glucocorticoid hormones, mainly cortisol.

Zona glomerulosa (glo-mer′-ū-LŌ-sa) The outer zone of the adrenal cortex, directly under the connective tissue covering, consisting of cells arranged in arched loops or round balls that secrete mineralocorticoid hormones, mainly aldosterone.

Zona pellucida (pe-LOO-si-da) Clear glycoprotein layer between a secondary oocyte and the surrounding granulosa cells of the corona radiata.

Zona reticularis (ret-ik′-ū-LAR-is) The inner zone of the adrenal cortex, consisting of cords of branching cells that secrete sex hormones, chiefly androgens.

Zygote (ZĪ-gōt) The single cell resulting from the union of male and female gametes; the fertilized ovum.

Credits

Illustration Credits

Chapter 1 CO art by Keith Kasnot. Table 1.2: Keith Kasnot. 1.1: Tomo Narashima. 1.2–1.4: Jared Schneidman Design. 1.5, 1.6: Kevin Somerville. 1.7: Lynn O'Kelley. 1.8–1.13: Kevin Somerville.

Chapter 2 CO art by Keith Kasnot. 2.1: Imagineering. 2.2, 2.3: Jared Schneidman Design. 2.4: Imagineering. 2.5: Jared Schneidman Design. 2.6–2.11: Imagineering. 2.12–2.15: Jared Schneidman Design. 2.16: Imagineering. 2.17: Jared Schneidman Design. 2.18: Imagineering. 2.19–2.21: Jared Schneidman Design. 2.22: Imagineering. 2.23: Jared Schneidman Design. 2.24: Imagineering. 2.25: Jared Schneidman Design.

Chapter 3 CO art by Keith Kasnot. 3.1, 3.2: Tomo Narashima. 3.3–3.7: Imagineering. 3.8: Jared Schneidman Design. 3.5: Adapted from Bruce Alberts et al., Essential Cell Biology, F12.5, p375 and F12.12, p380 (New York: Garland Publishing Inc., 1998). ©1998 Garland Publishing Inc. 3.9–3.16: Imagineering. 3.17–3.21: Tomo Narashima. 3.22: Imagineering. 3.23–3.25: Tomo Narashima. 3.26–3.33: Imagineering. 3.34: Hilda Muinos.

Chapter 4 CO art by Keith Kasnot. Table 4.1, Table 4.2: Nadine Sokol, Kevin Somerville. Table 4.3: Nadine Sokol, Kevin Somerville, Imagineering. Table 4.4, Table 4.5: Nadine Sokol. 4.1: Kevin Somerville. 4.2: Imagineering. 4.3: Kevin Somerville. 4.4, 4.5: Imagineering.

Chapter 5 CO art by Keith Kasnot. 5.1–5.7: Kevin Somerville. 5.9: Imagineering.

Chapter 6 CO art by Keith Kasnot. 6.1: Leonard Dank. 6.2: Lauren Keswick. 6.3–6.9: Kevin Somerville. 6.10: Leonard Dank. 6.11: Kevin Somerville. 6.12: Jared Schneidman Design. 6.13: Kevin Somerville.

Chapter 7 CO art by Keith Kasnot. Table 7.1: Nadine Sokol. 7.1–7.24: Leonard Dank.

Chapter 8 CO art by Keith Kasnot. Table 8.1, 8.1–8.17: Leonard Dank.

Chapter 9 CO art by Keith Kasnot. 9.1–9.4, 9.11–9.14: Leonard Dank.

Chapter 10 CO art by Keith Kasnot. 10.1, 10.3: Kevin Somerville. 10.4, 10.6: Imagineering. 10.7: Hilda Muinos. 10.8–10.10: Imagineering. 10.11: Kevin Somerville. 10.12: Imagineering. 10.13–10.15: Jared Schneidman Design. 10.16, 10.18: Imagineering. 10.19: Beth Willert. 10.20: Kevin Somerville.

Chapter 11 CO art by Keith Kasnot. Table 11.1: Kevin Somerville. 11.1–11.17: Leonard Dank. 11.18: Leonard Dank, Kevin Somerville. 11.19–11.23: Leonard Dank.

Chapter 12 CO art by Keith Kasnot. Table 12.1: Kevin Somerville. Table 12.3: Jared Schneidman Design. 12.1: Kevin Somerville. 12.2: Jared Schneidman Design. 12.3: Kevin Somerville. 12.4, 12.5: Imagineering. 12.6: Kevin Somerville. 12.7: Sharon Ellis. 12.8–12.12: Imagineering. 12.13: Jared Schneidman Design. 12.14: Imagineering. 12.15: Jared Schneidman Design. 12.16: Nadine Sokol. 12.17: Imagineering.

Chapter 13 CO art by Keith Kasnot. 13.1–13.5: Sharon Ellis. 13.6–13.9: Leonard Dank. 13.10: Kevin Somerville. 13.11: Sharon Ellis. 13.12, 13.13: Steve Oh, Myriam Kirkman-Oh. 13.14: Imagineering. 13.15, 13.16: Steve Oh, Myriam Kirkman-Oh. 13.17: Imagineering.

Chapter 14 CO art by Keith Kasnot. Table 14.1, Table 14.3: Hilda Muinos. 14.1, 14.2: Kevin Somerville. 14.3: Sharon Ellis. 14.4: Kevin Somerville, Imagineering. 14.5–14.8: Sharon Ellis. 14.9: Imagineering. 14.10, 14.11, 14.13–14.15: Sharon Ellis. 14.17: Hilda Muinos. 14.18–14.24: Sharon Ellis. 14.25, 14.26: Kevin Somerville.

Chapter 15 CO art by Keith Kasnot. Table 15.3: Kevin Somerville. 15.1: Imagineering. 15.2, 15.3: Kevin Somerville. 15.4: Leonard Dank. 15.5: Imagineering. 15.6: Kevin Somerville. 15.7: Jared Schneidman Design. 15.8: Kevin Somerville. 15.9, 15.10: Sharon Ellis. 15.11: Imagineering. Adapted from Purves et al., Neuroscience 2e, F26.1 and F26.2, p498 (Sunderland, MA: Sinauer Associates, 1997). ©1997 Sinauer Associates.

Chapter 16 CO art by Keith Kasnot. Table 16.1: Kevin Somerville. Table 16.2: Imagineering. 16.1: Tomo Narashima. 16.2: Molly Borman. 16.3: Imagineering. Adapted from "Making Sense of Taste" by David V. Smith and Robert F. Margolskee, *Scientific American,* March 2001, page 38. 16.5: Sharon Ellis. 16.6: Tomo Narashima . 16.9: Lynn O'Kelley. 16.10: Tomo Narashima. 16.11, 16.12: Jared Schneidman Design. 16.13: Lynn O'Kelley. 16.14: Jared Schneidman Design. 16.15: Lynn O'Kelley. 16.16: Imagineering. Adapted from Seeley et al., Anatomy and Physiology 4e, F15.22, p480 (New York: WCB McGraw-Hill, 1998) ©1998 The McGraw-Hill Companies. 16.17–16.21: Tomo Narashima. 16.22, 16.23: Tomo Narashima, Sharon Ellis. 16.24, 16.25: Kevin Somerville.

Chapter 17 CO art by Keith Kasnot. 17.1: Jared Schneidman Design. 17.2, 17.3: Imagineering. 17.4: Kevin Somerville. 17.5: Sharon Ellis. 17.6: Imagineering.

Chapter 18 CO art by Keith Kasnot. Table 18.4: Nadine Sokol, Imagineering. Table 18.5–18.8: Nadine Sokol. Table 18.9: Imagineering. Table 18.10: Nadine Sokol. 18.1: Steve Oh. 18.2: Jared Schneidman Design. 18.3, 18.4: Imagineering. 18.5: Lynn O'Kelley. 18.6: Imagineering. 18.7: Jared Schneidman Design. 18.8: Lynn O'Kelley. 18.9: Jared Schneidman Design. 18.10: Molly Borman. 18.11, 18.12: Jared Schneidman Design. 18.13: Molly Borman. 18.14: Jared Schneidman Design. 18.15: Molly Borman. 18.16, 18.17: Jared Schneidman Design. 18.18: Molly Borman. 18.19: Jared Schneidman Design. 18.20: Nadine Sokol. 18.21: Kevin Somerville.

Chapter 19 CO art by Keith Kasnot. Table 19.3: Jared Schneidman Design. 19.1: Hilda Muinos, Nadine Sokol. 19.3, 19.4: Nadine Sokol. 19.5, 19.6, 19.8: Jared Schneidman Design. 19.9: Nadine Sokol. 19.11: Imagineering. 19.12: Jean Jackson. 19.13: Nadine Sokol.

Chapter 20 CO art by Keith Kasnot. 20.1, 20.2: Kevin Somerville, Imagineering. 20.3–20.6: Kevin Somerville. 20.7: Nadine Sokol, Imagineering. 20.8, 20.9: Kevin Somerville. 20.10: Kevin Somerville, Imagineering. 20.11, 20.12: Burmar Technical Corp. 20.13, 20.14: Imagineering. 20.15: Imagineering, Kevin Somerville. 20.16: Hilda Muinos. 20.18: Kevin Somerville. 20.20: Hilda Muinos.

Chapter 21 CO art by Keith Kasnot. Exhibits 21.2–21.12: Keith Ciociola. Table 21.2: Imagineering. 21.1: Kevin Somerville. 21.2: Hilda Muinos. 21.3: Nadine Sokol, Imagineering. 21.4: Kevin Somerville. 21.6, 21.7: Jared Schneidman Design. 21.8: Imagineering. 21.9: Kevin Somerville. 21.10–21.12: Jared Schneidman Design. 21.13: Kevin Somerville. 21.14–21.16: Jared Schneidman Design. 21.17–21.30: Kevin Somerville. 21.31: Kevin Somerville, Keith Ciociola. 21.32: Kevin Somerville.

Chapter 22 CO art by Keith Kasnot. Table 22.3: Jean Jackson. 22.1: Molly Borman. 22.2: Sharon Ellis. 22.3: Molly Borman. 22.4: Nadine Sokol. 22.5: Steve Oh. 22.6: Molly Borman. 22.7: Steve Oh. 22.8: Kevin Somerville. 22.9, 22.10: Molly Borman. 22.11–21.20: Jared Schneidman Design. 22.21: Nadine Sokol, Imagineering.

Chapter 23 CO art by Keith Kasnot. 23.1: Molly Borman, Kevin Somerville. 23.2, 23.4, 23.5: Molly Borman. 23.6: Steve Oh. 23.8, 23.10: Molly Borman. 23.11, 23.12: Kevin Somerville. 23.13: Jared Schneidman Design. 23.14: Kevin Somerville. 23.15: Jared Schneidman Design, Imagineering. 23.16–21.24: Jared Schneidman Design. 23.25: Imagineering. 23.26: Jared Schneidman Design. 23.27: Kevin Somerville. 23.28: Jared Schneidman Design. 23.29: Kevin Somerville.

Chapter 24 CO art by Keith Kasnot. 24.1–24.3: Steve Oh. 24.4: Nadine Sokol. 24.5: Molly Borman. 24.6: Steve Oh. 24.7, 24.8, 24.10: Nadine Sokol. 24.11: Steve Oh. 24.12: Kevin Somerville. 24.13: Imagineering. 24.14–24.16: Jared Schneidman Design. 24.17: Steve Oh, Jared Schneidman Design. 24.18: Jared Schneidman Design. 24.19: Kevin Somerville. 24.20, 24.21: Jared Schneidman Design. 24.23: Kevin Somerville. 24.25, 24.26: Jared Schneidman Design. 24.27: Molly Borman. 24.28: Kevin Somerville.

Chapter 25 CO art by Keith Kasnot. 25.1–25.20: Imagineering.

Chapter 26 CO art by Keith Kasnot. Table 26.1: Nadine Sokol. 26.1, 26.2: Kevin Somerville. 26.3, 26.4: Steve Oh. 26.5: Imagineering. 26.6: Kevin Somerville. 26.7: Nadine Sokol. 26.8: Kevin Somerville. 26.9: Imagineering. 26.10–26.20: Jared Schneidman Design. 26.21: Steve Oh. 26.22: Kevin Somerville.

Chapter 27 CO art by Keith Kasnot. 27.1–27.8: Jared Schneidman Design.

Chapter 28 CO art by Keith Kasnot. Table 28.1, 28.1, 28.2: Imagineering. 28.3–28.6: Kevin Somerville. 28.7: Jared Schneidman Design. 28.8: Kevin Somerville. 28.9, 28.10–28.15, 28.17: Jared Schneidman Design. 28.18, 28.21–28.24: Kevin Somerville. 28.25–28.28: Jared Schneidman Design. 28.29, 28.30: Kevin Somerville.

Chapter 29 CO art by Keith Kasnot. Table 29.2, 29.1–29.15: Kevin Somerville. 29.16: Jared Schneidman Design. 29.17, 29.18: Kevin Somerville. 29.19–29.26: Jared Schneidman Design.

Focus on Homeostasis icons Imagineering.

Photo Credits

Chapter 1 Figure 1.1: John Wilson White. Figure 1.8a: From Stephen A. Kieffer and E. Robert Heitzman, *An Atlas of Cross–Sectional Anatomy.* Harper & Row, Publishers, New York, 1979. Figure 1.8b: Lester V. Bergman/Project Masters, Inc. Figure 1.8c: Martin Rotker. Figure 1.11: Mark Nielsen. Page 21 (top left): Biophoto Associates/Photo Researchers. Page 21 (top right): Scott Camazine/Photo Researchers. Page 21 (bottom left): Simon Fraser/Photo Researchers. Page 21 (bottom right): Courtesy Andrew Joseph Tortora and Damaris Soler. Page 22: Howard Sochurek/Medical Images, Inc.

Chapter 3 Figure 3.6: Andy Washnik. Figure 3.17c: Courtesy Kent McDonald, UC Berkeley Electron Microscope Laboratory. Figure 3.20b: D. W. Fawcett/Photo Researchers. Figure 3.21b: Biophoto Associates/Photo Researchers. Figure 3.23b: Courtesy Daniel S. Friend, Harvard Medical School. Figure 3.24b: D.W. Fawcett/Photo Researchers. Figure 3.25c: CNRI/Photo Researchers. Figure 3.33: Courtesy Michael Ross, University of Florida.

Chapter 4 Page 109 (top): Biophoto Associates/Photo Researchers. Pages 109 (bottom), 110, 111 (bottom), 112, 113 and 114 (bottom), 121, 122, 123 (top), 124 (bottom), 125, 126 (top), 127 and 130: Courtesy Michael Ross, University of Florida. Page 111 (top): Biophoto Associates/Photo Researchers. Page 114 (top): Lester V. Bergman/The Bergman Collection. Page 123 (bottom): Courtesy Andrew J. Kuntzman. Page 124 (top): Ed Reschke. Page 126 (bottom): John Burbidge/Photo Researchers. Page 131: Biophoto Associates/Photo Researchers. Page 132: © Ed Reschke.

Chapter 5 Figure 5.3b: © L.V. Bergman/Bergman Collection. Figure 5.4b: Science Photo Library/Photo Researchers. Figure 5.8a: Alain Dex/Photo Researchers. Figure 5.8b: Biophoto Associates/Photo Researchers.

Chapter 6 Figure 6.1b: Mark Nielsen. Figure 6.8a: The Bergman Collection. Figure 6.8b: Courtesy Michael Ross, University of Florida. Figure 6.14: P. Motta, Dept. of Anatomy, University La Sapienza, Rome/Photo Researchers.

Chapter 9 Figures 9.5, 9.6, 9.8, 9.7, 9.9 and 9.10: John Wilson White.

Chapter 10 Figure 10.2: Courtesy Fujita. Figure 10.5: Courtesy Denah Appelt and Clara Franzini-Armstrong. Figure 10.17: © John Wiley & Sons. Page 296: Biophoto Associates/Photo Researchers.

Chapter 12 Figure 12.3c: Science VU/Visuals Unlimited. Figure 12.6c: Dennis Kunkel/Phototake. Figure 12.6d: Martin Rotker/Phototake.

Chapter 13 Figure 13.1b: Mark Nielsen. Figure 13.3b: Jean Claude Revy/Phototake. Figure 13.10b: Copyright by Richard Kessel and Randy Kardon, *Tissues and Organs: A Text-Atlas of Scanning Electron Microscopy,* W. H. Freeman and Company, 1979. All rights reserved. Reprinted by permission.

Chapter 14 Figure 14.1b: Mark Nielsen. Figure 14.9e: From Stephen A. Kieffer and E. Robert Heitzman, *An Atlas of Cross-Sectional Anatomy,* Harper and Row, Publishers, 1979. Reproduced with permission. Figure 14.12: From N. Gluhbegovic and T.H. Williams, *The Human Brain: A Photographic Guide,* Harper and Row, Publishers, 1980. Reproduced with permission. Figure 14.16: From *Nature,* November 26, 1992, Vol. 360, page 340. Reproduced with permission from *Nature* and Robert Zatorre, Department of Neuropsychology, McGill University.

Chapter 16 Figure 16.4: John Moore. Figure 16.8: Courtesy Michael Ross, University of Florida. Figure 16.16a: From N. Gluhbegovic and T. H. Williams, *The Human Brain: A Photographic Guide,* Harper and Row, Publishers, 1980.

Chapter 18 Figure 18.5: Mark Nielsen. Figures 18.10b, 18.13b, 18.15b and 18.18c: Courtesy Michael Ross, University of Florida. Figure 18.22a: From *New England Journal of Medicine,* February 18, 1999, vol. 340, No. 7, Page 524. Photo provided courtesy of Robert Gagel, Department of Internal Medicine, University of Texas M.D. Anderson Cancer Center, Houston, Texas. Figures 18.22b,c and d: © The Bergman Collection/Project Masters, Inc. Figure 18.22e: Biophoto Associates/Photo Researchers.

Chapter 19 Figure 19.2: From Lennart Nilsson, *Our Body Victorious,* Boehringer Ingelheim International GmbH. Reproduced with permission. Figure 19.7: John Cunningham/Visuals Unlimited. Figure 19.10: From Lennart Nilsson, *The Incredible Machine,* Boehringer Ingelheim International GmbH. Reproduced with permission. Figure 19.14: Lewin/Royal Free Hospital/Photo Researchers.

Chapter 20 Figures 20.3b, 20.4b, 20.6e and 20.8c: Mark Nielsen. Figure 20.17: Gregg Adams/Stone. Figure 20.19a: © Vu/Cabisco/Visuals Unlimited. Figure 20.19b: W. Ober/Visuals Unlimited.

Chapter 21 Figure 21.1d: Dennis Strete. Figure 21.1e: Courtesy Michael Ross, University of Florida. Figure 21.5: Mark Nielsen.

Chapter 22 Figures 22.5, 22.6 and 22.7: Courtesy Michael Ross, University of Florida. Figure 22.9b: National Cancer Institute/Photo Researchers.

Chapter 23 Figures 23.1b and 23.9: Mark Nielsen. Figure 23.3: Courtesy Lynne Marie Barghesi. Figure 23.7: John Cunningham/Visuals Unlimited. Figure 23.11: Biophoto Associates/Photo Researchers.

Chapter 24 Figures 24.5b, 24.9, 24.19c, 24.24c–d and Figure 24.28c,d: Courtesy Michael Ross, University of Florida. Figures 24.11b and 24.22b: Mark Nielsen. Figure 24.12c: Ed Reschke. Figure 24.24a: Fred E. Hossler/Visuals Unlimited. Figure 24.24b: Willis/Biological Photo Service.

Chapter 26 Figure 26.3b: Mark Nielsen. Figure 26.6b: Dennis Strete. Figure 26.8b: Courtesy Michael Ross, University of Florida.

Chapter 28 Figures 28.3, 28.5b–c and 28.13b: Mark Nielsen. Figure 28.6a: Ed Reschke. Figure 28.16: Biophoto Associates/Photo Researchers. Figure 28.19: P. Motta/Photo Researchers. Figure 28.20: Courtesy Andrew J. Kuntzman, Wright State University.

Chapter 29 Figure 29.1b: David Phillips/Photo Researchers. Figure 29.1c: Myriam Wharman/Phototake. Figure 29.11b: Siu, Biomedical Comm./Custom Medical Stock Photo. Figures 29.14a, g and h: Photo provided courtesy of Kohei Shiota, Congenital Anomaly Research Center, Kyoto University, Graduate School of Medicine. Figures 29.14b–e: Courtesy National Museum of Health and Medicine, Armed Forces Institute of Pathology. Figure 29.14f: Photo by Lennart Nilsson/Albert Bonniers Förlag AB, *A Child is Born,* Dell Publishing Company. Reproduced with permission.

Index

Note Page numbers in **boldface** type indicate a major discussion. A *t* following a page number indicates a table, an *f* following a page number indicates a figure, and an *e* following a page number indicates an exhibit.

EPONYMS USED IN THIS TEXT

In the life sciences, an eponym is the name of a structure, drug, or disease that is based on the name of a person. For example, you may be more familiar with the Achilles tendon than you are with its more anatomically descriptive term, the calcaneal tendon. Because eponyms remain in frequent use, this listing correlates common eponyms with their anatomical terms.

EPONYM	ANATOMICAL TERM
Achilles tendon	calcaneal tendon
Adam's apple	thyroid cartilage
ampulla of Vater (VA-ter)	hepatopancreatic ampulla
Bartholin's (BAR-tō-linz) gland	greater vestibular gland
Billroth's (BIL-rōtz) cord	splenic cord
Bowman's (BŌ-manz) capsule	glomerular capsule
Bowman's (BŌ-manz) gland	olfactory gland
Broca's (BRŌ-kaz) area	motor speech area
Brunner's (BRUN-erz) gland	duodenal gland
bundle of His (HISS)	artrioventricular (AV) bundle
canal of Schlemm (SHLEM)	scleral venous sinus
circle of Willis (WIL-is)	cerebral arterial circle
Cooper's (KOO-perz) ligament	suspensory ligament of the breast
Cowper's (KOW-perz) gland	bulbourethral gland
crypt of Lieberkühn (LE-ber-kyūn)	intestinal gland
duct of Santorini (san′-tō-RĒ-nē)	accessory duct
duct of Wirsung (VĒR-sung)	pancreatic duct
Eustachian (yoo-STĀ-kē-an)	auditory tube
Fallopian (fal-LŌ-pē-an) tube	uterine tube
gland of Littré (LĒ-tra)	urethral gland
Golgi (GOL-jē) tendon organ	tendon organ
Graafian (GRAF-ē-an) follicle	mature ovarian follicle
Hassall's (HAS-alz) corpuscle	thymic corpuscle
Haversian (ha-VĒR-shun) canal	central canal
Haversian (ha-VĒR-shun) system	osteon
Heimlich (HĪM-lik) maneuver	abdomial thrust maneuver
islet of Langerhans (LANG-er-hanz)	pancreatic islet
Kupffer (KOOP-fer) cell	stellate reticuloendothelial cell
Leydig (LĪ-dig) cell	interstitial endocrinocyte
loop of Henle (HEN-lē)	loop of the nephron
Luschka's (LUSH-kaz) aperture	lateral aperture
Magendie's (ma-JEN-dēz) aperture	median aperture
Meibomian (mi-BŌ-mē-an) gland	tarsal gland
Meissner (MĪS-ner) corpuscle	corpuscle of touch
Merkel (MER-kel) disc	tactile disc
Müllerian (mil-E rē-an) duct	paramesonephric duct
organ of Corti (KOR-tē)	spiral organ
Pacinian (pa-SIN-ē-an) corpuscle	lamellated corpuscle
Peyer's (PĪ-erz) patch	aggregated lymphatic follicle
plexus of Auerbach (OW-er-bak)	myenteric plexus
plexus of Meissner (MĪS-ner)	submucosal plexus
pouch of Douglas	rectouterine pouch
Purkinje (pur-KIN-jē) fiber	conduction myofiber
Rathke's (rath-KĒZ) pouch	hypophyseal pouch
Ruffini (roo-FĒ-nē) corpuscle	type II cutaneous mechanoreceptor
Sertoli (ser-TŌ-lē) cell	sustentacular cell
Skene's (SKĒNZ) gland	paraurethral gland
sphincter of Oddi (OD-dē)	sphincter of the hepatopancreatic ampulla
Volkmann's (FŌLK-manz) canal	perforating canal
Wernicke's (VER-ni-kēz) area	auditory association area
Wharton's (HWAR-tunz) jelly	mucous connective tissue
Wolffian duct	mesonephric duct
Wormian (WER-mē-an) bone	sutural bone

COMBINING FORMS, WORD ROOTS, PREFIXES, AND SUFFIXES

Many of the terms used in anatomy and phsiology are compound words; that is, they are made up of word roots and one or more prefixes or suffixes. For example, *leukocyte* is formed from the word roots *leuk-* meaning "white", a connecting vowel (o), and *cyte* meaning "cell." Thus, a leukocyte is a white blood cell. The following list includes some of the most commonly used combining forms, word roots, prefixes, ad suffixes used in the study of anatomy and physiology. Each entry includes a usage example. Learning the meanings of these fundamental word parts will help you remember terms that, at first glance, may seem long or complicated.

COMBINING FORMS AND WORD ROOTS

Acous-, Acu- hearing Acoustics.
Acr- extremity Acromegaly.
Aden- gland Adenoma.
Alg-, Algia- pain Neuralgia.
Angi- vessel Angiocardiography.
Anthr- joint Arthropathy.
Aut-, Auto- self Autolysis.
Audit- hearing Auditory canal.

Bio- life, living Biopsy.
Blast- germ, bud Blastula.
Blephar- eyelid Blepharitis.
Brachi- arm Brachial plexus.
Bronch- trachea, windpipe Bronchoscopy.
Bucc- cheek Buccal.

Capit- head Decapitate.
Carcin- cancer Carcinogenic.
Cardi-, Cardia-, Cardio- heart Cardiogram.
Cephal- head Hydrocephalus.
Cerebro- brain Cerebrospinal fluid.
Chole- bile, gall Cholecystogram.
Chondr-, cartilage Chondrocyte.
Cor-, Coron- heart Coronary.
Cost- rib Costal.
Crani- skull Craniotomy.
Cut- skin Subcutaneous.
Cyst- sac, bladder Cystoscope.

Derma-, Dermato- skin Dermatosis.
Dura- hard Dura mater.

Enter- intestine Enteritis.
Erythr- red Erythrocyte.

Gastr- stomach Gastrointestinal.
Gloss- tongue Hypoglossal.
Glyco- sugar Glycogen.
Gyn-, Gynec- female, woman Gynecology.

Hem-, Hemat- blood Hematoma.
Hepar-, Hepat- liver Hepatitis.
Hist-, Histio- tissue Histology.
Hydr- water Dehydration.
Hyster- uterus Hysterectomy.

Ischi- hip, hip joint Ischium.

Kines- motion Kinesiology.

Labi- lip Labial.
Lacri- tears Lacrimal glands.
Laparo- loin, flank, abdomen Laparoscopy.
Leuko- white Leukocyte.
Lingu- tongue Sublingual glands.
Lip- fat Lipid.
Lumb- lower back, loin Lumbar.

Macul- spot, blotch Macula.
Malign- bad, harmful Malignant.
Mamm-, Mast- breast Mammography, Mastitis.
Meningo- membrane Meningitis.
Myel- marrow, spinal cord Myeloblast.
My-, Myo- muscle Myocardium.

Necro- corpse, dead Necrosis.
Nephro- kidney Nephron.
Neuro- nerve Neurotransmitter.

Ocul- eye Binocular.
Odont- tooth Orthodontic
Onco- mass, tumor Oncology.
Oo- egg Oocyte.
Opthalm- eye Ophthalmology.
Or- mouth Oral.
Osm- odor, sense of small Anosmia.
Os-, Osseo-, Osteo- bone Osteocyte.
Ot- ear Otitus media.

Palpebr- eyelid Palpebra.
Patho- disease Pathogen.
Pelv- basin Renal pelvis.
Phag- to eat Phagocytosis.
Phleb- vein Phlebitis.
Phren- diaphragm Phrenic.
Pilo- hair Depilatory.
Pneumo- lung, air Pneumothorax.
Pod- foot Podocyte.
Procto- anus, rectum Proctology.
Pulmon- lung Pulmonary.

Ren- kidneys Renal artery.
Rhin- nose Rhinitis.

Scler-, Sclero- hard Atherosclerosis.
Sep-, Spetic- toxic condition due to micoorganisms Septicemia.
Soma-, Somato- body Somatotropin.
Sten- narrow Stenosis.
Stasis-, Stat- stand still Homeostasis.

Tegument- skin, covering Integumentary.
Therm- heat Thermogenesis.
Thromb- clot, lump Thrombus.

Vas- vessel, duct Vasoconstriction.

Zyg- joined Zygote.

PREFIXES

A-, An- without, lack of, deficient Anesthesia.
Ab- away from, from Abnormal.
Ad-, Af- to, toward Adduction, Afferent neuron.
Alb- white Albino.
Alveol- cavity, socket Alveolus.
Andro- male, masculine Androgen.
Ante- before Antebrachial vein.
Anti- against Anticoagulant.

Bas- base, foundation Basal ganglia.
Bi- two, double Biceps.
Brady- slow Bradycardia.

Cata- down, lower, under Catabolism.
Circum- around Circumduction.
Cirrh- yellow Cirrhosis of the liver.
Co-, Con-, Com with, together Congenital.
Contra- against, opposite Contraception.
Crypt- hidden, concealed Cryptorchidism.
Cyano- blue Cyanosis.

De- down, from Deciduous.
Demi-, hemi- half Hemiplegia.
Di-, Diplo- two Diploid.
Dis- separation, apart, away from Dissection.
Dys- painful, difficult Dyspnea.

E-, Ec-, Ef- out from, out of Efferent neuron.
Ecto-, Exo- outside Ectopic pregnancy.
Em-, En- in, on Emmetropia.
End-, Endo- within, inside Endocardium.
Epi- upon, on, above Epidermis.
Eu- good, easy, normal Eupnea.
Ex-, Exo- outside, beyond Exocrine gland.
Extra- outside, beyond, in addition to Extracellular fluid.

Fore- before, in front of Forehead.

Gen- originate, produce, form Genitalia.
Gingiv- gum Gingivitis.

Hemi- half Hemiplegia.
Heter-, Hetero- other, different Heterozygous.
Homeo-, Homo- unchanging, the same, steady Homeostasis.
Hyper- over, above, excessive Hyperglycemia.
Hypo- under, beneath, deficient Hypothalamus.

In-, Im- in, inside, not Incontinent.
Infra- beneath Infraorbital.
Inter- among, between Intercostal.
Intra- within, inside Intracellular fluid.
Ipsi- same Ipsilateral.
Iso- equal, like Isotonic.

Juxta- near to Juxtaglomerular apparatus.

Later- side Lateral.

Macro- large, great Macrophage.
Mal- bad, abnormal Malnutrition.
Medi-, Meso- middle Medial.
Mega-, Megalo- great, large Magakaryocyte.
Melan- black Melanin.
Meta- after, beyond Metacarpus.
Micro- small Microfilament.
Mono- one Monounsaturated fat.

Neo- new Neonatal.

Oligo- small, few Oliguria.
Ortho- straight, normal Orthopedics.

Para- near, beyond, beside Paranasal sinus.
Peri- around Pericardium.
Poly- much, many, too much Polycythemia.
Post- after, beyond Postnatal.
Pre-, Pro- before, in front of Presynaptic.
Pseudo- false Pseudostratified.

Retro- backward, behind Retroperitoneal.

Semi- half Semicircular canals.
Sub- under, beneath, below Submucosa.
Super- above, beyond Superficial.
Supra- above, over Suprarenal.
Sym-, Syn- with, together Symphysis.

Tachy- rapid Tachycardia.
Trans- across, through, beyond Transudation.
Tri- three Trigone.

SUFFIXES

-able capable of, having ability to Viable.
-ac, -al pertaining to Cardiac.
-algia painful condition Myalgia.
-an, -ian pertaining to Circadian.
-ant having the characteristic of Malignant.
-ary connected with Ciliary.
-asis, -asia, -esis, -osis condition or state of Hemostasis.
-asthenia weakness Myasthenia.
-ation process, action, condition Inhalation.

-centesis puncture, usually for drainage Amniocentesis.
-cid, -cide, -cis, cut, kill destroy Spermicide.
-ectomize, -ectomy excision of, removal of Thyroidectomy.
-emia condition of blood Anemia.
-esthesia sensation, feeling Anesthesia.

-fer carry Efferent arteriole.

-gen agent that produces or originates Pathogen.
-genic producing Pyogenic.
-gram record Electrocardiogram.
-graph instrument for recording Electroencephalograph.

-ia state, condition Hypermetropia.
-ician person associated with Pediatrician.
-ics art of, science of Optics.
-ism condition, state Rheumatism.
-itis inflammation Neuritis.

-logy the study or science of Physiology.
-lysis dissolution, loosening, destruction Hemolysis.

-malacia softening Osteomalacia.
-megaly enlarged Cardiomegaly.
-mers, -meres parts Polymers.

-oma tumor Fibroma.
-osis condition, disease Necrosis.
-ostomy create an opening Colostomy.
-otomy surgical incision Tracheotomy.

-pathy disease Myopathy.
-penia deficiency Thrombocytopenia
-philic to like, have an affinity for Hydrophilic.
-phobe, -phobia fear of, aversion to Photophobia.
-plasia, -plasty forming, molding Rhinoplasty.
-pnea breath Apnea.
-poiesis making Hemopoiesis.
-ptosis falling, sagging Blepharoptosis.

-rrhage bursting forth, abnormal discharge Hemorrhage.
-rrhea flow, discharge Diarrhea.

-scope instrument for viewing Bronchoscope.
-stomy creation of a mouth or artificial opening Tracheostomy.

-tomy cutting into, incision into Laparotomy.
-tripsy crushing Lithotripsy.
-trophy relating to nutrition or growth Atrophy.

-uria urine Polyuria.